Crystallization Handbook

Crystallization Handbook

Edited by **Garry Hollis**

NY RESEARCH
PRESS

New York

Published by NY Research Press,
23 West, 55th Street, Suite 816,
New York, NY 10019, USA
www.nyresearchpress.com

Crystallization Handbook
Edited by Garry Hollis

International Standard Book Number: 978-1-63238-104-0 (Hardback)

Printed in China.

Contents

Preface

This book has been a concerted effort by a group of academicians, researchers and scientists, who have contributed their research works for the realization of the book. This book has materialized in the wake of emerging advancements and innovations in this field. Therefore, the need of the hour was to compile all the required researches and disseminate the knowledge to a broad spectrum of people comprising of students, researchers and specialists of the field.

Crystals are of various kinds, both organic and inorganic, and the process of crystallization is an age-old multidimensional subject of research study. Crystals can be formed from solids, melts, liquids and even vapors. Though complex in its characteristics, the process of crystallization forms a part of our regular lives from making ice cubes in our refrigerators to latest and advanced chemical and electronic industry processes. This book is aimed at bringing forth novel perspectives while presenting important data for experts in the related fields of crystallography. It includes topics on applications, techniques and mineral formation, dedicated to the practical application of crystallization processes and concepts. This book also attempts to be a wholesome document giving detailed accounts about the process of crystallization which will prove to be a valuable reference source to people with diverse backgrounds.

At the end of the preface, I would like to thank the authors for their brilliant chapters and the publisher for guiding us all-through the making of the book till its final stage. Also, I would like to thank my family for providing the support and encouragement throughout my academic career and research projects.

Editor

Applications, Techniques and Mineral Formation

1

Crystallization of Iron-Containing Oxide-Sulphide Melts

Evgeniy Selivanov and Roza Gulyaeva
Institute of Metallurgy of the Ural Branch of the Russian Academy of Sciences
Russia

1. Introduction

The processing of the sulphide raw materials (ores, concentrates and mattes) of non-ferrous metallurgy is related to the formation of a large amount of iron containing slags. The initial product of the oxidation of sulphides in real commercial plants is an oxide–sulphide melt, in which decomposition under the action of fluxes is accompanied by matte and slag formation (Selivanov et al., 2009a). The fraction of oxygen in a sulphide melt and the fraction of sulphur in an oxide melt are each controlled by the contents of silicon dioxide and iron oxides in a slag and the contents of non-ferrous metals in a matte. According to modern concepts, the heterogeneity of slags is caused by mechanical matte, magnetite and spinel inclusions, where the spinel inclusions form during oxidation processes (Selivanov et al., 2000; Spira & Themelis, 1969; Tokeda et al., 1983; Vanyukov & Zaitsev, 1969, 1973). The cooling (i.e., the crystallization) of a slag leads to the formation of new oxide and sulphide phases within it. Information on the available forms of the useful components is important for the reduction of metal loss through a slag and for the selection of their re-extraction methods.

A number of works are devoted to the study of the kinds of copper existing in slags. Major results are generalized in the monographs of (Ruddle, 1953; Vanyukov et al., 1988; Vanyukov & Zaitsev, 1969; 1973). Phase equilibria in the systems relevant to copper pyrometallurgy have been discussed mostly for molten states (Elliott, 1976; Kopylov, 2001; Yazawa, 1974). It is considered that the loss of non-ferrous metals through slags is caused by their oxide, sulphide and metal solubility. It was discovered that a part of copper is presented in the crystallized slag by matte mechanical inclusions (Vanyukov & Zaitsev, 1969; 1973). Data on the copper sulphide solubility in a slag was reported by (Mohapatra, 1994; Nagamori, 1974; Vanyukov et al., 1988; Vaysburd, 1996). There is no valid confirmation of the presence of individual copper oxide inclusions or copper silicates and ferrites in a slag. Information on the existence of other metals (Zn, Pb, As, etc.) in a slag needs to be specified more exactly in each separate case. The bulk of the zinc is transferred into the slag during the smelting of sulphide copper-zinc concentrates in the Vanyukov furnace for a rich matte and crude metal (Vanyukov et al., 1988). It is assumed herein that zinc is present in a slag in the form of an oxide. Some questions concerning the constituent phases of crystallization during the rapid cooling of a non-ferrous metallurgy slag are partially disclosed by (Cardona et al., 2011). However, no task-oriented studies devoted to the estimation of the cooling rate's effect on the formation of phases and the presence of different kinds of non-ferrous metals in a non-ferrous metallurgy slag have been found.

The goal of this work is to study phase composition and the kinds of metals present in the slag samples of copper-zinc concentrates of pyrometallurgical processing and the nickel oxide ores of smelting. The main task of the study lies in the estimation of the cooling rate and iron's oxidation level's influence on the phase composition, structure, thermal properties and forms of non-ferrous metals extant in the crystallized oxide-sulphide systems FeO_x-SiO_2-FeS-Cu_2O-ZnO and SiO_2-FeO_x-MgO-CaO-NiO-FeS.

2. Methods of investigation

The chemical analysis accuracy resolution is 0.1% for the elements' content in the slag samples over 1% (Fe, S, Zn, SiO, CaO, Al_2O_3). It is equal to 0.02% when the elements concentration in the slag is less than 1% (Cu, Sb, Pb, As). The phase composition of the samples has been determined by using an X-ray diffractometer (Cu–K_α - radiation). The temperatures and heats of the phase transformations are determined by means of differential-scanning calorimetry with a Netzsch STA 449 C Jupiter thermo-analyser with a heating rate of 20 °C/min in an argon flow. The determination of phase element composition is performed with a JSM-59000LV raster electronic microscope (ESM) and an Oxford INCA Energy 200 dispersion X-ray spectrometer (EDX). The results of the X-ray spectrum microanalysis have (EPMA) a relative error of 2% where the content of the elements is greater than 10%. The relative error is close to 5% at concentrations of elements are from 1% to 10%. This relative error is 10% than concentrations of elements are less 1%. The microstructure of the samples is studied by an Olympus optical microscope using the Simagic application program.

The analysis of the gases evolving in the heating of materials was carried out by a QMS 403C Aëolos mass - a spectrometer connected with the thermo analyser. To perform the thermodynamic simulation (TDS) of the equilibrium phases during the cooling of working bodies whose compositions corresponded to the initial slag samples, we used the HSC 5.1 Chemistry (Outokumpu) software package based on the minimization of the Gibbs energy and variational thermodynamics principles (HSC Chemistry, 2002; Moiseev & Vyatkin, 1999).

The initial slags were put in Al_2O_3 crucibles and melted (1300 °C) in a resistance furnace with an electrographite heater for the investigation of the cooling rate's influence on the crystallization of melts. The direct cooling of the slag was carried out in a furnace and it provided for a decrease of the temperature rate up to crystallization (solidus) at about 0.3 °C/s; in the air after removing the crucible from the furnace - 1.7 °C/s; by means of the pouring of the melt from a crucible into a water basin - 900 °C/s. With the water granulation of the slags, we fabricated particles with an average size of 1.5–2.0 mm. The calculation of the cooling time of these particles was carried out using the expression (Naboichenko et al., 1997):

$$\tau_{cool} = d_d \, (c_p \, \rho_{sl} / \, 6 \, \alpha) \, \ln[(T_m - T_s)/(T_d - T_s)], \tag{1}$$

where τ_{cool} is the drop solidification time; d_d is the drop size; c_p is the heat capacity; ρ_{sl} is the melt density; α is the heat-transfer coefficient of the melt–water system; T_m, T_d and T_s are the temperatures (K) of the slag melt, drop and vapour, respectively. The granules obtained from the slag were subjected to isothermal annealing in an electric resistance furnace (for 5 and 60 minutes) at temperatures of 750 °C and 1000 °C in an inert atmosphere.

The overall strategy of this investigation is presented schematically (Fig. 1).

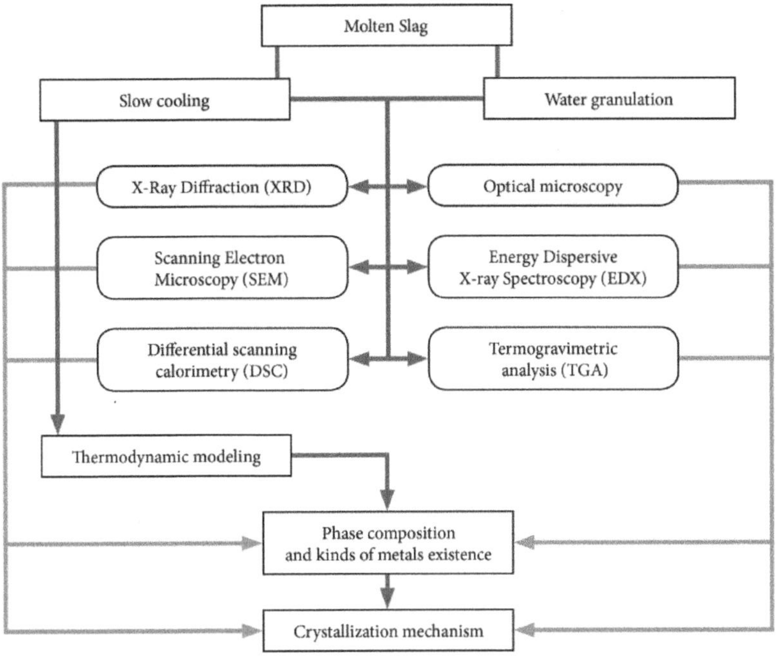

Fig. 1. Overall strategy of this investigation

3. Effect of the cooling rate on the structure of slag from the melt of copper-zinc concentrates in a Vanyukov furnace

The autogenous smelting technology of sulphide copper–zinc concentrates in a Vanyukov furnace was developed in "Sredneuralsky Copper Smelter Plant" JSC (Russia, Ural) (Vanyukov & Zaitsev, 1969, 1973; Vanyukov et al., 1988). Concentrates (14 - 16% Cu) are melted for the mattes contents with 45 – 55% copper.

The degree of copper concentration, defined as the ratio of metal content in the matte to its content in the charge, is within the range 3.0 - 4.0. The relatively low quality of the incoming concentrate and the desire to increase the copper content in the matte predetermine the high flow of the oxygen-air mixture and the large amount of slag which is produced. The slag contains iron oxide (III) in the form of magnetite, which largely determines the matte-slag emulsion delamination.

A large number of studies (Jalkann, 1991; Rüffler & Dávalos, 1998; Selivanov et al., 2000, 2004; Vanyukov & Zaitsev, 1969, 1973) have been devoted to the evaluation of slag structure and the metal forms of these of non-ferrous metals are presented in the literature. However, a common law for such complex systems as metallurgical slags does not allow us to extrapolate the known data on the studied samples because new objects require additional study.

The object of the research is the slag from the melting of copper-zinc concentrates in a Vanyukov furnace which contains, %: 40.5 Fe, 2.4 S, 0.8 Cu, 3.9 Zn, 32.1 SiO_2, 2.8 CaO, 0.8 MgO, 2.6 Al_2O_3, 0.1 Sb, 0.5 Pb, 0.1 As (Selivanov et al., 2009b, 2010).

A slag sample is taken from the furnace slag siphon at its overflow into the drain trough. The slag was in contact with a matte containing, %: 44.9 Cu, 23.8 Fe, 2.3 Zn, 22.8 S, 0.1 Sb, 2.0 Pb, until its discharge.

Reflexes which correspond to (Fe_2SiO_4) fayalite, (Fe_3O_4) magnetite and zinc sulphide (sphalerite) are identified in the initial slag through X-ray analysis (Fig. 2). The melting of the slag followed by its cooling reduces the intensity of the X-ray reflexes of the identified phases. Amorphization (glass formation) is reached throughout the mass of the sample when the cooling rate of the slag is equal to 900 °C/s.

Thermograms (Selivanov et al., 2009) of the samples (Fig. 3) allow us to estimate the melting and crystallization temperatures of the samples. Two endothermic effects are observed with the heating of the initial slag, which is begun at 972 °C and 1067 °C.

The first of these characterizes the melting of the eutectic and the second of the entire mass of the slag. The temperature of initial crystallization is equal to 1021 °C. According to the mass spectrometry data an evolution of a certain amount of SO_2 occurred under the sample's heating (from 300 °C - 400 °C). This evolution is caused by interaction of sulphides with iron oxides of higher valence. Slag mass loss does not exceed 1.0% with heating up to 1200 °C. The view of the sample thermogram crystallized at 0.3 °C/s is essentially identical to the results obtained for the original slag.

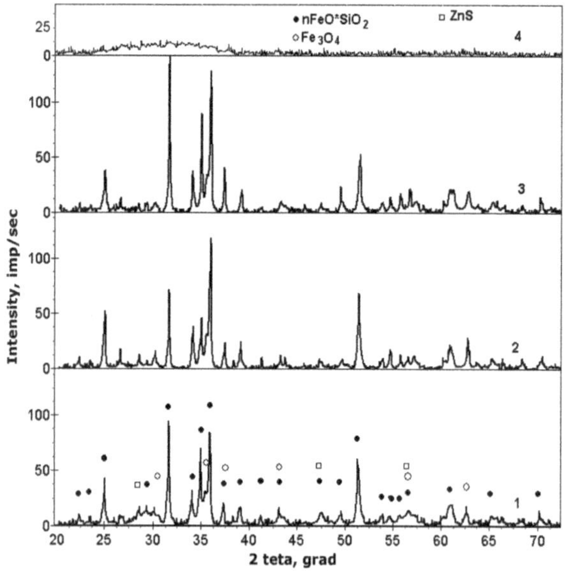

Fig. 2. Diffractograms of the initial (1) slag of melting of copper-zinc concentrates and samples obtained after their melting and cooling rates: 0.3 (2), 1.7 (3), 900 °C/s (4).

We may note the proximity of the starting temperatures of the thermal effects associated with melting (1062 °C) and melt crystallization (1045 °C). The appearance of the effect on the DSC curve is characterized for a sample cooled at a rate of 900 °C/s (Fig. 3). The effect starts from 507 (T_{ons}) and its middle is at 533 °C (Tg), which is connected with a second-order phase transition and the resulting process of slag devitrification (Mazurin, 1986) for the sample cooled at 900 °C/s (Fig. 3). Two exothermal heating effects are revealed on further heating with the onset/maximum at 541/577 °C and 628/644 °C. Apparently, the "cold" crystallization of the slag - the ordering of its structure - takes place at these temperatures and the presence of a doublet of peaks is caused by its two phases.

Devitrification observed on heating the sample containing glass and further "cold" crystallization is connected with the formation of magnetite (exothermal effect) and the isolation of crystals of the iron-silicate phase with a slightly lower (in comparison with glass) quantity of iron oxide:

$$Fe_{1+x}SiO_{3+x+y} = Fe_{1+x-3y/4}SiO_{3+x} + y/4Fe_3O_4 . \qquad (2)$$

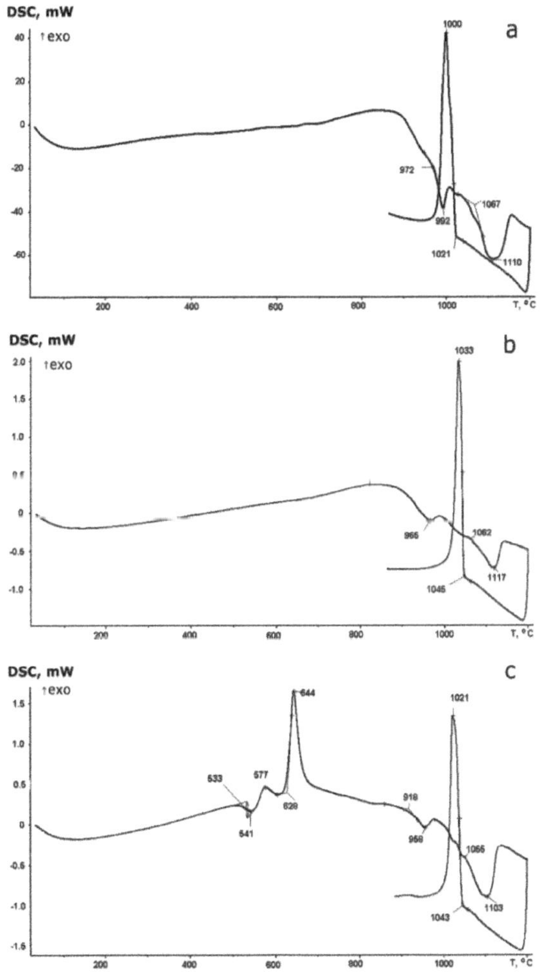

Fig. 3. Thermograms (20 °C/min, argon): of initial slag (a) from the melt of copper-zinc concentrates and the samples obtained after its melting and cooling at the rates of : 0.3 (b) и 900 °C/s (c)

The endothermic effects which started at 918 and 1055 °C point to the melting of the phase components of the slag. The temperature of the initiation of crystallization melting is 1043 °C which agrees with the temperature determined for the sample cooled at a rate of 0.3 °C/s. The temperature values and the change of heat capacity at devitrification (Δc_p) calculated from the experimental data and from the heat values of the 'cold' ($L_{c.cr.}$) and high ($L_{h.cr.}$) temperature crystallization of the hardened sample of slag are given in table 1. According to the data obtained, the heating of high-iron vitreous slag completely transforms it from an amorphous state to a crystalline state. The heat of slag melting is 165 J/g (Selivanov et al., 2009b).

Devitrification			$L_{c.cr.}$, J/g		$L_{m.}$, J/g		$L_{h.cr.}$, J/g
T_{ons}, °C	T_g, °C	Δc_p, J/(g·K)	1 peak	2 peak	1 peak	2 peak	
507	533	0.756	15	98	13	152	164

Table 1. The values of heat effect enthalpies of slags from a melt of copper-zinc concentrates at samples cooled at a rate of 900 °C/s.

The microstructure (Fig. 4) of the initial slag is represented by iron-silicates, magnetite and matte particles. Magnetite has been formed as fine-dispersed branchy dendrites. The isolated coarse matte particles are mechanically carried out together with the slag, which reach a size of up to 150 μm. The silicate constituent of the slag has small sulphide patches, which reach a size of 1.0-2.0 μm; they are concentrated along the boundaries of large iron-silicate aggregates. According to the X-ray spectral microanalysis data, the iron-silicate phase (Table 2) is heterogeneous, both in the main elements (silicon and iron) and the impurities dissolved in it. The calculated composition of the iron-silicates ranges from $Fe_2Si_3O_8$ to $Fe_3Si_2O_7$. With the elevation of the Fe/Si proportion in the iron-silicate phases, the content of calcium, sulphur, lead and zinc oxides in them decreases:

Fe	Ca	S	Pb	Zn	
$Fe_3Si_2O_7$	43.4	0.5	0.1	-	3.2
$FeSiO_3$	34.5	1.7	0.8	0.3	3.6
$Fe_2Si_3O_8$	22.1	7.9	1.6	0.6	4.4

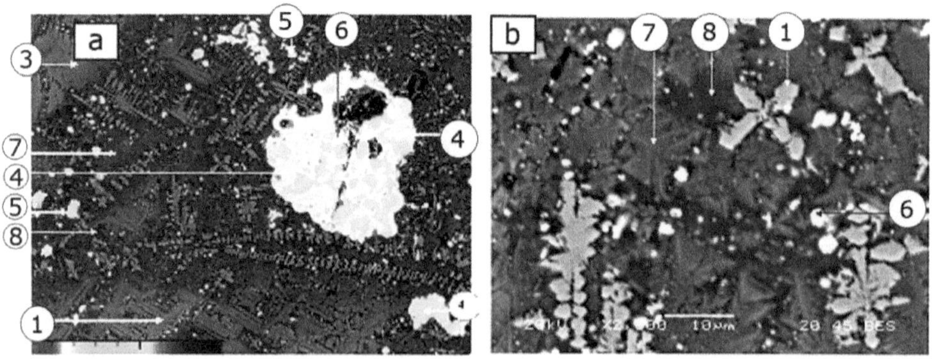

Fig. 4. The microstructure of the initial slag taken from smelting of copper-zinc concentrates and the point of local phase probing

The magnetic crystals (60.5 – 61.9% Fe) which are in the plane of the section also contain impurity elements in %: 1.0 Al; 2.5 - 3.0 Si; 0.3 - 0.4 Ti; 0.1 Cr; 0.1 Mg; 2.2 Zn and up to 0.2 Cu. The sulphide constituents of the slag are represented by the matte particles (48% Cu) with inclusions of zinc and lead sulphides. Solid solutions on the base of a ZnS-FeS system have a composition within the limits of $Zn_{0.24}Fe_{0.76}S$ to $Zn_{0.45}Fe_{0.55}S$ and, apart from the primary elements, contain 4.7-6.1% Cu and up to 0.5% Pb.

The iron-silicate phases which have different compositions in which magnetite and sulphides are found (Fig. 5) also constitute the base of the sample cooled at the rate of 0.3 °C/s. In comparison with the initial slag, the enlargement of iron, magnetite and sulphide silicate crystals has been marked. The main area of the matte is occupied by the conglomerates of coarse crystals of the iron-silicate phase, having a composition close to $Fe_3Si_2O_7$, and the spaces among them are filled up by $Fe_2Si_3O_8$ with small dendrites of $FeSiO_3$. As well as for the initial slag, the content of the impurity elements correlates with the iron content in the silicate:

Fe	Ca	S	Pb	Zn	Mg	
$Fe_3Si_2O_7$...	44.8	0.3	-	-	2.8	1.1
$FeSiO_3$	36.9	1.7	0.3	-	3.2	0.2
$Fe_2Si_3O_8$	23.5	4.3	1.0	0.4	4.0	-

№	Content, mas.%											Composition
	Mg	Al	Si	S	Ca	Ti	Fe	Cu	Zn	Pb	O	
1	0.1	1.0	2.5-3.0	0.1	0.3	0.3-0.4	60.5-61.9	0.2	2.2	-	31.5-31.7	Fe_3O_4
2	0.9-1.0	0.3-0.4	14.0-14.7	0.1	0.4-0.5	-	42.8-44.0	-	3.1-3.3	-	36.8-37.3	$Fe_3Si_2O_7$
3	-	-	0.2	27.1		-	20.6	48.8	3.0	-	-	$Cu_{2.1}FeS_{2.3}$ (matte)
4	-	-	0.2-0.3	32.4-32.5	0.1	-	38.2-38.6	25.1-26.0	2.7-3.2	0.5	-	$CuFe_{1.6}S_{2.4}$
5	-	-	0.5 1.4	33.2 32.6	0.2 0.5	-	43.0-30.8	6.1-4.7	16.9-29.8	0.5	-	$Zn_{0.24}Fe_{0.76}S$ $Zn_{0.45}Fe_{0.55}S$
6	-	-	0.2-0.5	20.0-23.3	-	-	12.7-19.5	19.1-23.8	0-4.8	33.3-41.9	-	(Pb,Cu,Fe)S
7	0.4	1.2	17.5	0.7-0.9	1.4-1.9	-	33.9-35.0	0.2	3.3-3.9	0.2-0.3	39.1-39.8	$FeSiO_3$
8	0.1	2.0-2.5	19.7	1.5-1.7	7.1-8.7	0.3	21.9-22.2	0.2	4.4	0.5-0.6	40.7-41.0	$Fe_2CaSi_3O_8$

Table 2. EPMA data on the phase composition of the initial slag from the smelting of copper-zinc concentrates (according to Fig. 4)

Matte particles with sizes from 1 to 15-30 μm are found mainly between iron-silicate blocks of a $Fe_3Si_2O_7$ composition. Matte decomposition into sulphides (bornite, sphalerite, galenite) occurs at cooling and their compositions - according to the analysis data - fluctuate widely (Table 3). More easily melted is lead containing a sulphide alloy from the margin along the surfaces of the matte particles. High-ferrous sphalerite ($Zn_{0.4}Fe_{0.6}S$) has been revealed both as an independent phase of around 2-10 μm in size and the inside of matte particles.

Magnetite (60.0-60.4% Fe) takes the form of both geometrical crystals and the form of dendrites arranged between $Fe_3Si_2O_7$ blocks and in direct contact with $Fe_2Si_3O_8$, both as in the initial slag and in the magnetite apart from the iron, which have revealed zinc, silicon and aluminium impurities as well as titanium (0.6%) and chromium (0.1%).

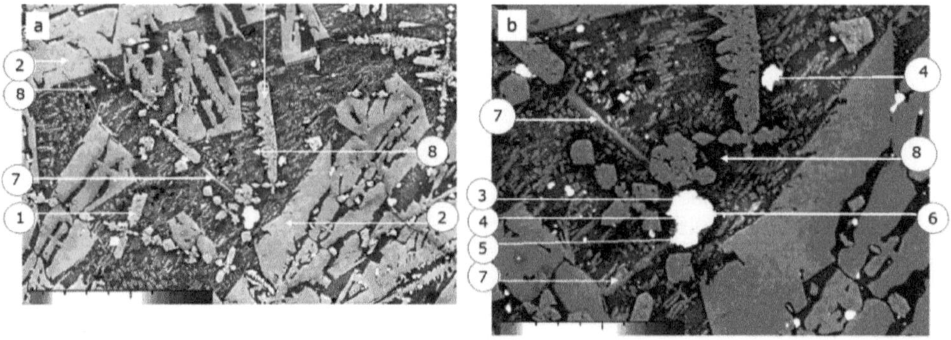

Fig. 5. Microstructure of the slag of copper-zinc concentrates' melting cooled from the melt at the rate of 0.3 °C/sec and the points of local phases' probing: an increase of x200 (a) and x500 (b)

The structure of a slag sample cooled at a rate of 900 °C/s is represented (Fig. 6) by glass and sulphide inclusions (up to 20 μm) with round forms. According to EPMA, the glass contains about 30% SiO_2 and 50% FeO_{1+x} (Table 4). The sulphide phase (inclusion of more than 15 μm in size) is inhomogeneous in its composition - its central part closely corresponds to Cu_5FeS_4. Lead and zinc sulphides are revealed inside matte particles.

№	Content, mas.%									Composition
	Al	Si	S	Ca	Fe	Cu	Zn	Pb	O	
1	2.5-2.6	1.7-2.0	-	0.2	60.0-60.4	-	2.8	-	31.4-31.6	Fe_3O_4
2	-	13.9-14.3	-	0.3	44.3-45.2	-	2.8-2.9	-	36.7-37.0	$Fe_3Si_2O_7$
3	-	0.3	26.7	-	18.9	53.0	0.6	0.8	-	$Cu_5Fe_2S_{4.5}$ (matte)
4	-	0.3-0.5	32.5	0.1	37.8	27.6	1.4	0.6	-	$Cu_2Fe_3S_{4.5}$ (matte)
5	-	0.6	27.8	0.1	35.7	6.1	26.7	3.0	-	$Zn_{0.4}Fe_{0.6}S$
6	-	0.3	18.2	-	13.0	17.0	1.0	50.6	-	$PbS-Cu_2S-FeS$
7	1.1-2.3	16.0-18.1	0.2-0.4	1.1-1.4	34.7-39.0	-	3.0-3.4	-	38.1-39.5	$FeSiO_3$
8	4.0-4.3	20.3-20.6	0.9-1.1	4.2-4.4	23.4-23.6	-	3.8-4.1	0.4	41.4-41.6	$Fe_2Si_3O_8$

Table 3. EPMA data on the composition of slag sample phases after their melting and cooling at a rate of 0.3 °C/s (according to Fig. 5)

The phase close to fayalite (Fe_2SiO_4) has not been revealed in any of study samples. All of the complexes of iron-silicate phases that were formed during slag cooling correspond to the atomic relations of Fe/Si within the limits of 0.7-1.5. Slag cooling at the high (900 °C/s) rate results in the formation of glass with the proportion of Fe/Si equal to about 1.4, without isolating magnetite in a self-dependent phase.

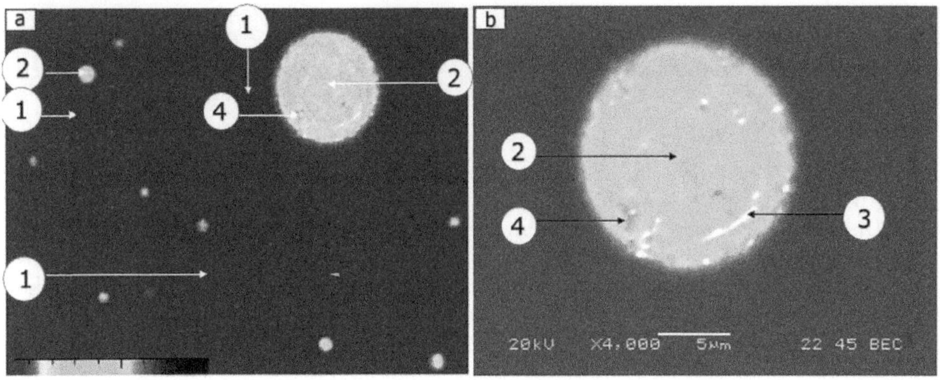

Fig. 6. Microstructure of the slag of copper-zinc concentrates' melting cooled from the melt at the rate of 900 °C/s and EPMA points

№	Content, mas.%										Composition
	Al	Si	S	Fe	Cu	Zn	Pb	O	Mg	Ca	
1	1.8-2.1	13.9-14.0	0.8	38.4-38.9	0.2-0.6	3.7-3.9	0.3-0.4	37.5-37.6	0.4	1.85	$Fe_{1.4}SiO_{3.4}$ (glass)
2	-	0.3-2.0	21.9-23.7	12.5-14.6	59.6-61.4	0.3-0.6	0.8-1.7	-	-	-	$Cu_5FeS_{3.3}$
3	-	0.4	23.3	11.6	44.7	16.3	1.7	-	-	-	Cu_5FeS_4-(Zn,Fe)S
4	-	0.3	21.1	9.7	57.9	0.3	8.0	-	-	-	Cu_5FeS_4-PbS

Table 4. EPMA data on the phase composition of the slag samples cooled from the melt at the rate of 900 °C/s (according to Fig. 6)

Proceeding from the fact that oxide phases with a high content of iron contain a lower quantity of CaO, one can draw a conclusion about the influence of a lime flux on phase formation (Selivanov et al., 2009a). For those slags with a Fe/SiO_2 ratio higher than 1, the increase of calcium oxide will not cause its solution in $Fe_3Si_2O_7$ but rather will favour the decomposition of this compound, which proceeds - in the limiting case - with the formation of calcium silicate and iron oxides. If the Fe/SiO_2 ratio in the slag is less than 1, then the CaO will dissolve in iron-silicate phases ($FeSiO_3$ and $Fe_2Si_3O_8$), reducing their melting temperature. One should bear in mind that these points are applied to those oxide melts which do not contain iron oxides of the highest valency. As has been shown in the works of (Okunev & Galimov, 1983; Tokeda et al., 1983), in oxide melts with a high degree of iron oxidation, CaO and Fe_2O_3 interaction establishes the formation of calcium ferrites.

Slow slag precipitation leads to the concentration of the matte particles among large grains of $Fe_3Si_2O_7$. On cooling, sulphide phases - the bulk of which is bornite - and crystallize from the matte. This is besides the fact that small crystals of sulphides form in the course of slag cooling, which can be explained by the peculiarities of the segregation of the oxide-sulphide system and by the change of the sulphides' solubility in iron-silicate melts.

The zinc on slag crystallization is distributed between oxide and sulphide phases. A (Zn,Fe)S independent phase containing 17-38% zinc has been revealed only at low rates of slag cooling. Lead in the slag takes both oxide and sulphide forms. The lead content in the iron-silicate phase increases as the content of silicon dioxide grows in it. Lead forms sulphide phases with 33-51% Pb which precipitate out of the sulphide melt (matte) on cooling (Selivanov et al., 2009b).

Thus, the rate of cooling of the melted slag influences the size and the number of forming phases, which defines the copper, zinc and lead distribution between oxide and sulphide forms. Changing the content of the calcium oxide in the slag and the rate of cooling, one can provide the preparation of the material for the subsequent redistribution of the precious metal's re-extraction. For example, in order to finish the slag by the floatation method (Dovshenko et al., 1997; Korukin et al., 2002; Sarrafi, 2004) it is necessary to form rather large sulphide particles, which is achieved by the familiar processes of reducing the cooling melts rate. The isolated concentrates besides copper will contain other non-ferrous metals, which designates their accumulation and concentration in the semi-products of closed-circuit processing schemes. The calcium oxide content and the rate of slag cooling also influence the composition of iron-silicate forming phases, the properties of which determine the expenditure of energy on slag grinding. The information about the structure of the high-ferriferous slag cooled at the high rate allows us to define characteristics of the glassy condition, including the devitrification temperature, heat and temperature, meaning the solid and liquid phases' crystallization of the quenched slag.

Oxide materials are important for pyrometallurgical processes for the production of non-ferrous metals, which are characterized by composition complexity and - apart from iron, silicon, calcium and aluminium oxides - contain the impurity elements Cu, S, Zn, Pb etc. Depending upon the composition and cooling rate from the melted state, the solid samples of oxide systems can be singled out both in crystalline and amorphous states. It is known that in oxide systems containing more than 50% SiO_2 and up to 20% of Fe_2O_3, a glassy state is formed even at a low cooling rate (Karamanov & Pelino, 2001). It is shown (Selivanov et al., 2009b, 2010) that all of the complex of iron-silicate phases generated during sample slow cooling corresponds to a molar Fe/Si ratio within the limits of 0.9-1.6. Sample cooling with a high (900 °C/s) rate results in the increase of this ratio by up to 3.4. It should be noted that a decrease of the cooling rate results in the increase of the portion and size of magnetite crystals and sulphide particles.

In connection with the study, the conditions of crystals' formation from amorphous high iron oxides based on FeO_x-SiO_2-MeS (Me – Cu, Zn) systems and the determination of the forming phases' composition are of great interest.

The sample of slag from the melt of copper-zinc concentrates in the Vanyukov furnace was melted in the furnace at a temperature of 1250 °C and cooled down using the water granulation method. The cooling rate of the oxide melt calculated from equation 1 is 900 °C/s. The granules obtained are isothermally annealed in the resistance electric furnace (during 5 and 60 min) at a temperature of 750 °C (Gulyaeva et al., 2011).

X-ray analysis results (Selivanov et al., 2009b) have shown that with the cooling of the oxide melt at a rate of about 900 °C/s, an amorphous product is formed (Fig. 2). The reflexes corresponding to the crystal phase are virtually absent at the diffractogram; however, the diffusion scattering characterizing the availability of a short-range order in the formed glass was observed.

The thermogram received by means of a quenched sample heating is shown in Fig. 3. Based on the data received, the heating of high iron vitrified oxides up to 1230 °C and their cooling at a rate of 20 °C/min results in their transition from an amorphous to a crystalline state. The stability factor of the vitreous state is shown in the following equation (Biswas et al., 2010):

$$\Delta T = T_g - T_{c.cr.}, \tag{3}$$

where T_g and $T_{c.cr.}$ - devitrification temperatures (533 °C) and the onset of "cold" crystallization (555 °C). For the sample under study, the ΔT value amounts to 22 degrees, which points to the non-stability of the amorphous state.

The annealing of sampled granular particles of slag at a temperature of 750 °C does not result in the external change of their forms. The data (Gulyaeva et al., 2011) of the X-ray analysis of amorphous oxide samples after annealing for durations of 5 and 60 min showed that the material heating results in the formation of a crystalline state. In diffractograms (Fig. 7), reflexes of annealed samples typical for fayalite, quartz and magnetite (very weak) were detected. Fayalite reflection (d = 2.50 Å) intensity increases from 74 (Fig. 7 a) to 86 pulses/s (Fig. 7 b) with the growth of the annealing duration, whereas magnetite reflection (d = 1.48 Å) intensity does not change and is equal to 11 pulses/s.

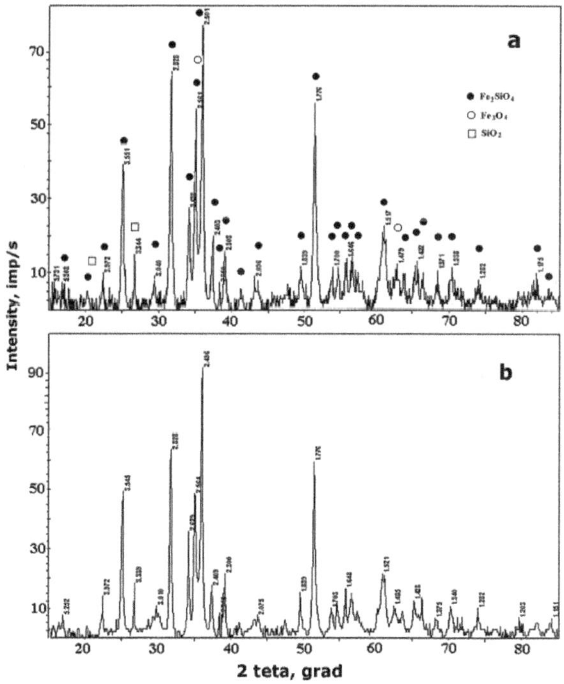

Fig. 7. X-ray spectra of granular slag samples from the melts of copper-zinc concentrates annealed at 750 °C within 5 (a) and 60 min (b)

The microstructure analysis of the samples (Fig. 8) received after annealing showed that initially (τ = 5 min) single faceted crystals of magnetite are presumably formed. Over the entire area of the metallographic section, a thin grid of crystal is formed with a cell size of 0.3 - 0.5 μm. It should be noted that an increase of the annealing duration (τ = 60 min) of an amorphous sample results in formation of certain particles of larger dendrites with the axes of the second- and third-order in the area close to the surface. Closer to the particles' border, the dendrites' axes' length decreases, the phases having - in the metallographic section plane - a structure in the form of triangles and stars sized of 5-10 μm form around them. Proceeding from the thermal analysis data (Sycheva & Polyakova, 2004), one can evaluate the viscosity crystallization criterion, which is equal to:

$$\alpha = T_g/T_m, \quad (4)$$

where T_m - temperature of the dissolution of the crystalline phases in the melt (1103 °C), defining together with the forces of surface tension the capacity of glasses to volume crystallization. The value of α calculated by this equation reached 0.6, which shows the inclination of the studied oxide glass-to-surface crystallization. The mechanism of the initial formation of crystalline phases is apparently explained by the annealing temperature (750 °C) which is higher than that of cold crystallization. During the heating of a quenched sample from an oversaturated silicate phase forming the glass diffusion separation of magnetite and the formation of iron-silicate phase crystals, there occurs a little less (compared to glass) iron content through equation 2.

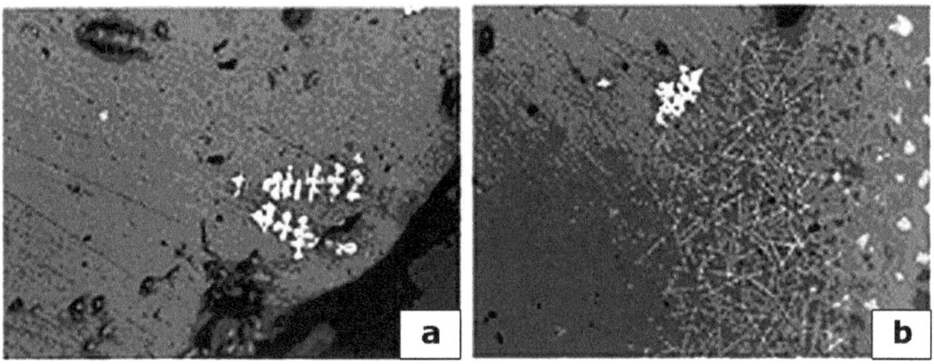

Fig. 8. Microstructure of the slag sample (x500) annealed at 750 °C during 5 (a) and 60 min (b)

The analysis of the sample annealed within 60 min by the X-ray spectral microanalysis of the characteristic radiation of the elements showed that the basis of the dendrite phase is formed by iron oxides while the basis of glass is formed by iron-silicates. The composition of the dendrite phases depending on crystal forms is variable - one of the phases in the form of triangles and stars is richer in iron and close to magnetite in its composition. They contain fewer impurity elements of silicon (0.9%) and calcium (0.2%) (Fig. 9, Table 5). Copper (0.2-0.7%) and zinc (2.7-3.7%) concentrations in magnetite are virtually unconnected with the geometric form of crystals. The silicate phase is close in composition to $Fe_{1.36}SiO_{3.36}$: the content of copper in it reaches 0.8% and that of zinc reaches 4.1%. The presence of sulphur determines the formation of fine bornite phases of non-spherical forms (1-6 μm) in the sample and larger chalcosine phases with a size of 15 - 25 μm.

Thus, the granulation of iron-silicate oxide melts containing more than 30% SiO_2 results in the formation of glass. The annealing of the granuled slag of the autogenous smelting of copper-zinc concentrates at 750 °C results in the initial separation of magnetite in the form of geometric crystals from an oversaturated iron-silicate matrix and - furthermore - in the form of a dendrite structure. The diffusion mechanism of magnetite crystals' growth has a superficial character. The composition of the crystallizing phases and the dissolubility of nonferrous metals in them have been established.

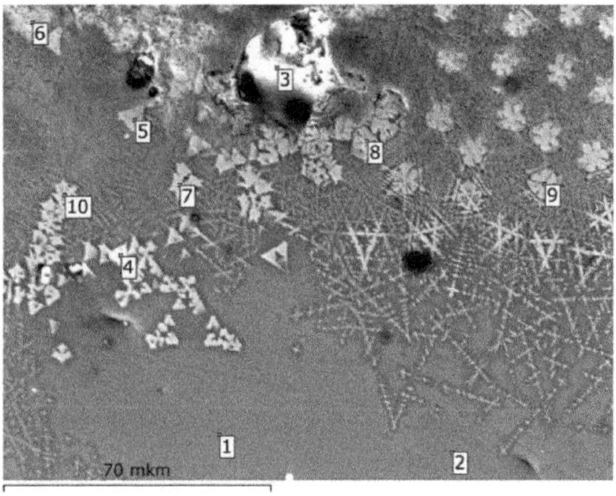

Fig. 9. Microstructure of the sample after annealing at 750 °C within 60 min and EPMA points

№	Content, mass, %					
	Si	Fe	S	Cu	Zn	Al
1	14.9	40.4	0.5	0.8	3.8	2.7
2	14.6	41.1	0.4	0.5	4.1	2.8
3	0.2	3.5	21.5	76.5	-	0.5
4	5.7	26.2	7.9	20.9	2.7	1.7
5	0.9	69.6	-	0.4	2.7	1.9
6	4.1	61.9	0.3	0.6	3.7	2.0
7	4.9	60.7	0.1	0.2	2.6	2.8
8	2.3	64.5	0.1	0.7	3.8	2.2
9	3.1	63.3	-	0.2	3.7	2.7
10	2.4	64.7	-	0.4	3.4	2.7

Table 5. EPMA data on the phase composition of a quenched slag sample annealed at 750 °C during 60 min (Fig. 9)

4. Effect of the cooling rate on the phase composition and structure of copper matter converting slags

As was mentioned (Selivanov et al., 2009a, 2009b), the distinguishing feature of the production process in the "Sredneuralsky Copper Smelter Plant" lies in the fact that mattes containing 45–55% copper recovered upon the smelting of copper–zinc concentrates in the

Vanyukov furnace are converter. Apart from copper and precious metals, the mattes concentrate zinc, lead, arsenic and antimony. During the conversion, a part of these metals passes into a gas phase and dust, and the other part of the metals is redistributed between white matte and slag and then between copper and slag. Thus, the precipitating converter slags have a high content of precious metals, the re-extraction of which increases the coefficient of an integrated approach to the raw materials' use. The choice of the slag processing method is determined based upon the forms of the metals.

A thermodynamic simulation (TDS) was performed for the working body of the following composition,%: 21.0 FeO; 16.0 Fe_2O_3; 20.3 SiO_2; 5.1 ZnO; 11.0 CuO; 2.6 Al_2O_3; 2.0 CaO; 1.2 MgO; 2.0 Sb_2O_3; 1.2 Pb, 3.1 S. The degree of iron oxidation (α) in the working body determined Fe^{3+}/Fe^{2+} the ratio amounts as 0.4. During the thermodynamic simulation, we used the compounds inherent in the FeO_x-SiO_2-FeS-Cu_2O-ZnO system. The calculations were carried out for the 100 kg working body at a gas phase (nitrogen) volume of 2.24 m^3 over the slag and at a pressure in the system equal to 0.1 MPa.

According to the TDS results, the main components of the equilibrium system which was cooled under equilibrium conditions were: Fe_3O_4, Fe_2SiO_4, SiO_2, ZnS, Cu_2S, Cu_5FeS_4, $CuFeS_2$, as well as metallic copper and Cu_2Sb; the possibility of the solidification of the last two components in such systems was noted earlier (Selivanov et al., 2000, 2004). A change in the degree of iron oxidation in the working body does not introduce qualitative changes in the phase composition of the slag but it does affect the interfacial distribution of non-ferrous metals. Copper occurs predominantly in the form of a metal or a sulphide (Fig. 10), and the temperature changes influence their mass ratios. As such, a larger amount of Cu_2S forms within the temperature range of 700-1100 °C, whereas the production of metallic copper in this temperature range decreases substantially. The reduction of the temperature favours the formation of complex copper sulphides, mainly bornite.

Zinc at high temperatures is represented by the oxide compounds ZnO, $ZnFe_2O_4$, $ZnSiO_3$ and $ZnAl_2O_4$, the fractions of which decrease as the slag cools down. At low temperatures, the probability of the solidification of zinc sulphide - ZnS - is high.

The samples for investigation have been prepared from a converter slag containing %: 40.7 Fe, 3.0 S, 8.5 Cu, 4.0 Zn, 19.5 SiO_2, 1.9 CaO, 0.7 MgO, 2.3 Al_2O_3, 0.2 Sb, 0.9 Pb, 0.1 As. According to the data of the X-ray diffraction phase analysis, the base phases in the initial converter slag are a fayalite of Fe_2SiO_4 and a magnetite of Fe_3O_4. Copper is mainly detected as bornite and copper sulphide. Zinc is revealed in the form of a sulphide with a sphalerite structure. The melting of the samples followed by rapid cooling decreases the diffraction reflex intensities and preserves the non-equilibrium high-temperature phases. With the cooling of the sample at the rate of 900 °C/s, a significant amount of amorphous phase forms. No reflections of the ZnS phase have been detected at a high cooling rate.

When the initial converter slag sample is subjected to differential thermal analysis during heating in an argon atmosphere (Selivanov et al., 2009a), the following melting-induced endothermic heat effect is observed: it begins at 1079 °C and has a maximum value at 1122 °C (Fig. 11). During its cooling, we established the solidification temperature of the slag melt, which is 1108/1078 °C. According to the mass-spectrometry data, a certain amount of SO_2 precipitates in the heating of the sample (beginning from 300–400 °C), which results from the interaction of sulphides with the oxides of iron of the highest valence. The slag weight loss upon heating up to 1300 °C is 1.2%.

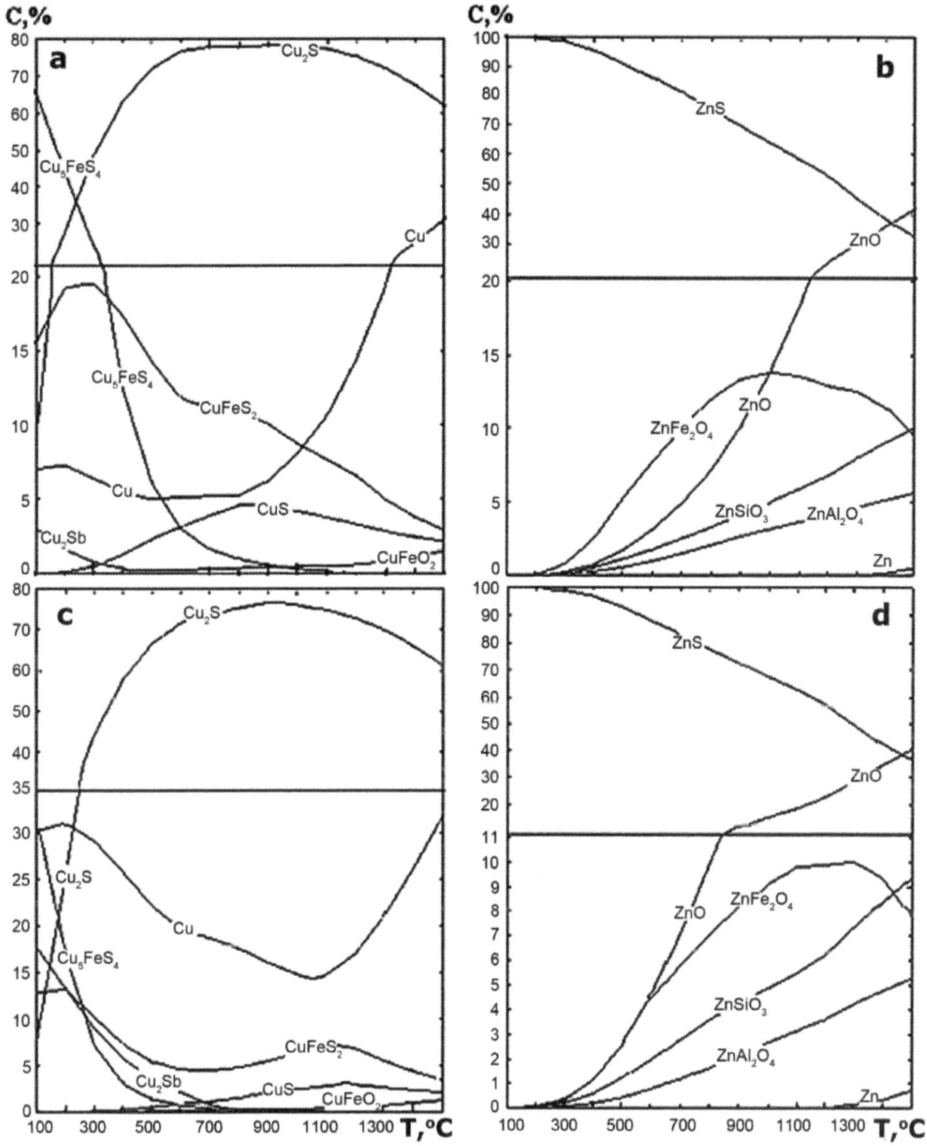

Fig. 10. Distributions (relative %) of copper (a, b) and zinc (c, d) in the components of a condensed phase, depending upon the temperature at the oxidizing degree of iron in the converter slag: 0.4 (a, b) and 0.1 (c, d)

A comparison of the thermograms of the initial slag sample and a sample cooled at a rate of 0.3 °C/s indicates that they are identical. However, a sample cooled at a rate of 900 °C/s is characterized by the appearance of an effect at 533 °C in the DSC curve, which is caused by a second-order phase transformation during devitrification. When this sample is heated further, we detect an exothermic heat effect with an onset/maximum at 608/635 °C. This effect is interpreted as "cold" slag crystallization (ordering its structure). The endothermic effects at 946/963 °C and 1064/1127 °C point to the melting of the sulphide and oxide components of the sample. On the DSC curves of the samples cooled at the rates of 0.3 and 900 °C/s, the solidification effects were equal to 1055/1049 °C and 1085/1072 °C, correspondingly.

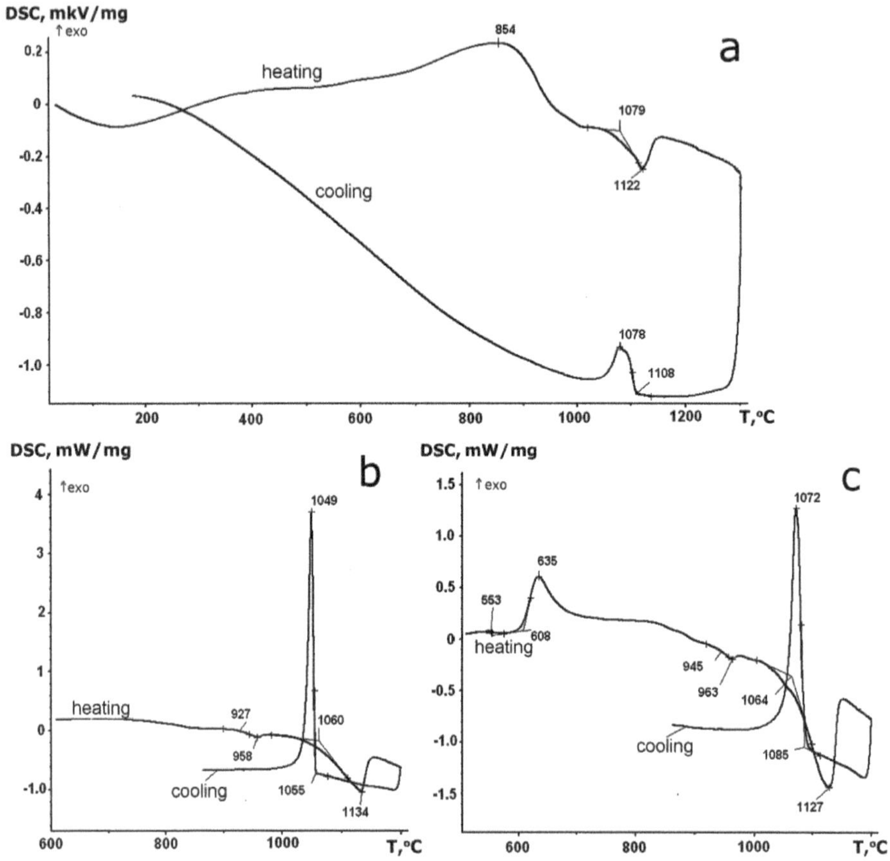

Fig. 11. Sample thermograms of (a) the initial converter slag and the slags that form upon the cooling of the melt at the rates of (b) 0.3 and (c) 900 °C/s

In essence, the temperatures and enthalpies of the thermal effects of the heating of the granulated slags from melts of copper-zinc concentrates and the converting of the matte have similar values (Table 6).

Devitrification			$T_{c.cr.}$, °C	$L_{c.cr.}$, J/g	L_m, J/g		$L_{h.cr.}$, J/g
T_o, °C	T_g, °C	Δc, J/g K			1 peak	2 peak	
533	553	0.150	608	66	4	164	157

Table 6. Temperatures and enthalpies of the thermal effects at the heating of the granulated and converter slags

The microstructure of the initial converter slag is represented by iron-silicates and matte particles (Fig. 12). The slag contains a large number of 100 µm magnetite crystals of a regular shape and spherical matte particles smaller than 300 µm. The matte particles have an eutectic structure (copper sulphides, bornite, metallic copper). The silicate constituent of the slag has a small amount of metallic and sulphide copper. These inclusions have sizes equal to 0.1 - 2.0 µm and are concentrated along the boundaries of large iron-silicate aggregates.

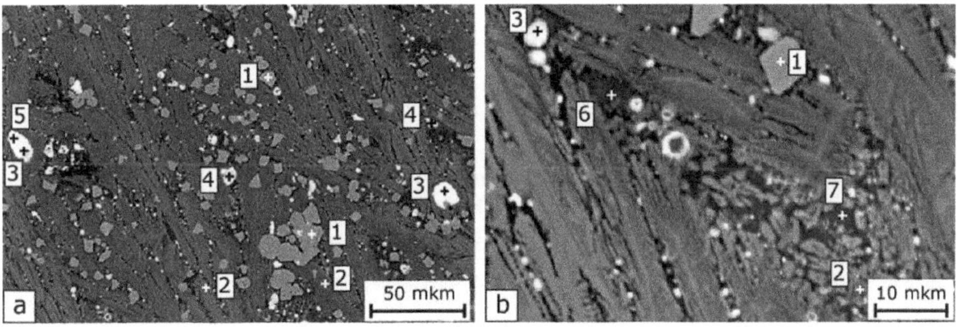

Fig. 12. Microstructure of the sample of industrial converter slag and EPMA points

According to the electron-probe microanalysis of the commercial sample, the iron-silicate phase is heterogeneous in terms of both major (iron, silicon) and dissolved impurities (Table 7). The calculated composition of the iron-silicates ranges from $Fe_{0.94}SiO_{2.94}$ to $Fe_{1.59}SiO_{3.59}$. The iron-silicates contain 0.5–2.0 Al_2O_3, 0.5–1.0 MgO, 0.2–0.5 K_2O, 0.2–1.1 CaO, 5.3–6.6 Zn, and 0.1–0.6% S. Moreover, we detected a silicate phase that corresponds to the SiO_2–FeO–CaO–Al_2O_3–Zn(Pb)O system and has a low iron content with a high calcium content. The magnetite crystals (61.4-63.3% Fe) located in the plane of the section contain the following impurity elements: 1.2-1.3 Al, 0.4–1.2 Si, 0.3–0.2 Ti, 0.1–2.5 Cr and 2.7–3.2% Zn. The sulphide phases in the slag are represented by bornite- and sphalerite-based solid solutions. The copper content in the bornite solid solution is lower than the stoichiometric copper content. The ZnS-based phase (sphalerite) contains 37.9–53.3 Zn, 0.9–7.3 Cu, 11.5–21.6% Fe, and a near-stoichiometric sulphur content. The regions of the PbS–Cu_2S–FeS solid solution are located along the periphery of the sulphide phases.

The base of the slag sample cooled at the rate of 0.3 °C/s also consists of iron-silicate phases, magnetite and sulphides (Fig. 13). The re-melting and slow cooling of the slag result in a significant coarsening of the formed crystals. The area of its polished section is a mainly occupied silicate phase with a composition close to $Fe_{1.59}SiO_{3.59}$; between them there are small $Fe_{1.12}SiO_{3.12}$ dendrites and calcium- and silicon-rich phases (Table 8). Apart from iron, the magnetite also contains zinc, titanium, silicon and aluminium impurities as in the initial converter slag.

№ point	Content, mas.%													Phases
	Mg	Al	Si	S	K	Ca	Ti	Cr	Fe	Cu	Zn	Pb	O	
1	-	1.2-1.3	0.4-1.2	-	-	0.1	0.2-0.3	0.1-2.5	61.4-63.3	-	2.7-3.3	-	30.4-30.8	Fe_3O_4
2	0.3-0.6	0.3-1.0	13.8-17.0	0.1-0.6	0.1-0.3	0.1-0.8	-	-	38.1-43.9	-	4.3-5.3	-	36.5-38.3	Iron-silicate
3	-	-	0.3	26.2-27.6	-	0.1	-	-	17.0-20.8	50.1-55.1	0.3-0.4	1.0-1.2	-	Cu_5FeS_4 solid solution
4	-	до 0.5	0.4-3.4	30.0-32.7	0.1	0,2	-	-	11.5-21.6	0.9-7.3	37.9-53.3	-	-	(Zn,Fe,Cu)S
5	-	-	0.2	16.6	-	-	-	-	9.4	23.4	0.6	49.8	-	(Pb,Cu,Fe)S
6	-	3.0	17.7	0.4	0.7	0.9	0.4	-	33.2		3.5	0.6	39.4	$Fe_{0.94}SiO_{3.47}$
7	0.3	4.6	21.9	1.6	0.2	6.3	0.4	-	14.6	0.5	4.4	3.1	42.1	Iron-silicate

Table 7. EPMA data on the phase composition of the initial converter slag (according to Fig.12)

The coarse sulphide particles of a size of 15–30 μm consist of bornite - the composition of which varies from $Cu_{7.2}FeS_{6.4}$ to $Cu_{3.2}FeS_{3.3}$ - and a PbS–Cu_2S–FeS alloy (Fig. 13). The bornite is located in the centres of the particles, while the lead-containing sulphide alloy with a lower melting point forms on the fringes on its surface.

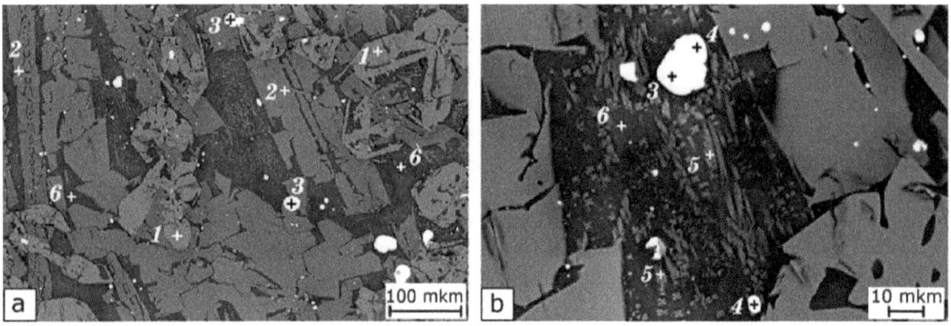

Fig. 13. The sample microstructure of the converter slag cooled at a rate of 0.3 °C/s and EPMA points

The structure of the slag sample cooled at a rate of 900 °C/s is represented by glass, magnetite, iron-silicate phase crystals and sulphide inclusions of a spherical form up to 10 μm (Fig. 14). According to the EPMA data (Table 9), the glass has about 36% SiO_2 and 51% FeO_{1+x}. Acicular crystals 5–15 μm long and about 1 μm thick are clearly visible against the background of the glass; their composition is close to that of iron-silicate of $Fe_{3.4}SiO_{4.4}$. Magnetite (60.4–61.6% Fe) is present in the form of dendrites. The sulphide phase (its coarsest particle is 6.4 μm in size) is inhomogeneous and its central portion corresponds to the formula of $Cu_{5.4}FeS_{3.4}$. The distribution of nonferrous metals in the sulphide particle is also non-uniform: the centre contains 6.2% Zn, 1.6% Pb, and 1.4% As, and the periphery contains 1.3% Zn, 2.0% Pb and 4.7% As.

№ point	Content, mas.%									Phases
	Al	Si	S	Ca	Fe	Cu	Zn	Pb	O	
1	2.4-2.5	0.4	-	0.1	63.3	-	2.4-2.5	-	30.8	Fe_3O_4
2	-	15.0	-	0.3	42.1-42.9	-	3.4	-	37.2-37.3	Iron-silicate
3	-	0.2-0,5	21.5-28.5	0.1	8.2-14.9	55.4-69.2	0.3	0.5-0.6	-	Cu_5FeS_4 solid solution
4	-	0.3-0.8	18.8-18.9	0.1-0.2	9.3-13.1	23.4-25.3	0.5-0.8	40.6-47.8	-	(Pb,Cu,Fe)S
5	2.4-2.7	16.4-16.9	0.4	1.9-2.1	32.0-33.4	< 0.2	5.3-5.5	0.9-1.0	38.1-38.5	Iron-silicate
6	5.8-7.2	19.9-22.5	0.6-1.1	5.7-6.9	13.0-15.3	-	3.9-6,2	2.3-2.7	40.7-42.5	Iron-silicate

Table 8. EPMA data on the phase composition of the converter slag cooled at a rate of 0.3 °C/s

This data indicates that an increase in the cooling rate leads to vitrification. However, even at a cooling rate of 900 °C/s an iron-silicate phase and magnetite solidify and sulphides precipitate. The compositions of the iron-silicate crystalline phases vary over wide limits: as the cooling rate increases, high-iron modifications form and the fraction of magnetite

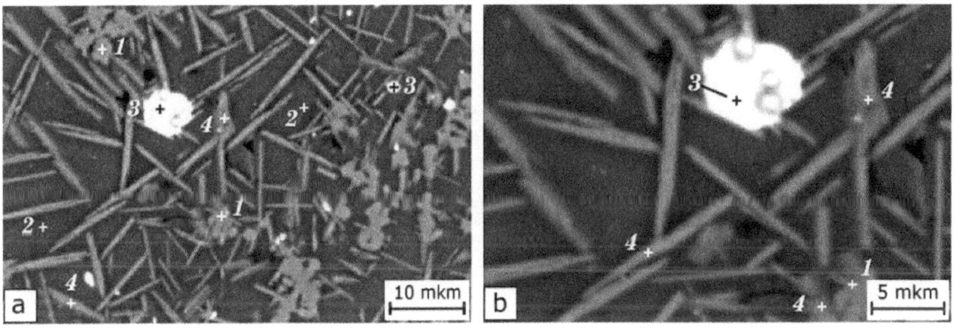

Fig. 14. Microstructure of the converter slag cooled at a rate of 900 °C/s and EPMA points

№ point	Content, mas.%										Phases
	Al	Si	S	Ti	Fe	Cu	Zn	Pb	O	As	
1	2.7-3.1	1.4	-	0.2	60.4-61.1	-	2.4-2.5	0.1	31.4-31.5	-	Fe_3O_4
2	2.3-2.6	15.5-17.3	0.8-0.9	0.1	34.2-41.0	0.2	4.7-5.4	0.9	28.0-38.1	0.3	Glass
3	-	0.3-0.9	17.2-20.3	-	9.8-10.8	62.6-64.8	0.4-6.2	0.9-2.0	-	0.6-4.7	Cu_5FeS_4
4	1.5-1.9	7.4-7.8	0.2-0.3	0.1	51.1-51.3	0.2-0.03	3.2-3.4	0.2-0.3	34.0-34.1	<0.1	Iron-silicate

Table 9. EPMA data on the phase composition of the converter slag cooled at rate of 900 °C/s (according to Fig. 14)

crystals decreases. A decrease in the cooling rate of the slag is accompanied by magnetite formation (endothermic effect) and the precipitation of iron-silicate crystals with a lower (compared to glass) iron content.

According to the EPMA data, none of the samples contains a phase close to stoichiometric fayalite (Fe_2SiO_4). The whole set of the iron-silicate phases that form during the slow cooling of the converter slag corresponds to an Fe/Si atomic ratio of 0.9–1.6. The cooling of the slag at the high (900 °C/s) rate is a result of the increase in this ratio to 3.4. A decrease in the cooling rate of the molten slag favours an increase in the fraction and sizes of magnetite crystals and sulphide particles. The distribution of non-ferrous metals between phases changes according to the fraction of the sulphides.

Copper is concentrated in the bornite-based solid solution, which forms in all the samples. The bornite content increases as the cooling rate decreases, which can be explained by the specific features of the solidification and separation of oxide–sulphide systems that are related to changes in the sulphides' solubilities. The results obtained agree with the TDS data with regard to the predominant formation of copper sulphides during the cooling of the slag. We failed to detect metallic copper in real slags, irrespective of the cooling rate (see Fig. 10). During slag solidification, zinc is distributed between oxide and sulphide phases. An individual (Zn,Fe)S phase containing 38–53% zinc was revealed only at the low cooling rate. Lead in the slag is present in both oxide and sulphide forms. Its content in the iron-silicate phase correlates (increases) with the silicon dioxide content. Lead forms the regions of a sulphide phase (40–48% Pb) of 1–2 μm in size at the low cooling rate. The formation of the oxide compounds of zinc and lead supports the absence of an equilibrium state in all of the investigated slag samples. Sulphide phases have a high arsenic content. The arsenic content in a bornite-based solid solution reaches 4.7% in a slag sample cooled at a rate of 900 °C/s and the arsenic content in iron-silicates is 0.1%.

Thus, when changing the cooling rate of the slag, we can affect the forms of copper, zinc, arsenic and lead in it in order to prepare the slag for the additional recovery of precious metals (Selivanov, 2009a). Moreover, the slag cooling conditions affect the composition of the iron-silicate forming phases, and its properties control the energy consumed for grinding as well as the possibility of using magnetic separation methods for the precipitation of iron oxides, and so on. The converter slag contains both mechanically-introduced coarse matte particles and fine sulphides, which precipitate during the solidification of an oxide–sulphide melt. The cooling rate of the molten slag controls both the phase composition and the particle size of the oxide and sulphide forming phases. The oxide component transforms into a glassy state at a high (900 °C/s) cooling rate of the slag. The devitrification and cold crystallization of the glass falls within the temperature range of 533 - 635 °C.

The copper in the slag is mainly represented by the bornite-based solid solution, the content of which increases as the cooling rate decreases. Zinc and lead are distributed between the oxide and sulphide components. The individual sulphides of these metals are only revealed with the low cooling rates of the slag.

5. Forms of metals finding in the slag of combined melting – The converting of copper concentrates

The autogenous processes of the converting of copper-containing raw materials, including the application of both «Noranda» and combined melting-converting (CMC) units have been widely used in non-ferrous metallurgy. However, the copper content in the slags which are formed during the smelting of concentrates in these units is rather large. In order to decrease the loss of metals and to choose the methods of the processing of slags, it is of the prime importance to reveal the forms of the existence of the precious components in them. The molten slags are - in their compositions - close to the FeO_x-SiO_2 system and, at the temperatures corresponding to pyrometallurgical processes, agree with homogeneic melts (Vanyukov & Zaitsev, 1969, 1973). During slag cooling, a number of micro-processes connected with compound crystallization, liquation phenomena and the change of the detection of forms of non-ferrous metals occurs (Kukoev et al., 1979). The last of these, in turn, determine the choice of methods for the re-extraction of precious components from the slag.

As a starting sample – the slag of a pilot unit of CMC cooled at a rate of about 0.5 °C/min (Selivanov et al., 2004). According to the data of the chemical analysis, the slag contained, %: 1.2 Cu, 55.9 Fe_{total}, 4.9 Fe^{3+}, 53.3 Fe^{2+}, 0.4 Fe_{met}, 3.1 S, 16.0 SiO_2, 3.6 Zn, 0.1 Pb, 0.1 As, 0.1 Sb, 0.5 CaO and 0.5 Al_2O_3. The studied slag sample in its chemical composition is close to the slag of copper matte converting. The relatively high sulphur content in the slag allows it to be referred to the oxide-sulphide melts class, the crystallization of which must be accompanied by a number of complex interactions changing the form of the metals' detection (Kukoev et al., 1979; Selivanov et al., 2000).

The thermodynamic modelling of the processes (Moiseev & Vyatkin, 1999; Selivanov et al., 2004) occurring during slag cooling was carried out for the working body and is in substantial agreement with the slag composition taken for the investigation. The thermodynamic functions of elements and compounds in the condensed (Cu, Cu_2O, CuO, $CuFe_2O_4$, $Cu_2Fe_2O_4$, Cu_2S, $CuFeS_2$, Cu_5FeS_4, Cu_3As, Cu_2Sb, Zn, ZnO, ZnS, $ZnSiO_3$, Zn_2SiO_4, $ZnAl_2O_4$, $ZnFe_2O_4$, Fe_{met}, FeO, Fe_3O_4 , Fe_2O_3, FeS, FeS_2, $FeSiO_3$, Fe_2SiO_4, $FeAl_2O_4$, Fe_2ZnO_4, Pb, PbO, Pb_2O_3, PbS, $PbSiO_3$, As, As_2O_3, As_2O_5, As_2S, As_2S_3, Sb, Sb_2O_3, Sb_2O_5, Sb_2S, Al_2O_3, Al_2SiO_5) and gaseous (S_2, SO_2, SO_3, Zn, ZnO, N_2, O_2, Pb, PbS, As_2O_3, As_2O_5, Sb_2O_5, Sb_2O_3, etc.) states have been used for the calculations. The modelling was carried out during the changing of the temperature from 1520 to 25 °C with steps of 50 degrees.

According to the TDS data, slag cooling leads to the changing of the parts of phases and the forms of the metals which exist (Fig. 15). Accordingly, the working body temperature decrease increases the crystallization probability of $FeSiO_3$ and Fe_2SiO_4 iron-silicate compounds and favours magnetite formation, which can be explained by the disproportionate amount of iron oxide (II) and by the interactions between non-ferrous metals oxides and iron oxide:

$$4FeO = Fe_3O_4 + Fe, \tag{5}$$

$$FeO + MeO = Me + Fe_3O_4. \tag{6}$$

For the non-ferrous metals in the slag, one would expect the changing of their forms of existence at the expense of reactions between sulphides and oxides (Belyaev et al., 2001; Spira & Themelis, 1969). If, at a high temperature, the copper in the slag is preferably in the form of sulphide, then cooling can lead to its transition into its metallic state:

$$3Cu_2S + FeS = Cu_5FeS_4 + Cu, \tag{7}$$

$$Cu_2S + 4FeO \rightarrow FeS + 2Cu + Fe_3O_4. \tag{8}$$

The working body (slag) cooling favours a ZnO → ZnS transformation according to the reaction:

$$ZnO + FeS \rightarrow ZnS + FeO. \tag{9}$$

This means that copper, antimony and arsenic are the most electropositive metals for the problem at hand and one should also expect the formation of Cu-Sb-As alloys as well as copper.

Iron-silicates and oxides, as well as bornite, were found in the slag by way of X-ray diffraction analysis (Fig. 16) and they were identified on the basis of the data from (PC-PDF, 2003) for Fe_2SiO_4, FeO, Cu_5FeS_4 and Fe_7SiO_{10}. Some discrepancies in the X-ray reflexes between their meanings for pure compounds is suggestive of the formation of solid solutions which distort the minerals' lattice.

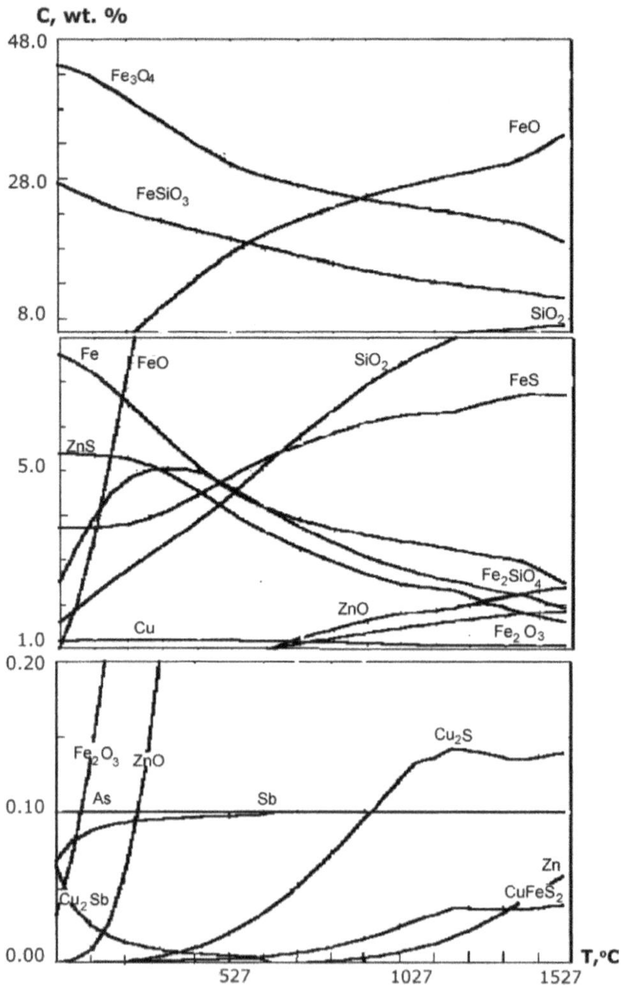

Fig. 15. Changing of the forms of metals present in the CMC slag according to the TDM data

The main structural components of a slag are fayalite and an iron-silicate phase with a high FeO content (Fig. 17). Fayalite, which is close to Fe_nSiO_{2+n} in composition, is presented by a solid solution of iron, zinc and calcium silicates. The iron-silicate phase (83% FeO) conforms to the $nFeO \cdot mSiO_2$ formula. The micro-hardness of fayalite varies within the limits of 5300 to 8400 MPa but that of the micro-hardness of the iron-silicate phase within the limits of 4100 to 6100 MPa. Apart from the main phases in the slag, we have revealed iron oxides with a micro-hardness of 5100 to 8400 MPa, which exceeds the values characteristic of pure wustite (~ 4300 MPa), iron sulphides which deposit in the form of FeS troilite (2440-2800 MPa) and FeS-FeO eutectic (3400 to 4900 MPa) of different dispersivity, zinc ferriferous sulphides (christophite) and, in lesser amounts, bornite and a solid solution on its base.

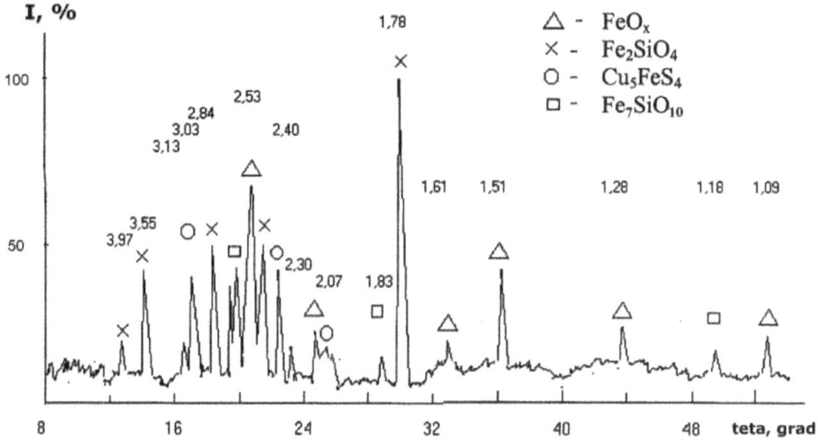

Fig. 16. X-ray diffractogram of the CMC slag

The crystallization of slag containing 16% silica, according to the diagram of the $FeO\text{-}SiO_2$ condition, proceeds from a temperature of 1250 °C and is accompanied by the rejection of iron-silicate phase crystals and excess wustite (Selivanov et al., 2004). X-ray spectral microanalyses of the slag (Fig. 18) enable us to establish the composition of the phase components (Table 10). As follows from the data obtained, the oxide and iron-silicate phases do not contain copper. A small quantity of non-ferrous metals (Zn, As - 0.2%) was dissolved in the FeO_x phase. Unlike FeO_x, the silicate phases (the first - with 13.4% and the second - with 5% of Si) have a zinc content of up to 3 - 4%. The silicate phases conform to compositions of Fe_2SiO_4 (fayalite) and Fe_7SiO_{10} (a solid solution with a high content of FeO).

Non-ferrous metals are preferably concentrated in metallic (a size up to 40 μm) and sulphide (up to 100 μm) inclusions. The metallic phase is represented by a Cu-Sb alloy with Sn (3-5%), As (5 - 10%) and Ni (1 - 7%) dissolved in it. The sulphide phases are formed with the participation of Cu, Zn and small quantities of Pb (0.2-0.8%) and As (0.2%). A conglomerate of sulphides and wustite occurs between the crystals of fayalite, and this conglomerate is revealed in the part of the section which is sized 200 μm (Fig. 18). There is a Cu-Sb particle in the centre of the conglomerate. The composition of this particle changes from the surface to the centre. The surface of the particle contains 50% copper and about 30% of Sb, whereas the internal part is more than 50% of the antimony. The heterogeneity of the particle is seen in the micro-structure obtained by the absorbed electrons during X-ray spectral microanalysis. The coefficients of distribution during liquation (K_l - the proportion of elements contents in copper $[C]_{Cu}$ and antimony $[C]_{Sb}$ parts) have the following meanings:

Element	Cu	Fe	Sb	Zn	Pb	Sn	Ni	As
$K_l=[C]_{Cu}/[C]_{Sb}$	10	0.2	0.5	0.5-1.0	0.5	1.2-2.5	0.2-0.4	0.3-0.5

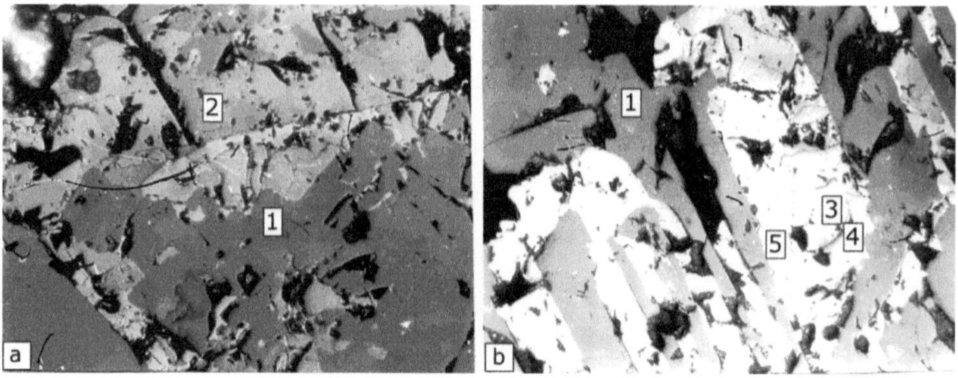

Fig. 17. Microstructure of the CMC slag : a) – x100; б) – x 200: 1 – fayalite, 2 – iron-silicate phase, 3 – wustite, 4 – christophite, 5 – eutectic of FeS-FeO

Scanning the area revealed that the sulphide-metallic part in the section space (Fig. 18) consists of a conglomerate of copper, zinc and iron sulphides as well as an intermetallic phase, closely conforming to the Cu_6Sb composition. The metallic phases stand out in close proximity or else together with wustite and sulphides. The phases are basically solid solutions and contain a significant proportion of impurities. Solid solutions of iron and zinc sulphides are enriched with copper (0.3-3.1%) in contrast with oxide-silicate phases.

As is known, zinc, iron sulphides and FeO are isolated during cooling from FeO-ZnS melts, having unlimited solubility in the liquid state (Kopilov et al., 2002; Toguzov et al., 1982). Apart from this, the double FeO-ZnS eutectic and below at a temperature 920 °C, a threefold FeO-FeS-ZnS eutectic crystallizes from the high-sulphurous residual melt. The last eutectic has the theoretical composition,%: 61.0 FeS, 36.0 FeO, 2.5 ZnS. During slow cooling, the formation of both the threefold and the double eutectic FeS-FeO and FeO–ZnS is possible. In the investigated sample, the FeO-ZnS eutectic is represented by christophite and wustite, which were evolved in turn, and the FeS-FeO eutectic by the primary troilite and wustite of a different dispersion (Fig. 17). Close contact with christophite and wustite in the field of the section confirms the progress of the reaction (9).

Complete (solidus) sulphide crystallization initiates at a temperature below 850 °C, with the formation of another threefold eutectic (30% Cu_2S, 45% FeS and 25% FeO) consisting of FeO, FeS and a solid solution of bornite. In cooling, the solid solution of bornite dissociates - partially or fully - forming a lattice structure of a chalcopyrite decomposition in bornite. The conglomerate illustrated in Fig. 18 is demonstrated by a bornite solid solution, a ferrous sulphide of zinc and copper, troilite, wustite and christophite, which provides evidence of both threefold eutectics. The phase components of the threefold eutectics are more dispersed than those of the double ones.

From the phase diagram of FeO-Cu_2S it follows that copper sulphide is soluble in those oxide melts containing FeO, but if the temperature is below 1100 °C then the interaction of the components with the formation of a solid solution of bornite, metallic copper and iron oxide occurs, which is due to the course of the reaction:

$$4Cu_2S + 4FeO = Cu_5FeS_4 + Fe_3O_4 + 3Cu. \quad (10)$$

In Fig. 18 Sulphide and metallic phases are represented by metallic copper as Cu_6Sb, with two kinds of sulphides containing iron, bornite and wustite as well.

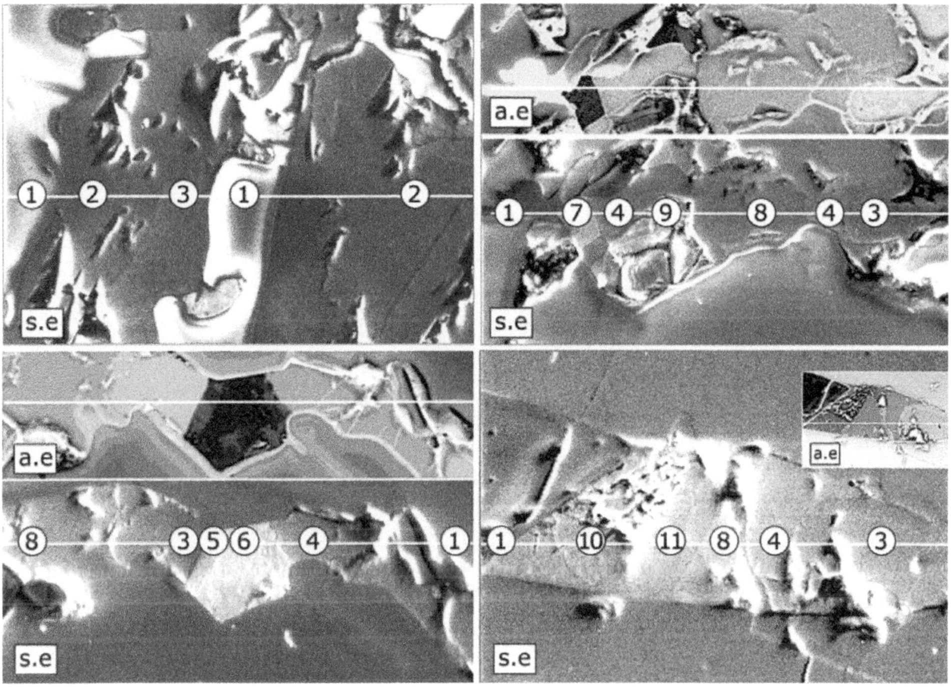

Fig. 18. Structure of the CMC slag in the absorbed (a.e.) and secondary (s.e.) electrons in the sections of 200 µm: 1 - Fe_2SiO_4; 2 - Fe_7SiO_{10}; 3 - FeO_x; 4 - FeS; 5 - Cu-Sb alloy (the particle surface), 6 - Sb-Cu alloy (the particle centre), 7 - Cu_6Sb; 8 - (Fe, Zn)S; 9 - Cu_5FeS_4, 10 - a solid solution on the base of the bornite, 11 - a solid solution of copper and zinc sulphides

Proceeding from the composition (according to the EPMA data) of clearly marked crystals, the phases containing copper correspond to $Cu_{4.4}FeS_3$, 2 and Cu_5FeS_4, the phases containing zinc to $Zn_{0.4}Fe_{0.6}S$ and the iron sulphide phase to FeS. The coefficient of the elements' distribution between the sulphide phases has the following meanings ($K_{b/t}$ and $K_{c/t}$ - the coefficients of the distribution are determined as the ratio of the metals' content in the bornite/troilite and christophite/troilite phases):

Element	Cu	Zn	Pb	Ni	As
$K_{b/t}$	104 - 149	2.5 – 5.0	1.8	2 - 5	1 - 2
$K_{c/t}$	1 - 3	1.5 - 2.5	0.6 - 0.9	1 - 3	1 – 3

Thus, the slag melts of the integrated melting-converting of copper concentrates refer to oxide-sulphide systems' crystallization, which is accompanied by a phase transformation characteristic of both oxide and sulphide systems. Besides this, the phase changes fraught with the interaction of oxides and sulphides proceed in oxisulphide systems, resulting in the extraction of metals and oxide-sulphide eutectics.

The oxide component of slowly cooled slag is represented by the solid solutions of iron oxides and iron-silicates. The number of non-ferrous metals dissolved in them (apart from zinc) is less than a tenth part of one percent. The cooling and crystallization of slag leads to the concentrating of non-ferrous metals in metallic and sulphide phases. The metallic phase is represented by Cu-Sb alloy with Fe, Ni, Sn, As and other elements dissolved in it. The metallic phase serves as the collector of non-ferrous metals and contains up to 50% antimony. In separating the metal from the solid slag by the methods concentration (flotation, gravitation, etc.) one should take into account the formation of copper alloys but not copper, as has been assumed before.

№ point	Content, mas, %											Phases
	Fe	Cu	Ni	Zn	S	Sb	Pb	Sn	Si	As	Ca	
1	50.9	0	0	3.5	0	0	0	0	13.4	0	0.9	Fe_2SiO_4
2	64.7	0	0	3.3	0	0	0	0	5.0	0	0.1	Fe_7SiO_{10}
3	76.4	0	0	0.2	0	0	01	0	0.2	0.1	0.1	FeO_x
4	62.5	0.4	0	0.1	36.4	0	0.3	0	0.1	0	0	FeS
5	3.7	50.9	0.9	0.2	0.2	29.2	0.1	5.5	0.1	5.7	0.1	Cu-Sb (surface)
6	21.7	4.8	7.2	0.4	0.1	50.2	0.3	2.7	0.1	10.7	0.1	Sb-Cu (centre)
7	0.9	60.5	3.3	0.4	0.1	23.8	0.1	6.7	0.3	3.0	0.1	Cu_6Sb
8	37.3	0.3	0.1	27.2	34.2	0	0.3	0	0	0	0	(Zn,Fe)S
9	12.6	62.6	0.1	0.3	23.3	0.1	0.6	0	0	0.2	0.1	Cu_5FeS_4
10	21.5	56.4	0.1	0.4	20.8	0	0.8	0	0.1	0.1	0.1	Cu_5FeS_4 solid solution
11	27.8	11.0	0.1	27.4	31.2	0	0.3	0	0	0.1	0.1	Solid solution of copper and zinc sulphides

Table 10. The composition of phases identified in the CMC slag (Fig. 18)

The sulphide phases are represented mainly by solid solutions containing copper, iron, zinc and small amounts of lead and arsenic as well. The main phases are bornite, christophite, troilite and the products of the eutectics' decomposition. The size of the metallic and sulphide phases in the slag amounts to 10 - 100 μm, which will allow their breakdown to be carried out by way of grinding with the standard equipment.

6. Peculiarities of the crystallization of high-magnesian iron-silicate slags

The processing of oxide nickel ores with the smelting technique for mattes is connected with the formation of a great amount of high-magnesian silicate slags. The investigations presented here are devoted to the study of forms for finding metals and structures of granulated slags of nickel production which are in substantial agreement with the SiO_2–

FeO_X-MgO-CaO system, with the phase transformations' peculiarities occurring during their heating as well as during their crystallization during the course of annealing.

The commercial slag sample of the shaft smelting of oxide nickel ores from "Ufaleynickel" JSC (Russia, Ural), obtained by way of the granulation of an oxide melt in the water pond and containing, %: 13.0 MgO, 42.1 SiO_2, 20.8 Fe, 7.2 CaO, 5.8 Al_2O_3, 0.2 Ni, 0.3 S, was taken for investigation (Sergeeva et al., 2011). The sample slag annealing was carried out in a furnace by heating it up to 1000 °C and upon the subsequent maintenance of this temperature for 60 minutes.

The results of the X-ray-phase analysis showed that the initial granulated sample of slag almost completely consisted of glass (Fig. 19). There are no reflexes conforming to crystal phases in the diffractogram. Responses conforming to $Ca(Mg,Fe)Si_2O_6$ diopside and $(Fe,Mg)_2SiO_4$ fayalite, with the structure of olivine, were revealed in the slag diffractogram after the annealing.

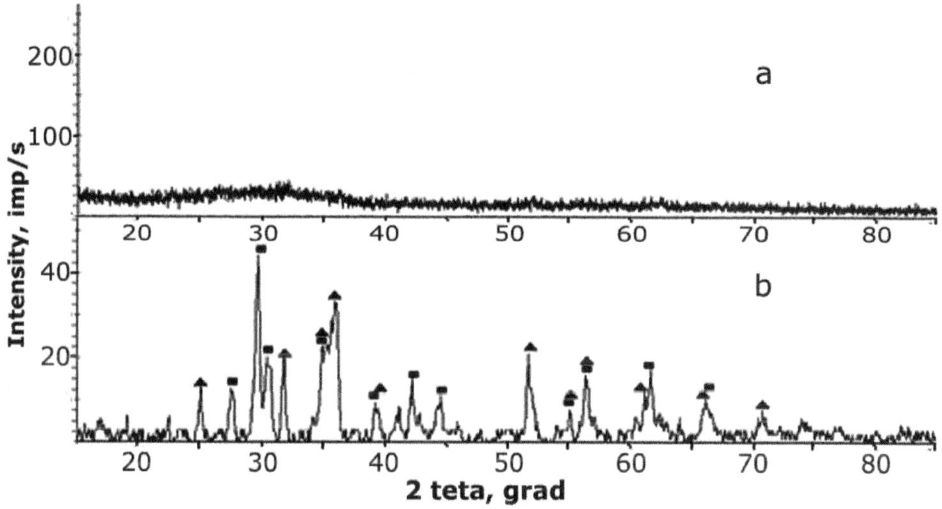

Fig. 19. Slag diffractograms: a – initial, granulated; b – after annealing: ■ - $Ca(Mg,Fe)Si_2O_6$; ▲ - $(Fe,Mg)_2SiO_4$

During the heating (20 °C/min) of the granulated slag in the argon flow on the curve of the heat flow, we revealed the effect at 646 °C due to the phase transition of the second sort's conformity to the process of devitrification (Fig. 20). The heat capacity change of devitrification was equal to 0.513 J/(g·K). A complex exothermal effect beginning at 744 °C and with the maximum meaning 766 °C and 786 °C was due to the "cold" phases' crystallization - probably to the formation of diopside and olivine crystals. These two peaks show that the crystallization temperatures of these phases are shifted with respect to each other. The endothermic effect with a beginning/maximum of 1102/1171 °C is connected with the slag melting. During the cooling of the melted sample, the temperature of the crystallization is determined - it is 1159 °C. The practically complete amorphous state of the phases in the initial granulated slag confirms the relationship of the heat meanings of "cold" crystallization (153 J/g), melting (152 J/g) and melt crystallization at cooling (154 J/g). According to the data obtained, the heating of high-magnesian glassy slag up to 1250 °C and its cooling at a rate of 20 °C/min lead to its transition from an amorphous state to a crystal state. The value of the stability factor of the glassy state of ΔT for the investigated slag sample is equal to 98 °C, which points to the stability of the amorphous state.

The microstructure of the initial granulated slag is represented mainly by glass (Fig. 21). There are multiphase areas on the surface of the section and globular particles with a size of 5 - 15 μm are concentrated in them. The representation of the section in the characteristic emission of elements is evidence of the fact that the main slag component consists of magnesium, silicon, iron, aluminium and calcium oxides. The other part is represented by small insular edged parts (10 - 20 μm) with a high iron, silicon and magnesium oxide content. The globular phases (less than 1% of the section area) consist of solid solutions of iron and nickel sulphides and their compounds. The results of the determination of the elements' composition of the parts of the granulated slag showed (Table 11) that the main part of it is represented by glass, containing up to 23.0% Fe, 20.0% Si, 6.3% Ca, 6.8% Mg and 2.7% Al. Another silicate phase with a high magnesium (14.0 – 23.0%), silicon (16.6 - 18.3%) and iron (18 - 30%) oxides content is in close agreement with the ferrous olivine.

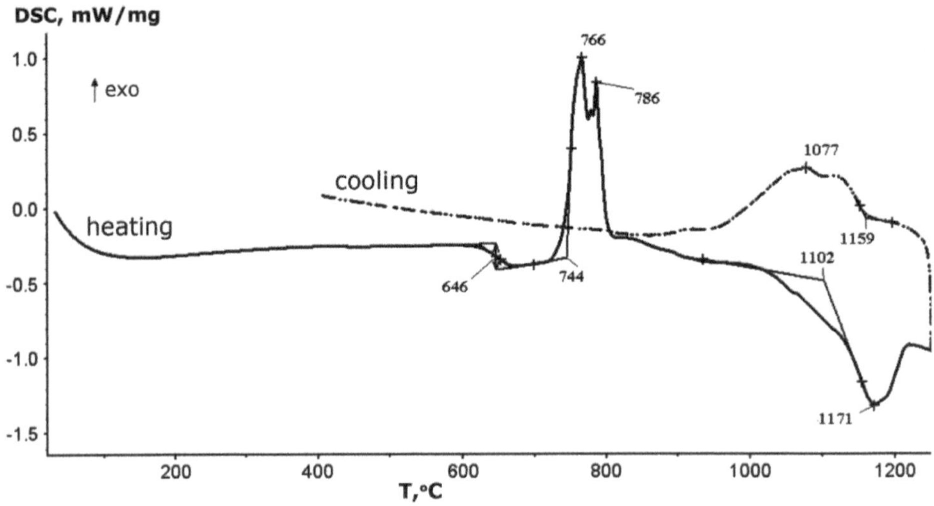

Fig. 20. DCK curves of granulated high-magnesian iron-silicate slag under heating and cooling at a rate of 20 °C/min

The nickel content in the glass is negligible (less than 0.1%), whereas its concentration in olivines reaches 0.2 - 1.2%. Sulphide metal shorts are either a mechanical suspension of matte or else are those solid solution phases which were isolated as the result of unbalanced slag cooling. The detailed analysis of the impurities' composition showed their heterogeneity. The grains consisting of the sulphide phases (27.8 - 29.1% S) concentrated in both nickel (up to 30.9%) and iron (up to 54.0%) are revealed in the impurities. The sulphide grain boundaries are metalized and the sulphur content in them is within the limits of 7.1 - 19.1%, Fe - 31.0 - 43.9%, Ni - 43.2 - 45.0% and Co - 1.6 - 2.6%.

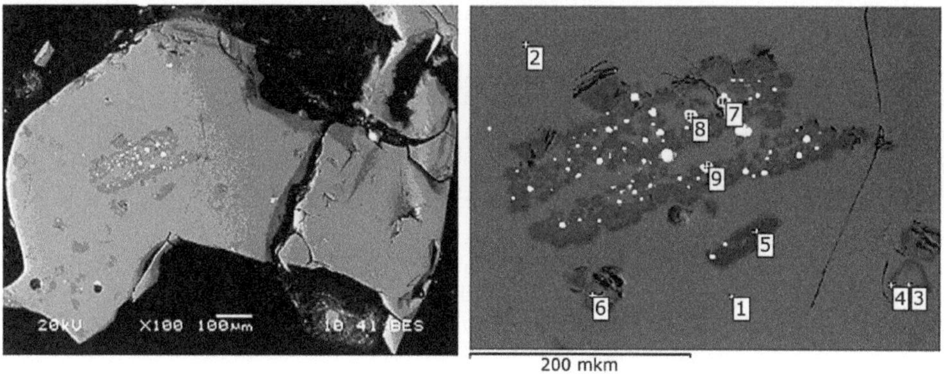

Fig. 21. Microstructure of the particle of high-magnesian iron-silicate granulated slag and EPMA points

The annealing at 1000 °C leads to the change of samples microstructure; the starting processes of solid crystallization can be seen in them, in contrast with the ferrous magnesian-free slags (Gulyaeva et al., 2011) where these phenomena proceed on the surface. The phases of the needle forms (Fig. 22) form in the heterogeneous part of the section - they have an average composition of $(Mg_{0.79}Fe_{0.71}Al_{0.09}Ca_{0.16})Si_{1.10}O_4$. Thus, the cut of olivine

№	Composition in the points of zonding	Content, %							
		Ni	Mg	Si	Al	Fe	S	Co	Ca
Before annealing (Fig. 21)									
1	- $Fe(Mg_{0.67}Ca_{0,37})(Si_{1.74}Al_{0.24})O_6$	0	6.7	20.3	2.7	23.0	0.4	0.2	6.3
2	$Fe(Mg_{0.67}Ca_{0,37})(Si_{1.74}Al_{0.23})O_6$	0.1	6.8	20.1	2.6	22.3	0.4	0.2	6.1
3	$(Mg_{1.03}Fe_{0,97})Si_{1.05}O_4$	1.2	19.9	17.8	0.1	30.4	0	0.5	0.3
4	$(Mg_{1.40}Fe_{0,70})Si_{1.09}O_4$	0.4	14.0	16.6	0	22.9	0	0.2	0.5
5	$(Mg_{1.54}Fe_{0,53})Si_{1.07}O_4$	0.2	22.7	18.3	0	18.0	0	0.2	0.7
6	$(Mg_{1.20}Fe_{0,84})Si_{1.03}O_4$	0.5	17.1	17.1	0	26.3	0	0.4	0.4
7	Matte	16.8	0.2	0.4	0.1	51.0	26.8	1.4	0.1
8	Matte	21.3	0.2	0.4	0.1	47.0	26.2	1.5	0.2
9	Matte	33.0	0.2	0.4	0.1	38.2	24.7	1.3	0.2
After annealing (Fig. 22)									
10	- $(Mg_{1.53}Fe_{0.52})Si_{1.05}O_4$	0.2	23.0	18.1	0.1	18.0	0	0.1	0.3
11	$(Mg_{1.46}Fe_{0.55})Si_{1.03}O_4$	0.2	22.1	17.8	0.1	18.9	0	0.2	0.6
12	$(Mg_{1.37}Fe_{0.67})Si_{1.05}O_4$	1.6	18.9	17.5	0.1	23.0	0	0.2	0.5
13	$(Mg_{1.15}Ca_{0.11}Fe_{0.64})Si_{1.07}O_4$	0.1	14.6	18.2	1.1	22.4	0.2	0.2	3.0
14	$Fe_{0.97}(Mg_{0.74}Ca_{0.49})(Si_{1.80}Al_{0.24})O_6$	0.1	7.2	20.2	2.7	21.3	0.4	0.2	7.8
15	Matte	23.3	0.2	0.4	0.1	40.7	31.1	1.2	0.2
16	$Fe_{0.75}Ni_{0.42}S$ $(Mg_{1.42}Fe_{0.59})Si_{1.04}O_4$	0.3	21.0	17.7	0	20.2	0	0.1	0.8
17-20	$Fe_{0.94}(Mg_{0.74}Ca_{0.37})(Si_{1.75}Al_{0.24})O_6$	0.1	7.5	20.4	2.7	21.8	0.3	0	6.2
21-23	$Fe_{1.09}(Mg_{0.70}Ca_{0.39})(Si_{1.73}Al_{0.22})O_6$	0	6.7	19.6	2.5	24.5	0.5	0.1	6.3

Table 11. Elemental phase composition of the samples of granulated and annealed high-magnesian slags at EPMA points

crystals is shown clearly and their sizes increase to 60 μm. The homogeneous glass breaks down into fine-dispersed (less than 0.5 μm) phases in which the $Fe_{1.09}(Mg_{0.70}Ca_{0.39})(Si_{1.73}Al_{0.22})O_6$ composition differs slightly from the matrix, with a slight deviation in the iron (Table 11). The annealing also leads to changes in the microstructure of sulphide impurities, increasing their inhomogeneity. Fine-dispersed metallic phases containing 52.9 - 57.6% Ni, 36 - 38.9% Fe, 2.6 - 2.9% Co and 1.2 - 4.6% S also precipitate from an unbalanced matte and a monosulphide solid solution close in the composition to $Fe_{1.03}Ni_{0.06}S$ forms.

Thus, the processes at the beginning of glass crystallization and the extraction of diopside and olivin are registered as a result of the annealing of the granulated slag of the shaft melting of oxide nickel ores. The main part of the nickel and cobalt is in a granulated slag as globules of a matte, while the sample annealing leads to the precipitation of fine-dispersed metallic phases enriched in nickel. Thermic effects conforming to devitrification processes, "cold" crystallization, melting and melting crystallization are determined and so they can be taken as basic effects for metallurgical calculation.

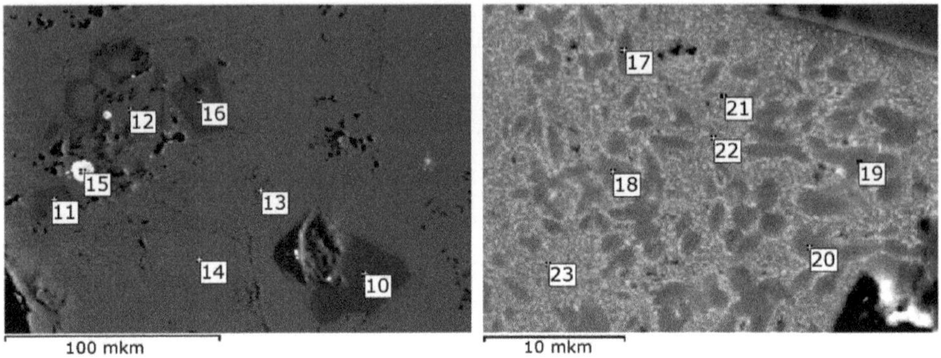

Fig. 22. Microstructure of particles of high-magnesian granulated slag after annealing at 1000 °C and EPMA points

7. Conclusion

The effect of melts' cooling rates on the phases' structure, composition and size, as well as the inter-phase distribution of impurity elements, have been estimated using exemplary slag samples taken from copper-zinc concentrates smelted in a Vanyukov furnace ("Sredneuralsky Copper Smelter Plant" JSC), combined smelting-converting ("Svyatogor" JSC), copper matte converting ("Sredneuralsky Copper Smelter Plant" JSC) and oxide ores smelting for matte ("Ufaleynickel" JSC). The cooling rate of iron oxide-sulphide melts affects the crystal structure, particle size and the number of forming phase. The cooling of the melt by using water granulation leads to the formation of glassy phases, the number of which increases with the content of SiO_2. The iron-silicate phases of variable composition, magnetite and sulphide components constitute the basis of the samples under study. Sulphides are represented by matte mechanical losses (particle size 50 μm) and fine inclusions (2-10 μm) formed from the melt during crystallization.

The amorphous constituent (glass) was observed in all the rapidly quenched samples. The devitrification temperature and the "cold" crystallization of the glassy phases of slags from the melting of copper-zinc concentrates and the converting of a copper matte are within the range of 530-550 °C and 630 °C, respectively. The devitrification and cold crystallization of the granulated and high manganese slag processes from nickel production occur around temperatures of 650 to 740 °C.

The modes of the cooling of the melted slag determine the distribution of non-ferrous metals (copper, zinc, lead, nickel) between oxide and sulphide forms. The possibility of christophite Zn(Fe)S formation under rapid and slow slag cooling has been demonstrated for the first time. It was shown that the content of impurity elements in the phases arising depends upon the cooling rate and upon the Fe/SiO_2 ratio in these phases. The metal, copper-based, component accumulating antimony and arsenic was revealed in the slag sample taken from the copper matte conversion.

The forms of non-ferrous metals in the crystallized slag are able to be regulated by changing the Fe^{+3}/Fe^{+2} and CaO/SiO_2 ratios, the amounts of iron, CaO and MgO oxides. The data obtained is useful for the justification of processes for the re-extracting of precious metals.

8. References

Biswas, K., Sontakke, A.D., Majumder, M. & Annapurna, K. (2010). Nonisothermal crystallization kinetics and microstructure evolution of calcium lanthanum metaborate glass. *J. Therm. Anal. Calorim*, Vol.101, pp. 143–151, ISSN 1388-6150.

Cardona, N., Coursol, P., Vargas, J. & Parra, R. (2011). The Physical Chemistry of Copper Smelting Slags and Copper Losses at the Paipote Smelter. Part 2 - Characterization of industrial slags, *Canadian Metallurgical Quarterly*, Vol.50, No.4, pp. 330-340, ISSN 0008-4433.

Dovchenko, V.A., Paretskii, V.M., Chakhotin, V.S. et al. (1997). Flotation Depletion of the Self-Decomposing Slags of the Slags of Melting of Copper Concentrates for Producing White Matte, *Tsvetn. Met.*, No.1, pp. 27–31, ISSN 0372-2929.

Elliott, J. (1976). Phase relationships in the pyrometallurgy of copper. *Metallurgical Transactions*, Vol.7B, pp. 17-33, ISSN: 1073-5623.

Gulyaeva, R.I., Selivanov, E.N. & Selmenskikh, N.I. (2011). Crystallization of oxide high iron melts. EPJ Web of Conferences, Vol.15, 01010. Published online 18.05.2011, available from http://www.epj-conferences.org.

Karamanov, A. & Pelino, M. (2001). Vitrification of copper flotation waste. *J. Non-Crystalline Solids*, Vol.281, No.1-2, pp. 333-339, ISSN 0022-3093.

Kopylov, N.I., Lata, V.A. & Toguzov, M.Z. (2001). *Interactions and Phase States in Molts Sulfides Systems*, ISBN 5-628-02752-9, Gilim, Almaty, in Russian.

Koryukin, E.B., Litovskikh, S.N. & Kireeva, O.V. (2002). Flotation-Magnetic Scheme of Processing of Converter Slags. *Tsvetn. Met.*, No.8, pp. 18–20, ISSN 0372-2929.

Mazurin, O.V. (1986). *Vitrification*, Nauka, Moscow, Russia.

Mohapatra, B.K., Nayak, B.D. & Rao, G.V. (1994). Microstructure of and metal distribution in Indian copper converter slags - study by scanning-electron microscopy. *Mineral Processing and Extractive Metallurgy*, Vol.103, pp. C217-C220, ISSN 0371-9553.

Moiseev, G.K. & Vyatkin, G.P. (1999). *Thermodynamic Simulation in Inorganic Systems*, YuUrGU , Chelyabinsk, Russia.

Naboichenko, S.S., Nichiporenko, O.S., Murashev, I.B. et al. (1997). *Nonferrous Metal Powders*, Metallurgiya, Moscow, Russia.

Nagamori, M. (1974). Metal loss to slag: Part I. Sulfidic and oxidic dissolution of copper in fayalite slag from low grade matte, *Metallurgical Transactions*, Vol.5, pp. 531-538. ISSN 1073-5623.

Okunev, A.I. & Galimov, M.D. (1983). *Oxidation of iron and sulfur in the oxide-sulfide systems,* Nauka, Moscow, Russia.
Roine, A. (2002). *Outokumpu HSC Chemistry for Windows,* (Outokumpu Research OY), Pori, www.outotec.com.
Ruddle, R.W. (1953). *The Physical Chemistry of Copper Smelting,* Institution of mining and metallurgy, London.
Rüffler, R. & Dávalos, J. (1998). Phase Analytical Studies of Industrial Copper Smelting slags: Part I: Silicate slags, *Hyperfine Interac,*. Vol.111, pp. 299–307, ISSN 0304-3843.
Sarrafi, A., Rahmati, B., Hassani, H.R. & Shirazi, H.A. (2004). Recovery of Copper from Reverbatory Furnace Slag by Flotation. *Mineral Engineering,* Vol.17, No.3, pp. 457–459, ISSN 0892-6875.
Selivanov, E.N., Okunev, A.I. & Moiseev, G.K. (2000). Phase Transformations during Cooling of the Slags of Melting of Copper Concentrates for Producing High-Grade Matte. *Rasplavy,* No.2, pp. 18–24, ISSN 0235-0106.
Selivanov, E.N., Belyaev, V.V., Selmenskikh, N.I. & Pankratov, A.A. (2004). Forms of Metals in the Slag of Combined Melting–Converting of Copper Concentrates. *Rasplavy,* No.1, pp. 33–41, ISSN 0235-0106.
Selivanov, E.N., Gulyaeva, R.I., Udoeva, L.Yu. et al. (2009a). Effect of the Cooling Rate on the Phase Composition and Structure of Copper Matte Converting Slags. *Russian Metallurgy (Metally),* No.4, pp. 281–288, ISSN 0036-0295.
Selivanov, E.N., Gulyaeva, R.I., Zelytin D.I. et al. (2009b). Effect of speed of cooling on structure of slag from of copper-zinc concentrates in Vanukov furnace. *Tsvetnye Metally,* No.12, pp. 27-31, ISSN 0372-2929.
Selivanov, E.N., Gulyaeva, R.I. & Selmenskich, N.I. (2010). Phase Formation in FeO_x-SiO_2-Cu_2O-ZnO-FeS System during Melts Crystallization. *Defect and Diffusion Forum,* Vols. 297-301, pp. 602-607, ISSN 1662-9507.
Sergeeva, S.V., Gulyaeva, R.I. & Selivanov, E.N. (2011). Features of crystallization of high-magnesian silicate slag. *13th Russian Conference on Composition and Properties of Melts of Metals and Slags.* Vol.3, pp. 99-102, Yekaterinburg, Russia, September 12-16, 2011.
Spira, P. & Themelis, N.J. (1969). The Solubility of Copper in Slags. *Journal of Metals,* Vol.29, No.4, pp. 35-42, ISSN 0148-6608.
Sycheva, G.A. & Polyakova, I.G. (2004). Nucleation of crystals in glasses based on blast furnace slag, *11th Russian Conference on Composition and Properties of Melts of Metals and Slags.* Vol.3, pp. 186-190, Yekaterinburg, Russia, September 14-16, 2004.
Tokeda, Y., Ishiwata, S. & Yazawa, A. (1983). Distribution Equilibria of Minor Elements between Liquid Copper and Calcium Ferrite Slag. *Trans. Japan Inst. Metals.* Vol.24. №10. pp. 518–528, ISSN 0021-4434.
Vanyukov, A.V., Bystrov, V.P., Vaskevich, A.D. et al. (1988). *Melting in the Liquid Bath,* Metallurgiya, ISBN 5-229-00062-7, Moscow, Russia.
Vanyukov, A.V. &. Zaitsev, V.Ya. (1969). *Slags and Mattes of Nonferrous Metallurgy,* Metallurgiya, Moscow, Russia.
Vanyukov, A.V. &. Zaitsev, V.Ya. (1973). *Theory of Pyrometallurgical Processes,* Metallurgiya, Moscow, Russia.
Vaysburd, S.E. (1996). *The Physical Chemistry Properties and Structural Features of Sulfide Melts.* Metallurgiya, ISBN 5-229-00903-9, Moscow, Russia.
Yazawa, A. (1974). Thermodynamic considerations of copper smelting, *Canadian Metallurgical Quarterly,* Vol.13, No.3, pp. 443-455, ISSN 0008-4433.

2

Preparation of Selected Ceramic Compounds by Controlled Crystallization Under Hydrothermal Conditions

Juan Carlos Rendón-Angeles[1], Zully Matamoros-Veloza[2] and Kazumichi Yanagisawa[3]
[1]*Research Institute for Advanced Studies of the NPI Saltillo Campus*
[2]*Technological Institute of Saltillo*
[3]*Research Laboratory for Hydrothermal Chemistry, Kochi University*
[1,2]*México*
[3]*Japan*

1. Introduction

The main aim of this chapter is to review some particular aspects related with the hydrothermal crystallization process for preparing some selected oxide and non-oxide powders, namely functional compound with electric, piezoelectric, ionic conducting and catalytic properties. The literature on the hydrothermal crystallization of oxide and non-oxide powders is vast; the references cited here are those more appropriated to illustrate this review. A particular attempt is made to broaden the traditional concepts of processing perovskite powders with controlled chemical composition and particle morphology, those aspects will be discussed based on the chemical reactivity of the precursor reactants (gels). Furthermore, an additional approach that takes into account the solubility of the solid species (mineral reagents) that are employed as a precursor in the hydrothermal systems on controlling the crystallization process of oxide particles was further investigated for preparing titanates oxide ceramic. The specific reaction pathways and kinetic aspects are discussed and illustrated by experimental setups for the solution of selected problems in hydrothermal crystallization. Also the chapter includes recent work on the formation of inorganic salts of Sr or Ba, under ordinary hydrothermal treatments via the anionic replacement in sulphate minerals, because these particular reactions have promoted peculiar microstructure on the crystallized material that preserves the bulk geometrical features of particular mineral specie. Therefore, this method could be attractive for preparing net-shaped materials with controlled porosity, with optimized functional properties, because can be used as gas sensor, substrates for porous catalytic materials, filters, among other potential applications.

2. Hydrothermal crystallization as technique for oxide and non-oxide synthesis

2.1 Brief history of the hydrothermal technique development

The studies recorded at the scientific annals indicate that the pioneering research on hydrothermal systems was initiated in the middle of the 19th century (Schafthaul, 1845, as cited in Suchanek et. al., 2004). The term of hydrothermal is purely of geological origin; this was firstly used by the geologist Sir Roderick Murchinson (1792–1871), whom analysed the reactivity of water at elevated pressure and temperature, to explain the mineral formation of several rocks and minerals (Byrappa & Yoshimura, 2001). Continuous progress in material synthesis was initially accelerated by notorious developments in hydrothermal pressure vessels apparatus. Hence, at the first decade of the 20th century, geologist achieved a marked technological expansion on the hydrothermal research field, namely efforts were directed towards for designing the first metal vessels and carried out preliminary experiments at laboratory scale related to material synthesis area, mainly in the field of inorganic single crystal growth (Riman et. al., 2002; Suchanek et. al., 2004). However, the development of chemical reactions via hydrothermal processing was further limited because of the severe treatment (supercritical) conditions, which were normally required for single crystal growth. This resulted on discourage extensive research and commercialization of various materials. Hydrothermal epitaxial growth is one example, it was popular during the 1970s, but it did not reach commercial success due to the high temperatures (> 500 °C) and pressures (> 100 MPa), which were involved to achieve the epitaxial crystal growth process. In the middle of the 1980s, the commercial interest in the hydrothermal technology was revived, because the steadily increasing of a large group of materials, manly ceramic powder, had emerged that can be produced under more environmentally friendly conditions (T < 350 °C, P < 100 MPa). At present, the major developments on the hydrothermal synthesis technology including in particular the hydrothermal crystallization has been accomplished in several countries around the world like: China, Japan, USA, UK, Germany and some others with minor contributions (Byrappa & Yoshimura, 2001).

2.2 Definition of hydrothermal synthesis

The term *"hydrothermal"* is difficult to define, based in the etymologic root of the Greek word, *"hydrous"* means water and *"thermal"* means heat. One of the accepted statements for hydrothermal defines it as any heterogeneous chemical reaction that occurs in the presence of a solvent media at above the room temperature (> 25 °C) and pressure levels greater than 0.1 MPa in a closed system, at these conditions does not matter whether the solvent is aqueous or non-aqueous. Hitherto, there is still a bit of confusion regarding the correct use of the term hydrothermal, because in the case of the chemistry field, the chemists prefer to use a different term, namely *solvothermal*, which means any chemical reaction conducted with a non-aqueous solvent or solvent in supercritical conditions. Additionally, another similar related terms widely used amongst the physicist, chemist and material scientist communities are: *glycothermal*, *alcothermal*, *ammonothermal*, and so on. However, some researchers also use hydrothermal for describing processes conducted at ambient conditions. The *crystallization* process of solid phases under hydrothermal conditions is usually conducted at autogeneous pressure, achieving a particular saturated vapour pressure of the solution at the specified temperature and composition of the hydrothermal solution. In this concern, in terms of industrial commercial processes mild operating conditions are preferred, for example treatment temperatures below than 350 °C and pressures > 50 MPa (Byrappa, 2005). The limit that indicates the transition from mild to severe conditions during a hydrothermal treatment is normally determined by the strength of the inner vessel materials, which at severe treatment conditions undergoes into a corrosion process. The continuous research in this field has led the way to a better understanding for controlling chemical reactions in various hydrothermal medias, which has remarkably reduced the processing parameters such as: reaction time, temperature, and pressure for hydrothermal crystallization of several oxide and non-oxide materials (T<200 °C, P<1.5 MPa). At present, the recent scientific and technological achievements have made hydrothermal synthesis more economical, because powder preparation can be carried out by a single step cost-effective process, in advanced pressure reactor technology coupled with processing methodologies proposed for a wide number of inorganic compounds (Suchanek et al., 2004).

2.3 Hydrothermal processes for synthesis of inorganic compounds

2.3.1 Hydrothermal single crystal growth

Regardless of the numerous investigations on hydrothermal single crystal growth, quartz crystal is one of the materials extensively investigated up to now. Nowadays, the electronic industry requires the use of pure large single crystals of synthetic origin, because natural quartz crystals are generally irregular in shape, and is difficult to obtain large-scale single crystals wafers by automatic cutting. The crystal growth of oxide species, namely α-quartz; is conducted by the conventional hydrothermal temperature gradient method, thus, the autoclaves frequently used consist of two chambers were different process occurs. One of the important parameter to consider for the single crystal growth of α–quartz is the selection of the proper nutrient material; among the most employed are small particle size α–quartz, silica glass, high quality silica sand, or silica gel (Byrappa 2005). The nutrient reactant is placed at the liner chamber (vessel bottom, see Figure 1) made up of iron, silver or titanium that is less employed, with a suitable baffle and a frame (at the top of the vessel) holding the seed of the material that is been grown. The mineralizer solution is other factor to select coupled with a definite molarity; this is poured into the liner to make the required filling volume and achieves the dissolution of the nutrient. Under these conditions, matter transport proceeds from the nutrient chamber and the crystallization and growth of the crystals is achieved due to the temperature difference at the top-seeded part of the autoclave. The particular optimum conditions determined at Bell Laboratories for growing quartz single crystals are dissolution temperature of 425 °C at a pressure in the range of 100–175 MPa, the crystallization and simultaneous growth proceeds at 375 °C whilst the temperature gradient from the nutrient to the growing chamber was of 50 °C. The mineralizer employed was an alkaline solution of NaOH with concentration varying between 0.5 up to 1 M, and the volume of poured mineralizer solution varied in the range of 78–85 % of the total volume of the vessel (Laudies, 1970; as cited in Byrappa, 2005).

The crystallization and growth of other single crystals rather than SiO_2 was firstly investigated for oxide species, namely TiO_2, ZrO_2, HfO_2 and some related perovskite oxides $PbTiO_3$ and $PbZrO_3$. Experimental results evidenced that the mineralizer employed for dissolving and transport the nutrient reactant markedly affected the crystallization of the oxides and the crystal growth. Thus, the single oxide species were found to dissolve and recrystallize faster in fluoride solutions (NaF, KF and NH_4F) in comparison with alkalis, KOH and K_2CO_3, because of the high chemical reactivity of these nutrient oxides in fluoride solutions. However, the complete dissolution and transport of the nutrient was preferentially achieved in NH_4F solutions rather than NaF or KF mineralizers, even at a low temperature of 470 °C for a reaction interval up to 6 days (Kuznetzov, 1968), and the TiO_2, ZrO_2, HfO_2 single crystal growth was achieved at a temperature of 520 °C with a positive temperature gradient

(30 °C). This is a popular method that promotes the crystal growth and has been widely used on the preparation of berlinite ($AlPO_4$) crystals (Byrappa & Yoshimura 2001).

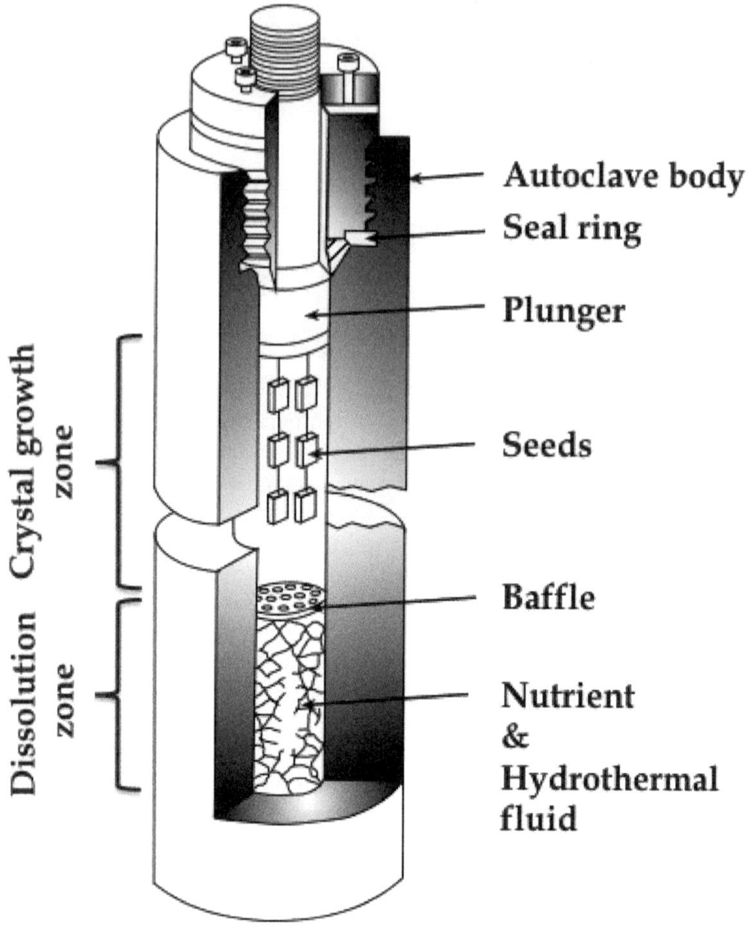

Fig. 1. Scheme of the conventional autoclave employed for single crystal growth under hydrothermal conditions (Schubert, 2000; as cited in Byrappa & Yoshimura, 2001).

A particular emphasize on the hydrothermal crystal growth research has been focused for the crystallization of materials that melts incongruently because such materials cannot be grown with compositional and phase uniformity. The relevant examples of compounds that melts incongruently are those belonging to lead family oxides titanate ($PbTiO_3$), zirconate ($PbZrO_3$), single crystals with millimetre size of both compounds were found to grown satisfactorily in NH_4F hydrothermal solution by seedless grown crystallizing conditions at temperatures between 500-600 °C for several days (Kuznetzov, 1968). However, the hydrothermal crystal grown of some lead related solid solutions, $PbZr_{1-x}Ti_xO_3$ (PZT), $PbMg_{1/3}Nb_{2/3}O_3$ (PMN) and $PbSc_{0.5}Nb_{0.5}O_3$ (PSN), is limited due to the low chemical stability of the perovskite structure even in KF solution, and results on the grown of small single crystals with pyrochlore structure $Pb_{1.83}Mg_{0.29}Nb_{1.71}O_{6.39}$ and $Pb_{1.83}Sc_{0.29}Nb_{1.71}O_{6.39}$, as shown in Figure 2 (Yanagisawa et. al., 1999, 2000). This incongruence has not been determined thermodynamically and experimentally for the PZT, because this compound has been used to grown thin and thick single crystals on different seeds ($SrTiO_3$) with excellent compositional control (Oledzka et. al., 2003). In general, the crystallization and crystal growth of a specific compound can be carried out using different mineralizer or solvents, e.g., water, soluble salts, acid solutions, non aqueous solutions; but some physicochemical factors must be taking into account to fulfil suitable conditions that facilitate both processes.

Fig. 2. Pyrochlore single crystals, of (a) PMN, (b) PSNT and (c) PSN; grown at the top of platinum capsule at 600 °C with a gradient temperature gradient of 40 °C in KF solutions 5, 4.2 and 6.2 M for reaction intervals of 3 (a) and 5 days (c,d), respectively (Yanagisawa et. al., 1999).

2.3.2 Advanced hydrothermal processing methods

Remarkable achievements for developing the *Hydrothermal Microwave Assisted Synthesis* process have been conducted at Pennsylvania State University (Komarneni et. al., 1992); this method enhances solid crystallization kinetics 1-4 order of magnitude faster than that occurring on the conventional hydrothermal processing for a wide solutions. In addition, other advantages of the hydrothermal microwave assisted technique are very high heating rates and the synthesis of novel phases. A great variety of ceramic powders with particular morphologies and controlled particle size, have been produced, e.g. TiO_2, ZrO_2, Fe_2O_3, $BaTiO_3$, $Ca_{10}(PO_4)_6(OH)_2$, etc (Komarneni et. al., 1992; Roy, 1994; Komarneni et. al., 2002).

Another variation implemented to the hydrothermal conventional process resulted on the development of a different hybrid method denominated *Hydrothermal-Electrochemical Synthesis*. This technique was tailored to deposit polycrystalline oxide films on reactive metal substrates. This technique becomes very important when the crystallization of oxide products from supersaturated hydrothermal solutions is hinder in absence of an applied electrical potential. Nowadays, highly crystallized ceramic thin films, such as a $BaTiO_3$, $SrTiO_3$, $LiNiO_2$, $PbTiO_3$, $CaWO_4$ and $BaMoO_4$ can be deposited on metal substrates from aqueous solutions at relatively low temperature 50-200 °C for several hours under a continuous applied voltage charge (Yoshimura & Suchanek, 1997). Additionally, semiconductor thin layers of GaAs, CdTe, CdSe and CdS have been successfully prepared by electrochemical atomic layer epitaxy growth, this mechanism is analogous to molecular epitaxy, however, the crystallization and growth of the layer is enhanced from a saturated hydrothermal media instead of a vapour phase that transport the growing species (Colletti et. al., 1998; as cited in Suchanek et. al., 2004).

Among other tailored techniques, the high energy milling technique has been challenged to hybridize the conventional hydrothermal processing, resulting in a coupled process that involves the classical powder mechanochemical and hydrothermal syntheses. The *Mechanochemical-Hydrothermal* route utilizes the solvency of an aqueous solution, which facilities the crystallization of solid species due to the pressure environment generated inside the mechano-chemical autoclave; it mainly accelerates the rate-determining steps of those factors, for instance: interfacial reaction, solute dissolution or dehydroxylation; that limit hydrothermal chemical reactions at low temperature. These reactions enhance the crystallization to occur locally at the particle surface because of the perturbation of superficial bonded species in the solid coupled with high temperature gradients (400-700 °C) and pressure localized zones that are generated during the mechanical activation of slurries. This is mainly promoted by the friction and adiabatic heating of gas bubbles while maintaining the average temperature close to room temperature (Kosova et. al., 1997; as cited in Suchanek et. al., 2004). Hitherto, this technique has been used for preparing bioceramic materials such as hydroxyapatite (Suchanek et. al., 2002; Chen et. al., 2004), another tailored material like $PbTiO_3$ was also crystallized by this technique.

Since Rustum Roy reported that the use of ultrasonic devices are feasible for improving low temperature inorganic syntheses because reaction kinetics is two orders of magnitude faster than that for the conventional wet chemistry synthesis methods (Roy, 1994). Some attempts to adapt ultrasonic devices emitting acoustic signals of 20 kHz up to 10 MHz have been conducted, because the acoustic signal produce very sharp temperature gradients with localized peak temperature zones, speculated as high as 5000 K, as well as, localized pressure zones of up to 100's MPa. The sonochemical environment also alters the molecular chemistry (chemical bond scission, generate excited states and accelerate electron transfer steps in chemical reactions), and enhances mass transport and crystallization kinetics due to the high convection of the fluid. (Peters, 1996, as cited in Riman et. al., 2002). *Hydrothermal-sonochemical synthesis* method have been used for preparing various ceramic powders

($Ca_{10}(PO_4)_6(OH)_2$, $AlPO_4$, InSb, CdS) and thin films ($Li_2B_4O_7$, $Ba_2TiSi_2O_8$).while the average temperature of the reactors is maintained close to room temperature.

The advanced hydrothermal techniques discussed in this section, even though wet chemical synthesis is offer in conjunction with so many significant advantages over the conventional method, might be able to have wide application in the industry. Compared to solid-state materials processing, these technologies might be more facile for scaling up. This situation, however, still cannot guarantee successful application of all these technologies to industries. Hence, much effort is needed to obtain a more comprehensive understanding of several physicochemical phenomena related to each of the new technologies hybridizing the conventional hydrothermal synthesis, in order to establish a relationship between science and technology that could lead to optimize these methods for their employment at a industrial scale (Shi & Hwang, 2003).

2.4 Conventional hydrothermal crystallization process for advanced ceramic materials

2.4.1 Factors that affect optimizing hydrothermal crystallization experiments

In general, a majority of the hydrothermal synthesis research work that has been done in the past was based on Edisonian trial and error design for process development, but this is not the best experimental approach for discerning between processes that are controlled by either thermodynamics or kinetics. In contrast, much effort has been paid to use thermodynamic modelling for processing design; this approach is based on fundamental physicochemical principles instead of the Edisonian methods (Riman et. al., 2002). Therefore, a great number of hydrothermal fundamental works for some particular solid-aqueous systems have provided sufficient experimental hydrothermal physical chemistry data. An important point derived from these works is related with the behaviour of the solvent under hydrothermal conditions, because it has a relationship with aspects like structure at critical, supercritical and sub-critical conditions, solution dielectric constant, pH variation, viscosity, expansion coefficient, density, etc., all these parameters depend markedly on thermodynamic variables such as pressure and temperature. At present, hydrothermal crystallization process is the only one where a fundamental understanding of kinetics is lacking due to the absence of physicochemical data of the intermediate phases forming in specific aqueous solutions. Although, fundamental research works related to synthesis of specific compounds demonstrated the importance of crystallization kinetics, however, a better understanding of crystallization kinetics still in an early stage of development. In this case, due to the absence of predictive methodology models, ones must estimated on terms of chemical equilibrium of the reaction the effect of temperature, pressure, precursor, and time to achieve solid crystallization and improve the reaction kinetics. Insight into this would enable us to understand how to control the formation of ionic species in the solution, the crystallization of solid phases and the rate of their growth.

In the last decade, much effort has directed towards for developing thermochemical models based in fundamental knowledge of thermodynamic and the Hegelson-Kirkham-Flowers-Tanger equation of state, which allows to represent standard-state properties across substantial temperature and pressure ranges in order to estimate chemical reactions path ways under hydrothermal conditions (Lencka & Riman, 2002). Modelling can be successfully applied to very complex aqueous electrolytes over specific ranges of temperature and reactant and solvent concentration and non-aqueous systems can also be model as well. Practical computer software for conducting thermochemical modelling studies was recently developed by OLI System Inc., USA. This software can be used to conduct studies for chemical reactions for hydrothermal systems within the temperature range of -50 to 300 °C, at pressures ranging from 0 to 150 MPa and concentration of 0 up to 30 m in molar ionic strength; for the non-aqueous systems the temperature range covered is from 0 to 1200 °C and pressure from 0 to 150 MPa with species concentration from 0 to 1.0 mole fraction. Major predictions using the thermodynamic model have been done to determine the optimum hydrothermal conditions for achieving the crystallization of a wide variety of oxides such as, $BaTiO_3$, $PbTiO_3$, $CaTiO_3$, $SrZrO_3$ and $(BaSr)TiO_3$ (Lencka & Riman, 1993, 1995; Gersten et. al., 2004).

Currently, the research efforts on the thermochemical-modelling topic are being focused to establish an overall rational engineering-based methodology that will speed up process development. The proposed methodology for conducting this study involves the following four steps:

1. Compute thermodynamic equilibrium as a function of chemical processing variables.
2. Generate equilibrium diagrams to draw the process variable space for the phases of interest.
3. Design hydrothermal experiments to test and validate the computed diagrams.
4. Utilize the processing variables to explore opportunities for controlling reactions and crystallization kinetics.

Recently, population balance modelling has received much attention from both academic and industrial areas because of its applicability to a wide variety of particular processes. In general, a population balance model can be proposed by the collective phenomenology contained in entities displacement through their state space and the birth-and-death processes that terminate existing entities and produce new entities. The phenomenology concerns the behaviour of any single entity in conjunction with other entities, which is for the population balance modelling a reasonable description of the system (Ramkrishna & Mahoney, 2002). Regarding the hydrothermal crystallization process, a rigorous kinetic model for the solution precipitation and hydrothermal synthesis of $BaTiO_3$ particles based on mass and population balances has been recently proposed (Testino et. al., 2005). The population balances considered three elementary chemical kinetic processes: primary nucleation, secondary nucleation, and diffusion-controlled particle growth. Secondary nucleation accounts for the acceleration of the formation kinetics after an initial slow crystallization of $BaTiO_3$ particles. The time evolution of yield and particle size is represented by means of discretized mass and population balance equations. Hence, the algorithm is capable for calculating a generic number of conditions involved in the chemical reaction, which can optimize the crystallization and control the growth of $BaTiO_3$ particles. Another different approach based in a bi-variate population balance equations have prove to give more accurate results for the modelling of the barium titanate hydrothermal crystallization. The results obtained from the proposed population balance mathematical model clearly showed that such an approach can overcome the limitations of previous modelling work, and provide a useful tool for more detailed kinetic parameters estimation.

Moreover the model shows that secondary nucleation is indeed very important but particle aggregation cannot be neglected, these variables were not considered in the previous model (Marchizo, 2009).

Despite of the remarkable achievement to develop new strategies based on fundamental principles of thermodynamics and chemistry (reaction equilibrium and kinetics), caution should bear in mind for applying the proposed models to diverse hydrothermal systems and experimental situations. Hence, the estimation of kinetic parameters using population balance derived models could not be correct, these parameters can exhibit variations due to differences on temperature and concentration in the fluid media, which would lead to erroneous conclusions, because the reaction mechanism at certain conditions might change. Likewise, the population balance derived models assume that the system is well mixed and the crystallization rate is uniquely controlled by the chemical kinetics. These assumptions are generally valid for some small-scale laboratory reactors, but they fail in larger industrial plants where mixing is not well controlled. A mixing-limited precipitation rate is one of the problems commonly encountered in the scale-up of solid precipitation processes.

2.4.2 Hydrothermal crystallization process of selected oxide compounds

In the past, the term *hydrothermal crystallization* was used to referred it as nonconventional chemical process, this process involves heating an aqueous suspension of insoluble salts in an autoclave at temperatures and pressures greater than 100 °C and 0.1 MPa, respectively; resulting in the crystallization of the desired well-crystallized phases. This process, however, can be analogous to the term *hydrothermal synthesis* broadly used for physicists and chemists involved in this type of research, because both terms are related with the *genesis* of a specific compound. The crystallization process can mainly occur under very precise conditions of reaction, i.e. temperature, pressure, pH, concentration mineralizer or solvent solution. Thus, the hydrothermal crystallization method using inexpensive and chemicals easy to handle, can produce single or multicomponent oxides. The advantages of hydrothermal crystallization are the reduced energy costs due to the mild temperatures sufficient to achieve chemical reactions, less pollution, simplicity in the process equipment, and the fast rate of solid precipitation reactions (Vivekanandan & Kutty, 1989; Yoshimura & Suchanek, 1997; Riman et. al., 2002).

There are thousands of research works in the literature that include a vast experimental data related with the hydrothermal crystallization of metal oxides or inorganic compounds. The most popular among the metal oxides are those of perovskite oxides group, because of their wide application at the electronic industry. Hence, fundamental principles of crystallization of these materials are discussed in detail and compared with the present research experimental of the present authors regarding the crystallization of perovskite particles of $SrTiO_3$, $BaTiO_3$ and $La_{1-X}M_XCr_{1-Y}N_YO_3$.

2.4.3 Hydrothermal crystallization of perovskite derived oxides from gel precursors

In this section, we focus on the hydrothermal crystallization of perovskite-type structure oxide materials, because this particular group of oxides offers stupendous functional properties for a wide number of applications, in particular electronic applications. Perovskite compounds have the general formula ABO_3, and Perovskite is cubic structure in nature which is shown in Figure 3, where the large A cations (Ba^{2+}, Sr^{2+}, Ca^{2+}, Pb^{2+}, La^{3+}, Bi^{3+} and K^{+}), but low in electric charge, are surrounded by 12 oxygens, whilst B ions relatively small in size (Ti^{4+}, Zr^{4+}, Sn^{4+}, W^{6+}, Nb^{5+}, Mn^{3+}, Mg^{2+}) are coordinated by 6 oxygens. The crystallization of the most representative compounds in this group, namely $BaTiO_3$ (BT), $SrTiO_3$ (ST) and solid solutions of $Ba_{1-X}Sr_XTiO_3$ (BST) or $PbZr_{1-X}Ti_XO_3$ (PZT) has been extensively studied under hydrothermal conditions.

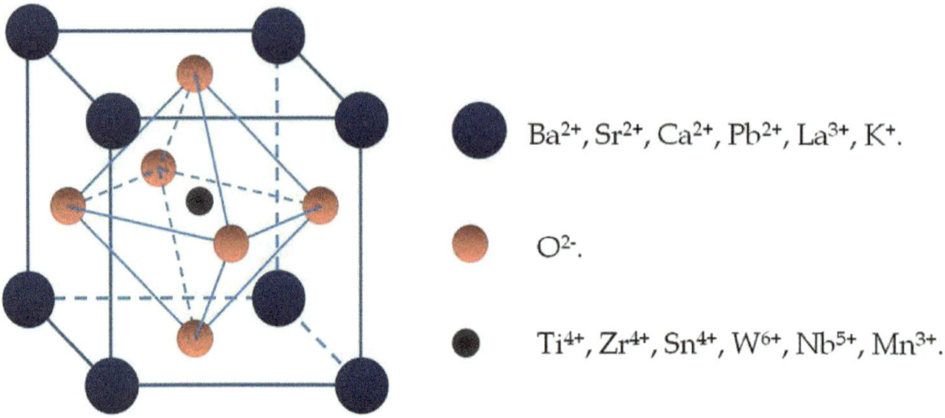

Fig. 3. Schematic representation of the unit lattice cell of the Perovskite cubic structure.

2.4.3.1 Hydrothermal crystallization of perovskite barium and strontium titanate oxides

$BaTiO_3$ (BT) and $SrTiO_3$ (ST) are the metal earth alkaline perovskite-like structures most prepared species under hydrothermal conditions, these compounds has similarity on the crystallization process as that proposed for PT. In terms of the chemical reaction equilibrium that can be produced on the hydrothermal system, and in accordance with the chemical reagent reactants that have been used the chemical reactions that are related with the crystallization of BT and ST powders are as follows:

Where M on chemical equations 3 and 4 is related to earth alkaline metals of the group IIA of the periodic table of elements, *viz.* Ca, Sr and Ba. The chemical reactions from 1 to 9 represent all the reaction equilibrium that are able to proceed in the system M-Ti-Na-H_2O and these reactions serve to produce the crystallization process, because they stem from the principal chemical reactions 1–12 that have widely investigated on the crystallization of BT under hydrothermal alkaline conditions and can operate on the case of the ST compound as well.

$$H_2O = H^+ + OH^- \quad (1)$$

$$H_2O(g) = H_2O \quad (2)$$

$$MTiO_3(s) + H_2O = M^{2+} + 2OH^- + TiO_2(s) \quad (3)$$

$$MOH^+ = M^{2+} + OH^- \quad (4)$$

$$Ti^{4+} + H_2O = TiOH_3^+ + H^+ \quad (5)$$

$$TiOH^{3+} + H_2O = Ti(OH)_2^{2+} + H^+ \quad (6)$$

$$Ti(OH)_2^{2+} + H_2O = Ti(OH)_3^{+} + H^+ \quad (7)$$

$$Ti(OH)_3^{+} + H_2O = Ti(OH)_4(aq) + H^+ \quad (8)$$

$$Ti(OH)_4(aq) = TiO_2(s) + 2H_2O \quad (9)$$

$$BaCl_2(aq) + TiCl_4(aq) + 6NaOH(aq) = BaTiO_3(s) + 6NaCl(aq) + 3H_2O(l) \quad (10)$$

$$Ba(OH)_2(aq) + TiO_2{\bullet}xH_2O(s) + NaOH(aq) = BaTiO_3(s) + NaOH(aq) + xH_2O(l) \quad (11)$$

$$Ba(OH)_2(aq) + Ti(OH)_4{\bullet}xH_2O(aq) + NaOH(aq) = BaTiO_3(s) + NaOH(aq) + 4xH_2O(l) \quad (12)$$

The chemical synthesis of BT particles has been studied over a broad range of experimental hydrothermal conditions; this compound is preferentially formed in highly concentrated alkaline solutions (pH > 10), as indicated by thermodynamic calculations in BT phase stability diagram proposed elsewhere (Lencka et. al., 1993; Riman et. al., 2002). The normal intervals of reaction time and temperature are 2–96 h and 120–250 °C where the crystallization is normally achieved, thus, the preparation of this type of powder is carried out in typical stainless steel 304 Teflon-lined autoclaves. Additional parametric variable studies recently conducted accounts for other variables that also have an effect on the formation of the BT pure phase like the molar ratio of precursor alkaline (KOH or NaOH) solvent (Vivekanandan & Kutty, 1989; Lee et. al, 2003) and the reactants ($Ba(OH)_2$ or titania source) (Moon et. al., 2003, Qi et. al., 2004), the Ba/Ti molar ratio of the raw materials (Wada et. al., 1995). Another specific studies recently conducted lead to remarkable analysis of the kinetic process related to the crystallization of BT particles and their growth (Walton et. al., 2001), and the control of the particle size and shape of the BT particles have been carried out by stirring the autoclave (Kubo et. al., 2009) and microwave assisted (Moreira et. al., 2008) hydrothermal methods.

Hitherto, in accordance with the analyses conducted based on experimental observations, indicate that the crystallization mechanism of BT particles is dissolution-precipitation in nature, and this mechanism is mainly controlled by the dissolving rate of the reactant solid specie, namely titania (oxide or amorphous gel). Thus, the presence of a high concentration of OH^- ions in the hydrothermal system is required to produce the hydrolysis of aqueous species, and these ions also seemed to act as catalysts to accelerate the transition from Ba-OH bonds, usually resulting on the crystallization of BT via precipitation that proceeds via the chemical reaction (13), resulting from the previous reaction (10–12) that is carried out during the earlier and intermediate stages of the hydrothermal treatment and are linked with the dissolution process.

$$BaOH^+(aq) + Ti(OH)_4^0(aq) = BaTiO_3(s) + 2H_2O + H^+(aq) \quad (13)$$

Once the solution is supersaturated due to dissolution of the precursor, precipitation of BT from the homogeneous solutions spontaneously occurs, yielding an abundant number of nuclei. Therefore, the dissolution of either titania hydrous amorphous or crystalline powders is the rate limiting step that controls the early stage of BT nucleation and growth during the hydrothermal crystallization process, for anhydrous TiO_2 precursor, Ti-O bonds must be broken via hydrolytic attack to for hydroxyl-titanium complexes ($Ti(OH)_X^{4-X}$) capable of dissolution and reaction with barium ions or complexes (Ba^{2+} or $BaOH^+$) in solution to precipitate BT. Subsequently, the nuclei grow rapidly, resulting in an accelerated crystallization kinetics rate at the initial step of the process. In the solvent, the solute concentration decreases to below the supersaturation point as a result of this event, but remains sufficiently high for the particles to growth, avoiding a secondary nucleation. The particle coarsening proceeds during intermediate and final steps of the crystallization process, and mainly depends on the dissociation of the remaining reactants, or in some cases the separation of terminal organic groups that are liked to the Ti or Ba, e.g. acetiylacetone or acetate, these crystallization barriers serves to slow, if not halt, the kinetic rate in the final stage of the hydrothermal treatment (Eckert et. al., 1996). The crystallization studies indicates that the activation energy fro barium titanate crystallization under hydrothermal conditions varies on the range of 21-105 $kJmol^{-1}$, and the value of activation energy specifically depends on the type of compound used as titania source (Eckert et. al.; 1996; Walton et. al., 2001), but these values of activation are within the range of activation energy values for chemical reaction that proceed on the liquid stage.

On the other hand, major attempts recently conducted to investigate the crystallization of $SrTiO_3$ particles under hydrothermal conditions, have been designed by considering the fundamental principles derived from the synthesis of its related compounds, *viz.* BT or PT perovskite-like structure. A broad type of strontium titanate particle shapes, e.g. nanotube, dendrite, cuboidal and spherical; had been produced through optimizing the hydrothermal crystallization conditions by controlling parameter such as: the pH of the hydrothermal alkaline media in the range between 10–12 (Wendelbo et. al, 2006), reaction temperature and time. Furthermore, different chemical reagents including inorganic and organometallic had been employed as source of Sr^{2+} ($Sr(NO_3)_2$, $SrCl_2$, $Sr(OH)_2 \bullet 6H_2O$) and Ti^{4+} ($TiCl_4$, $TiO_2 \bullet H_2O$, $Ti(SO_4)_2$, $Ti(OC_4H_9)_4$, $Ti(OPr^i)_4$) for producing ST particles (Moon et. al., 1999; Zhang et. al., 2001; Wang et. al., 2009). Another factor that has been studied is related with the use of organic dispersants, namely polyvinyl alcohol, for controlling the crystallization of fine regular shaped ST particles. The dispersant that have large polymeric changes can operate as micelles that attract the hydrolysed ionic species that are formed during the hydrothermal treatment, this lead to and homogeneous nucleation and limited particle growth, resulting in the preparation of nanometer sized ST regular particles (Wei et. al., 2008). Regarding the mechanism correlated with the crystallization of ST particles, one model derived from further experimental data states that the dissolution-crystallization coupled with a second aggregative growth-recrystallization mechanisms are related with the bulk crystallization stage of ST particles under alkaline hydrothermal conditions, but is similar to that proposed for the formation of BT particles (Fig. 4). The second mechanism is achieved when heterogeneous nucleation promotes the formation of ST particles, the surface of the TiO_2 particles acts as the precipitation site for ST nuclei, and when the size of TiO_2 particles is reduced by the progressive dissolution, the aggregation growth of the ST particles proceeds in the reaction media. This phenomenon produces a marked particle agglomeration and also the coarsening of the particles can be promoted by the Ostwald recrystallization mechanism (Zhang et. al., 2004).

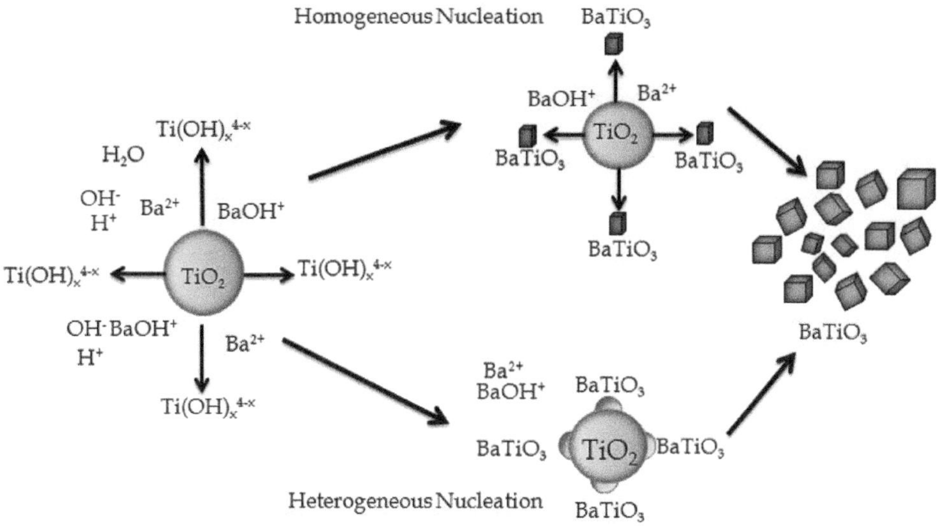

Fig. 4. Schematic sketch of the dissolution/precipitation mechanism that conduces to the hydrothermal crystallization of $BaTiO_3$ (Ecker et. al. 1996, Zhang et. al., 2004).

2.4.3.2 Single step hydrothermal crystallization of perovskite strontium and barium oxides using mineral precursors

In accordance to the former literature, in the last decade, the interest for using pure mineral species to produce pure synthetic inorganic compounds has been rising. In general, the synthesis of $SrTiO_3$ or $BaTiO_3$ has been broadly conducted by employing $Ti(OH)_4$ gel (Ti-gel), while several strontium soluble salts had been used as a source of Sr^{2+} ions. However, the challenge for employing low-grade chemical reagent precursors for preparing ABO_3 particles has not been considered yet. The approach for employing a pure mineral ore, like celestite ($SrSO_4$) or barite ($BaSO_4$) with a low grade of impurities; to produce strontium or barium compounds under hydrothermal conditions, was considered based on the analysis of the proposed chemical methods for producing functional ceramic materials and the former information related to the ionic substitution on mineral species, which are analogous to the mechanistic principles of hydrothermal crystallization discussed in this section. The employment of a low cost precursor may provide an additional advantage in order to propose an economical effective processing method.

Recently, the present authors have conducted an exhaustive efforts to develop a simple single step reaction method for the preparation of $SrTiO_3$, which involves the employment of a mineral $SrSO_4$ crystal plate (0.2 ± 0.0010 g, 6 ± 1 mm side and 2 ± 0.5 mm thick) with $Ti(OH)_4 \cdot 4.5H_2O$ gel (1 g, stoichiometric ratio Sr/Ti=1) under hydrothermal conditions, at various temperatures (150-250 °C) for different reaction intervals (0.08–96 h) in KOH solutions with different concentrations (5–10 M). The hydrothermal treatments were carried out in stainless steel micro-autoclaves lined with Teflon with a filling volume ratio of 50% of the total inner volume (30 ml). This process involves a complex solute dissolution stage because $SrSO_4$ is chemically stable even at acid, neutral and mild basic conditions. However, the use of a mineral single crystal favours to analyse systematically the effect of the dissolution of this reactant coupled with the $Ti(OH)_4$ dehydration and serve to control the synthesis and crystallization kinetics of ST particles. Thus, the complete crystallization of ST particles in a single step of reaction occurred with the complete dissolution of the $SrSO_4$ crystal obtained at 250 °C for 96 h in a 5 M KOH solution, resulting in the formation of $SrTiO_3$ particles with two different shapes (peanut-like and cubic) as is shown in Figure 4. The parameter that has a marked effect on control the particle size and morphology is the temperature rather than the interval of reaction. ST particles having a bimodal size distribution (0.5–1 μm and 0.2–0.5 μm) were prepared at different temperatures for 96 h in a 5 M KOH solution from $SrSO_4$ crystals. The ST powders produced at mild temperatures (200 °C) were constituted by a large amount of agglomerated peanut-like shape irregular particles (Fig. 5a), together with a small amount of bulky crystals with a regular pseudo-cubic shape (1–4 μm). In contrast, at high temperature (250 °C), the pseudo-cubic (3–6 μm) ST particles exhibited a coarsening, while the amount of peanut-like shape (length = 1–5 μm and width = 0.5–1 μm) particles was significantly reduced (Fig. 5b). Hence, the coarsening of pseudo-cubic shaped particles is attributed to the Ostwald ripening particle growth mechanism (Peterson & Slamovich, 1999). Indeed, the growth of the aggregated particles is due to a recrystallization mechanism, which involves the dissolution of the primary small ST peanut-like particles produced at lower temperatures (< 200 °C). This process is promoted due to lower chemical stability that exhibits the strontium titanate oxide in highly concentrated (> 5 M) KOH solvent solutions at 250 °C. Another point that deserves to be emphasized is that related with the structural features of either peanut-like or cubic-like shaped particles. EDX spectra obtained on both particles did not show a marked difference on the molar Sr/Ti ratio, because the particles contain similar Sr and Ti amounts as it is suggested by the corresponding EDX spectrum in Fig. 5c (Rangel-Hernandez et. al., 2009).

In contrast, the increase on the surface area of the precursor $SrSO_4$ produces the control of morphology and size homogeneity for the ST particles. In general, the use of $SrSO_4$ powders with a narrow particle size distribution in the range from 25–38 μm, favoured the crystallization of very fine cubic ST particles, even at very short reaction intervals (0.08 h, Fig. 6a). However, ST powders with a bimodal size distribution, consisting of a large amount of pseudo cubic particles (0.75 μm) and a small quantity of large cubic particles (1.5 μm) were produced during intermediate reaction intervals between 3-12 h. In contrast, at the longest reaction interval of 24 h, the preferential formation of very fine pseudo-spherical ST particles (average size of 0.4 μm) was observed (Fig. 6b). The formation of the cubic particles is due to the crystallographic habit growth, which proceeded when the alkaline solvent reaches a supersaturation steady-state with the ionic species $Sr(OH)^+$ and $Ti(OH)_4^0$. Hence, a massive homogeneous nuclei formation and fast growth of the cubic particles proceeds in the hydrothermal system during intermediate reaction intervals. However, the ST cubic particles underwent into a preferential dissolution by increasing the reaction interval at long reaction period (24 h). The ST particle dissolution proceeds at high-energy faceted edges on the fine cubic particles, because of the relative low chemical stability of the ST perovskite structure in the concentrated alkaline media. This fact leads to further dissolution of the cubic particles and the recrystallization of the pseudo-spherical shaped particles from the solution.

Fig. 5. Morphologies of ST particles obtained after hydrothermal treatments of $SrSO_4$ crystal plates, carried out at temperatures of (a) 200 °C, (b) 250 °C; and (c) EDX spectra of ST (*i*) peanut-like and (*ii*) cubic particles shown in Fig. 3b (Rangel-Hernandez et. al., 2009).

Fig. 6. $SrTiO_3$ particles hydrothermally produced using $SrSO_4$ powders at 250 °C in 5 M KOH solution for intervals of (a) 0.08 and (b) 24 h (Rangel-Hernandez et. al., 2009).

The bulk crystallization process of ST particles is dissolution-precipitation in nature, and the reaction path involving this mechanism that produces nuclei formation and crystal growth, is represented physically in the Figures 7a and 7b. These Figures gave the set of the original

$SrSO_4$ crystal plate embedded in the $Ti(OH)_4 \bullet 4.5H_2O$ gel prior the hydrothermal treatment and after 6 h of reaction at 200 °C. In terms of fundamental chemistry the reaction path way occurs trough the chemical equations 14 and 15:

$$SrSO_4(crystal) + Ti(OH)_4 \bullet 4.5H_2O(gel) + 2KOH \\ = SrOH^+(aq) + K_2SO_4(aq) + Ti(OH)_4^{\;0}(aq) + 4.5H_2O + OH^-(aq) \tag{14}$$

$$SrOH^+(aq) + Ti(OH)_4^0(aq) + OH^-(aq) = SrTiO_3(s) + 3H_2O \tag{15}$$

Macroscopic aspects related with the single-step synthesis process were revealed during the early stages of the reaction at 200 °C for 6 h in a 5 M KOH solution (Fig. 7b). In general, the reaction gradually proceeds by the $SrSO_4$ crystal dissolution, the consumption of $SrSO_4$ occurs locally on those crystal surfaces exposed to the alkaline fluid and also on those crystal surfaces in contact with the Ti-gel. This phenomenon agrees with those results previously reported and provided evidences of the solubility of $SrSO_4$ compound in alkaline hydrothermal solutions (5 and 10 M NaOH) (Rendón-Angeles et. al., 2006). Hence, the continuous dissolution of the $SrSO_4$ crystal must yield the formation of $SrOH^+$ species (Eq. 14), because of the saturated alkaline conditions of the solvent. Indeed, the solubility of the Ti-gel is low in high concentrated KOH (> 0.1 M) solutions (Wang et. al., 2009, Rangel-Hernandez et. al., 2009). However, the consumption of the $SrSO_4$ crystal dissolution (α) was markedly reduced due to the presence of the Ti-gel (Fig. 7c). This fact let us to conclude that the reactivity of Ti-gel is the limiting rate variable for crystallization, because reduces the dissolution rate of the $SrSO_4$ phase. Thus, the equation (4) is proposed for the crystallization of ST and is similar to that discussed at the early part of the present section for PT and BT compounds. The nucleation and growth processes of the ST particles occur when the alkaline (KOH) solution reaches a supersaturation steady-state of the species $Sr(OH)^+$ and $Ti(OH)_4^0$. The ST particles precipitation locally occurred at Ti-gel surface and between the spaces with the remaining $SrSO_4$ crystal, because the mass transport was limited during the treatments, which were conducted under neither agitation nor thermal gradients. Additionally, kinetic data obtained from $SrSO_4$ consumption curves depicted that the activation energy required for the synthesis of $SrTiO_3$ powders from the complete consumption of an $SrSO_4$ crystal plate under hydrothermal conditions, is 27.9 kJ mol^{-1}. One particular fact that must be emphasised is related with the incorporation of the major impurities on the ST particles during the synthesis process, namely Ba (5.7 wt.%) and CO_3^{-2} (0.6 wt.%), contained in the $SrSO_4$ crystals, does not proceed during the crystallization event because neither the presence of barium nor carbonate compounds were found on the compositional analysis conducted by EDX and DRX techniques in the ST particles.

On the other hand, the crystallization of perovskite related compounds like ST or BT via the transformation of sulphate alkaline earth metal mineral species is hinder for the barite mineral. The crystalline transformation of $BaSO_4$ to $BaTiO_3$ was studied using the single step reaction route aforementioned; and the high chemical stability (low solubility) of the orthorhombic structure limited the dissolution of the barite crystals under hydrothermal conditions even in highly concentrated KOH solutions (>5 M). The reactivity of the orthorhombic structure with the alkaline solvent can be influenced by the size of the alkaline earth metal ion incorporated in the structure, these results are in agreement with those early reported on the transformation of Ca, Sr and Ba-chlorapatite crystals into their respective hydroxyapatite species (Rendón-Angeles et. al., 2000a). This factor reduces the rate of the BT particle crystallization process, but the reaction was conducted by increasing the concentration of the KOH solution up to 10 M, and the complete dissolution of the $BaSO_4$ crystal occurred for large reaction intervals of 144 h above 200 °C. Optimum steady-state supersaturation of the solvent media with Ba^{2+} and Ti^{4+} ions that achieve the conditions for homogeneous nucleation of BT particles were found to proceed at temperatures below 200 °C (Figure 8). Fine BT particles with pseudo-cubic, star-like and dendrite shapes were preferentially formed in 10 KOH. At sever treatment conditions of temperature 250 °C for 144 h, the $BaSO_4$ crystals were completely dissolved, and the formation of reaction by-products Ba_2TiO_4 (needle crystals in Figure 8c) and a few amount of TiO_2 simultaneously occurred with the excessive growth of polyhedral aggregated BT crystals. This particular behaviour is attributed to the differences in the chemical reactivity of the alkaline solution resulting in different dissolving rates for solid species, this can shift the reaction equilibrium coupled with the saturation conditions. Therefore, other crystalline phases are able to precipitate because thermodynamically are more stable than BT.

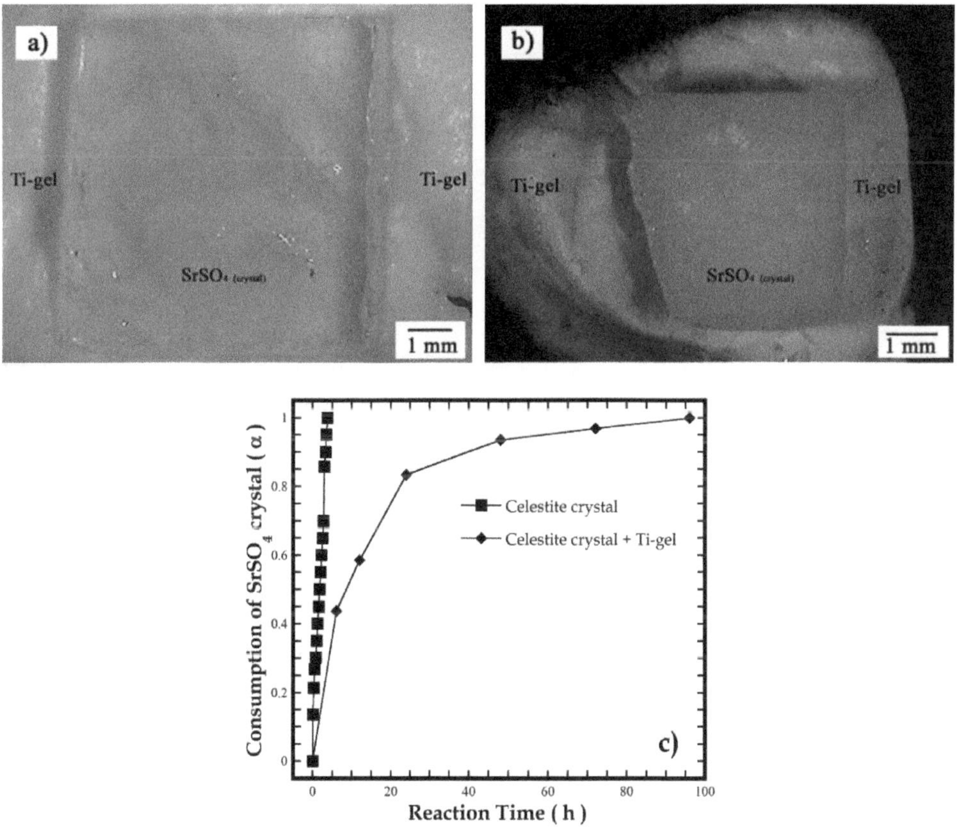

Fig. 7. Aspects of (a) original $SrSO_4$ crystal plate embedded in Ti-gel, and (b) the partially reacted $SrSO_4$ crystal and Ti-gel at 200 °C for 6 h in a 5 M KOH solution. (c) Consumption curves (α) of $SrSO_4$ crystal vs. the reaction interval obtained at 250 °C in a 5 M KOH solution, (■) without Ti-gel and (◆) with Ti-gel additions (Rangel-Hernandez et. al., 2009).

2.4.3.3 Elimination of mineral impurities during the synthesis of strontium titanate particles under hydrothermal conditions

Regarding the presence of high content of impurities during the crystallization of ST particles from mineral celestite ores. The experimental research work conducted recently by the authors of the present chapter; clearly contribute to demonstrate that refining of the major metal ion contained in the mineral ore proceeds during the hydrothermal crystallization of ST powders. Thus, the chemical stability of the barite-celestite mineral specie was investigated to elucidate the feasibility to conduct simultaneously the crystallization of ST particles and the release of major impurities such as Ba, and the preparation of BT powders as well. In particular, the barite-celestite mineral consists in a solid solution with a general chemical formula of $Sr_{0.70}Ba_{0.30}SO_4$, which was determined by wet chemistry using ICP, the total content of major constituents $SrSO_4$ and $BaSO_4$ was 64.7 wt.% and 35.3 wt.%, respectively. Additional microprobe wavelength X-ray diffraction analyses conducted by scanning electron microscope observations indicated that the microstructure of the mineral consists of two major solid solutions, one rich in strontium $Sr_{0.95}Ba_{0.05}SO_4$ and a less amount of an intermediate $Sr_{0.75}Ba_{0.25}SO_4$. The hydrothermal treatments of barite-celestite crystal plate samples were conducted in accordance with the experimental procedure explained before for high pure celestite crystals.

Fig. 8. $BaTiO_3$ particles crystallized from barite mineral plates in a 10 M KOH solution for 24 h at temperatures of (a) 150, (b) 200 and (c) 250 °C.

The synthesis of ST particles was preferentially obtained using the barite-celestite crystals in a feedstock alkaline 5 M KOH solution; the transformation of the mineral specie into the perovskite oxide is strongly affected by the temperature and concentration of the solvent media rather than the reaction interval. Marked morphological and particle size differences were derived due to the increase of reaction temperature, a mixture of fine particles resembling cubic and start-like shapes were formed at mild temperatures (150 and 200 °C) for 12 h, increasing the temperature at 250 °C bulky aggregated cubic ST particles (20 µm size) were crystallized (Figure 9). During the hydrothermal crystallization of the ST compound the incorporation of the Ba^{2+} ions was avoided to occur in the structure of ST particles, this is supported by the EDX spectra of the cubic and star-like ST particles where no traces of Ba are visible (Fig. 9d).

Fig. 9. $SrTiO_3$ particles crystallized from barite-celestite mineral plates in a 5 M KOH solution for 12 h at different temperatures of (a) 150, (b) 200 and (c) 250 °C. (d) EDX spectra of the ST particles shown in (b).

Two main factors are associated with; one is related to the marked compositional gradient differences of the metal ions Sr^{2+} and Ba^{2+} produced in the solvent media during mineral dissolution, coupled with the stability of Ba^{2+} to form complexes ions in alkaline solutions ($BaOH^{+}$, Eq. 4). Indeed, wet chemical analyses of the remaining solutions after the treatments gave evidences that support the above inference, because the Ba^{2+} ions content gradually increased as far as the complete mineral dissolution was concluded for 96 h at 250 °C. The content of Ba determined by in the solution was 20.0 ± 0.8 wt%, which is nearly the same, measured in the original barite-celestite crystals (20.8 wt%). The ST particles crystallization proceeds by the same model established previously on the case of the transformation of high pure celestite crystal to ST particles, the control of morphology and particle size for the ST powders can be optimized by limiting the dissolution-recrystallization mechanism that operates at sever treatment conditions of temperature (T> 200 °C) and long periods (t> 72 h). Therefore, these results probe that the hydrothermal technique combined with the use of mineral (pure and contaminated) as precursor for the synthesis of inorganic materials can be an attractive technique to be explored at an industrial scale.

2.4.3.4 Hydrothermal crystallization of perovskite lanthanum chromite oxides

Lanthanum chromite ($LaCrO_3$, LC) powders substituted with alkaline metals (Ca or Sr) have been widely accepted as the candidate for interconnection materials in Solid Oxide Fuel Cells (SOFCs) and also particular solid solutions can be used as oxygen sensor at high temperature (Chakraborty et. al., 2000) The partial substitution of lanthanum ions with alkaline metal ions, Ca or Sr in the A site of La, increases the chemical stability and electric conductivity, whilst Al or Ni in the B site reduces the thermal expansion coefficient at high temperature, when compared with the properties of pure lanthanum chromite (Ianculescu, et. al., 2001). Hitherto, various chemical routes have been used to process lanthanum chromite powders (Bliger. et. al., 1997). However, these chemical processes involve heat treatments at temperatures beyond 700 °C, in order to obtain the crystalline phase pure phase and their solid solutions with Ca or Sr in the A site and Ni or Al in the B in the sites of the perovskite-like structure ABO_3. Hence, the present authors recently determined through exhaustive research work, the aspects related to the crystallization of precursor complex gel of the $La_{1-X}M_XCr_{1-Y}N_YOH_{6-x}$ compound into their respective perovskite solid solutions $La_{1-X}M_XCr_{1-Y}N_YO_3$ (where M = Ca or Sr and N = Al or Ni) with orthorhombic crystalline structure under hydrothermal conditions, at a temperature range between 350–500 °C for short reaction intervals (0.5–2 h).

The precursor complex gel was prepared by the alkaline coprecipitation method widely used to prepare this type of materials; the details of the chemical preparative method are given elsewhere (Inagaki et. al., 1990). Precursor lanthanum chromite complex gels were prepared by employing reagent grade chemicals of: $LaCl_3 \cdot 7H_2O$ (99.998%), $Cr(NO_3)_3 \cdot 9H_2O$ (99.9%), $CaCl_2 \cdot 2H_2O$ (99%), $SrCl_2 \cdot 2H_2O$ (99%), $Ni(NO_3)_2 \cdot 6H_2O$, $Al(NO_3)_3 \cdot 9H_2O$ and NaOH (99.998%) (Wako Pure Chemical Industries, Ltd., Japan). Aqueous solutions with a concentration 0.05 M of $LaCl_3$, $Cr(NO_3)_3$ and $CaCl_2$ were prepared with deionized water, and a solution of 0.5 M of NaOH was employed as coprecipitation media. In a typical procedure, a volume of 475 ml of the precipitating solution (NaOH) was poured in a biker, and chromium or the mixtures of Cr+Al, Cr+Ni; solution (500 ml) was then mixed, which results in the formation of an opaque whitish green precipitate ($Cr(OH)_3$, or $Al(OH)_3$, $Ni(OH)_2$), which was subsequently dissolved by vigorous stirring. Finally, the coprecipitation of the complex gel was carried out by the addition of the same volume (500 ml) of the solution containing the other elements, La or the mixture of La+Ca, La+Sr. The solutions were mixed in different volumetric ratios, La:M:Cr:N, 1:0:1:0, 0.9:0.1:1:0 and 0.8:0.2:1:0, 1:0:0.95:0.05; 1:0:0.9:0.1; 0.8:0.2:0.95:0.05, 0.8:0.2:0.9:0.1; which matches the compositional stoichiometric of the solid solutions, $LaCrO_3$, $La_{0.9}Ca_{0.1}CrO_3$, $La_{0.8}Ca_{0.2}CrO_3$, $La_{0.8}Sr_{0.1}CrO_3$, $La_{0.8}Sr_{0.2}CrO_3$, $LaCr_{0.95}Al_{0.05}O_3$, $LaCr_{0.9}Al_{0.1}O_3$, $LaCa_{0.2}Cr_{0.95}Al_{0.05}O_3$, $LaCa_{0.2}Cr_{0.9}Al_{0.1}O_3$, $LaSr_{0.2}Cr_{0.95}Al_{0.05}O_3$, $LaSr_{0.2}Cr_{0.9}Al_{0.1}O_3$, $LaSr_{0.2}Cr_{0.95}Ni_{0.05}O_3$, $LaSr_{0.2}Cr_{0.9}Ni_{0.1}O_3$. The coprecipitated gel was centrifuged and a volume of 20 ml was then poured into a hydrothermal Hastelloy C-lined microautoclave (40 ml capacity, Figure 10). The vessel was heated at a constant rate of 20 °C/min up at various temperatures (350–500

°C) for a reaction interval between 0.5 and 2 h. After the treatments, the precipitates were well washed with distilled water, decanted and then dried in an oven at 100 °C overnight (Rivas-Vázquez et. al., 2004, 2006; Rendón-Angeles et. al., 2009).

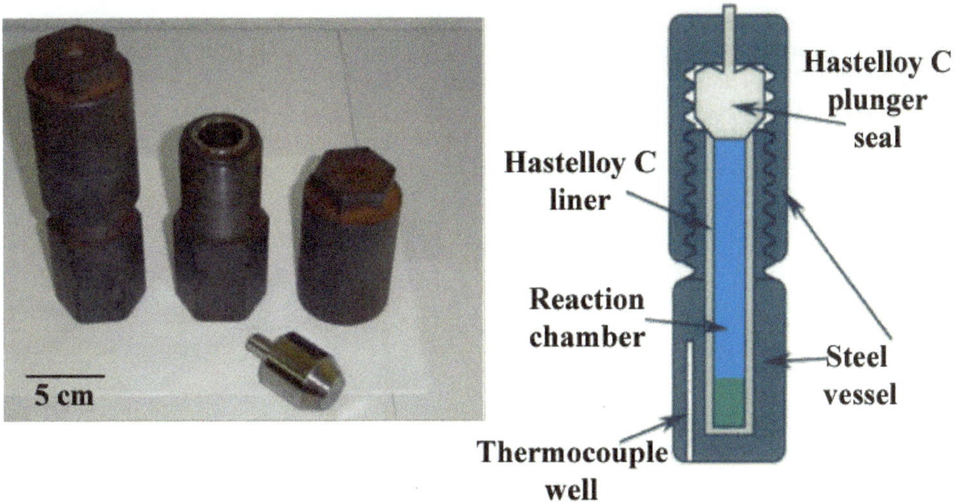

Fig. 10. Typical Hastelloy C type lined microautoclave and scheme of the cross section of the vessel, which was employed for conducting the synthesis of perovskite lanthanum chromite powders under hydrothermal conditions.

The minimum temperature for conducting the crystallization of the pure LC powders was 375 °C for a reaction interval of 1 h. This parameter, however, is mainly affected by the incorporation of the metal ions in both A and B sites of the perovskite structure ABO_3, and also with the amount of metal dopant ion inserted. Thus, the crystallization of powders corresponding to the solid solutions $La_{1-X}M_XCr_{1-Y}N_YO_3$ (where M = Ca or Sr and N = Al or Ni) was observed to proceeds at temperatures above 400 °C, when only a 10 mol% of Ca^{2+} or Sr^{2+} were partially substituting La^{3+} site. Thus, in the particular case of the $La_{0.9}Ca_{0.1}CrO_3$ and $La_{0.8}Ca_{0.2}CrO_3$ solid solutions, the structural analyses in Figure 11a depicted that at a constant reaction interval 1 h, the powders of $La_{0.9}Ca_{0.1}CrO_3$ were produced at a temperature of 400 °C without contaminant formation, in contrast with the $La_{0.8}Ca_{0.2}CrO_3$ powders that were produced together with a marked amount of secondary crystalline phases of $La(OH)_3$, CrOOH, and $CaCrO_4$, these phases were eliminated by increasing the reaction temperature up to 425 °C resulting in the crystallization of the pure phase $La_{0.8}Ca_{0.2}CrO_3$. This behaviour was also determined for the crystallization of the $La_{0.9}Ca_{0.1}CrO_3$ and $La_{0.8}Ca_{0.2}CrO_3$ powders. The minimum temperature that achieves the perovskite powders formation was found to increase at 450 °C, when Al^{3+} partially substituted Cr^{3+} sites. This situation is more critical when two different metal dopants were simultaneously incorporated at the same time to form a complex oxide such as; $LaCa_{0.2}Cr_{0.95}Al_{0.05}O_3$ and $LaCa_{0.2}Cr_{0.9}Al_{0.1}O_3$, the powders corresponding to these LC solid solutions were obtained at minimum temperature of 475 °C, which is a significant increase (100 °C) when compared with the low temperature needed to form the pure $LaCrO_3$. The significant variation on the processing parameters experimentally determined, namely the temperature, can be associated with the thermodynamic fundamental principles, e.g. the Gibbs free energy. The calculus of the ΔG values for the formation of Ca and Sr doped LC solid solutions over a wide range of temperature (Fig. 11b), are in a good agreement with the trend of the experimental results aforementioned. Because, in terms of energy, the crystallization reaction that required less energy consumption is that leads to the formation of pure $LaCrO_3$, compound that exhibit the lowest ΔG values in comparison with those of $La_{0.9}M_{0.1}CrO_3$ and $La_{0.8}M_{0.2}CrO_3$ solid solutions (M= Ca^{2+} or Sr^{2+}).

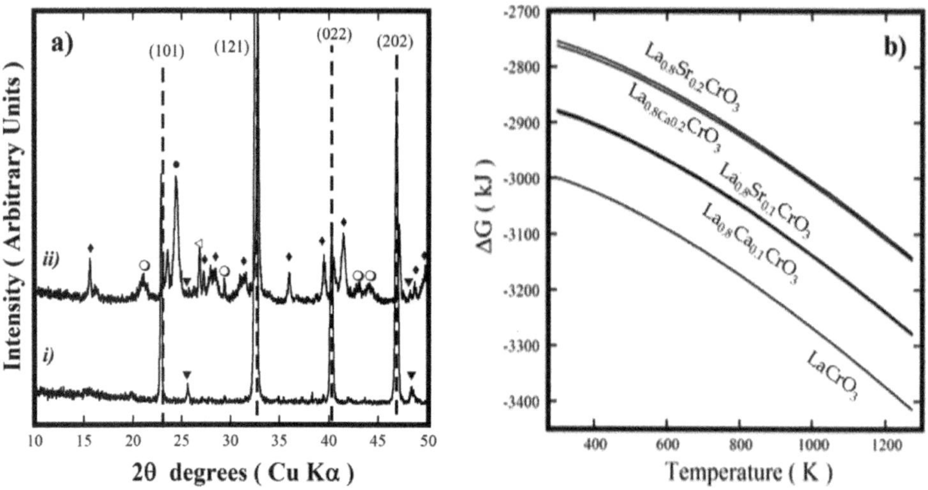

Fig. 11. (a) X-ray diffraction patterns of (*i*) $La_{0.9}Ca_{0.1}CrO_3$ and (*ii*) $La_{0.8}Ca_{0.2}CrO_3$ powders produced at 400 °C for 1 h. $LaCrO_3$ (JCPDS 33-701) compound doted line; (◆) $La(OH)_3$, (◁) CrOOH, (●) $CaCrO_4$, (▼) low intensity peaks of $LaCrO_3$. (b) Variation of the ΔG formation for pure $LaCrO_3$ and $La_{0.9}M_{0.1}CrO_3$ and $La_{0.8}M_{0.2}CrO_3$ solid solutions (M= Ca^{2+} or Sr^{2+}).

Morphological aspects of powders corresponding to the solid solutions of $La_{0.9}M_{0.1}CrO_3$ and $La_{0.8}M_{0.2}CrO_3$ (M= Ca^{2+} or Sr^{2+}) solid solutions obtained at 400 and 425 °C, respectively; are shown in Figure 12. Particles with submicron size and irregular morphology, which resembles a peanut-like shape, were preferentially formed on all cases investigated, it is also indicated that the particle morphology of these compounds is irrespective of the gel crystallization temperature. Furthermore, the particles showed a marked agglomeration due to its particular morphology and particle size aspect. One point that deserves emphasis is that related with the marked tendency for particle bounding that underwent the particles. These are reliable evidences that indicate the particles of the formed LC solid solutions were partially dissolved in the fluid at an intermediate reaction stage (>1 h) of the hydrothermal treatment, once the secondary crystalline phases were completely exhausted.

This phenomenon seems to occur markedly on the $La_{0.8}Sr_{0.2}CrO_3$ particles (Fig. 12d) hydrothermally synthesized 1 h at 425 °C, because these particles exhibit the smallest particle size (average 250 nm) in comparison with the other three powders that have an average particle size of 350 nm (Figs. 12a-12c). This fact can be explained based on the chemical stability of the crystalline phase, which must be low and therefore the dissolution proceeds rapidly. Indeed, this inference is supported by the fact that the amount of regular $La_{0.8}Sr_{0.2}CrO_3$ particles is further decreased in this particular powder. Although, the particles exhibited a non-peculiar morphology when compared with those used for preparing dense advanced ceramic materials, it was experimentally probed that these fine powders have better sinterability features even in oxidizing atmospheres (Rivas-Vázquez et. al., 2004, 2006; Rendón-Angeles et. al., 2009).

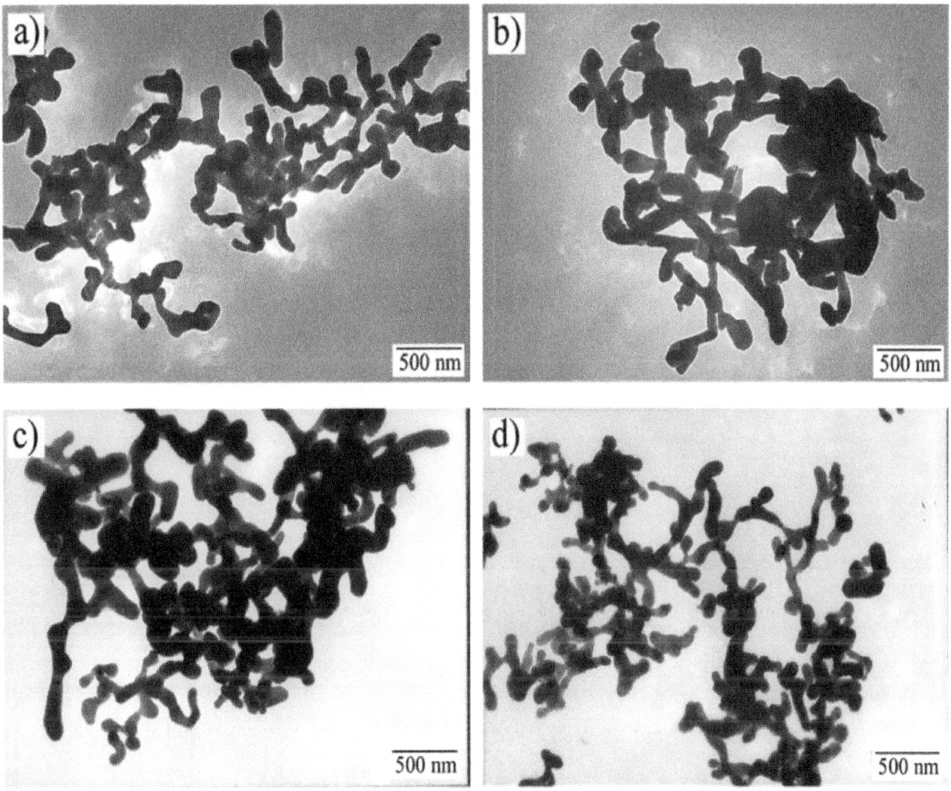

Fig. 12. Transmission electron micrographs of hydrothermally produced powders for 1 h at 400 °C (a) $La_{0.9}Ca_{0.1}CrO_3$ and (c) $La_{0.9}Sr_{0.1}CrO_3$; and 425 °C (b) $La_{0.8}M_{0.2}CrO_3$ and (d) $La_{0.8}M_{0.2}CrO_3$.

The details of the reaction pathway that are linked to the crystallization of the final tailored LC composition were determined during the preparation of $La_{0.9}Ca_{0.1}CrO_3$ solid solution. Preliminary observations conducted even during the heating stage of treatment confirmed that the hydrothermal crystallization involves a preliminary reaction stage, which is related with the dehydration process of the complex gel and proceeds at temperatures above 300 °C, resulting on the autogeneous formation of hydrothermal solvent inside the reaction vessel. The pH of the remaining solutions after the hydrothermal treatment varied in the range between 7.2-7.8, which confirms that the crystallization of the lanthanum chromite powders was conducted in low alkaline conditions. Once the treatment temperature is reached (400 °C), as a result of the complex gel dehydration process, the crystallization of intermediate secondary phases proceeded during the earlier stages of the reaction (10–30 min). The major stable secondary phases that were produced during this reaction interval range were $La(OH)_3$, CrOOH, which correspond to the needle and platelets, respectively (Figs. 13a–13b); these observations are in a good consistence with the X-ray diffraction patterns of the powders produced at 400 °C for different reaction intervals (Fig. 13d), these results also showed the presence of a slight amount of $CaCrO_4$. The dissolution of the reaction by-products is fast in such a diluted alkaline hydrothermal media. Hence, the nucleation of the oxide particles proceeds preferentially at the surface of the remaining complex gel during the dehydration process, because the crystallization by the process employed in this study proceeded at higher temperatures (300–450 °C) than those determined for $LaCrO_3$ and $La_{0.5}Sr_{0.5}MnO_3$ particles under concentrated alkaline hydrothermal conditions (8 M KOH), 240 and 260 °C, respectively (Zheng et. al., 1999; Spooren et. al., 2003). The complete crystallization of $La_{0.9}Ca_{0.1}CrO_3$ particles was even promoted for a shorter interval as 1 h, resulting in the formation of irregular particles resembling peanut-like shapes (Fig. 13c).

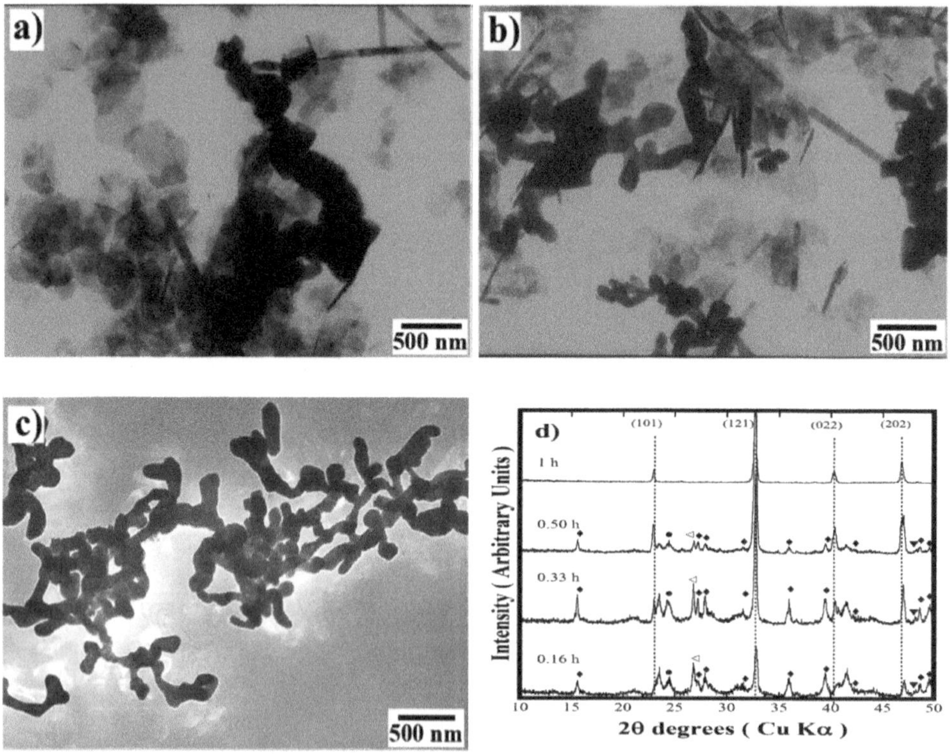

Fig. 13. TEM micrographs of the reaction products obtained under hydrothermal conditions at 400 °C for different intervals (a) 0.33, (b) 0.5 and (c) 1 h. (d) X-ray diffraction patterns of complex gel corresponding to the above micrographs. $LaCrO_3$ (JCPDS 33-701) compound doted line; (◆) $La(OH)_3$, (◁) CrOOH, (●) $CaCrO_4$, (▼) low intensity peaks of $LaCrO_3$.

Regarding the peculiar morphology of the $La_{0.9}M_{0.1}CrO_3$ and $La_{0.8}M_{0.2}CrO_3$ solid solutions (M= Ca^{2+} or Sr^{2+}). Two main factors promoted the formation of the peanut-like shaped particles (Figure 14); the first is related with the nucleation and growth process which proceeded very fast for reaction intervals of 0.5 up to 2 h at interval of temperatures (400-450 ◦C) and limited the growth of $La_{1-X}M_XCrO_3$ particles. The second one is associated with the concentration of the solvent media; the alkaline fluid autogenously formed during the hydrothermal treatment is not capable of achieving the actuated conditions that lead the crystallization of particles with cubic habit (Zheng et. al., 1999; Spooren et. al., 2003). However, the $La_{1-X}Sr_XCrO_3$ particles were partially dissolved in the solvent media at high temperature (400–450 °C), resulting in a marked particle joining by developing necks on the surfaces of primary synthesized particles in contact, as shown in Figure 14 for $La_{0.8}Sr_{0.2}CrO_3$ powders.

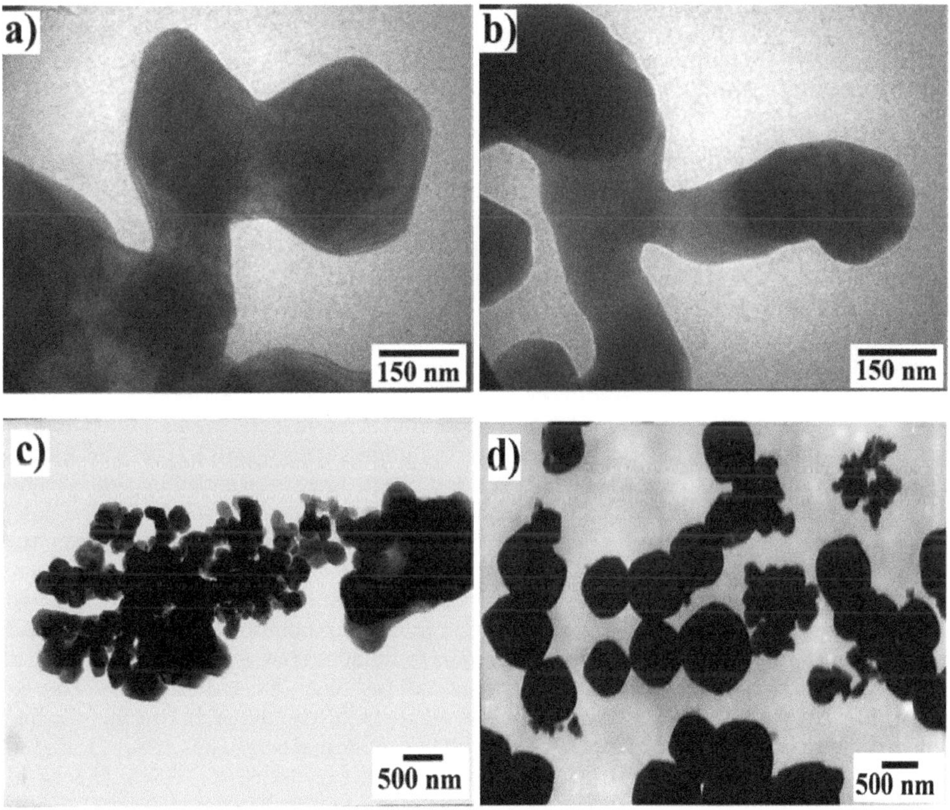

Fig. 14. Transmission electron micrographs of $La_{0.8}Sr_{0.2}CrO_3$ powders obtained at 450 °C for reaction intervals of (a) 1 and (b) 2 h. $La_{0.8}Sr_{0.2}Cr_{0.9}Al_{0.9}O_3$ powders prepared at 475 °C for 1 h in NaOH solutions of (c) 0.1 and (d) 5 M using as a precursor dried complex gel.

The local recrystallization of the solute is mainly promoted inside the reaction vessel, due to the fact that no fluid convection occurred during the hydrothermal treatment, because the treatments were conducted under static conditions at constant temperature, therefore, mass transfer due to convection was further limited (Bayrappa & Yoshimura, 2001). Hence, the crystallization mechanism is similar to that described previously for other perovskite species, but it differs on the rate of kinetics that lead to produce particles particular aspect. In contrast, the particle bonding was limited by modifying the treatment conditions, using a dried complex gel $La_{0.8}Sr_{0.2}Cr_{0.9}Al_{0.1}O_3$ and different aqueous solvent media (water, KOH, NaOH and KF), resulting in the optimum control of dissolution and crystallization of fine particles with regular. A marked particle growth, however, was determined as a result of increasing the concentration of the alkaline solution from 0.1 to 5 M (NaOH or KOH). The formation of submicron particles monodispersed (0.5 - 0.75 μm) was found to proceed on these solvents (Figs. 14c, 14d). Furthermore, the employment of the alkaline solvents NaOH and KF leads to the crystallization of pseudo-cubic shaped and hexagonal plate particles, respectively; the differences on the particle morphology control are due to the chemical reactivity of the different solutions with the dried complex gel.

2.5 Replacement reactions on minerals species under hydrothermal conditions as a new approach for preparing inorganic materials

Another different approach similar to the mineral transformation into perovskite oxide powders was investigated to establish an alternative processing route for mineral ores via controlled dissolution-precipitation. The precursor mineral considered were the alkaline earth sulphates of $SrSO_4$ and $BaSO_4$, for preparing high pure strontium inorganic compounds, namely $SrCO_3$, $BaCO_3$, $SrCrO_4$, SrF_2, $Sr(OH)_2$. This chemical preparative method involves the dissolution–recrystallization mechanism, similar to the process that achieves the ionic replacement reaction in mineral ores, and promotes the conversion of natural ores at the earth's crust into more chemically stable mineral species or inorganic compounds (Putnis, 2002, 2009). The development of more efficient and environmental friendly chemical routes has recently been under concern of some researchers. The new chemical techniques or the optimized conventional routes might lead to reduce the pollution grade of contaminants, which is produced during mineral processing stages. The related alkaline earth sulphate minerals, such as $SrSO_4$ and $BaSO_4$, have been exploited since several decades, because these are the main source of the alkaline earth metals elements, Sr and Ba, and are used as a main source for preparing inorganic compounds of Sr and Ba. During the last decade, some attention has been paid to the use of celestite ore for producing functional ceramic compounds with magnetic properties, e.g. strontium hexaferrite ($SrFe_{12}O_{19}$); two different methods were proposed, coprecipitation in aqueous solutions via mineral powder leaching and powder mechanochemical activation techniques (Hessien et. al., 2009; Tiwary, et. al. 2008). Hence, since a decade, the present authors have devoted efforts in order to investigate, from a different approach of preparative chemistry; the nature of the possible replacement reactions that can be achieved under hydrothermal conditions, as was found to proceed even in synthetic inorganic compounds (Rendón-Angeles et. al., 2000b). This could derive in an optimum method for preparing high grade Sr and Ba compounds. The most relevant aspects found regarding this topic are given in this section.

2.5.1 Aspects of the compositional and structural transformation on sulphate minerals under alkaline hydrothermal conditions

Hitherto, the ion exchange reaction of SO_4^{2-} ions with CO_3^{2-} ions has been one of the subjects of research work, which involves both mineral celestite ($SrSO_4$) and barite ($BaSO_4$) species.

This particular transformation involves the conversion of MSO_4 into MCO_3, M= Sr or Ba, because these carbonated compound are widely used as a precursors for the preparation of strontium and barium inorganic compounds (Suaréz-Orduña et. al., 2004a; Rendón-Angeles et. al., 2008). Preliminary evidences of the ion exchange replacement process were investigated on the mineral celestite specie, thermodynamic and kinetic details were reported elsewhere (Yoshino et. al., 1985). The exchange of SO_4^{2-} ions with CO_3^{2-} ions was investigated on large mineral $SrSO_4$ bulky crystal plates, which were leached at low temperature (55 °C). The ion replacement was achieved by two reaction mechanisms, at initial and intermediate stages, the superficial reaction and the diffusion of SO_4^{2-} ions produced a dense $SrCO_3$, the complete ion replacement process proceeded by a second mechanism by a solid-state ion exchange mechanism even under hydrothermal conditions (Yoshino et. al., 1985).

Recently, a better approach conducted to elucidate the source of the conversion $M^{2+}SO_4$ to $M^{2+}CO_3$, M= Sr or Ba; was proposed based on the crystalline structural differences associated to the physical bulk molar volume change, which also must occur due to the replacement of a large anion SO_4^{2-} by a smaller one CO_3^{2-}. Thus, the study was conducted in large single crystals of celestite $SrSO_4$ and barite $BaCO_3$, which were treated under alkaline hydrothermal conditions using high concentrated carbonated solutions of Na_2CO_3 and K_2CO_3. The mineral celestite single crystals ($SrSO_4$, square plates 10 mm wide and 3 mm thick) were topotaxially converted to strontianite ($SrCO_3$) under alkaline hydrothermal conditions. The reaction was completed in a short reaction time (such as 24 h) at a temperature of 250 °C. Increasing the treatment temperature and the molar ratio CO_3^{2-}/SO_4^{2-} accelerated the exchange of SO_4^{2-} with CO_3^{2-} ions. Under these conditions the topotaxial hydrothermal conversion to strontianite is carried out with the formation of intermediate CO_3^{2-}-rich solid solutions in the system $SrCO_3$–$SrSO_4$, this is shown in the Figure 15; and these solid solutions were formed by a controlled crystallization process achieved by the cluster dissolution–recrystallization. An anisotropic dissolution gave a characteristic texture in the converted crystals, and the difference on the reactivity of the celestite crystals in Na_2CO_3 and K_2CO_3 solutions resulted in a different texture inside the pseudomorphical converted strontianite crystals (Rendón-Angeles et. al., 2000b).

Despite the aspect of the converted crystals remained without any change after the conversion process (Fig. 15c), the formation of small holes randomly distributed were produced on outside layer produced as a result of the transformed into the MCO_3 phase Figure 16. However, the morphology of these holes underwent a change due to a simultaneous dissolution of the solid product, the recrystallization process of this phase also occurred at the same place where the previous dissolution occurred. One point that must be emphasized is that the two phases are separated by a sharp boundary in texture, as well as composition. From this observation it is clear that the replacement reaction begins from the surface of the mineral that was in contact with the hydrothermal media (Fig. 16a). In addition, the reaction proceeded by the incorporation of the solvent through the inner porosity. The holes form a zig-zag network inside the crystals which allows incorporating fresh solvent media at the reaction front (Suaréz-Orduña et. al., 2004a; Rendón-Angeles et. al., 2008). This reaction is similar to the ionic replacement process that was found for the conversion of chlorapatite and hydroxyapatite single crystals into fluorapatite single crystals under hydrothermal conditions (Rendón-Angeles et. al., 2000b). In terms of the macroscopic aspects and crystalline structural differences associated with the replacement of a large ion (SO_4^{2-} = 4.32 Å) by a smaller (CO_3^{2-} = 1.55 Å). However, the completely converted MCO_3 crystals are constituted for very tiny crystals randomly oriented which resemble a polycrystalline arrangement on the converted $M^{2+}CO_3$. Thus in this particular case, the conversion process proceeds with the formation of a converted layer that has a peculiar texture (holes, Fig. 16b) and a moving reaction front. Hence, the conversion of the $M^{2+}SO_4$ mineral species into their related carbonated inorganic compounds; is associated to a pseudomorphic replacement process rather than the ion-exchange process. In addition, at the hydrothermal conditions in which the mineral conversion can be achieved, the pseudomorphic replacement process is mainly achieved by a mechanism of coupled bulk dissolution and precipitation. It is well known that this mechanism promotes the formation of a great wide type of inorganic compounds (Putnis, 2002).

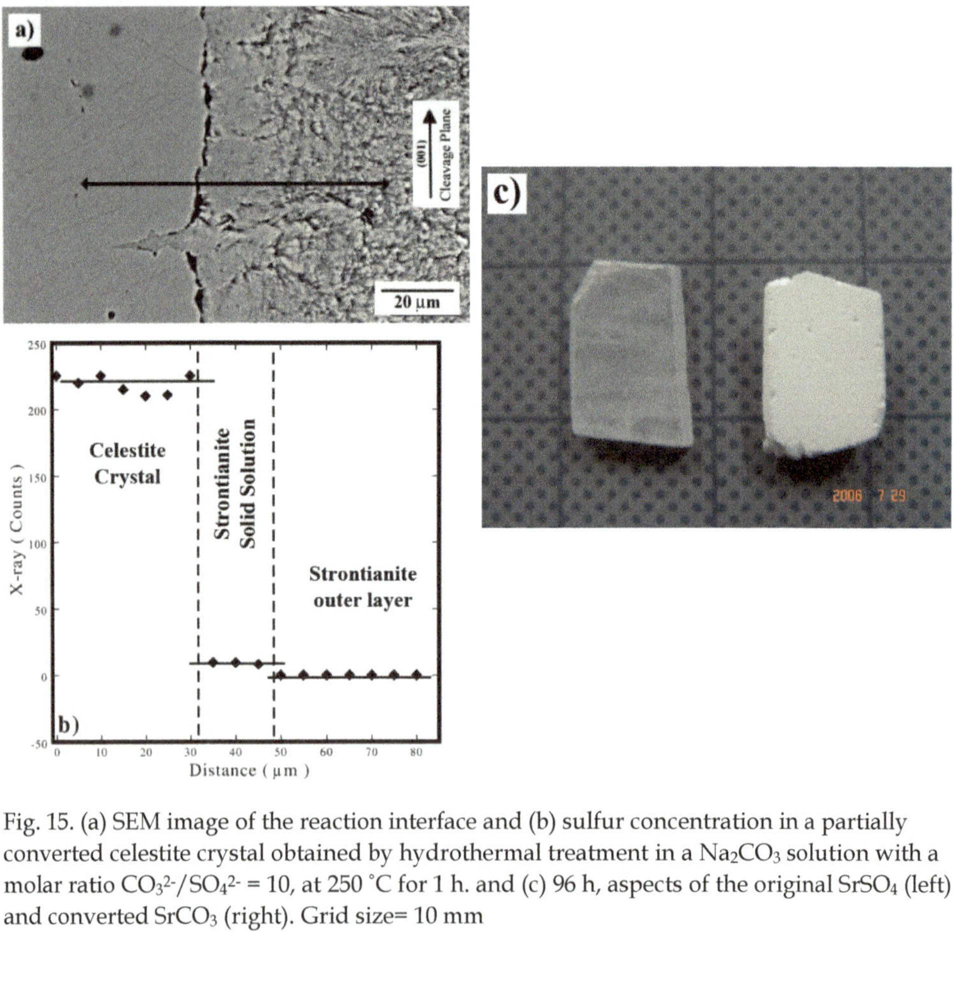

Fig. 15. (a) SEM image of the reaction interface and (b) sulfur concentration in a partially converted celestite crystal obtained by hydrothermal treatment in a Na_2CO_3 solution with a molar ratio CO_3^{2-}/SO_4^{2-} = 10, at 250 °C for 1 h. and (c) 96 h, aspects of the original $SrSO_4$ (left) and converted $SrCO_3$ (right). Grid size= 10 mm

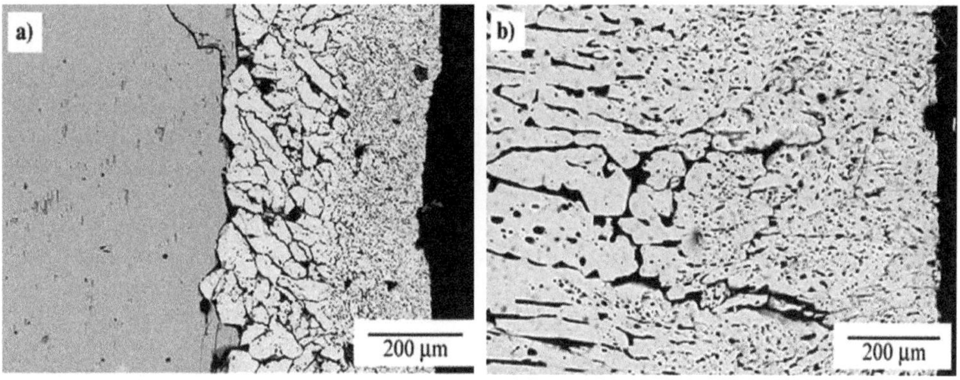

Fig. 16. SEM micrographs of barite crystals converted to barium carbonate after hydrothermal treatment at 250 °C in a Na_2CO_3 solution with a molar ratio CO_3^{2-}/SO_4^{2-} of 10 for different reaction intervals (a) 24 h (b) 192 h (Rendón-Angeles et. al., 2008).

2.5.2 Aspects associated with the formation and control of the residual porosity during the replacement reaction of mineral sulphate species

In terms of the porosity resulted as consequence of the crystalline transformation, this factor has also been the subject of controversy, because the causes that promote its formation are not clarified yet. However, we have recently found that the formation of the residual porosity depends strongly of two principal factors: i) the differences on the molar volume associated with the crystalline structural differences and ii) the chemical stability (solubility) of the new converted crystalline phase in the hydrothermal media. These inferences were established on partially and completely converted $SrCO_3$ specimens obtained by hydrothermal treatments at 250 °C for a interval of 24 h with a molar ratio $CO_3^{2-}/SO_4^{2-} = 10$. The inner volume on the converted crystals was determined by helium picnometry measurements and those results are giving in Figure 17. It is clear that the volume measurements conducted on the completed converted $SrCO_3$ plates at 250 °C for 96 h, revealed that the residual inner porosity value obtained on the specimens treated in Na_2CO_3 solutions is nearly similar to the theoretical value (dotted line in Fig. 17), this value (15.62 %) associated with the reduction of the molar volume, was calculated by considering the unit cell volume values of the parent (celestite, 307.06 Å^3) and the product (strontianite, 259.07 Å^3). This crystalline structural variation is related with the formation of the residual inner porosity, because macroscopically the crystal plate remains without any change regarding its shape and dimension, therefore, the bulk molar volume reduction does not proceed on the crystal plate and this must be compensated by which is like to proceed the residual porosity. Moreover, the control of the porosity depends on the chemical stability of the phase crystallized with the ion exchange media, once the replacement reaction was completed (Suaréz-Orduña et. al., 2004b). This inference is supported by the fact that when the conversion was carried out in K_2CO_3 solutions, the residual inner porosity on the completed converted $SrCO_3$ plate was markedly increased due to a further dissolution of the converted $SrCO_3$ crystal, in comparison with the porosity obtained the $SrCO_3$ plates transformed with highly concentrated Na_2CO_3 solutions at $CO_3^{2-}/SO_4^{2-} > 5$.

On the other hand, the remaining porosity is limited by the differences on the crystalline structure between the parent and the product. This is likely to occur when a large one replaces a small anion and a change on the structural structure also proceeds. One example was found in the case of the conversion of celestite to $SrCrO_4$, this reaction proceeds at low temperatures (200 °C) in relatively alkaline hydrothermal conditions (K_2CrO_4 solution), with the formation of a peculiar phase on the surface of the partially converted crystals. It is well known that dissolution of mineral species in hydrothermal fluids normally proceeds anisotropically, producing a peculiar texture with holes inside the recrystallized mineral specie (Suaréz-Orduña et. al., 2004a; Rendón-Angeles et. al., 2008). The holes might not be inherited from etch pits produced during the dissolution process, because they did not penetrate the crystals. The factor that has a marked influence in limiting the formation of a residual porosity is related with the replacement of SO_4^{2-} ions by CrO_4^{2-} ions in the $SrSO_4$ crystals. This fact is suggested from the structural change of the orthorhombic to monoclinic structure, which occurs during the conversion. In terms of the global unit cell volume, an expansion process is likely to proceed, therefore, in accordance with the differences on the unit cell volume between that for celestite (312.37 $Å^3$) and that of $SrCrO_4$ (354.11 $Å^3$), a volume increase of 41.74 $Å^3$ is attained in the converted new layer of $SrCrO_4$ as is seen in Figure 18. Hence, the global volumetric unit cell expansion must be compensated by the formation of a continuous solid phase and the formation of some microcracks in this new phase (Rendón-Angeles et. al., 2000b). The formation of a solid phase covering the partially reacted $SrSO_4$ might reduce the transfer of fresh ion exchange media, because under hydrothermal conditions the absence of a texture (small porosity) in the converted phase avoids the penetration of the hydrothermal fluid, coming to an abrupt halt of the replacement reaction

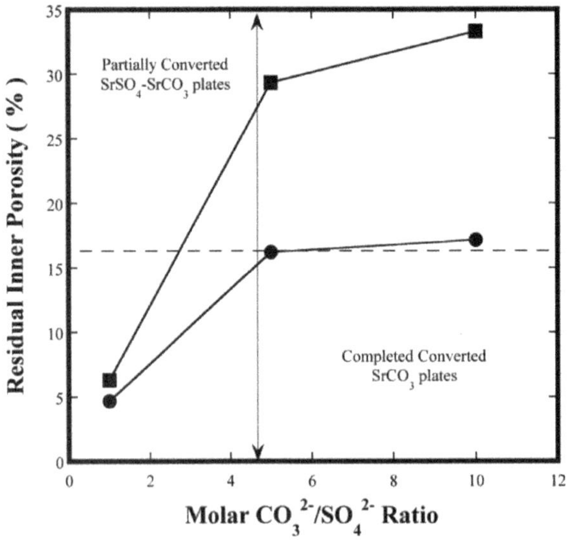

Fig. 17. Variation of the residual porosity on partial and completely converted $SrCO_3$ crystal plates, under hydrothermal conditions at 250 °C for 96 h in (●) Na_2CO_3 and (■) K_2CO_3 solutions with different molar CO_3^{2-}/SO_4^{2-} ratios. Dotted line= theoretical value calculated from the variation of the cell unit lattice associated with the structural conversion.

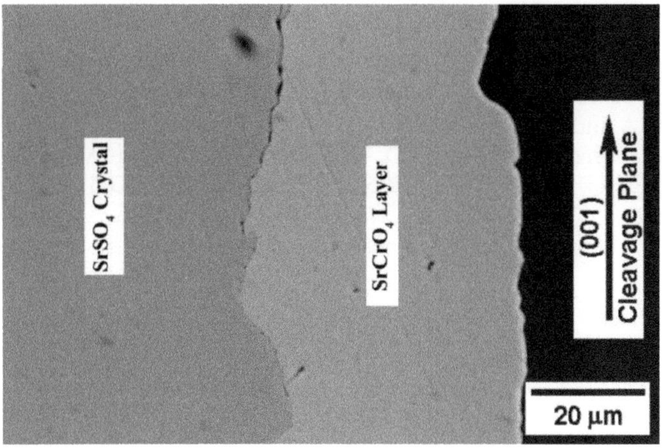

Fig. 18. Partially reacted $SrSO_4$ crystal at 200 °C for 96 h in a hydrothermal media with a molar ratio $CrO_4^{2-}/SO_4^{2-}=1$.

3. Conclusions

The hydrothermal crystallization of advanced materials is an important branch of chemical science and technology, this process has advantages over conventional technologies, namely the purity of products, quality, and performance, and can be consider as a green chemistry process because is environmentally friendly, because reactions consume lesser energy and these can be carried out under controlled parameters in a closed system. Although, the study of a wide number of oxides have provided physico-chemical experimental information related to solid-aqueous interactions, that affect the dissolution-precipitation mechanism main driving force for achieving bulk solid crystallization, coupled with thermodynamic modelling analysis; have contributed to establish the optimum hydrothermal processing conditions that favours the crystallization proceeds at high yield rates. However, still crystallization research work to carry out on those solid-aqueous systems associated to a specific compound that departs from the thermodynamic modelling limits. In addition, the employment of mineral ore as reactant precursors emerges as an interesting route to explore in order to produce other advanced ceramic compounds via the hydrothermal crystallization. Among the new research routes explored for hydrothermal processing, the ionic replacement is able to become common knowledge in the near future as alternative processing route, because this reaction promotes a peculiar microstructure in the transformed material that preserves the bulk original geometrical aspects during a single step hydrothermal crystallization reaction. This processing route can lead to prepare pore net-shaped materials with controlled porosity, which depending on their functional properties of the compound, can be used as gas sensor, substrates for porous catalytic materials, filters, and other applications. The conversion process focused toward a particular preparation of a functional inorganic compound from the mineral transformation requires a proper appreciation of the physico-chemical aspects involved for the solid solution-aqueous solution system. Likewise, to evaluate the likely porosity development in any mineral replacement reaction requires knowledge of the coexisting solid and fluid phases involved and their relative solubility.

4. Acknowledgment

One of the authors JCRA wish to acknowledge CONACYT and COECYT for the financial support trough research grants (Project 34830-U) and (FOMIX-COAH 2003-C02-02), respectively. JCRA and ZMV are indebt to the SNI for the financial support. Many thanks are particularly offered to former Ph D. students that collaborate with the experimental work, Roberto Suarez-Orduña, Laura Patricia Vazquez-Rivas and Yadira Marlen Rangel-Hernandez, also to technicians MSc. Martha Rivas-Aguilar and Eng. Felipe de Jesus Márquez-Torres, whom helped in the preparation of samples and its observation by SEM.

5. References

Bilger, S.; Blab, G. & Förthmann, R. (1997). Sol-gel Synthesis of Lanthanum Chromite Powder, *Journal of the European Ceramic Society*, Vol. 17, No. 8, pp. 1027-1031, ISSN 0955-2219

Byrappa, K. & Yoshimura, M. (2001). *Handbook of Hydrothermal Technology*, William Andrew Publishing/Noyes, ISBN 0-8155-1445-X, New York, USA

Byrappa, K. (2005). *Hydrothermal Processing*, Kirk-Othmer Encyclopedia of Chemical Technology, pp. Copyright John Wiley & Sons, Inc., ISBN 9780471238966

Chakraborty, A.; Basu, R. N. & Maiti, H. S. (2000). Low Temperature Sintering of La(Ca)CrO_3 Prepared by an Autoignition Process, *Materials Letters*, Vol. 45, No. 3-4, (September 2000), pp. 162-166, ISSN 0167-577X

Chen, C. W.; Riman, R. E; TenHuisenb, K. S. & Brown. (2004). Mechanochemical–Hydrothermal Synthesis of Hydroxyapatite from Nonionic Surfactant Emulsion Precursors, *Journal of Crystal Growth*, Vol. 270, pp. 615–623, ISSN 0022-0248

Eckert J. O.; Hung-Houston, C. C.; Gersten, B. L.; Lencka, M. M. & Riman, R. E. (1996). Kinetics and Mechanism of Hydrothermal Synthesis of Barium Titanate, *Journal of the American Ceramic Society* Vol. 79, No. 11, (November 1996), pp. 2929-2939, ISSN 0002-7820

Gersten, B. l.; Lencka, M. M. & Riman R. E. (2004). Low Temperature Hydrothermal Synthesis of Phase-Pure (Br,Sr)TiO_3 Perovskite Using EDTA, *Journal of the American Ceramic Society*, Vol. 87, No. 11, (January 2004), pp. 2025-2032, ISSN 0002-7820

Hessien, M. M.; Hassan, M. M. & El-Barawy, K. (2009). Synthesis and Magnetic Properties of Strontium Hexaferrite from Celestite Ore, *Journal of Alloys and Compounds*, Vol. 476, No. 1-2, 12 (May 2009), pp. 373-378, ISSN 0925-8388

Ianculescu, A.; Braileanu, A.; Pasuk, I. & Zaharescu, M. (2001). Phase Formation Study of Alkaline Earth-doped Lanthanum Chromites, *J. Therm. Anal. Calorim.*, Vol. 66, No. 2, (November 2001), pp. 501-507, ISSN: 1388-6150

Inagaki, M.; Yamamoto, O. & Hirohara, M., (1990). Synthesis of $LaCrO_3$ from Complex Precipitation and its Electrical Conductivity, *Journal of the Ceramic Society of Japan*, Vol. 98, pp. 675–678, ISSN 1882-0743

Kashkurov, K. F., Nikitichev, P. I., Osipov, V. V., Sizova, L. D., & Simonov, A. V. (1968). Growth of Large Corundum Crystals by the Hydrothermal Method, *Soviet Physics Crystallografia*, Vol. 12, pp. 837– 839, ISSN 0038-5638

Kobe K. A. & Deiglmeier N. J. (1943). Strontium Carbonate, Conversion from Strontium Sulfate by Metathesis with Alkali Carbonate Solution, *Industrial and Engineering Chemistry*, Vol. 35, No. 3, pp. 323-325, ISSN 1226-086X

Komarneni S., Roy R. & Li, Q. H. (1992). Microwave-Hydrothermal Synthesis of Ceramic Powders, *Materials Research Bulletin*, Vol. 27, pp. 1393-1405, ISSN 0025-5408

Komarneni, S. ; Komarneni Y. S.; Newalkar, B.; Stour, S. (2002). Microwave-Hydrothermal Synthesis of Al-Sustituted Tobermorite from Zeolites. *Materials Research Bulletin*, Vol. 3, (March 2002), pp. 1025-1032, ISSN 0025-5408

Kubo, T.; Hogiri, M.; Kagata, H. & Nakahira, A. (2009). Synthesis of Nano-Sized $BaTiO_3$ Powders by The Rotatory-Hydrothermal Process, *Journal of the American Ceramic Society*, Vol. 92, No. S1, (January 2009), pp. S172-S176, ISSN 0002-7820

Lee, S. K.; Choi, G.J.; Hwang, U. Y.; Koo, K. K. & Park, T. J. (2003). Effect of Molar Ratio of KOH to Ti-Isopropoxide on the Formation of $BaTiO_3$ Powders by Hydrothermal Method, *Materials Letters*, Vol. 57, No. 15, (April 2003), pp. 2201-2207, ISSN 0167-577X

Lencka, M. M. & Riman, R.E . (1993). Thermodynamic Modeling of Hydrothermal Synthesis of Ceramic Powders, *Chemistry of Materials*, Vol. 5, (October 1993), pp. 61-70, *ISSN* 0897-4756

Lencka, M.M. & Riman, R. E. (1995). Thermodynamics of the Hydrothermal Synthesis of Calcium Titanate with Reference to Other Alkaline-Earth Titanates, *Chemistry of Materials*, Vol. 7, (September 1995), pp. 18-25, *ISSN* 0897-4756

Lencka, M.M. & Riman, R.E. (2002). *Intelligent Systems of Smart Ceramics*, Encyclopedia of Smart Materials, Vol. 1, John Wiley & Sons, Inc., ISBN 9780-471216278, N.J, USA

Marchisio, D. L. (2009). On the Use of Bi-Variate Population Balance Equations for Modelling Barium Titanate Nanoparticle Precipitation, *Chemical Engineering Science*, Vol. 64, (February 2009), pp. 697-708, ISSN 0009-2509

Moreira, M. L.; Mambrini, G. P.; Volanti, D. P.; Leite, E. R.; Orlandi, M. O.; Pizani, V. R. Mastelaro, P.S.; Paiva-Santos, C. O.; Longo, E. & Varela, J. A. (2008). Hydrothermal Microwave: A New Route to Obtain Photoluminescent Crystalline $BaTiO_3$ Nanoparticles, *Chemistry of Materials*, Vol. 20, pp. 5381–5387, *ISSN* 0897-4756

Moon, J.; Kerchener, J. A.; Krarup, H. & Adair, H. (1999). Hydrothermal Synthesis of Ferroelectric Perovskites from Chemically Modified Titanium Isopropoxide and Acetate Salts, *Journal of Materials Research*, Vol. 14, No. 2, (February 1999), pp. 425-434, ISSN 0884-2914

Oledzka, M.; Lencka, M. M.; Pincelup, M. P.; Mikulka-Bulen, K.; McCandlish, L. & Riman, R. (2003). Influence of Precursor on Microstructure and Phase Composition of Epitaxial Hydrothermal $PbZr_{0.7}Ti_{0.3}O_3$ Films, *Chemistry of Materials*, Vol. 15, (March 2003), pp. 1090-1098, ISSN 0897-4756

Peterson, C. R. & Slamovich, E. B. (1999). Effect of Processing Parameters on the Morphology of Hydrothermally Derived $PbTiO_3$ Powders, *Journal of the American Ceramic Society*, Vol. 82, No. 7, (July1999), pp. 1702-1710, ISSN 0002-7820

Putnis, A. (2002). Mineral Replacement Reactions: From Macroscopic Observations to Microscopic Mechanisms, *Mineral Magazine*, Vol. 66, No.5, (October 2002), pp. 689-708, ISSN 0026-461X

Putnis A. (2009). Replacement Reactions, *Review Mineral Geochemistry*, Vol. 70, pp. 87-124, ISSN 1529-6466

Qi, L.; Lee, B. I.; Badheka, P.; Yoon, D. H.; Samuels, W. D. & Exarhos, G. J., (2004). Short-range Dissolution–Precipitation Crystallization of Hydrothermal Barium Titanate, *Journal of the European Ceramic Society*, Vol. 24, No. 13, (October 2004), pp. 3553–3557, ISSN 0955-2219

Ramkrishna, D. & Mahoney, A. W. (2002) Population Balance Modeling Promise for the Future, *Chemical Engineering Science*, Vol. 57, (February 2002), pp. 595 - 606, ISSN 0009-2509

Rangel-Hernandez, Y. M.; Rendón-Angeles, J. C.; Matamoros-Veloza, Z.; Pech-Canul, M. I.; Diaz-de la Torre, S. & Yanagisawa, K. (2009). One-step Synthesis of Fine $SrTiO_3$ Particles Using $SrSO_4$ Ore Under Alkaline Hydrothermal Conditions, *Chemical Engineering Journal*, Vol. 155, No. 2, (January 2009), pp. 483–492, ISSN 1385-8947

Riman R.E .; Suchanek, W. L. & Lencka, M. M. (2002). Hydrothermal Crystallization of Ceramics, *Annales de Chimie Science des Materiaux*, Vol. 27, No. 6, (November 2002), pp. 15-36, ISSN 0151-9107

Rivas-Vázquez, L. P.; Rendón-Angeles, J. C.; Rodríguez-Galicia, J. L.; Zhu, K. & Yanagisawa, K. (2004). Hydrothermal Synthesis and Sintering of Lanthanum Chromite Powders Doped With Calcium, *Solid State Ionics*, Vol. 172, No. 1-4, (August 2004), pp. 597-600, ISSN 0167-2738

Rivas-Vázquez, L. P.; Rendón-Angeles, J. C.; Rodríguez-Galicia, J. L.; Gutiérrez-Chavarria, C. A.; Zhu, K. & Yanagisawa, K. (2006). Preparation of Calcium Doped $LaCrO_3$ Fine Powders by Hydrothermal Method and its Sintering, *Journal of the European Ceramic Society*, Vol. 26, No. 1-2, (October 2006), pp. 81-88, ISSN 0955-2219

Rendón-Angeles, J. C.; Yanagisawa, K.; Ishizawa, N. and Oishi, S. (2000a). Effect of Metal Ions of Chlorapatites on the Topotaxial Replacement by Hydroxyapatite under Hydrothermal Conditions, *Journal of Solid State Chemistry*, Vol. 154, No. 2, (November 2000), pp. 569-578, ISSN 0022-4596

Rendón-Angeles, J. C.; Yanagisawa, K.; Ishizawa, N. and Oishi, S. (2000b). Topotaxial Conversion of Chlorapatite and Hydroxyapatite to Fluorapatite by Hydrothermal Ion Exchange, *Chemistry of Materials* Vol. 12, No. 8, pp. 2143-2150, (June 2000), *ISSN* 0897-4756

Rendón-Angeles, J. C.; Pech-Canul, M. I.; López-Cuevas, J.; Matamoros-Veloza, Z. & Yanagisawa., K. (2006). Differences on the Conversion of Celestite in Solutions Bearing Monovalent Ions Under Hydrothermal Conditions, *Journal of Solid State Chemistry*, Vol. 179, No. 12, (December 2006), pp. 3645-3652, ISSN 0022-4596

Rendón-Angeles, J. C.; Matamoros-Veloza, Z.; López-Cuevas J.; Pech-Canul, M. I & Yanagisawa, K. (2008). Stability and Direct Conversion of Mineral Barite Crystals in Carbonated Hydrothermal Fluids, *Journal of Material Science*, Vol. 43, pp. 2189, ISSN 0022-2461

Rendón-Angeles, J. C., Yanagisawa, K.; Matamoros-Veloza, Z.; Pech-Canul, M. I.; Mendez-Nonell, J. & Diaz-de la Torre, S. (2010). Hydrothermal Synthesis of Perovskite Strontium Doped Lanthanum Chromite Fine Powders and its Sintering, *Journal of Alloys and Compounds*, Vol. 504, No. 1, (August 2010), pp. 251-256, ISSN 0925-8388

Roy, R. (1994). Accelerating the Kinetics of Low-Temperature Inorganic Syntheses, *Journal of Solid State Chemistry*, Vol. 111, (July 1994), pp. 11-17, ISSN 0022-4596

Schubert, U. and Hüsing, N., 2000. Synthesis of Inorganic Materials

Shi, S. &. Hwang, J. Y. (2003). Microwave-Assisted Wet Chemical Synthesis: Advantages, Significance, and Steps to Industrialization, *Journal of Mineral & Materials & Characterization & Engineering*, Vol. 2, No.2, pp. 101-110, ISSN: 1539-2511

Suárez-Orduña, R; Rendón-Angeles, J. C.; López-Cuevas, J; & Yanagisawa, K. (2004a). The Conversion of Mineral Celestite to Strontianite Under Alkaline Hydrothermal

Conditions, *Journal of Physics: Condensed Matter*, Vol. 16, pp. S1331-S1344, ISSN 0953-8984

Suárez-Orduña, R.; Rendón-Angeles, J. C.; Matamoros-Veloza, Z. & Yanagisawa, K. (2004b). Exchange of SO_4^{2-} Ions With F^- Ions In Mineral Celestite Under Hydrothermal Conditions, *Solid State Ionics*, Vol. 172, pp. 393-393, ISSN 0167-2738

Suchanek, W. L.; Shuka, P.; Byrappa, K.; Riman, R. E.; TenHuisen, K. S. & Janas, V. F. (2002). Mechanochemical–hydrothermal synthesis of Carbonated Apatite Powders at Room Temperature, *Biomaterials*, Vol. 23, (April 2001), pp. 699–710, ISSN 0142-9612

Suchanek, W. L.; Lencka M. M; & Riman, R. E. (2004). Hydrothermal Synthesis of Ceramics Materials, *Aqueous System at Elevated Temperature and Pressure Physical Chemistry in Water, Steam and Hydrothermal Solutions*, Chapter 18, ISBN 0-12-544461-3, Elsevier Ltd., New Jersey, USA

Shi, S. &. Hwang, J. Y. (2003). Microwave-Assisted Wet Chemical Synthesis: Advantages, Significance, and Steps to Industrialization, *Journal of Mineral & Materials & Characterization & Engineering*, Vol. 2, No.2, pp. 101-110, ISSN 1539-2511

Spooren, J.; Rumplecker, A.; Millange, F. & Walton, R. I. (2003). Subcritical Hydrothermal Synthesis of Perovskite Manganites: A Direct and Rapid Route to Complex Transition-Metal Oxides, *Chemistry of Materials* Vol. 15, (March 2003), pp. 1401-1403, ISNN 0897-4756

Testino, A.; Buscaglia, V.; Buscaglia, M. T.; Viviani, M. & Nanni, P. (2005). Kinetic Modeling of Aqueous and Hydrothermal Synthesis of Barium Titanate ($BaTiO_3$), *Chemistry of Materials*, Vol. 17, (September 2005), pp. 5346-5356, *ISSN* 0897-4756

Tiwary, R. K.; Narayan, S. P. & Pandey, O.P. (2008). Preparation of Strontium Hexaferrite Magnets from Celestite and Blue Dust by Mechanochemical Route, *Journal of Mining and Metallurgy, Section B*, Vol. 44, (February 2008), pp. 91 - 100, ISSN 1450-5339

Vivekanandan, R. & Kutty, T. R. N. (1989). Characterization of Barium Titanate Fine Powders Formed From Hydrothermal Crystallization, *Powder Technology*, Vol. 57, No. 3, (March 1989), pp. 181 – 192, ISSN 1570-1468

Wada, S. Suzuki T. & Noma T. (1995). Preparation of Barium Titanate Particles by Hydrothermal Method and Their Characterization, *Journal of the Ceramic Society of Japan*, Vol. 103, No. 12, (December 1995) pp. 1220-1227, ISSN 1882-0743

Walton, R. I.; Millange, F. Dmith, R. I., Hansen, T. C. & O'hare, D. (2001). Real Time Observation of the Hydrothermal Crystallization of Barium Titanate Using In Situ Neutron Powder Difracction. *Journal of the American Chemical Society*, Vol. 123, pp. 12547-12555, ISSN 0002-7863

Wang, Y.; Xu, G.; Yang, L.; Ren, Z.; Wei, X.; Weng, W.; Du, P.; Shen, G. & Han, G. (2009). Formation of Single-Crystal $SrTiO_3$ Dendritic Nanostructures via a Simple Hydrothermal Method, *Journal of Crystal Growth*, Vol. 311, No. 8, (April 2009), pp. 2519-2523, ISSN 0022-0248

Wei, X.; Xu, G.; Ren, Z.; Xu, C.; Sheng, G. & Han, G. (2008). PVA-Assited Hydrothermal Sinthesis of $SrTiO_3$ Nanoparticles with Enhanced Photocatalytic Activity for Degradation of RhB, *Journal of the American Ceramic Society*, Vol. 91, No. 11, (November 2008), pp. 11-17, ISSN 0002-7820

Wendelbo, R.; Akporiaye, D. E.; Karlsson, A.; Plassen, M. & Olafsen, A. (2006). Combinatorial Hydrothermal Synthesis and Characterization of Perovskites, *Journal of the European Ceramic Society*, Vol. 26, No. 6, pp. 849-859, ISSN 0955-2219

Yanagisawa, K.; Rendón-Angeles, J. C.; Kanai H. & Yamashita, Y. (1999). Stability and Single Crystal Growth of Dielectric Materials Containing Lead Under Hydrothermal Conditions, *Journal of the European Ceramic Society*, Vol. 19, No. 6-7, (June 1999), pp. 1033-1036, ISSN 0955-2219

Yoshimura, M. & Suchanek, W. (1997). In Situ Fabrication of Morphology-Controlled Advanced Ceramic Materials by Soft Solution Processing, *Solid State Ionics*, Vol. 98, No. 3-4, (June 1997) pp. 197–208, ISSN 0167-2738

Yoshino, K.; Nishino, T.; Yoshino, M. & Yoshimura, S. (1985). Exchange Reaction between Alkaline Earth Carbonate and Sulfate under Hydrothermal Condition, *Yogyo-Kyokai-Shi*, Vol. 93, No. 6, pp. 334-337, ISSN 0372-7718

Zhang, S.; Han, Y.; Chen, B. & Song, X. (2001). The Influence of $TiO_2 \bullet H_2O$ Gel on Hydrothermal Synthesis of $SrTiO_3$ Powders, *Materials Letters*, Vol. 51, No. 4, (November 2001), pp. 368-370, ISSN 0167-577X

Zhang, S.; Liu, J.; Han, Y.; Chen, B. & Li, X. (2004). Formation Mechanisms of $SrTiO_3$ Nanoparticles Under Hydrothermal Synthesis, *Materials Science and Engineering B*, Vol. 110, No. 1, (June 2004), pp. 11-17, ISSN 0921-5107

Zheng, W.; Pang, W.; Meng, G. & Peng, D. (1999). Hydrothermal Synthesis and Characterization of $LaCrO_3$, *Journal of Materials Chemistry*, Vol. 9, pp. 2833–2836, ISNN 0959-9428

3

Fe-Based Nanocomposite Formed by Thermal Treatment of Rapid-Quenched $Fe_{81}B_{13}Si_4C_2$ Alloy

Dragica M. Minić[1], Vladimir A. Blagojević[1] and Dušan M. Minić[2]
[1]*University of Belgrade, Faculty for Physical Chemistry,*
[2]*Military Technical Institute, Belgrade,*
Serbia

1. Introduction

Amorphous alloys have been a focus of considerable scientific interest, both from fundamental and practical point of view, ever since the first of its kind ($Al_{75}Si_{25}$) was produced by Klement, Willens and Duwez in 1960 (Klement et al., 1960). It has been shown that the amorphous alloys have features that are different from those of crystalline alloys in both alloy compositions and atomic configurations. This enabled the exhibition of various characteristics which were not obtained for conventional crystalline alloys. Their soft ferromagnetic properties (saturation magnetization, high permeability, low coercivity and loss), high corrosion resistance and good mechanical properties make them suitable for use in a variety of applications, such as power devices, information handling technology, magnetic sensors, anti-theft security systems and construction materials (Minić et al., 2007). Since amorphous alloys are meta-stable, elevated temperature or prolonged performance can induce a transformation into a crystalline state, which could lead to a loss of their advantageous physical properties limiting them to single-use applications. Commercial soft magnetic nanocrystalline materials have recently been successfully obtained by crystallization of amorphous precursors. Materials like this are characterized by a microstructure of nanocrystals embedded into an amorphous matrix, exhibiting superior soft magnetic and mechanical properties to both amorphous and crystalline magnetic alloys (Blagojević et al., 2011; Minić et al., 2011a). This dependence of functional properties on microstructure can be used to produce functional materials with tailored properties (Maričić et al., 2012).

The production of early amorphous alloys required very high cooling rates (as much as 10^6 K/s) to avoid crystallization. This limited the form in which they could be produced, as one dimension had to remain small enough to allow sufficiently rapid heat extraction, in order to achieve the necessary high cooling rates. The result was that the thickness of amorphous metal specimens was limited to less than 100μm. In 1976, Liebermann and Graham developed a new method of manufacturing thin ribbons of amorphous metals on a supercooled rapidly rotating disc - the melt-spinning method (Liebermann & Graham, 1976). For the next thirty years, with the production of new materials, the required cooling rate diminished, until, in 1990s, materials were developed, whose production required cooling rates as low as 1 K/s, allowing these alloys to be cast into metallic moulds to produce specimens up to 100mm thick (Ponnambalam et al., 2004). The alloys that require cooling rates below 10^3 K/s are known as bulk amorphous alloys, as the production process allows for higher specimen thickness (above 1mm), classifying them as bulk materials.

2. Experimental procedures

The ribbon shaped samples of $Fe_{81}B_{13}Si_4C_2$ amorphous alloy were obtained using the standard procedure of rapid quenching of the melt on a rotating disc (melt-spinning method). The obtained ribbon was 2 cm wide and 35 μm thick. During the preparation process of the amorphous alloy ribbon, one of the sides was in direct contact with the cooled rotating disc, while the other was in inert atmosphere. As a result, the two sides of the ribbon show an easily observable difference in reflectivity, surface morphology and structure, as can be seen in X-ray diffraction (XRD) spectra, Fig.1, as well as SEM images of the sample in the as-prepared alloy (Fig.2). The side that was in the contact with the cooled spinning disc is usually labeled as fishy or matte side and the other side, free of contact, as shiny side.

DSC was obtained using SHIMADZU DSC-50 analyzer. In this case, samples weighting several milligrams were heated in the DSC cell from the room temperature to 650°C in a stream of nitrogen with nitrogen flowing at a rate of 20 mL min^{-1} at the heating rates of 5, 10, 20 and 30°Cmin^{-1}.

Mössbauer spectra were taken in the standard transmission geometry using a $^{57}Co(Rh)$ source at room temperature. The calibration was done against α-Fe foil data. "CONFIT" program package was used for the spectra fitting and decomposition (T. Žák, 1999).

The X-ray diffraction (XRD) patterns were recorded on an X'Pert PROMPD diffractometer from PANalytical with CoKα radiation operated at 40kVand 30mA. For routine characterization, diffraction data was collected in the range of 2θ Bragg angles (15–135°, step 0.0081). For a quantitative analysis and determination of crystallite size from XRD spectra, TOPAS V3 general profile and structure analysis software for powder diffraction data was used (BrukerAXS, general profile and structure analysis software for powder diffraction data, Karlsruhe, 2005). Dislocation density was obtained from the Rietveld analysis, while microstrain was calculated using Williamson-Hall method (Williamson, Hall 1953), using the XRD data. Lattice parameters obtained through XRD spectra were used to calculate the unit cell volumes, which were then compared to the standard values in JCPDS database.

Vickers microhardness tests were performed using MHT-10 (Anton Paar, Austria) microhardness tester, with loads of 0.4N and loading time of 10s (Minić, et al. 2011c). Up to seven measurements were performed on each individual sample, using the average value of microhardness for each sample. The measurements were performed on the cross-section of the ribbons, rather than on any of the sides. Error was calculated as standard deviation for each series of measurements.

2.1 Microstructure of as-prepared $Fe_{81}B_{13}Si_4C_2$ amorphous alloy

XRD spectra of the as-prepared alloy ribbon, Fig.1, showed that the as-prepared alloy already had a degree of crystallinity caused by presence of α-Fe phase (JCPDS-PDF 06-0696). The degree of crystallinity is much higher on the shiny side, with peak intensity in XRD spectrum being about 8 times higher on the shiny side.

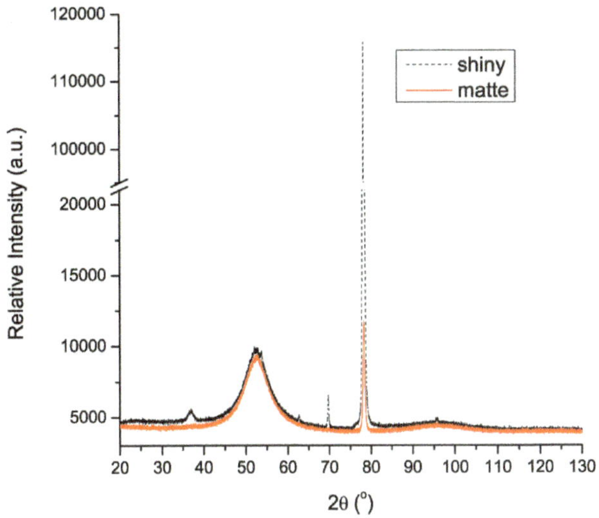

Fig. 1. X-ray diffraction spectra of as-prepared alloy

The nanocrystalline phase can be identified as α-Fe on matte side and a mixture of α-Fe and Fe_3Si on shiny side, with α-Fe being the major component. In addition to sharp crystalline peaks (37, 69 and 78°) in the XRD spectra of the as-prepared alloy, a broad spread halo around 53°, corresponding to domains of short-range ordering in the sample, was also observed. The position of the spread halo corresponds to approximate position of Fe_3Si peak and, using the Scherrer equation, we estimated the size of these domains to be 1-1.5 nm. The entire structure could best be described as a combination of nanocrystals and short-range ordered domains embedded in an amorphous matrix. Recent theoretical studies of iron-based binary systems predict existence of short-range ordering in iron-based amorphous alloys (Lass et al., 2010). Therefore, appearance of domains of short-range ordering in $Fe_{81}B_{13}Si_4C_2$ amorphous alloy could be expected.

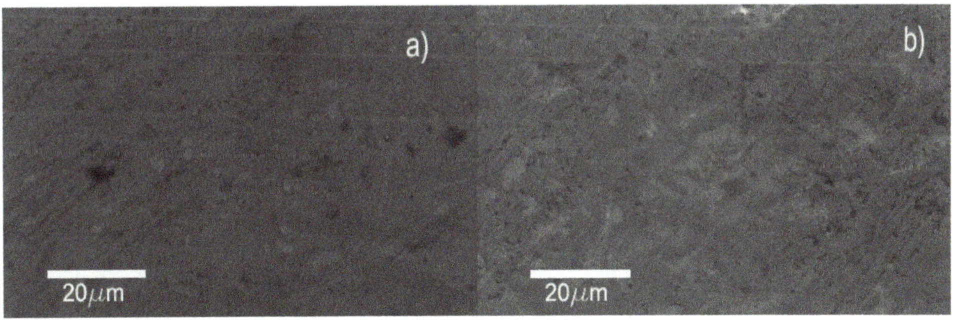

Fig. 2. SEM of as-prepared $Fe_{81}B_{13}Si_4C_2$ amorphous alloy (a – shiny; b – matte side)

Crystal structures of α-Fe and Fe_3Si are closely related, as Fe_3Si crystal system is cubic, same as α-Fe crystal system. Fe_3Si lattice is composed of four sub-lattices: three composed of iron atoms and one of silicon atoms, and this leads to doubling of the unit cell (when compared to α-Fe). Fe_3Si lattice is, also, slightly distorted, so the value of its lattice parameter is slightly higher than twice the value of lattice parameter of α-Fe.

2.2 Thermal stability of alloy

The thermal stability of the alloy was investigated using differential scanning calorimetry (DSC) in a nitrogen atmosphere (Minić, et al., 2009a). Typical DSC scan obtained during heating and cooling cycle is presented in Fig. 3. DSC scan involves a series of endothermic and exothermic peaks indicating a stepwise process of structural stabilization of the alloy in the temperature range 170-560°C. A broad exothermic peak, indicated as (T_{sr}), in temperature range 170-400°C, coresponding to structural relaxation, is followed by endothermic hump (temperature of glass transition T_g) and a short supercooled liquid region before the sharp exothermic crystallization peak (T_k) in temperature range 500-540°C. The enthalpy of crystallization is 83.5 J/g as determined from area of corresponding peak obtained at heating rate of 20 °Cmin^{-1}.

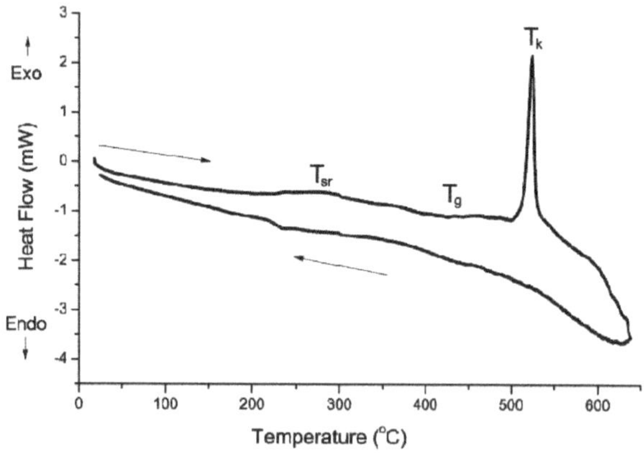

Fig. 3. DSC scan of heating and cooling cycle in nitrogen atmosphere; heating rate 10°C/min.

2.3 Magnetic properties

2.3.1 Thermo-magnetic behavior

Thermally induced processes were also studied using the thermo-magnetic scan, where the sample is heated, annealed and then cooled in vacuum furnace at low magnetic field of 4 kA m^{-1} while its magnetic moment is monitored (Minić, et al., 2011b). Both heating and cooling rate were 4°C min^{-1}, dwell time at maximum temperature of 800°C was 30 minutes. The shape of the thermo-magnetic curve, Fig. 4, reflects changes in the magnetic moment of the sample, caused by phase or structural transitions.

Most pronounced change represents the Curie point (T_c = 420°C), where the magnetization of the respective phase drops to almost zero, because the thermal motion overcomes magnetic interaction. Annealing at the temperature near 200°C is sometimes called stress-relieving (or structural relaxation). Temperatures marked by arrows in Fig. 4 have been identified as points of interest for further study of structural transformations.

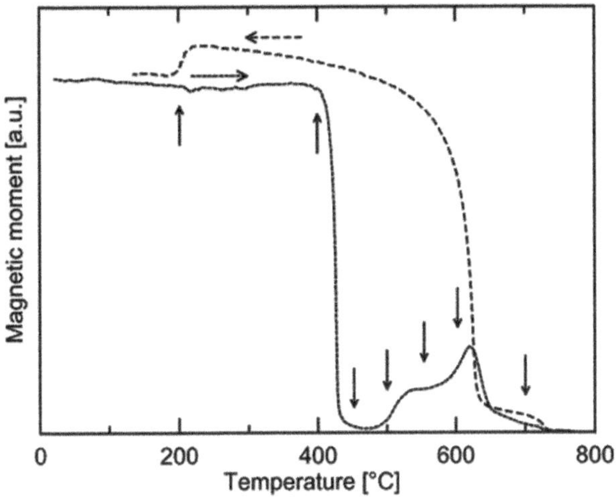

Fig. 4. Thermomagnetic scans for increasing (dotted line) and decreasing (dashed line) temperature.

2.3.2 Magnetic susceptibility

Measurements of relative magnetic susceptibility were performed using a modified Maxwell method, based on the action of an inhomogeneous field on the magnetic sample. The magnetic force measurements were performed with a sensitivity of 10^{-6}N in an argon atmosphere (Minić, et al. 2009b).

The temperature dependence of the relative magnetic susceptibility of the as-prepared $Fe_{81}B_{13}Si_4C_2$ amorphous alloy during three thermal treatments to different temperatures is presented in Fig. 5. During the first and second treatment, the decrease in the magnetic susceptibility in the temperature region from 320°C to 380°C is the result of proximity to the Curie temperature of the amorphous alloy. Before the start of the second treatment it was observed that magnetic susceptibility increased slightly. This was caused by the structural relaxation of an amorphous structure during the first treatment. During this process, internal strains and the free volume are reduced in the starting material. These changes are accompanied by subtle inter-atomic movements, causing the changes in the electron structure and leading to an increase in the number of electrons with unpaired spin in the direction of the outer magnetic field (Minić et al., 2009b). This also leads to a decrease in the number of electrons spinning in the reverse direction and causes an increase in the magnetic susceptibility upon cooling. At the same time, strains and decrease in the free volume enable greater mobility of the walls of the magnetic domains and this behavior further contributes to the increase in the magnetic susceptibility.

During the second treatment, the alloy loses its ferromagnetic properties in the temperature region from 400°C to 470°C. With further heating, the magnetic susceptibility starts to rise, and the alloy regains its ferromagnetic properties once the crystallization process starts at about 490°C. After the second heating to 490°C, the magnetic susceptibility decreases by 23 % when compared to the value of as-prepared alloy and to the value of the relaxed state of the lattice after the first heating cycle. During the third treatment, above the crystallization temperature, the alloy maintains its ferromagnetic features in the whole temperature region, whereas the maximum change in the magnetic susceptibility occurs at about 190°C as a consequence of further phase transformation of the crystallized alloy.

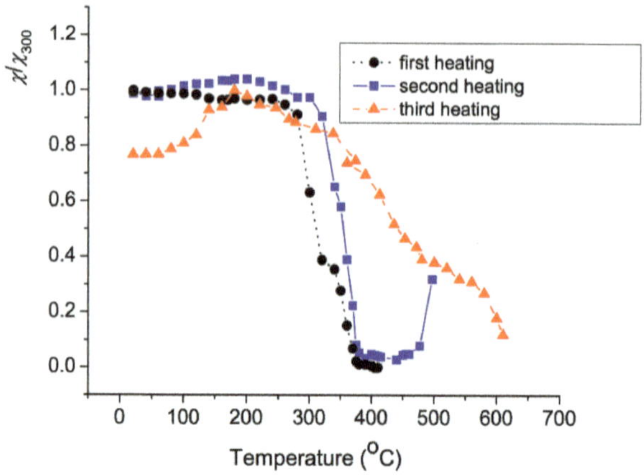

Fig. 5. Temperature dependence of relative magnetic susceptibility of as-prepared $Fe_{81}B_{13}Si_4C_2$ amorphous alloy during three thermal treatments up to different temperatures: a) 420°C; b) 500°C; c) 630°C.

2.3.3 Mössbauer spectra

In the Fig. 6, spectra illustrate the ability of Mössbauer effect to distinguish between individual iron-containing phases of different structure. Broad-line components (Fig. 6a) are

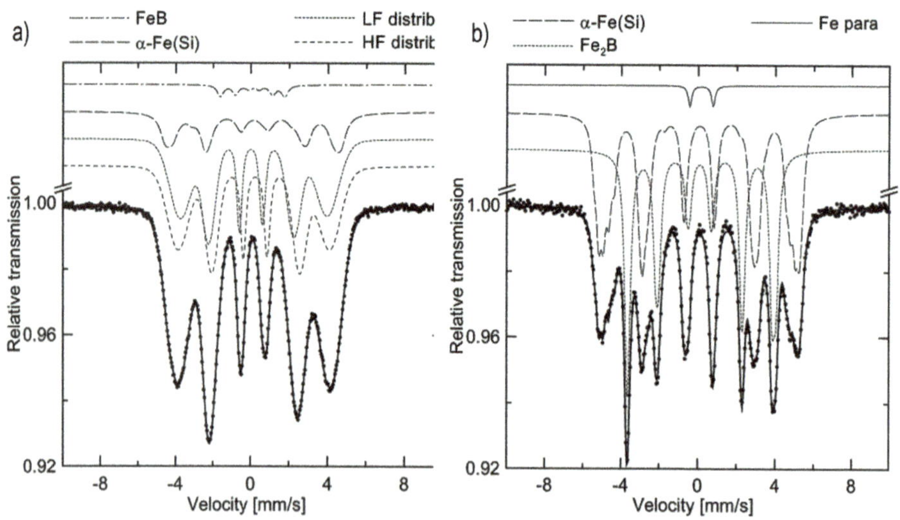

Fig. 6. Mössbauer spectra of the as-prepared material (left) and of material after final annealing at 700°C (right), including components of iron containing phases.

typical for the amorphous volume of the as-prepared sample, while the sharp lines (Fig. 6b) characterize the crystalline structure with well defined position of atoms, which results from the thermally induced crystallization process during annealing of the sample (Minić, et al. 2011b). The computer processing of Mössbauer spectra yielded intensities, *I*, of components, their hyperfine inductions, B_{hf}, isomer shifts, δ, and quadrupole splitting, σ (T. Žák & Y. Jirásková, 2006). The contents of the iron-containing phases are determined as proportional to the relative areas of the corresponding spectral components. The phase percentages correspond to the distribution of Mössbauer iron atoms among phases (Table 1).

Annealing temperature	Amorphous at%	α-Fe(Si) at%	Fe_2B at%	Fe_3B at%	FeB at%	α-Fe at%	Fe para at%
as-prepared alloy	0.95	0.03	–	–	0.02	–	–
200°C/30 min.	0.94	0.02	–	–	0.02	0.02	–
450°C/30 min.	0.83	0.14	–	0.03	–	–	–
500°C/30 min.	–	0.42	0.42	0.15	–	–	0.01
550 °C/30 min.	–	0.52	0.47	–	–	–	0.01
600°C/30 min.	–	0.54	0.45	–	–	–	0.01
700°C/30 min.	–	0.55	0.44	–	–	–	0.01

Table 1. Mössbauer tentative phase analysis (distribution of Mössbauer iron atoms among phases) (Minić, et al. 2011e)

In the as-prepared alloy, the amorphous structure, having a high-field and a low-field component is accompanied by a small amount of α-Fe(Si) solid solution and a FeB phase. Mössbauer phase analysis at higher temperatures reveals α-Fe(Si) solid solution and Fe_2B phase to be the most important final crystallization products, although metastable phase Fe_3B, was detected initially at 450°C and in higher percentage at 500°C. Amount of iron atoms in paramagnetic positions is almost below the sensitivity threshold. Content of silicon in the α-Fe(Si) solid solution seems to be about 9 at.%, which is close to 7 at.% published in (Saegusa & Morrish, 1982).

2.4 Electrical properties

2.4.1 Electrical resistance

The electrical resistance of the ribbon was measured using the four-point method within a temperature interval of 20–630°C in an argon atmosphere (Minić, et al., 2011d). Fig. 7a shows the temperature dependence of the electrical resistivity of the alloy in the temperature range of 25-630°C. The dependence clearly shows each structural stabilization step which causes the change in the ordering of the investigated material. These changes are more obvious in the derivative curve (Fig. 7b).

The slow increase of electrical resistivity was caused by the structural relaxation process in the temperature range of 200-380°C. This process is followed by an increase of electrical resistivity in the vicinity of Curie temperature T_c at 420°C, corresponding to the first maximum of the differential curve. At this point the effect that scattering of conductive electrons had on the magnons disappeared (I. Balberg & J. S. Helman, 1978; G. Bohnke et al., 1983) and the amorphous alloy loses its ferromagnetic properties (D. M. Minić et al. 2010). This is in excellent agreement with the results of the thermo-magnetic measurements (Fig.

4). The beginning of crystallization at about 520°C caused a sharp decrease of electrical resistivity. The appearance of two clearly separated maxima, T_{k1} and T_{k2} (490 and 510°C respectively) on the differential curve of electrical resistivity (Fig. 7b), suggests that crystallization of the amorphous alloy is a complex process, occurring in two steps, which appear as a single overlapping peak in the DSC scans, Fig. 3.

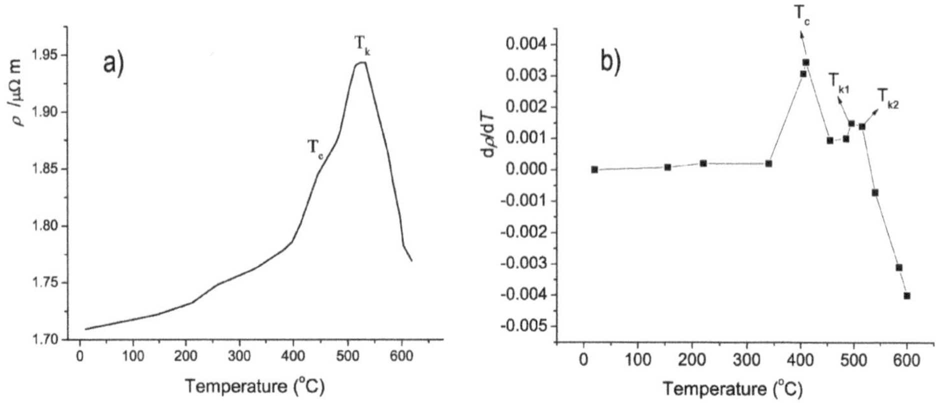

Fig. 7. Temperature dependence of the electrical resistivity of amorphous alloy.

The electrical resistivity of the crystalline alloy is lower than that of the amorphous alloy of the same composition, as a result of the increase in electron free path. The linear change of electrical resistivity with increasing temperature during the second treatment shows that crystallization was completed during the first heating cycle (Fig. 8).

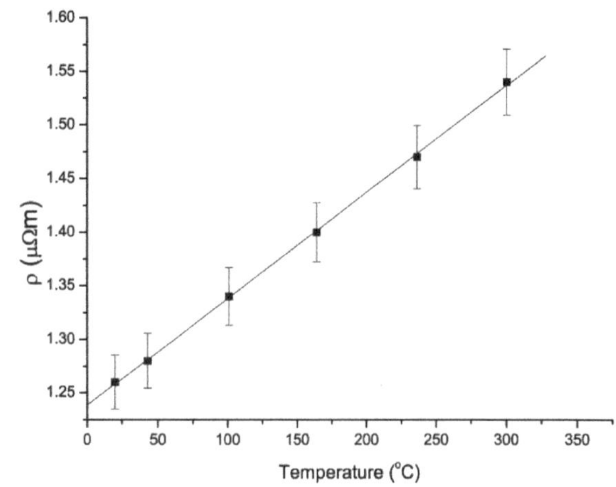

Fig. 8. Temperature dependence of the electrical resistivity of second thermal treatment

2.4.2 Thermo-electromotor force

The structural relaxation processes, as well as the crystallization, in the temperature interval of 25-680°C, were also investigated by measuring the thermo-electromotor force (TEMF) of a thermocouple made by coupling a copper conductor to the amorphous alloy (D.M. Minić, et al. 2009b). The alloy sample was mechanically attached to a copper conductor, forming the Cu - $Fe_{81}B_{13}Si_4C_2$ thermocouple, which was placed into a specially designed furnace, while the other end of the sample was submerged into a mixture of water and ice. The TEMF produced by the thermocouple during the heating process was measured by a voltmeter with the sensitivity of 10^{-5}V.

The temperature dependence of a thermo-electromotor force (Fig. 9) shows three linear regions corresponding to the structural transformations of the alloy. Different slopes of these linear dependences correspond to structural changes involving a structural relaxation, the loss of ferromagnetic properties, and the crystallization, respectively. The temperature coefficient of TEMF is a function of the electron state density at the Fermi level:

$$\alpha = \frac{k}{2e}\left(\frac{N_{1E_F}}{N_{2E_F}} - \frac{N_{2E_F}}{N_{1E_F}}\right) \quad (1)$$

where k is Boltzmann's constant, e is electron charge, $N_{1(Ef)}$ is the electron state density in copper and $N_{2(Ef)}$ is the electron state density in the alloy.

The electron density of states in copper remained unchanged during heating to 680°C, meaning that the change in the temperature coefficient during the heating of the thermocouple was caused only by the change of the electron density of states at the Fermi level of the alloy. Based on the slope of the first linear segment in Fig. 9, temperature coefficient a_1= 9.4 μV/K, and the relative change in the electron density of states of the alloy caused by the structural relaxation process was determined to be $\frac{\Delta N_1}{N} = 3.53\%$. The temperature coefficient for the second linear segment is a_2= 8.36 μV/K, and $\frac{\Delta N_2}{N} = 5.33\%$

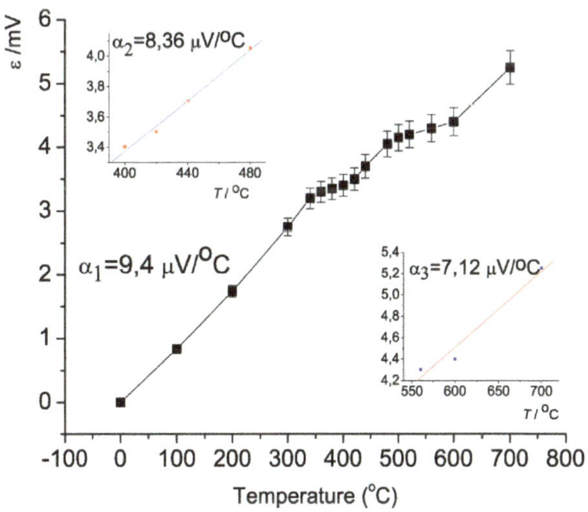

Fig. 9. Temperature dependence of thermo-electromotor force.

was determined to be 5.33% and for the third linear segment a_3= 7.12 μV/K and $\frac{\Delta N_3}{N} = 7.81\%$ was 7.81%. The overall change in the electron density of states at the Fermi level caused by the structural transformations during heating the alloy in temperature range 25-680°C is the sum of the three ΔN/N values and equals 16.67%.

The increase in the electron density of states at the Fermi level and the above mentioned increase in free path of the electron combine to diminish resistivity of the crystalline alloy (Fig. 8).

2.5 Structural transformations induced by thermal treatment

X-ray diffraction (XRD) spectra of the alloy ribbon samples (Fig. 10) show that thermal treatment (200-700°C) caused a series of structural transformations of the amorphous alloy leading to formation of more than one crystalline phase: the stable α-Fe, Fe_3Si and Fe_2B as well as metastable Fe_3B. Crystallization initially leads to formation of a nanocomposite structure of nanocrystals dispersed in the amorphous matrix. After thermal treatment at 700°C, the alloy ribbon sample was fully crystallized and composed of interdispersed nanocrystals of three crystalline phases: α-Fe, Fe_3Si and Fe_2B. The analysis of crystallite orientation on the two sides of the ribbon showed that Fe_3Si and α-Fe crystallites exhibit a degree of preferential orientation after thermal treatment. In the as-prepared alloy, Fe_3Si and α-Fe crystallites on the shiny side are oriented preferentially in [100] direction, while those on the matte side are not. This is probably the reason for higher reflectivity of the shiny side of the ribbon and the consistently high intensity of peak around 78° in the XRD spectra of the shiny side. After the treatment at 500°C, the degree of preferential orientation increases on the shiny side and forms on the matte side, suggesting asymmetric growth of the crystallites. After treatment at 700°C, amplitudes of the degree of preferential orientation on both sides decreased, suggesting that the crystallites were growing in more symmetrical manner. All of these changes are much more pronounced on the shiny side than on the matte side.

Rietveld analysis (Minić, et al. 2011c, Fig. 11) of XRD spectra yielded phase composition of the crystalline portion of the samples, average crystal sizes, dislocation density and microstrain for individual phases (Table 2). R^2 was greater than 0.98 for all XRD measurements. Individual phase contributions of α-Fe and Fe_3Si could not be completely separated, because of the overlap of their peaks. The phase composition data shows that shiny side has higher percentage of metastable Fe_3B phase, but lower percentage of Fe_2B phase. After heating at 700°C, the final phase composition on the matte side shows more crystalline Fe_2B than on the shiny side. The phase content of Fe_3B declines in two steps, with increase in heating temperature, to disappear completely after thermal treatment at 650°C. The first sharp decrease in percentage of Fe_3B (550°C) coincides with an increase in percentage of combined α-Fe and Fe_3Si. After thermal treatment at 650°C, Fe_3B phase disappears completely, and this coincides with increase in phase content of Fe_2B. Phase content of Fe_2B phase shows an increase after treatment at temperatures above 550°C.

The evolution of average crystal sizes for the respective phases shows that while there is some difference between shiny and matte side of the ribbon at the onset of crystallization, those differences become almost negligible after heating at 700°C. On closer examination, Fe_3Si and α-Fe have lower crystal sizes on the matte side, at heating temperatures of 500°C and below, and then grow faster than they do on the shiny side. As a consequence, crystal sizes of Fe_3Si and α-Fe, respectively, at the end of crystallization are almost the same on both sides of the ribbon. The evolution of average crystal size of Fe_2B shows the same trend of smaller initial crystal size and then faster crystal growth on the matte side, with increase in temperature. Metastable Fe_3B phase shows the same trend with regards to growth, except, it has a higher average crystal size on the matte side, after the samples are treated at 500°C.

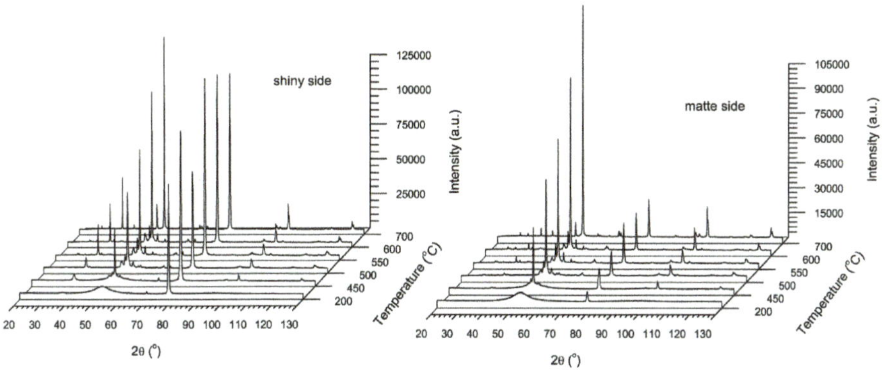

Fig. 10. XRD spectra of alloy samples after thermal treatment (left - shiny side; right - matte side)

In order to further comprehend the changes in phase composition, unit cell volumes for individual phases were determined using XRD data, and compared with standard values in JCPDS database (Minić, et al. 2011c). This way, we estimated the lattice distortion caused by the presence of boron in α-Fe and Fe_3Si lattices. The change in unit cell volume was negative for α-Fe and positive for Fe_3Si phase. The diagram (Fig 12) shows that the distortion of the unit cell of α-Fe was the greatest before the crystallization started; it decreased during the crystallization and remained relatively stable after treatment at 500°C and higher temperatures.

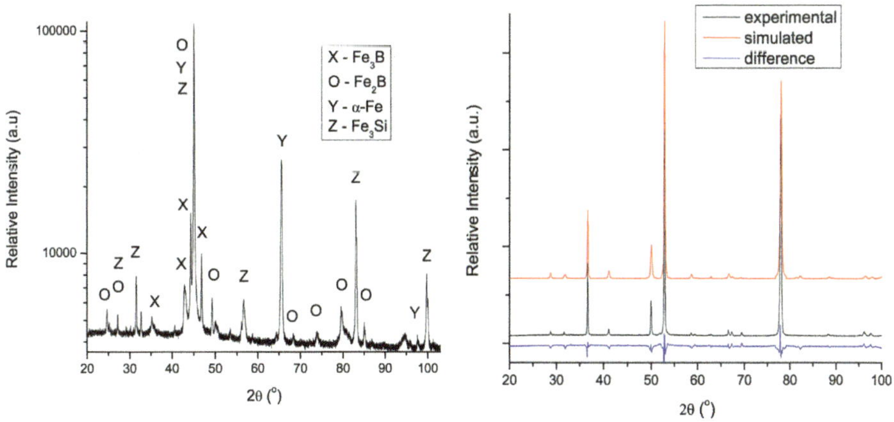

Fig. 11. XRD spectrum of alloy sample treated at 600°C with individual peak assignments (left); Rietveld analysis of XRD spectrum of alloy sample treated at 700°C (right)

The change in Fe_3B percentage after treatment at 550°C was accompanied by a decreased distortion of the unit cell of both α-Fe and Fe_3Si and increase in their combined phase content. This indicates that iron from Fe_3B was transformed into these two phases, with boron being incorporated back into the α-Fe/Fe_3Si matrix. As Fe_3Si showed greater change, Fe_3B probably transformed more to Fe_3Si than it did to α-Fe.

Temperature °C	Matte				Shiny			
	α-Fe	Fe_3Si	Fe_2B	Fe_3B	α-Fe	Fe_3Si	Fe_2B	Fe_3B
Phase composition (% mass of crystalline phase)								
500	84± 4		8± 2	8± 4	83± 5		6± 2	11± 2
550	88± 3		8± 2	4± 1	88± 3		8± 2	4± 1
600	86± 3		10± 2	4± 1	89± 3		8± 2	3± 1
700	87± 2		13± 2	-	90± 2		10± 2	
Average crystal size (nm)								
25	26.2				36.9			
200	24.6				36.5			
450	19.7				32.9			
500	26.7	22.3	49.3	16	33.1	35.1	37.3	15.8
550	32.1	35.4	90.6	46.4	36.4	45.6	82.7	44.8
600	38.6	51.6	97.7	56.6	39.0	52.7	111.1	67.7
700	44.6	71.9	107.5	-	44.9	67.5	133.5	
Dislocation density (10^{15} m^{-2})								
25	4.37				2.2			
200	4.96				2.25			
450	7.73				2.77			
500	4.21	6.03	1.23	11.72	2.74	2.43	2.16	12.02
550	2.91	2.39	0.37	1.39	2.26	1.44	0.44	1.50
600	2.01	1.13	0.31	0.94	1.97	1.08	0.24	0.66
700	1.51	0.58	0.26	-	1.49	0.66	0.17	
Microstrain (%)								
25	2.19				2.47			
200	1.85				2.63			
500	2.53				3.56	4.73		
550	2.00	4.27			2.26	3.76		
600	1.94	4.03			1.70	5.69		
700	1.26	3.47			1.37	2.66		

Table 2. Phase composition and dislocation density in the samples, obtained by analysis of XRD data

Reduced lattice distortion in these phases is probably due to the fact that mass percentage of boron in Fe_3B (6.06%) is lower than the mass percentage of boron in the as-prepared alloy (13%). Therefore, transformation of Fe_3B would lead to dilution of boron in α-Fe/Fe_3Si matrix, leading to stabilization of its crystal structure. The change in unit cell volumes of Fe_3Si and α-Fe showed increase at higher heating temperatures and it is possible that the distortion of the lattice in α-Fe and Fe_3Si is caused by the increasing occurrence of crystal/crystal interfaces between α-Fe and Fe_3Si on one side and Fe_2B on the other.

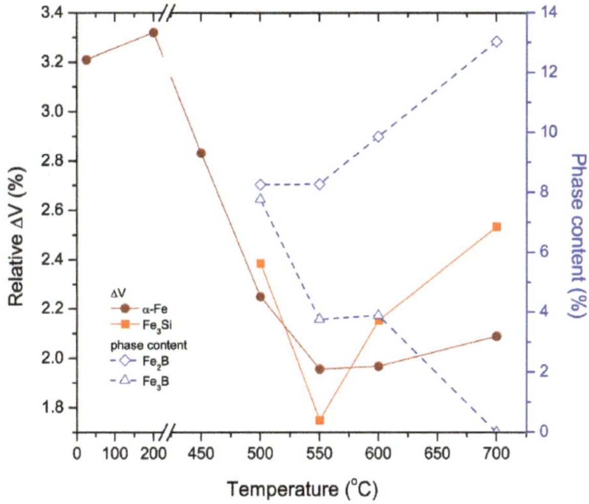

Fig. 12. Change in unit cell volumes of α-Fe and Fe_3Si against change in phase contents of Fe_2B and Fe_3B

2.6 Mechanical properties

Microhardness was measured on the cross-section of the alloy ribbon samples, rather than on any of the sides, giving a good measure of the average properties of the alloy samples. This was done due to extreme brittleness of the ribbon samples after crystallization. In order to clearly show the influence of the change in microstructure, induced by thermal treatment, on microhardness, we presented the microhardness data in combination with DSC scan (Fig. 12). In terms of microstructure, the as-prepared alloy contains a small percentage (less than 5%) of crystalline α-Fe phase in form of nanocrystals dispersed in the amorphous matrix (Fig. 1) and small domains (1-1.5 nm in size) of short-range crystalline ordering. This structure can be best described as a nanocomposite of nanocrystals and nanoclusters embedded in the amorphous matrix. The nanocomposite structure combines with chemical composition, involving significant percentage of boron, silicon and carbon, all of which are known to increase hardness in iron alloys, to produce high hardness of the as-prepared alloy sample (909HV).

The change of microhardness with respect to heating temperature showed three distinct temperature regions with completely different behavior. Before the onset of crystallization around 450°C, microhardness exhibited slight growth, from 909 to 931HV, which corresponds to lattice relaxation, as shown by the broad exothermic peaks in 200-400°C region in the DSC (Fig. 13a). In the second region, after thermal treatment at 450°C, it increased to 951HV and then to 1250HV, after thermal treatment at 500°C. This corresponds to crystallization involving formation of disordered Fe_3Si crystal phase (around 450°C), stable crystalline phases Fe_3Si, Fe_2B, α-Fe and metastable Fe_3B (exothermic peak around 512°C). These phases form a nanocomposite of nanocrystals dispersed in the amorphous matrix, which is the main cause of the increased microhardness. In the last temperature region, after the sample was treated at 700°C, microhardness decreased to 908HV, as a consequence of further crystal growth, leading to the change in the nature of the interfaces.

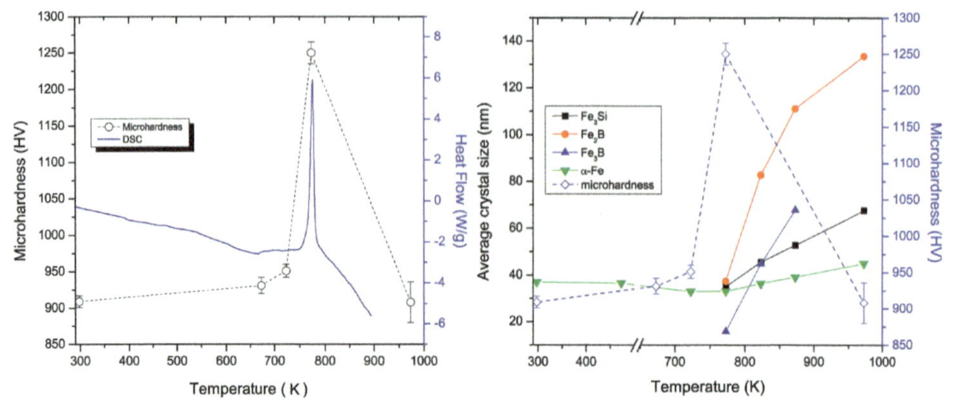

Fig. 13. Microhardness shown with DSC scan (a) and average crystal size on matte side (b).

These changes in microhardness can also be correlated with change in microstructure, induced by thermal treatment, through evolution of average crystal size of the crystalline phases in the alloy (Fig. 13b) (Minić, et al. 2011c). It can be seen that average crystal size has a significant influence on the change in microhardness. When the formed nanocrystals are relatively small (below 50 nm), microhardness remains high, while appearance of larger nanocrystals (over 100 nm) leads to a sharp decrease in microhardness. After thermal treatment at temperatures from 200 to 700°C, the alloy structure gradually transformed from a relatively homogeneous to a granulated structure with larger crystalline domains. This is consistent with creation, at the beginning of the crystallization, of a nanocomposite of small nanocrystals dispersed in amorphous matrix, which significantly increases microhardness, due to the fact that the dominant type of interface in the alloy is crystal/amorphous, as opposed to crystal/crystal in a completely crystalline alloy. As the average crystal size increased, crystal/crystal interfaces became dominant and microhardness decreased, due to increased interfacial energy and more successful propagation of shear bands and cracks along these interfaces. During the course of the observed structural transformations, the overall composition of the alloy did not change drastically, while microhardness showed significant fluctuation. This means that the changes of microhardness in our alloy were caused primarily by changes in its microstructure, rather than changes in its composition, with the main factor being the change in the average crystal size and creation of a granulated structure, as opposed to nanocrystals embedded in amorphous matrix (Table 2). The granulated structure, in addition to containing larger crystals and more crystal/crystal interfaces, is also much more porous than the original nanocomposite crystal/amorphous structure of the as-prepared alloy.

2.7 Kinetics of crystallization

All kinetic studies assume that the isothermal rate of conversion $d\alpha/dt$ is a linear function of the temperature-dependent reaction rate constant, $k\ (T)$, and a temperature-independent function of the conversion, $f\,(\alpha)$ (Vyazovkin, 2000)

$$\frac{d\alpha}{dt} = k(T)f(\alpha)\,, \tag{2}$$

where a is the fractional extent of reaction (conversion degree), t is time and the function $f(\alpha)$ depends on the particular crystallization mechanism.

The temperature dependence of the rate conversion is introduced by replacing $k(T)$ with the Arrhenius equation, which gives the relation

$$\frac{d\alpha}{dt} = A\exp\left(-\frac{E}{RT}\right)f(\alpha) \tag{3}$$

where A (pre-exponential factor) and E (activation energy) are the Arrhenius parameters and R is the gas constant.

Kinetic description of solid state transformations usually includes a kinetic triplet, involving Arrhenius parameters (activation energy, E and pre-exponential factor, A) as well as an algebraic expression of the conversion function, $f(a)$ (presented in Table 3), which describes the dependence of the reaction rate on the conversion degree, a.

In solid state reactions, the constant value of activation energy can be expected only for a single-step reaction, therefore E in equation (3) becomes an apparent quantity (E_a), based on a quasi-single-step reaction. In non-isothermal measurements at constant heating rate, β, the equation (3) transforms to:

$$\beta\frac{d\alpha}{dT} = A\exp\left(-\frac{E_a}{RT}\right)f(\alpha) \tag{4}$$

where $d\alpha/dt \equiv \beta\,(d\alpha/dT)$.

The integral form of the reaction model can be obtained by integration of equation (4)

$$g(\alpha) = \int_0^\alpha \frac{d\alpha}{f(\alpha)} = \frac{AE_a}{R\beta}p(x) \tag{5}$$

where $p(x)$ is the temperature integral for $x=E_a/RT$ which does not have analytical solution.

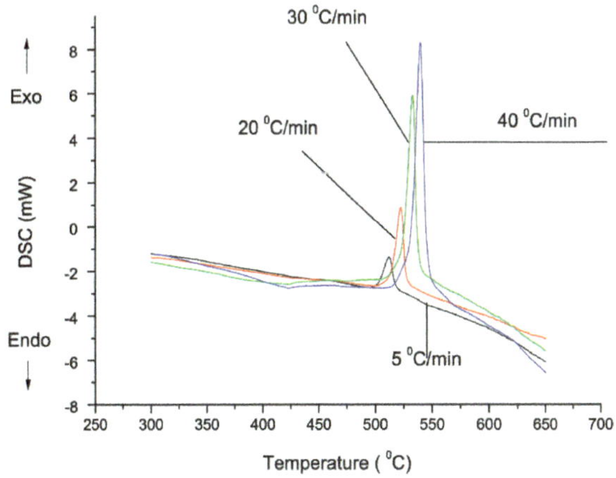

Fig. 14. DSC curves at different heating rates

The different algebraic expressions of conversion functions for solid state transformations are given in Table 3.

Fig. 14 shows the continuous DSC curves of $Fe_{81}B_{13}Si_4C_2$ ribbon in temperature range 300-650°C taken at four different heating rates. All curves contain a single well formed exothermic peak representing crystallization in the temperature range 500-560°C.

Mechanism label	$f(\alpha)$	$g(\alpha)$
Power law P4	$4\alpha^{3/4}$	$\alpha^{1/4}$
Power law P3	$3\alpha^{2/3}$	$\alpha^{1/3}$
Power law P2	$2\alpha^{1/2}$	$\alpha^{1/2}$
Power law P3/2	$3/2\alpha^{1/3}$	$\alpha^{2/3}$
Avrami-Erofeev A3/2	$3/2(1-\alpha)[-\ln(1-\alpha)]^{1/3}$	$[-\ln(1-\alpha)]^{2/3}$
Avrami-Erofeev A2	$2(1-\alpha)[-\ln(1-\alpha)]^{1/2}$	$[-\ln(1-\alpha)]^{1/2}$
Avrami-Erofeev A3	$3(1-\alpha)[-\ln(1-\alpha)]^{3/2}$	$[-\ln(1-\alpha)]^{1/3}$
Avrami-Erofeev A4	$4(1-\alpha)[-\ln(1-\alpha)]^{3/4}$	$[-\ln(1-\alpha)]^{1/4}$
Prout-Tompkins B1	$\alpha(1-\alpha)$	$\ln[\alpha(1-\alpha)^{-1}]$
Šesták-Berggren	$\alpha^N(1-\alpha)^M$	-
One dimensional phase boundary R1	1	α
Contracting cylinder R2	$2(1-\alpha)^{1/2}$	$1-(1-\alpha)^{1/2}$
Contracting sphere R3	$3(1-\alpha)^{2/3}$	$1-(1-\alpha)^{1/3}$
D1 One-dimensional diffusion	$1/2\alpha$	α^2
D2 Two-dimensional diffusion	$[-\ln(1-\alpha)]^{-1}$	$(1-\alpha)\ln(1-\alpha)+\alpha$
D3 three-dimensional diffusion	$3/2(1-\alpha)^{2/3}[1-(1-\alpha)^{1/3}]^{-1}$	$[1-(1-\alpha)^{1/3}]^2$
D4 Ginstling-Brounshtein	$3/2[(1-\alpha)^{-1/3}-1]^{-1}$	$(1-2\alpha/3)-(1-\alpha)^{2/3}$
F1 First-order	$1-\alpha$	$-\ln(1-\alpha)$
F2 Second-order	$(1-\alpha)^2$	$(1-\alpha)^{-1}$
F3 Third-order	$1/2(1-\alpha)^3$	$(1-\alpha)^{-2}$

Table 3. Algebraic expressions of conversion functions for solid state transformations

All values of initial (T_i), maximal (T_p) and final temperature (T_f) for both exo-peaks are shifted to higher values with increasing heating rate indicating the presence of kinetic effects (Table 4). The shape of DSC curves depends on the heating rate, because, as the values of shape factor S show, asymmetry of peaks increases with the decrease of heating rates indicating that the heating rate has a strong influence on the crystallization process. The values of shape factor S is obtained as the ratio of half-widths (left and right) for individual crystallization peaks for particular heating rate. The fractional extent of the sample transformed into crystalline phase, α, has been obtained from the DSC curve as a function of temperature (T) (Minić & Adnađević, 2008). At any temperature T, α is defined as $\alpha = S_T/S$, where S is the total area of the exotherm, between the temperature T_i, of the onset of crystallization and the temperature T_f, of the end of crystallization, and S_T is the area between the initial temperature T_i and a generic temperature, T, ranging between T_i and T_f (Friedman, 1964). The plots of α versus T at different heating rates for the considered crystallization process are shown in Fig. 15.

The sigmoid shape of fractional conversion curves in Fig. 15 indicates, for all heating rates, a slow initial period corresponding to nucleation with increasing rate caused by the increase of specific surface area of nuclei. The saturation part that follows is the consequence of nuclei merging together causing a decrease in their surface area.

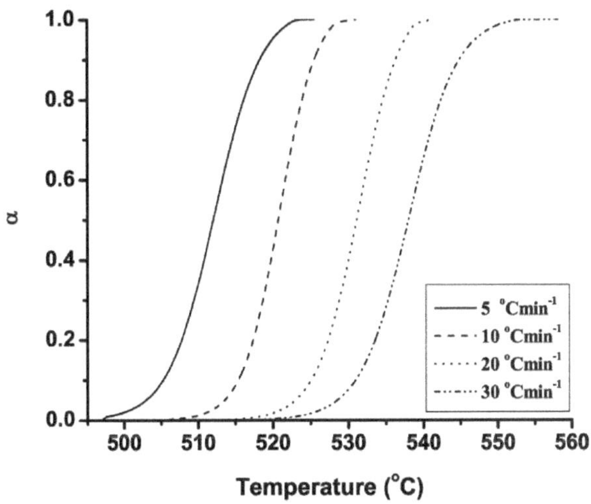

Fig. 15. Fractional conversion (α) as a function of temperature (T) for the crystallization alloy at different heating rates (5, 10, 20 and 30°Cmin^{-1}).

2.7.1 The overall apparent activation energy of crystallization

The overall activation energy of crystallization of an amorphous alloy under linear heating condition can be determined by the Kissinger as well as by the Ozawa peak method relating to the dependence of exothermic peak temperature Tp on heating rate β.

Kissinger (Kissinger, 1957) proposed that the activation energy can be determined according the equation

$$\ln\left(\frac{\beta}{T_p^2}\right) = \ln\left(\frac{AR}{E_a}\right) - \frac{E_a}{RT} \tag{6}$$

For the determination the activation energy in non-isothermal conditions Ozawa (Ozawa, 1970) proposed the equation:

$$\ln\beta = \ln\frac{AE_a}{R}C - 1.0516\frac{E_a}{RT_p} \tag{7}$$

The values of peak temperatures together with values of kinetic parameters (E_a and $\ln A$) calculated by both methods and the symmetry factors (SF) are given in the Table 4.

β °Cmin^{-1}	T_i °C	T_p °C	T_f °C	S	Ozawa		Kissinger	
					E_a kJmol^{-1}	A × 10^{22} min^{-1}	E_a kJmol^{-1}	A × 10^{21} min^{-1}
5	492	512	542	0.59	338.0 +1.8	1.10 ± 2.3	351.2±1.8	3.06 ±2.3
10	501	520	546	0.59				
20	509	531	554	0.70				
30	513	538	560	0.75				

Table 4. Values of T_i, T_p, T_f and S for crystallization of the amorphous $Fe_{81}B_{13}Si_4C_2$ alloy upon continuous heating at different heating rates.

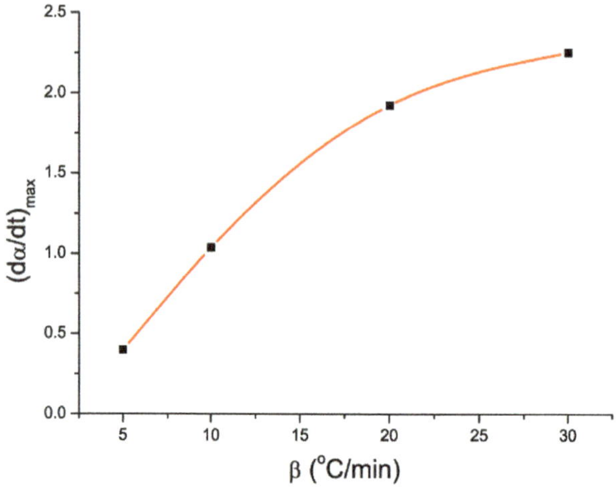

Fig. 16. The dependence of maximum rate of crystallization on heating rate.

The dependence of maximum of rate on rate of heating (Fig. 16) shows the growth of maximum of rate of crystallization with respect to the rate of heating. It indicates that the rate of crystallization reaches a saturation point at high heating rates.

The activation energy of crystallization process involving formation of nuclei and their growth, according to some opinions, has no physical meaning, just empirical significance and only establishes the dependence of the rate of conversion on temperature. This energy can be spent, not just for overcoming the activation barrier but also for its downturn due to cooperative displacement of groups of atoms. Finally, the crystallization of amorphous alloys is a very complicated process accompanied by nucleation and growth of various crystal phases under continuously varied conditions in the conversion zone. With the multitude of possible ways of conversions, only those mechanisms and activated complexes of the crystallization process will be realized that have the highest probability at a given temperature. Any change in crystallization conditions, such as heating rate, can result in change of the mechanism and main activation complex of the crystallization process. Thus, high values of activation energy of crystallization of amorphous alloys, first of all, indicate that a large number of atoms participate in an elementary act of structure reorganization, as well as high complexity of the transformation processes.

2.7.2 The dependence of apparent activation energy on range of conversion

On the basis of dynamic DSC measurements for different heating rates, isoconversional method of Kissinger-Akahira-Sunose method, also known as the „model-free method", enables the determination values of E_a over a wide range of α without the determination of the conversion function (Kissinger, 1957; Akahira, Sunose, 1971). This model involves measuring the temperatures T_a corresponding to fixed values of the crystallized volume fraction, a, for different heating rates, β, and application of the relation:

$$\ln\left(\frac{\beta_i}{T^2_{\alpha,i}}\right) = \ln\left(\frac{AR}{E_{a,\alpha}} F(\alpha)\right) - \frac{E_{a,\alpha}}{RT_{\alpha.i}} \quad (8)$$

where a subscript α designates values related to a given conversion degree, and i is a number of the non-isothermal experiment conducted at the heating rate β_i.

The left-hand side of Eq. (8) is linear with respect to the inverse temperature, $1/T_a$, and enables the activation energy to be evaluated using a linear regression method. In case of a single step process, a constant value of $E_{a(a)}$ is obtained. On the other hand, the dependence of $E_{a(a)}$ on α indicates complex process involving more than one step with different activation energies. It can be observed that the apparent activation energy for the considered crystallization process (Fig. 17) is practically constant in the $0.05 \leq \alpha \leq 0.7$ range indicating the existence of a single-step reaction. (Vyazovkin, 2000; Opfermann & Flammersheim, 2003). The average value of apparent activation energy was determined as $E_a = 356.5 \pm 5.5$ kJmol^{-1}.

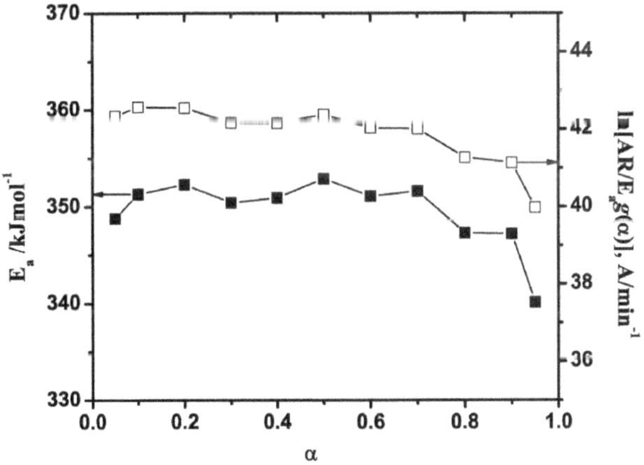

Fig. 17. Apparent activation energies (E_a) and the intercepts as function of the crystallized fraction α.

2.7.3 Preliminary determination of kinetic model

For the preliminary determination of kinetic model of the crystallization process, Dollimore's method was used (D. Dollimore, 1991, 1992; Lee, 1998).

β (°C/min)	α_{max}	$\frac{\Delta LoT}{\Delta HiT}$	Half-width (°C)	T_i	T_f
5	0.51	1.0	9.5	sharp	sharp
10	0.53	1.0	9.5	sharp	sharp
20	0.53	0.9	10.0	sharp	sharp
30	0.55	0.8	11.0	sharp	sharp

Table 5. Parameters describing the asymmetric DSC peaks of crystallization α-Fe phase in amorphous $Fe_{81}B_{13}Si_4C_2$ alloy

This model is based on the "sharpness" of initial and final temperature of differential rate curves, Fig. 18, as well as on its asymmetry. The "sharpness" of the initial and final temperatures is influenced by kinetic factors, and especially by the mechanism of the process. Certain kinetic models lead to an asymptotic or diffuse departure from the base line in the differential form of the thermal curve, while others produce a very sharp approach to the final plateau. The investigation of these parameters that describe geometry and asymmetry of the differential rate curves can be an indicator of the probable kinetic mechanism. According to these parameters, different types of kinetic mechanisms have been listed (Lee, 1998). So, when the crystallization process is not complex the qualitative approach to its kinetics may be obtained using parameters such as α_{max} or $(d\alpha/dt)_{max}$, the shape of the initial and final temperatures as well as peak temperature (T_p) or half-width from differential rate curves.

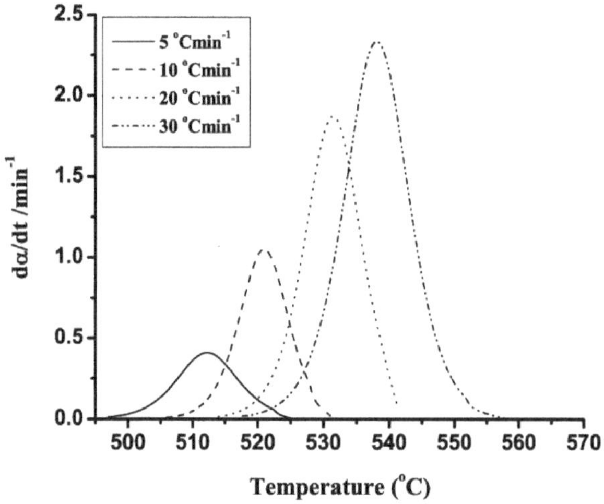

Fig. 18. Differential rate curves ($d\alpha/dt$ vs. T) for different heating rates (β = 5, 10, 20 and 30°Cmin^{-1}).

In this case, Dollimore's procedure was applied to the conversion and differential rate curves (Fig.18) whose slight asymmetry is observed between T_i and T_f. The other parameters such as the conversion at the rate of maximum crystallization, α_{max}, peak temperature, T_p, at $(d\alpha/dt)_{max}$, and the ratio ΔLoT/ΔHiT (shape factor), which is the ratio between the low and high temperature points at half-width of the differential rate curve peak are presented in

Table 5. It is clearly seen that the position of the broadening exotherm, which is connected with the crystallization, was shifted toward higher temperature with the increase of the heating rate as well as asymmetry of peaks. This suggests that the crystallization process should not be characterized by a definite critical temperature independent of the heating rate. The determined values of α_{max} for different heating rates were in the range from 0.51 to 0.55. These results indicate that the non-isothermal crystallization mechanism of α-Fe in amorphous $Fe_{81}B_{13}Si_4C_2$ alloy can not be fully described within the JMA (Johnson-Mehl-Avrami) models (A2, A3 and A4, Group A).

2.7.4 Determination of the kinetic model

General equation, enabling the analysis of conversion kinetics involving nucleation and growth in solid phase was proposed by Avrami (Avrami, 1939):

$$\alpha(t) = 1 - \exp[-(kt)^n], \tag{9}$$

where $\alpha(t)$ is conversion degree, n is kinetic exponent, $k=k_o\exp(-E_a/RT)$.

Differentiation of this equation with respect to time gives the rate equation, known as the JMA equation:

$$\left(\frac{d\alpha}{dt}\right) = kn(1-\alpha)[-\ln(1-\alpha)]^{1-1/n}. \tag{10}$$

The JMA equation is based on assumptions of isothermal crystallization, homogenous or heterogeneous nucleation at randomly dispersed particles of the second phase. The growth rate of new phase is independent of time and controlled by temperature and low anisotropy of growing crystals. However if the entire nucleation process takes place during the early stage of transformation and becomes negligible afterwards, JMA equation can also be applied to non-isothermal conditions (Henderson, 1979).

The validity of listed assumptions is not given a priori and simple and reliable testing methods were developed (Málek, 1992, 1995; Gotor et al, 2000; Criado et al, 2003). Once the apparent activation energy has been determined, it is possible to find the kinetic model which best describes the measured set of thermoanalytical data. It can be shown that, for this purpose, it is useful to define two special functions y(α) and z(α), which can easily be obtained by simple transformation of the experimental data. The conversions, in which the y(α) and z(α) functions exhibit the maximum values are designated as α_y^* and α_z^*, respectively. Under non-isothermal conditions these functions can be expressed as follows (Málek, 1992, 1995, 2000):

$$y(\alpha) = \left(\frac{d\alpha}{dt}\right)\exp\left(\frac{E_a}{RT}\right) = Af(\alpha) \tag{11}$$

$$z(\alpha) \approx \left(\frac{d\alpha}{dt}\right)T^2 \tag{12}$$

The maximum of the y(α) function for the JMA model depends on the value of the kinetic exponent:

$$\alpha_y^* = 0 \text{ for n} \le 1$$

$$\alpha_y^* = 1\text{-exp } (n^{-1}\text{-}1) \text{ for n} > 1$$

The value of α_y^* is always lower than the maximum of value for α_z^*. For JMA model, α_z^* =0.632. This value is characteristic "fingerprint" of the JMA model and it can be used as a simple test of the applicability of this model.

The obtained normalized functions y(α) and z(α), Fig 19, are independent on the heating rate (β) of the system, and the both functions exhibits the well-defined maxima which were located at exactly defined value of α (α_y^* for the y(α) function and α_z^* for z(α) function, respectively), Table 6.

β °C min^{-1}	α_y^*	α_z^*
5	0.41 ± 0.01	0.53 ± 0.01
10	0.42 ± 0.01	0.51 ± 0.01
20	0.42 ± 0.01	0.55 ± 0.03
30	0.41 ± 0.01	0.52 ± 0.01

Table 6. The maximum of α_y and α_z for the different heating rates

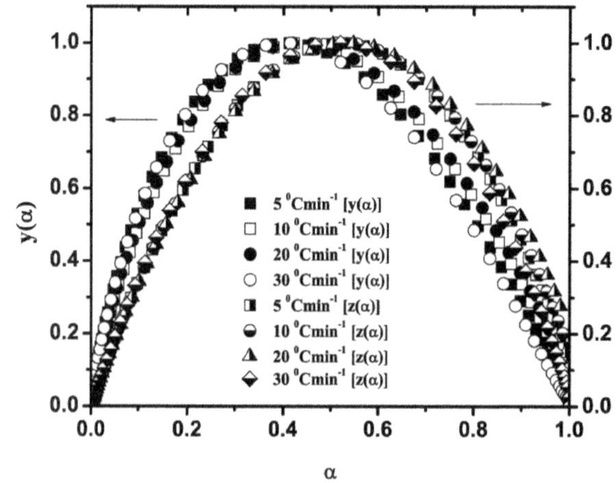

Fig. 19. Normalized *y*(*a*) and *z*(*a*) functions at the different heating rates.

From Table 6, it can be seen that the values of α_y^* fall into the range $\alpha_y^* \in (0, \alpha_z)$ ($0.41 \le \alpha_y^* \le 0.42$) and the values of α_z^* are less than 0.632 ($0.51 \le \alpha_z^* \le 0.55$). From the obtained results, it follows that the conditions of validity of the JMA model are not fulfilled for crystallization of *a*-Fe. The displacement of α_z^* in lower value range indicates complexity of the process and can be caused by the influence of surface nucleation or the effect of released crystallization heat on temperature distribution within the sample. However, the relatively high value of α_y^* indicates an increasing effect of the crystallized phase to overall crystallization kinetics, where the crystallized phase further increases the rate of the crystallization. Such autocatalytic behavior can be well described using an empirical two parameter Šesták-Berggren's kinetic model (Šesták, Berggren, 1971; Málek, et al. 1989). This model is based on the equation:

$$f(\alpha)=\alpha^M(1-\alpha)^N, \tag{13}$$

where M and N represents the kinetic exponents.

In this case the expression for reaction rate of the investigated crystallization process can be given as:

$$\frac{d\alpha}{dt}=A\exp\left(-\frac{E_a}{RT}\right)\alpha^M(1-\alpha)^N \tag{14}$$

For this model, the ratio of the kinetic exponents $p = M / N$ can be calculated from the maximum of the y (α) function (Málek, 2000):

$$p=\frac{M}{N}=\frac{\alpha_y^*}{\left(1-\alpha_y^*\right)}. \tag{15}$$

Introducing this equation in equation this (13) gives:

$$\ln\left[\left(\frac{d\alpha}{dt}\right)\exp\left(\frac{E_a}{RT}\right)\right]=\ln A+N\ln\left[\alpha^p(1-\alpha)\right]. \tag{16}$$

This equation describes the processes of nucleation and growth in non-crystalline solids very well. The parameters M and N define relative contributions of acceleratory and decay regions of the kinetic process. From the linear dependence $\ln\left[(d\alpha/dt)\exp(E_a/RT)\right]$=f($\ln\left[\alpha^p(1-\alpha)\right]$), it is possible to obtain the kinetic exponent N and the pre-exponential factor, ln A. The value of kinetic exponent M can be obtained directly from eq. (15).

β °C/min	M	N	ln$A$ (*10^{22})
5	0.75 ± 0.03	1.08 ± 0.10	9.0±0.2
10	0.66 ± 0.05	0.92 ± 0.05	10.7±0.3
20	0.64 ± 0.05	0.89 ± 0.07	9.4±0.2
30	0.81 ± 0.10	1.17 ± 0.04	10.6±0.2
Average	0.72 ± 0.06	1.02 ± 0.07	9.9±0.2

Table 7. Kinetic exponents M and N at different heating rates (Adnađević et al., 2010)

Table 7 lists the values of kinetic exponents M and N, as well as the values of ln$A$ obtained by the procedure described above, for the considered crystallization process at different heating rates.

The obtained values of kinetic exponents M and N vary a little with respect to heating rate β. The values of M vary in the range of $0.64 \leq M \leq 0.81$ with average value of $M_{av} = 0.72$. The values of N vary in the range of $0.89 \leq N \leq 1.17$ with average value of $N_{av} = 1.02$. The values of the pre-exponential factor (ln$A$) are independent on the heating rate (β), within the limits of the experimental error. It was shown that this two parameter autocatalytic model has physical meaning only for M<1(Gotor, et al. 2000).

In order to check the established kinetic model we applied the "Master-plot" method (Criado, et. al 2003; Gotor, 2000). Using the value at $\alpha = 0.5$ as a reference point, the following differential master equation is easily derived from eq. (4):

$$\frac{f(\alpha)}{f(0.5)} = \frac{d\alpha / dt}{(d\alpha / dt)_{0.5}} \frac{\exp(E_a / RT)}{\exp(E_a / RT_{0.5})} \tag{17}$$

where $(d\alpha/dt)_{0.5}$, $T_{0.5}$ and f (0.5) are the reaction rate, the temperature reaction and the differential conversion function, respectively at $a = 0.5$.

The left side of Eq. (17) is a reduced theoretical curve which is characteristic of each kinetic function. The right side of the equation is associated with the reduced rate and can be obtained from experimental data if the apparent activation energy is known and remains constant throughout the reaction. Comparison of both sides of Eq. (17) tells us which kinetic model describes an experimental reaction process. It can be seen in Fig. 20, using the average value for the apparent activation energy determined from Kissinger-Akahira-Sunose isoconversional method, the suggested kinetic model works very well in the entire conversion range. Therefore, the Šesták-Berggren autocatalytic model represents the best reaction model for describing the crystallization process of α-Fe in the amorphous $Fe_{81}B_{13}Si_4C_2$ alloy. The higher value of N exponent designates that the formed crystallized phase has the decisive influence on the kinetics of transformation and the rate of growth. In the propagation process, on account of overlapping of nuclei during growth, it acts to retard crystallization rate. Bearing this in mind, we can suppose that in the amorphous alloy, the α-Fe embryos

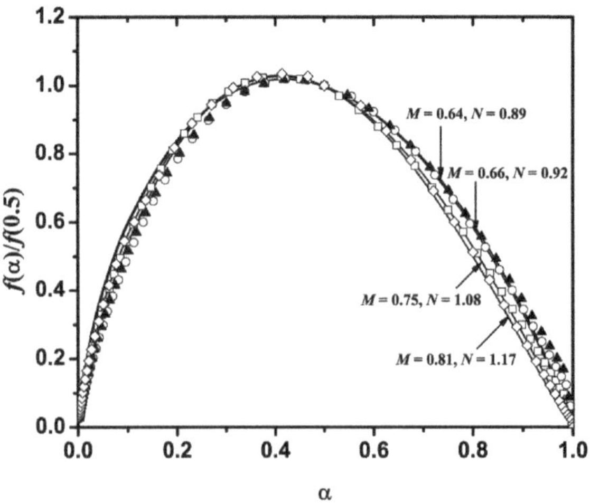

Fig. 20. The theoretical (solid line) and experimental differential master plots of $f(\alpha)/f(0.5)$ versus α for different heating rates: (□) 5 °Cmin^{-1}; (○) 10 °Cmin^{-1}; (▲) 20 °Cmin^{-1} and (◊) 30 °Cmin^{-1}.

already exist at $T \leq 500°C$ and at $T \geq 500°C$ these embryos are momentarily transformed into nuclei. The established acceleration of crystallization process is a consequence of significant increase of strain in alloy, which arises on account of formation of α-Fe.

2.7.5 Morphology of crystal growth

For non-isothermal crystallization, where the volume fraction of crystalline phase α precipitated in glass heated at a uniform heating rate β is related with the activation energy E_a, Matusita et al. proposed the following relation (Matusita & S. Sakka, 1979, 1980; Matusita, et al. 1984):

$$\ln[-\ln(1-\alpha)] = -n\ln\beta - \frac{1.052mE_a}{RT} + const, \quad (18)$$

where m and n are constants with values between one and four depending on the morphology and kinetics of the growth nuclei, Table 8.

Mechanism	n	m
Bulk nucleation		
Three dimensional growth	4	3
Two dimensional growth	3	2
One dimensional growth	2	1
Surface nucleation	1	1

Table 8. Values of constants n and m for different crystallization mechanisms

The values of n obtained from the slopes of linear plots ln[-ln (1-α)] versus $-\ln\beta$ at different temperatures for considered crystallization process are given in Table 9. For all considered temperatures, the value of n is ≈ 4.0, within the limits of experimental errors. It follows, then, that the kinetics of crystallization process is independent from the temperature.

The crystallization exponent n is connected with the number of growth dimensions (m) and the number of nuclei forming stages (s) (Matusita, Sakka, 1979) by the following equation

$$n = m + s \quad (19)$$

where m is the number of growth dimensions as defined in Table 8, s is the number of the nuclei forming stages ($s = 0$ – at instantaneously nucleation; $s = 1$ – at constant nucleation rate and $s > 1$ at self-acceleratory nucleation rate).

In order to describe the crystallization process in detail, the value of parameter m should be determined from the plot of ln [-ln (1-α)], because a function of reciprocal temperature is linear with a slope of $1.051\times(m+1)E_a/R$, using the value of activation energy determined above.

Temperature °C	n	β °C/min	m	s
518	3.92	5	2.84	1.16
520	4.08	10	3.08	0.92
522	4.07	20	3.22	0.78
		30	2.84	1.16

Table 9. The values of n at three temperatures and values of m and s for four different heating rates

The values of parameters *m* and *s* obtained at the different heating rates for the investigated crystallization process of α-Fe in $Fe_{81}B_{13}Si_4C_2$ amorphous alloy are given in Table 9. Based on the obtained values of parameters *m* and *s*, at the different heating rates, (Table 9), we asserted with high degree of reliability, that the nucleation process of α-Fe occurs within amorphous alloy with a constant rate, in three effective directions (three-dimensional growth) proceeding with constant nucleation rate.

3. Conclusion

Metallic amorphous alloy are a class of materials which has seen dramatic developments in recent times with design of materials stable enough to allow bulk production. While the functional properties of amorphous alloys allow for many possible fields of application, they can also be used as precursors in preparation of nanocomposite materials composed of nanocrystals in the amorphous matrix. These nanocomposites often exhibit superior mechanical, electrical and magnetic properties to both purely amorphous and purely crystalline materials. Thermal treatment of amorphous alloys can allow for controlled crystallization, leading to formation of nanocomposite materials with targeted properties.

$Fe_{81}B_{13}Si_4C_2$ alloy undergoes multi-step crystallization process as a result of thermal treatment. The changes in microstructure cause changes in electrical, magnetic and mechanical properties of the alloy as the alloy structure changes from predominantly amorphous to crystal/amorphous nanocomposite to nanocrystalline composite of α-Fe and Fe_2B phases.

4. References

Adnađević, B.; Janković B. & Minić D. M.; (2010), Kinetics of the apparent isothermal and non-isothermal crystallization of the α-Fe phase within the amorphous $Fe_{81}B_{13}Si_4C_2$ alloy, Journal of Physical Chemistry of Solids, vol. 71, pp. 927-934

Akahira, T. & Sunose, T. (1971) Trans. Joint Convention of Four Electrical Institutes, paper no. 246 (1969), Research Report, Chiba Institute of Technology, vol. 16, pp. 22-31

Avrami, M. (1939), Kinetics of Phase Change. I General Theory, Journal of Chemical Physics, vol. 7, No. 12, pp. 1103-1113

Blagojević, V. A.; Minić, D. M.; Žak, T.; Minić, D. M. (2011), Influence of thermal treatment on structure and microhardness of $Fe_{75}Ni_2Si_8B_{13}C_2$ amorphous alloy, Intermetallics vol. 19, pp. 1780-1785

Balberg, I. & Helman, J. S. (1978), Critical behavior of the resistivity in magnetic systems. II. Below Tc and in the presence of a magnetic field, Physical Review B, vol. 18, No. 1, pp. 303-318

Böhnke, G.; Kaul, S. N.; Kettler, W. & Rosenberg, M. (1983), Critical behaviour of the electrical resistivity in amorphous ferromagnetic alloys, Solid State Communications, vol. 48, No. 9, pp. 743-746

Criado, J. M.; Pérez-Maqueda, L. A.; Gotor, F. J.; Málek, J. & Koga, N. (2003), A unified theory for the kinetic analysis of solid state reactions under any thermal pathway, Journal of Thermal Analysis and Calorimetry, vol. 72, No. 3, pp. 901-906

Dollimore, D.; Evans, T.A.; Lee, Y.F. & Wilburn, F.W. (1991), Calculation of activation energy and pre-exponential factors from rising temperature data and the generation of TG and DTG curves from A and E values, Thermochimica Acta, vol. 188, No. 1, pp. 77-85

Dollimore, D.; Evans, T.A.; Lee, Y.F.; Pee, G.P. & Wilburn, F.W. (1992), The significance of the onset and final temperatures in the kinetic analysis of TG curves, Thermochimica Acta, vol. 196, No. 2, pp. 255-265

Friedman, H. L. (1964), Kinetics of thermal degradation of char-forming plastics from thermogravimetry. Application to a phenolic plastic, Journal of Polymer Science Part C: Polymer Symposia, vol. 6, No. 1, pp. 183-195

Gotor, F. J.; Criado, J. M.; Málek, J. & Koga, N. (2000), Kinetic Analysis of Solid-State Reactions: The Universality of Master Plots for Analyzing Isothermal and Nonisothermal Experiments, Journal of Physical Chemistry A, vol. 104, No. 46, pp. 10777-10782

Henderson, D. W. (1979), Experimental analysis of non-isothermal transformations involving nucleation and growth, Journal of Thermal Analysis and Calorimetry, vol. 15, No. 2, pp. 325-331

Kissinger, H. E. (1957), Reaction Kinetics in Differential Thermal Analysis, Analytical Chemistry, vol. 29, No. 11, pp. 1702-1706

Klement, W.; Willens, R. H & Duwez, P.O.L. (1960), Non-crystalline Structure in Solidified Gold-Silicon Alloys, Nature, vol. 187, No. 4740, pp. 869–870

Lass, E. A.; Zhu, A.; Shiflet, G. J. & Poon, S. J. (2010), A short-range ordering description of amorphous metal alloys using the central atoms model, Acta Materialia, vol. 58, No. 16 pp. 5460-5470

Lee, Y.F. & Dollimore, D. (1998), The identification of the reaction mechanism in rising temperature kinetic studies based on the shape of the DTG curve, Thermochimica Acta, vol. 323, No. 1-2, pp. 75-81

Libermann H. & Graham C. (1976), Production Of Amorphous Alloy Ribbons And Effects Of Apparatus Parameters On Ribbon Dimensions, IEEE Transactions on Magnetics, vol. 12, No. 6, pp. 921

Málek, J.; Criado, J. M.; Šesták, J. & Militky, J. (1989), The boundary conditions for kinetic models, Thermochimica Acta, vol. 153, pp. 429-432

Málek, J. (1992), The kinetic analysis of non-isothermal data, Thermochimica Acta, vol. 200, pp. 257-269

Málek, J. (1995), The applicability of Johnson-Mehl-Avrami model in the thermal analysis of the crystallization kinetics of glasses, Thermochimica Acta, vol. 267, pp. 61-73

Málek, J. (2000), Kinetic analysis of crystallization processes in amorphous materials, Thermochimica Acta, vol. 355, No. 1-2, pp. 239-253

Maričić, A.; Minić, D. M.; Blagojević, V. A.; Kalezić-Glišović, A.; Minić, D. M. (2012), *Effects of structural relaxation on functional properties of amorphous alloy* $Fe_{73.5}Cu_1Nb_3Si_{15.5}B_7$, Intermetallics vol. 21, pp. 45-49

Matusita, K. & Sakka, S. (1979), Kinetic study of the crystallisation of glass by differential scanning calorimetry, Physics and Chemistry of Glasses, vol. 20, pp. 81-85

Matusita, K. & Sakka, S. (1980), Kinetic study of crystallization of glass by differential thermal analysis—criterion on application of Kissinger plot, Journal of Non-Crystalline Solids, vol. 38-39, No. 2, pp. 741-746

Matusita, K.; Konatsu, T. & Yokota, R. (1984), Kinetics of non-isothermal crystallization process and activation energy for crystal growth in amorphous materials, Journal of Materials Science, vol. 19, No. 1, pp. 291-296

Minić, D. M.; Maričić, A., Dimitrijevic, R. Z.; Ristić, M. M. (2007), Journal of Alloys and Compdounds, vol. 430, pp. 241-245

Minić, D. M.; & Adnađević, B. (2008), Mechanism and kinetics of crystallization of α-Fe in amorphous $Fe_{81}B_{13}Si_4C_2$ alloy, Thermochimica Acta, vol. 474, No. 1-2, pp. 41-46

Minić, D. M.; Maričić, A. & Adnađjević, B. (2009), Crystallization of α-Fe phase in amorphous $Fe_{81}B_{13}Si_4C_2$ alloy, Journal of Alloys and Compounds, vol. 473, pp. 363-367

Minić, D. M.; Minić, D. G. & Maričić, A. (2009), Stability and crystallization of $Fe_{81}B_{13}Si_4C_2$ amorphous alloy, Journal of Non-Crystalline Solids, vol. 355, No. 50-51, pp. 2503-2507

Minić, D. M.; Plazinić, M. ; Živanić, J. & Maričić, A. (2010), Influence of Structural Transformations on Electric and Magnetic Properties of $Fe_{81}B_{13}Si_4C_2$ Amorphous Alloy, Science of Sintering, vol. 42, pp. 61-68

Minić, D. M.; Blagojević, V.; Minić, D. G.; Gavrilović, A.; Rafailović, L. (2011), Influence of thermally induced structural transformations on hardness in $Fe_{89.8}Ni_{1.5}Si_{5.2}B_3C_{0.5}$ amorphous alloy, Journal of Alloys and Compounds vol. 509, pp. 8350-8355

Minić, D. M.; Minić, D. M.; Žák, T.; Roupcova, P. & David, B. (2011), Structural transformations of $Fe_{81}B_{13}Si_4C_2$ amorphous alloy induced by heating, Journal of Magnetism and Magnetic Materials, vol. 323, No. 5, pp. 400-404

Minić, D. M.; Blagojević, V. A.; Minić, D. M.; Gavrilović, A. & Žak, T. (2011), Influence of microstructure on microhardness of $Fe_{81}Si_4B_{13}C_2$ amorphous alloy after thermal treatment, Metallurgical and Materials Transactions A, vol 42, pp. 4106-4112

Minić D. M; Blagojević V. A; Minić D. M.; Žák T(2011), Influence of microstructural inhomogeneouity of individual sides of $Fe_{81}Si_4B_{13}C_2$ amorphous alloy ribbon on thermally induced structural transformations, Materials Chemistry and Physics, vol. 130, pp. 980-985

Opfermann, J. & Flammersheim, H. J. (2003), Some comments to the paper of J.D. Sewry and M.E. Brown: "Model-free" kinetic analysis, Thermochimica Acta, vol. 397, No. 1-2, pp. 1-3

Ozawa, T. (1970), Kinetic analysis of derivative curves in thermal analysis, Journal of Thermal Analysis and Calorimetry, vol. 2, No. 3, pp. 301-324

Ponnambalam, V.; Poon, S. J. & Shiflet, G. J. (2004), Fe-based bulk metallic glasses with diameter thickness larger than one centimeter, Journal of Materials Research vol. 19, No. 5, pp. 1320-1323

Ribić-Zelenović, L.; Rafailović, L.; Spasojević, M. & Maričić, A. (2008), Correlation between electron state density change and the electrical resistivity and magnetic permeability changes in the nanostructured powder of the NiMo alloy, Physica B: Condensed Matter, vol. 403, No. 12, pp. 2148-2154

Saegusa, N. & Morrish, A. H. (1982), Atomic rearrangements and crystalline transformations in amorphous Fe81B13.5Si3.5C2, Physical Review B, vol. 26, No. 1, pp. 305-314

Šesták, J. & Berggren, G. (1971), Study of the kinetics of the mechanism of solid-state reactions at increasing temperatures, Thermochimica Acta, vol. 3, No. 1, pp. 1-12

Vyazovkin, S. (2000), Computational aspects of kinetic analysis.: Part C. The ICTAC Kinetics Project — the light at the end of the tunnel?, Thermochimica Acta, vol. 355, No. 1-2, pp. 155-163

Williamson, G. K. & Hall, W. (1953), X-ray line broadening from filed aluminium and wolfram, Acta Metallurgica, vol. 1, No. 1, pp. 22-31

Žák, T. (1999) In: Mössbauer spectroscopy in Material Science, Eds. M. Miglierini and D. Petridis, NATO Science Series, Kluwer, Dordrecht, 1999, pp. 385

Žák, T. & Jirásková, Y. (2006), CONFIT: Mössbauer spectra fitting program, Surface Interface Analysis, vol. 38, No. 4, pp. 710-714

4

Real-Time Analysis to Evaluate Crystallization Processes

João F.Cajaiba da Silva, Andréia P. M. da Silva and Rodrigo C. de Sena
Instituto de Química – Universidade Federal do Rio de Janeiro
Brazil

1. Introduction

Crystallization is one of the most important unit operations employed by the pharmaceutical, microelectronics, food and fine chemicals industries for the production of solid products with high added value. This operation can be used as a method to perform separation and purification of crystalline compounds such as proteins, polymers, pharmaceuticals, inorganic salts, etc. (Févotte & Klein, 1995; Feng & Berglund, 2002; Mersmann, 2001; Lewiner et al., 2001; Mullin, 2001; Liotta & Sabesan, 2004; Joung et al., 2005; Pelberg et al., 2005; Derdour et al., 2011).

The experimental conditions used during a crystallization process may alter the physical properties of the final product such as its chemical purity, crystal size distribution and morphology. These properties can have impacts on the subsequent purification operations such as filtration, washing and drying, and can also alter the bioavailability of pharmaceuticals (Sabesan & Liotta, 2004). Polymorphism is another issue related to the crystallization process that has profound importance in the pharmaceutical industry because it can alter the kinetics of a crystal's solubilization and in some cases polymorphs present problems of toxicity (Fujiwara et al., 2005).

The challenges involved in controlling crystallization are significant, since the kinetic parameters of the process are strongly affected by several factors such as the presence of impurities (Gunawan et al., 2002; Ma et al., 1999; Rauls et al., 2000; Poddar, 2002), breaking of crystals (Kougoulos et al., 2005; Gahn & Mersmann, 1995) and clustering (Yu et al., 2005; Paulaime et al., 2003) among other effects that are difficult to characterize.

The identification and control of factors that affect the final quality of crystals are essential to ensure uniformity among different batches and to improve product quality. The development of more accurate and sensitive sensors for real-time analysis of crystallization must allow significant advances in monitoring, control and optimization of crystallization processes (Liotta & Sabesan, 2004).

The accuracy of off line methods for evaluating crystallization processes is strongly dependent on sampling. When the collected samples do not represent the whole, the errors introduced may cause a misinterpretation of the crystallization process. By using real-time analysis these types of errors can be greatly reduced. The possibility of performing in situ analysis of crystal size distribution, crystal shape, crystal habit, agglomeration and breakage can indicate what changes should be made to the crystallization process parameters such as cooling and stirring rates and time for seeding. The optimization of these parameters allows crystals of the desired characteristics to be obtained (Yu et al., 2004).

This section is intended to give a brief overview of crystallization processes and the methods dedicated to monitor them in real-time. Additionally, a comparison between four in line methods to determine the onset of adipic acid crystallization was performed.

1.1 Solubility and supersaturation

The determination of the solubility of a solid in a specific solvent is a key step in the study of crystallization processes. The solubility curve is used as a benchmark to assess the degree of supersaturation and the metastable zone limits. The solubility or condition of saturation is determined experimentally by heating a suspension and observing the temperature at which the solid phase is completely dissolved. The cooling of a saturated solution results in a system that is not in thermodynamic equilibrium, a supersaturated solution (Mullin, 2001; Giulietti et al., 2001).

The supersaturation is the driving force for crystallization processes and can be defined as the difference between the chemical potential of a solute in a supersaturated solution and the chemical potential of the saturated solution. Supersaturation can be created by cooling, by adding an anti-solvent, by performing a chemical reaction that generates a product of lower solubility, by solvent evaporation, etc. Among the methods used to create supersaturation, cooling is the most used. The usage of this method is restricted to substances whose solubility changes significantly during a temperature variation. The expected properties of the solid material, as well as economic aspects, form the basis for making a decision about which method should be used to create supersaturation (Mullin, 2001; Giulietti et al., 2001).

The difference between the concentration of a compound in a supersaturated solution, c, and its concentration in a saturated solution, c*, is known as absolute supersaturation and is expressed by equation 1.

$$\Delta c = c - c^* \tag{1}$$

When Δc is greater than 0, the system is supersaturated. Another important parameter is the supersaturation ratio, S, which is defined by equation 2. In this case the system is said to be supersaturated when S is greater than 1.

$$S = \frac{c}{c^*} \tag{2}$$

1.2 Metastable zone width

Supersaturated solutions exhibit a metastable region, where despite the instability of the system, there is no separation of a solid phase. The determination of this region is, in general, the first phase in the design of a batch cooling crystallization process. The metastable zone width (MZW) is a property that depends on several characteristics of the system (cooling rate, solute concentration, stirring rate, thermal history of the solution, presence of impurities, etc.) (Liotta & Sabesan, 2004).

The metastable zone width is defined as the difference between saturation temperature and the temperature which is detected in the formation of the first crystals. This temperature difference is known as the maximum undercooling, ΔT_{max}. Figure 1 illustrates schematically the solubility curve and metastable zone boundary for a hypothetical case of a system cooled from an under saturated condition until the condition of supersaturation.

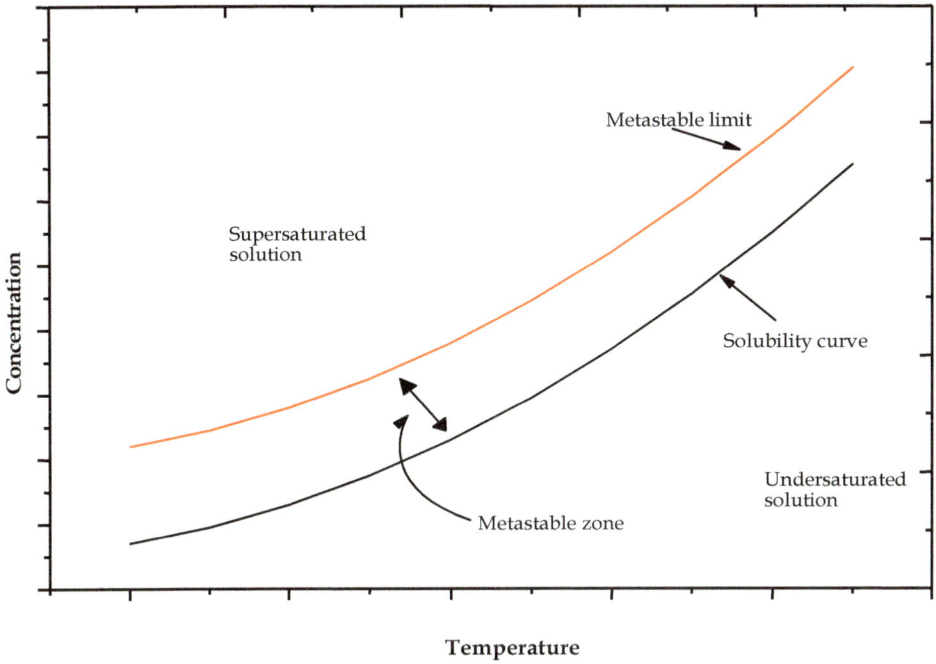

Fig. 1. Solubility curve and metastable zone.

In batch cooling crystallizations for example, maintaining the solution concentration profile within the metastable zone and close to the solubility curve promotes crystal growth and helps to avoid secondary nucleation (Beckmann, 2000). In the case of industrial crystallizers as a general rule, the level of supersaturation is maintained at about half the metastable zone (Marciniak, 2002).

1.3 Nucleation and crystal growth

The supersaturation is the main requirement for crystallization to occur and its creation does not imply the immediate separation of the phases. In a supersaturated solution, part of the dissolved solute tends to reorganize again to form the solid phase. However, the formation of the solid phase (positive energy) implies the generation of an interface (energetically unfavourable). Therefore, for the formation of nuclei to occur within the solution, it is necessary that this barrier is overcome (Mullin, 2001; Ulrich & Strega, 2002; Giulietti, 2001).

The nucleation rate depends on the supersaturation. When the supersaturation is extremely high, the nucleus formation is a random process and difficult to reproduce. For this reason, whenever possible, this condition is avoided in industrial applications (Mullin, 2001).

The crystal growth also depends on the supersaturation level. High growth rates lead to products with a degree of purity lower than those generated in conditions of lower supersaturation owing to the inclusion of liquids and other impurities. From the industrial point of view there is a compromise between the desired characteristics of the products and the economic efficiency of the process. This means that is not always possible to carry out crystallizations with a low rate of crystal growth because such a condition significantly increases the residence time of the product inside the crystallizer (Mullin, 2001; Ulrich & Strege, 2002).

1.4 Sensor technologies for monitoring crystallization processes

Continuous monitoring and control of crystallization processes in real-time require the use of sensors able to provide information regarding product quality and critical process variables.

The methods used to assess information of the products obtained from crystallization processes can be divided into four main groups (Yu et al., 2004):

- off line methods: the analysis is performed after sampling;
- on line methods: the sample stream is diverted from the crystallizer for analysis and subsequently returned to the system;
- in line methods: sensors are integrated into the crystallizer and provide real-time information about the process. The sensors are in direct contact with the material and can cause disturbances in the system;
- non-invasive methods: sensors are integrated into the crystallizer and provide real-time information of the process. In this case, the sensors do not come into direct contact with the material.

The real-time analyses described in this chapter can be included in the in line and the non-invasive methods.

1.4.1 Attenuated total reflectance Fourier transform infrared spectroscopy (ATR-FTIR)

ATR-FTIR is a consolidated technique in the monitoring of crystallization processes (Chen et al., 2009; Kadam et al., 2011; Pöllänen et al., 2006; Qu et al., 2009; Sheikhzadeh et al., 2008; Liotta & Sabesan, 2004). The method allows estimating the degree of supersaturation of the system by measuring the concentration of solute in the solution. The only prerequisite for application of the technique is the absorption of infrared radiation in the medium by the solute. The sensor used in this method has an element of internal reflection of high refractive index. The radiation passes through the element of reflection, being reflected when it encounters a material with lower refractive index. The amount of reflected radiation depends on the angle of incidence on the interface and when this angle is greater than a critical angle (depending on the ratio between the two refraction index) the refraction is attainable. However, radiation penetrates only a short distance into material of lower refractive index; this radiation is called the evanescent radiation (evanescent wave). Thus, if a sample is able to absorb infrared radiation, the beam is attenuated at frequencies absorbed by the sample (Man et al., 2010; Dunuwilaa et al., 1994).

1.4.2 Focused beam reflectance measurement (FBRM)

The FBRM measurement principle is based on backward light scattering. A laser beam is coupled to a probe via an optical fibre. This laser beam is deviated from the probe's central axis and focused into a disperse medium with an optical conduit. When this laser beam intersects with a particle, light scattering occurs. A certain fraction of the light is scattered back into the system. This back scattered light is coupled via a beam splitter to a second fibre and conduced to a detector. The rotational velocity of the laser is constant. The time span in which back scattering is detected is therefore directly proportional to the path length of the laser on the particle. It is assumed that the particle velocity is small compared to the laser rotational velocity. The length of the laser path on the particle is therefore proportional to the time span in which scattering is detected. This path length is called a chord length. Depending on the laser position, different chord lengths are measured even for a single particle. Those chord lengths are generally different from any characteristic particle length. In order to calculate the particle size distribution from the chord length distribution, a model is needed. This model has to cover all relevant aspects of the measurement technology (Kail et al., 2008; Barrett & Glennon, 2002). Figure 2 presents a schematic view of the Mettler Toledo FBRM probe.

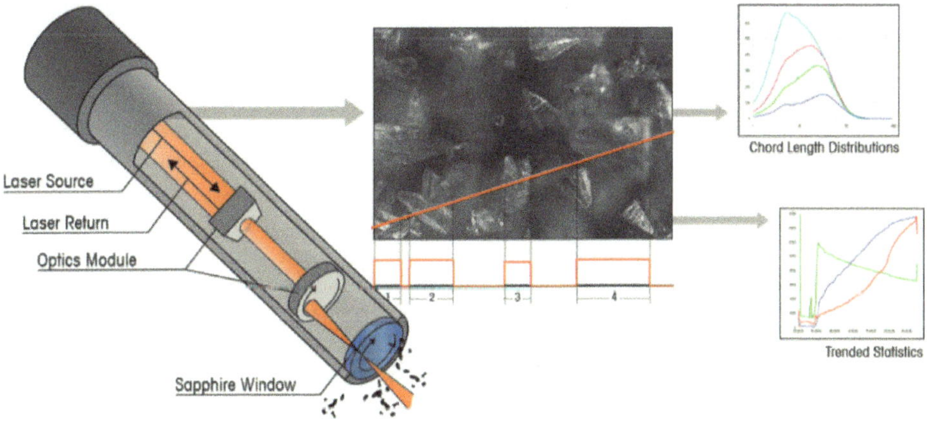

Fig. 2. Measurement principles of the Lasentec FBRM probe.

Just as with the ATR-FTIR technique, the FBRM is a well-established method of monitoring crystallization processes (Sun et al., 2010; Barrett & Glennon, 2002; Czapla et al., 2010; Sheikhzadeh et al., 2008; Nguyen & Kim, 2008).

1.4.3 Calorimetry

The principle of reaction calorimetry is based on the heat flow in processes involving changes in chemical or physical properties. The rate of release or absorption of heat is a parameter dependent on the kinetics and thermodynamics of the process under study (Gesthuisen et al., 2005; Mantel & Meyer, 2008). The combination of heat flow with the mass balance allows us to estimate the conversion of the instantaneous and cumulative process under study. Reaction calorimetry is non-invasive, fast, robust, relatively simple and its principle is based on the measurement of temperature differences (Gesthuisen et al., 2005; Mantel & Meyer, 2008). From the industrial point of view, calorimetry is a technique of great importance because it allows safely scaling up a process from pilot to industrial scale (Gesthuisen et al., 2005). The method finds application in the study of polymerization reactions (Benamor et al., 2002; Elizalde et al., 2005), biotechnological processes (Marison et al., 1985), study of supercritical fluids (Lavanchy et al., 2004; Mantel & Meyer, 2008), optimization of chemical reactions (Barton et al., 2003) and determination of kinetic parameters of chemical reactions (Silva et al., 2003; Seiceira et al., 2005) among other applications. There are few articles where heat flow calorimetry is applied to the study of crystallization processes (Févotte & Klein, 1995)

1.4.4 Image analysis

The use of Complementary Metal Oxide Semiconductor (CMOS) and Charge Coupled Device (CCD) cameras has been widely introduced in analytical chemistry for different reasons such as fast image capturing, stable background and good linearity (Jolling et al., 2007). These sensors are capable of converting the intensity of light that focuses on it in digital storable values as bits. The analytical response that generates an image representing the patterns of the colours Red (R), Green (G) and Blue (B). These patterns are known as RGB 8 bits for each channel, totalling 256 levels. The combination of the three matrices (R, G and B) allows the acquisition of 16 million colours (Gaiao et al., 2006; Safavi et al., 2007). Different methodologies employing this image analysis has been described in literature. A digital camera was used as a sensor for simultaneous determination of Al(III) and Fe(III) in alloys using the chrome azurol S(CAS) as chromogenic reagent (Maleki et al., 2004). An instrumental detection technique for titration based on digital images was proposed (Gaiao et al., 2006). A similar method for the measurement of lithium, calcium and sodium through the radiation emitted by the sample into an air-butane flame was developed (Lyra et al. (2009). Image analysis was also used for a real-time assessment of the coffee roasting process (Hernández et al., 2008). An approach employing a CCD camera as a sensor for recognizing volatile alcohols was described (Shirshov et al., 2007). A method based on external bulk video imaging was proposed for metastable zone identification in food and pharmaceutical crystallization processes, and showed good performance when compared to FBRM and ultra-violet visible spectroscopy (Simon et al., 2009). Additionally, CCD cameras have been used as detectors in clinical analysis and showed high detection sensitivity (Liang et al., 2004; Alexandre et al., 2001). The rapid improvements in digital camera technology provide the opportunity for the development of new methodologies employing digital cameras as an analytical sensor with high sensitivity, robustness, speed and low cost for implementation that reduces the analysis time.

2. Experimental part

In this section batch cooling crystallization of adipic acid will be used as a model to demonstrate the use of in line and non-invasive techniques for monitoring crystallization. For this purpose, four analytical tools with different physical principles were used. The procedures and techniques used in the experiments are described in the following subsections.

2.1 Materials and methods

2.2.1 Solubility

The solubility of adipic acid (99.8%) in ethanol (analytical grade) was determined for twelve different temperatures ranging from 16.0 to 64.0°C. The experiments were performed in a 1.8-L Hastelloy jacketed reactor vessel connected to an RC1e reaction calorimeter. The solutions were prepared by successive additions of adipic acid to a solution containing ethanol at a stirring rate of 300 rpm. For measurements at 44.0°C, the mass of ethanol used was 524.0g. The used mass for other temperatures are presented in Table 1. The solubilization temperature was maintained constant during the whole process.

Temperature (K)	Temperature (°C)	Ethanol mass (g)
289.15	16.0	590.3
293.95	20.8	582.7
298.15	25.0	574.7
303.95	30.8	564.3
306.15	33.0	556.3
308.25	35.1	550.8
312.15	39.0	537.0
317.15	44.0	524.0
323.15	50.0	492.4
330,15	57.0	459.5
334.15	61.0	428.1
337.15	64.0	405.3

Table 1. Experimental conditions for testing the solubility of the adipic acid in ethanol.

The ATR-FTIR measurements were performed by using a Mettler-Toledo ReactIR IC10 spectrometer. The base unit contains the Fourier transform mid-infrared source and the mercuric cadmium telluride (MCT) detector that should be cooled with liquid nitrogen. The sample interface module (SIM) is the interface on the instrument base unit where the K6 (16 mm diameter) conduit connects. It contains the optics that transfer the infrared source light from the base unit to the probe in contact with the chemical materials contained in the vessel and then back to the detector. Measurements are taken optically using a diamond sensing element that uses a multiple reflection ATR crystal and a gold seal between the metal housing and the sensor. Figure 3 illustrates schematically the K6 Mettler Toledo probe tip.

The focused beam reflectance measurements were obtained by using a Mettler-Toledo Lasentec D600L probe consisting of a Hastelloy C-22 tube with the sensor at one end with an optical diameter of 19 mm and a length of ~406 mm. The FBRM laser provides a continuous beam of monochromatic light with a wavelength of 780 nm. The beam is located approximately 3 mm to the focal point.

The ATR-FTIR and FBRM probes were kept immersed in the adipic acid solution 5 cm above the propeller stirrer. Infrared spectra obtained from 4000 to 650 cm^{-1} at 4 wave numbers resolution were collected at 15 s intervals with each spectrum averaged over 30 scans.

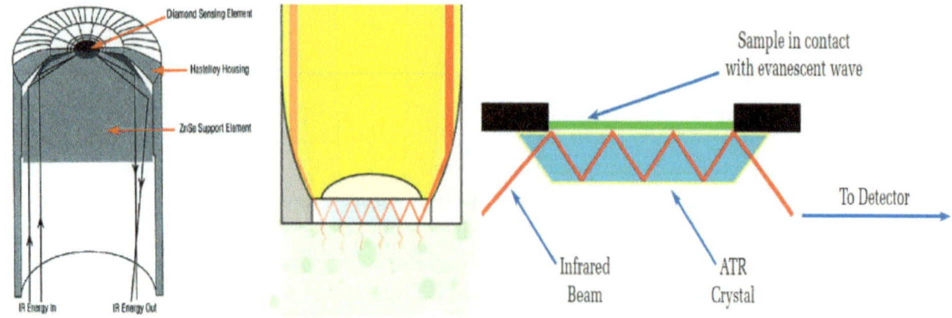

Fig. 3. Schematic draw of the Mettler Toledo K6 probe tip.

2.2.2 Batch cooling crystallization

The batch cooling crystallization experiments were carried out in a double walled glass reactor with a capacity of 2.0L. The solutions were stirred by a propeller stirrer at 300 rpm. The experimental setup is schematically presented in Figure 4.

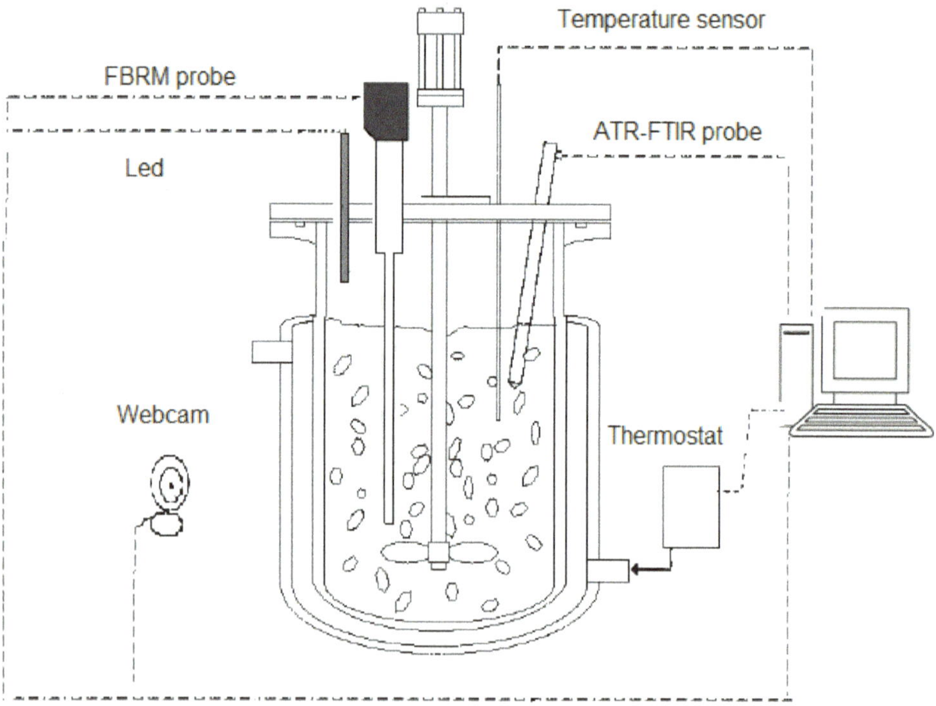

Fig. 4. Experimental setup.

The mass of ethanol used in each run was 600 g and the amount of adipic acid was varied from 91.2 to 279.9 g. The temperature was controlled by the precise RC1e thermostat. The temperature of solution was kept 5°C higher than that saturation condition in order to assure that no crystal was presented in the solution prior to starting the cooling process. After 0.5 h, the solution was cooled at constant rate of 0.2 and 1°C/min, until the onset of crystallization. The experimental conditions of experiments are presented in Table 2.

Concentration (mass %)	Temperature at the start of cooling (°C)	Final temperature (°C)	Mass of adipic acid (g)
13.2	30.0	-2.00	91.20
15.1	35.0	5.00	106.8
17.3	40.0	10.0	125.2
19.7	45.0	15.0	146.8
22.3	50.0	15.0	172.2
25.2	55.0	20.0	202.1
28.4	60.0	25.0	237.6
31.8	65.0	25.0	279.7

Table 2. Experimental conditions for crystallization experiments of adipic acid in ethanol.

In situ methods, FBRM and ATR-FTIR, were used to monitor adipic acid crystallization. Non-invasive measurements were carried out by monitoring the heat released and the image patterns of colours red, green and blue during the crystallization. Images of experiments were acquired by using a low cost PC webcam (Microsoft Life Cam VX-2000). The webcam was placed externally and, in order to avoid interferences by external light and to maintain the CCD noise under controlled conditions, the reactor was enclosed with a black box. The images were captured during whole experiments. A light-emitting diode (LED) was used as a source of illumination. The images acquired were analysed according to their patterns of colour red, green and blue, and for this purpose software was developed which allows the evaluation of alteration in these patterns of colours. The software enables the analysis of the whole image or the user can define a specific region previously selected from an image. The software automatically saves the coordinates of the delimited region for all digital images and calculates the R, G and B values averaging all pixels.

The heat released during the crystallization was monitored using the RC1e. To be able to calculate the heat flow during the crystallization, the total heat transfer coefficient (U) and the heat capacity (Cp) of the solution of adipic acid were measured.

3. Results and discussion

This section presents the results obtained in the study of solubilization of adipic acid in ethanol as well as its crystallization, evaluating the possibility of using calorimetry, infrared (ATR-FTIR) and FBRM as techniques for determining the width of the metastable zone.

3.1 Determination of the solubility of adipic acid

The solubility curve of adipic acid in ethanol, presented in Figure 5, was prepared according to procedure in literature that used ATR-FTIR and heat flow to calculate the solubility curve of adipic acid in acetone (Silva & Silva, 2011).

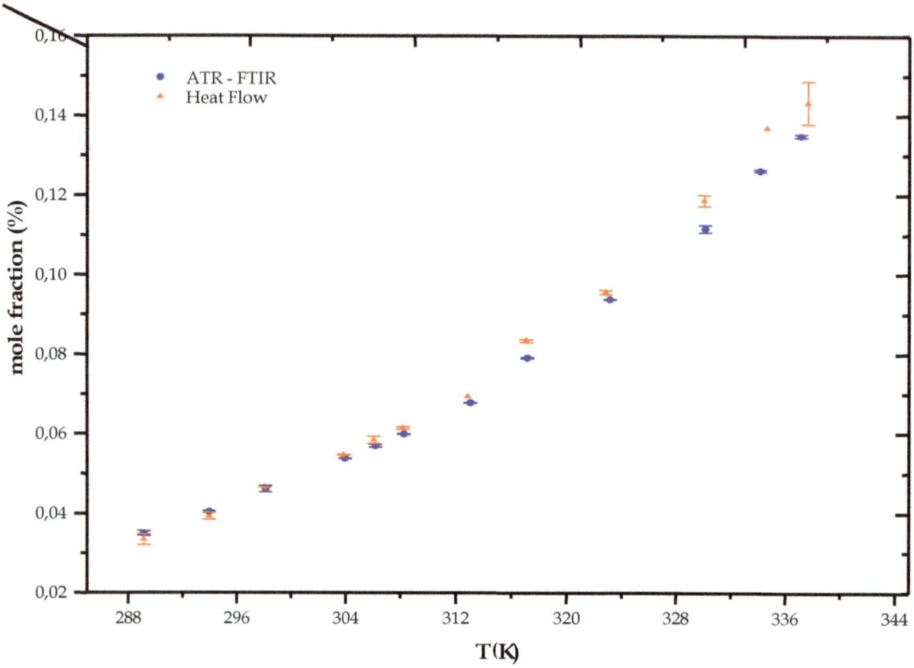

Fig. 5. Solubility curves of adipic acid in ethanol by using ATR-FTIR and heat flow calorimetry.

3.2 Determination of the onset temperature for the crystallization of adipic acid in ethanol by heat flow calorimetry

Since during crystallization there is a decrease in entropy, the second term of the Gibbs free energy (equation 3) becomes positive and therefore, for ΔG to be negative, it is necessary that the change in enthalpy is negative, which shows that the crystallization processes are exothermic.

$$\Delta G = \Delta H - T\Delta S \tag{3}$$

The heat flow for cooling at $1^{0}C/min$ is presented in Figure 6.

The analysis of Figure 6 shows that when the solution temperature was approximately 37.2°C, the jacket temperature was reduced by about 10°C, so that the cooling rate of 1°C/min could be maintained, meaning that an exothermic process had started. The heat flow measurement had a fast increase at the same moment indicating that the crystallization of the adipic acid (the exothermic process) had begun. In the initial moments of the crystallization there was an intense release of heat and soon after that the release decreased and became nearly constant. The region where heat release is constant may be associated with the growth of crystals (Riesen, 2005). The heat release continued until the moment when the reactor temperature was maintained at 15.0 °C. With the constant temperature the heat flow remained stable, indicating that the process that was responsible for the release of heat had ceased. As the only process that was taking place inside the reactor was the crystallization of adipic acid, this release is solely related to this process.

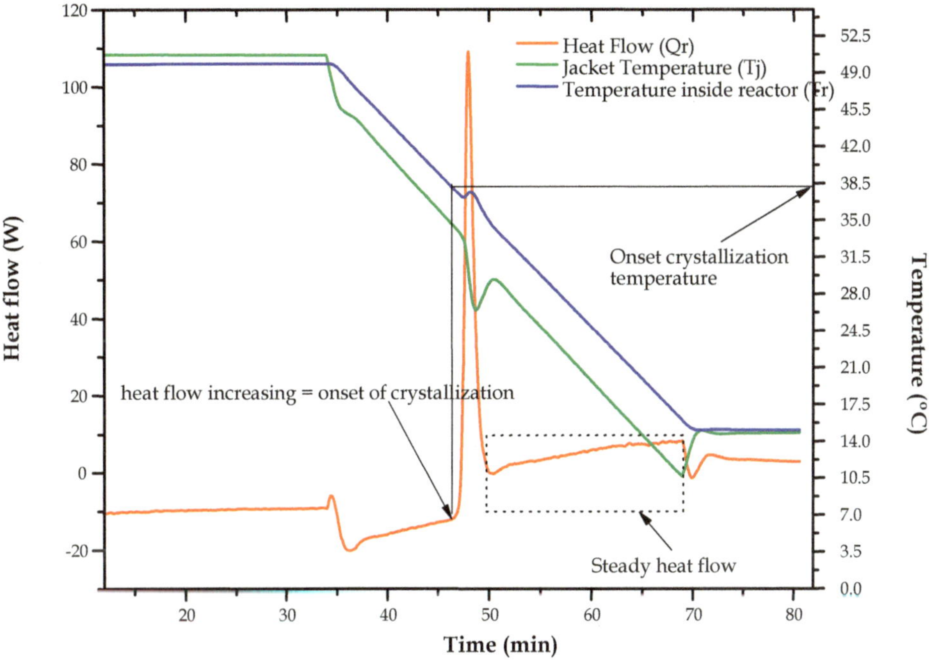

Fig. 6. Heat release curve (Qr), the temperature inside the reactor (Tr) and temperature of the jacket (Tj), obtained from crystallization experiments using an ethanol solution with initial concentration of 22%, cooled at a rate of 1°C/min and with a stirring rate of 300 rpm.

The maximum supersaturation achieved in the solution is a function of the maximum cooling achieved by the system. Thus, it is expected that the higher the maximum cooling achieved by a system, the greater the value of supersaturation in the medium. In general the smaller nucleation rates are obtained at lower cooling rates (Mullin, 2001). In order to check this information, the crystallization of adipic acid was performed at a lower cooling rate as presented in Figure 7.

The rate at which heat is released when employing a cooling rate of 1°C/min is higher than when using a cooling rate of 0.2°C/min.

The comparison between the results presented in Figures 6 and 7 confirms that the maximum heat released is a function of cooling rate.

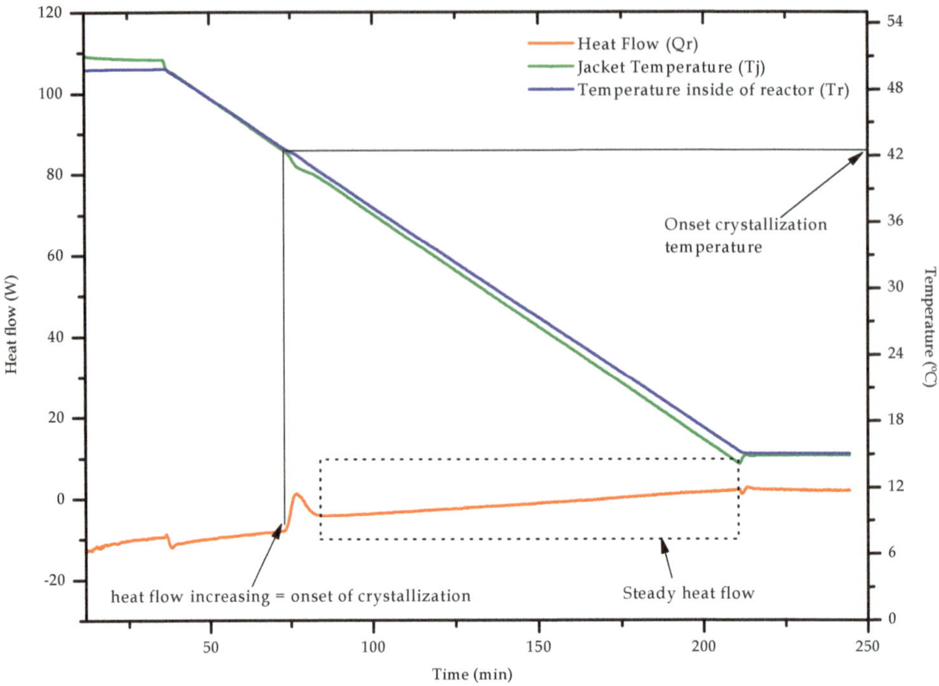

Fig. 7. Heat release curve (Qr), the temperature inside the reactor (Tr) and temperature of the jacket (Tj), obtained from crystallization experiments using an ethanol solution with initial concentration of 22%, cooled at a rate of 0.2°C/min and with a stirring rate of 300 rpm.

3.3 Determination of the onset temperature for the crystallization of adipic acid in ethanol by RGB image analysis

The image video analysis was performed by using software named MasterView RGB that captures images in real-time from a PC webcam by evaluating changes in the component's RGB colour pixel by pixel. The software automatically saves the coordinates of the delimited region for all digital images and calculates the RGB values averaging all pixels (de Sena et al., 2011).

Figure 8 shows the graphic interface of MasterView RGB for an adipic acid solution and Figure 9 presents the crystallization onset by measuring the RGB variation.

The method of image analysis is based on comparison of images. This procedure is comparable to human visual inspection and the method of image analysis has greater sensitivity and is not subject to misinterpretation.

The variation of the red channel during the crystallization of adipic acid is presented in Figure 10.

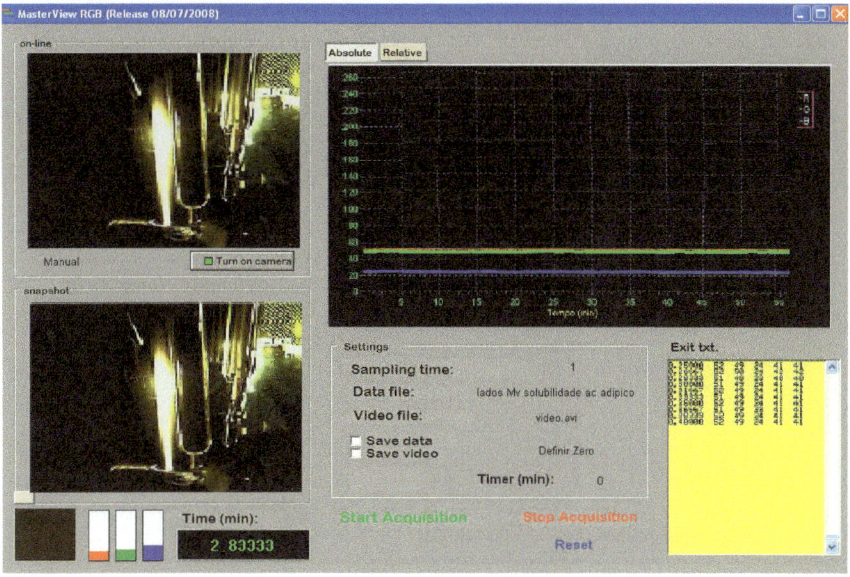

Fig. 8. Monitoring of the onset of the crystallization of adipic acid using the image analysis method (saturated solution).

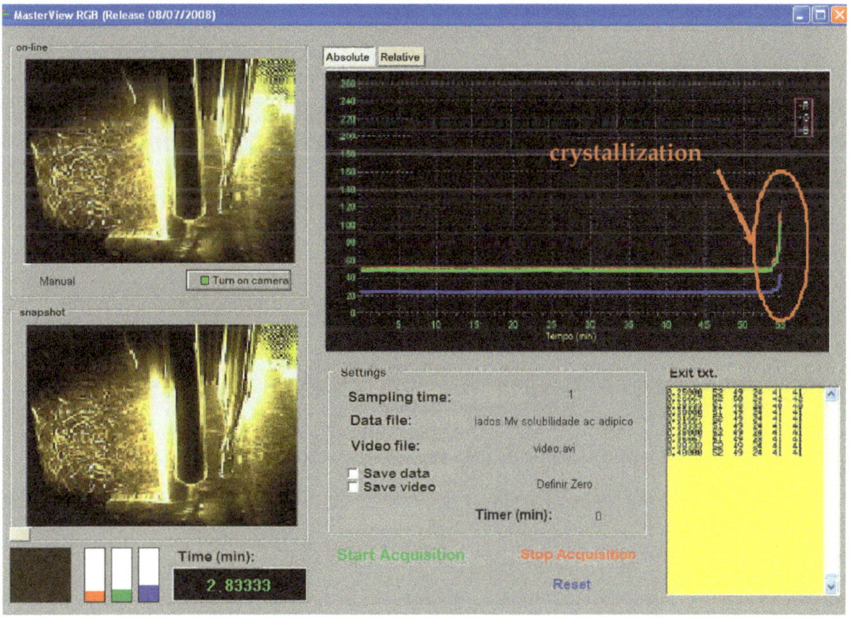

Fig. 9. Monitoring of the onset of the crystallization of adipic acid using the image analysis method (increasing the mass of adipic acid crystals due to cooling).

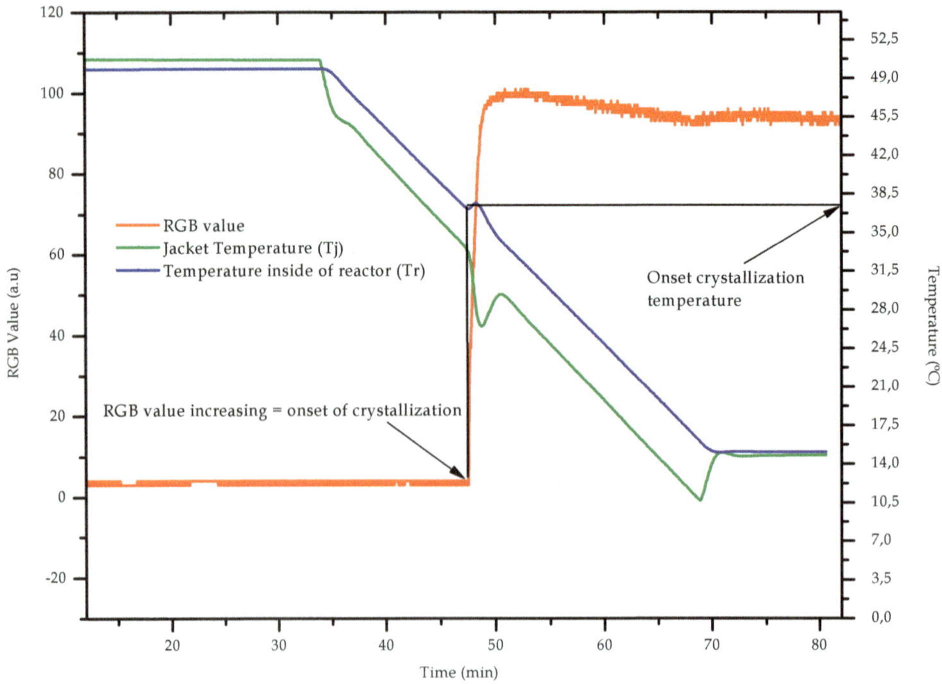

Fig. 10. Image analysis red channel, the temperature inside the reactor (Tr) and temperature of the jacket (Tj), obtained from crystallization experiments using ethanol solution with initial concentration of 22%, cooled at a rate of 1°C/min and with a stirring rate of 300 rpm.

As can be seen in Figure 8, before the crystallization of adipic acid (red signal channel close to zero), the red value increases immediately when the adipic acid crystallization starts. This result is in total accordance with the results obtained by heat flow calorimetry. It was established that the onset temperature of crystallization, determined by the variation in RGB signal, would be that corresponding to the time when the RGB signal began to increase.

3.4 Determination of the onset temperature of crystallization using infrared (ATR-FTIR)

The infrared absorption spectrum of adipic acid is presented in Figure 11.

The carbonyl stretching absorption region of carboxylic acids used in the ATR-FTIR analysis is indicated in Figure 11.

The analysis of crystallization of a solution of adipic acid in ethanol was accompanied by infrared (ATR-FTIR), through the decrease of the signal of the peak area for the absorption of carbonyl (C=O) in the medium. Figure 12 presents the carbonyl absorption variation during a crystallization process that started at 50 °C.

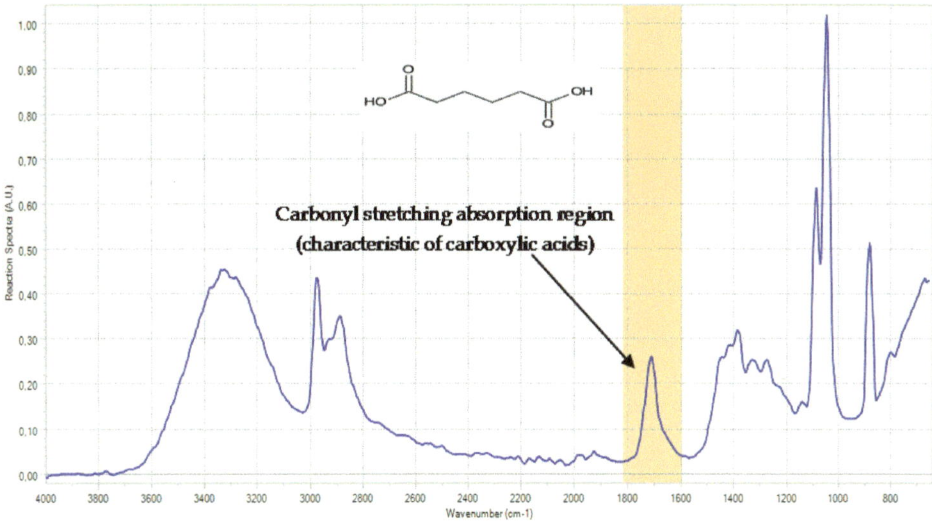

Fig. 11. ATR/FTIR spectra of adipic acid in the infrared region.

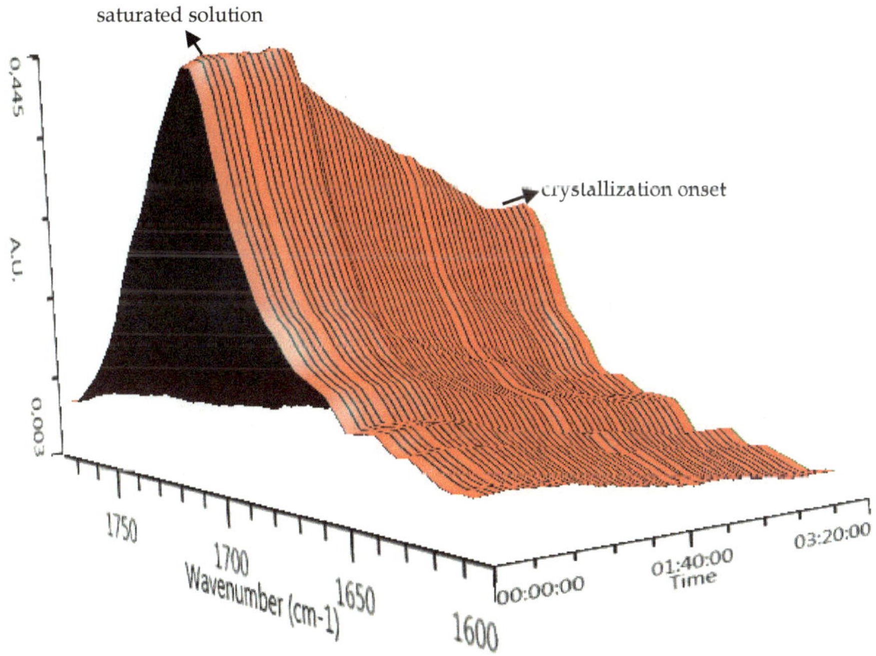

Fig. 12. Adipic acid carbonyl absorption variation measured in the interval of 1750-1600 cm^{-1}.

The carbonyl absorption trend related to the data presented in Figure 12 is presented in Figure 13.

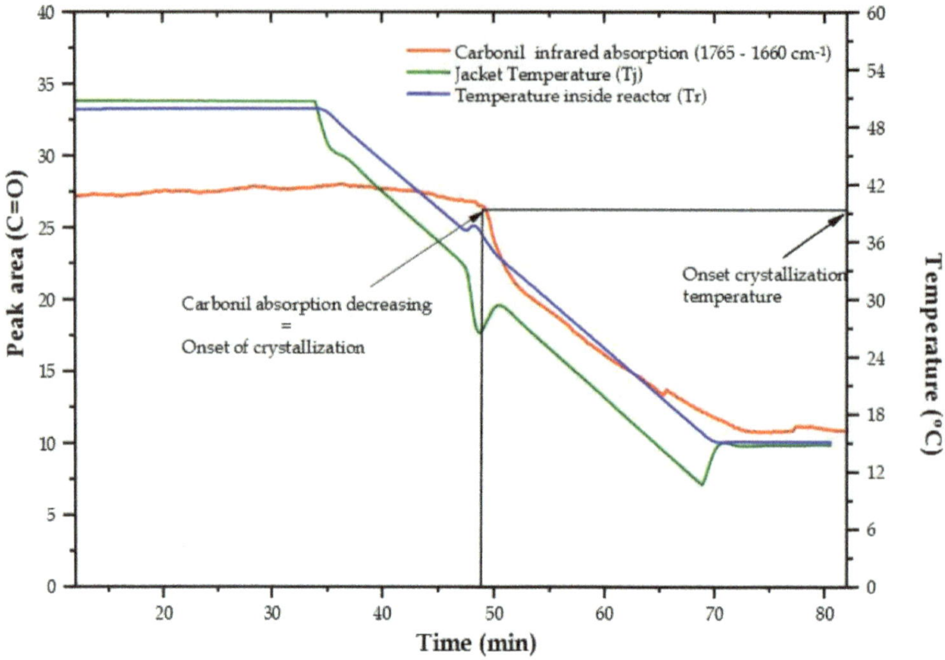

Fig. 13. Absorption curve of the carbonyl C=O (1765-1660 cm 1) in ethanol, the temperature inside the reactor (Tr) and jacket temperature (Tj), obtained from crystallization experiments using an ethanol solution with initial concentration of 22% , cooled at a rate of 1°C/min and stirring rate of 300 rpm.

The carbonyl absorption area presented a decrease at the precise moment the crystallization started. This observation is in total agreement with the fact that the ATR probe detects the concentration of adipic acid soluble in ethanol. A comparison between the infrared absorption for the cooling rates of 1 and 0.2ºC/min is presented in Figure 14.

The FTIR analysis may be considered as an indirect measurement of the kinetics of crystallization, since the ATR probe cannot detect the crystallized adipic acid. As expected, the diminution of the carbonyl absorption was more pronounced when the higher cooling rate was used.

3.5 Determination of onset temperature of the crystallization using Focused Beam Reflectance Measurement

The variation in the total chord counts obtained from the FBRM analysis can detect the crystallization onset. This signal remained in the form of a stable baseline, with a few counts per second before the formation of crystals. By the time the crystallization of adipic acid had started the chord counts increased dramatically. This behaviour can be seen in Figure 15.

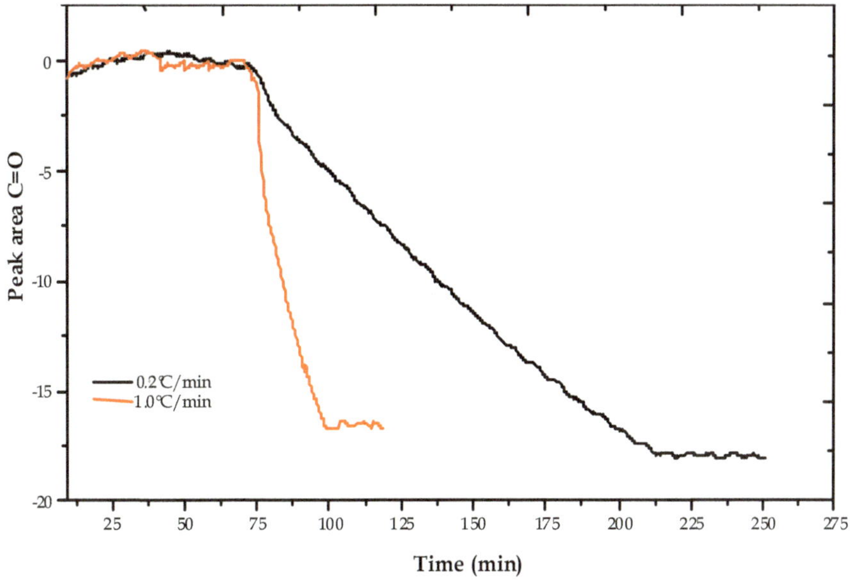

Fig. 14. Infrared absorption curve of the carbonyl C=O (1765-1660 cm^{-1}) of adipic acid during crystallizations performed at cooling rates of 1.0 and 0.2^{0}C/min.

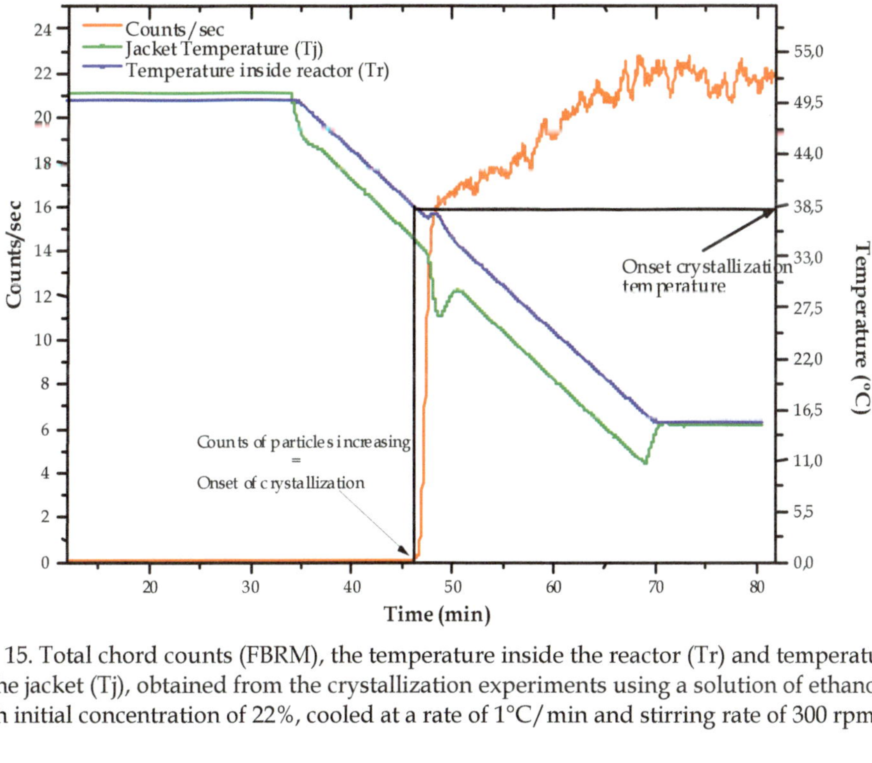

Fig. 15. Total chord counts (FBRM), the temperature inside the reactor (Tr) and temperature of the jacket (Tj), obtained from the crystallization experiments using a solution of ethanol with initial concentration of 22%, cooled at a rate of 1°C/min and stirring rate of 300 rpm.

The comparison between the chord length distribution in the crystallization onset with the values obtained in the end cooling, presented in Figure 16, can furnish information about crystal growth and this variation can also be detected by analysing the particle video microscopy images correspondent to these different moments that are shown in Figure 17.

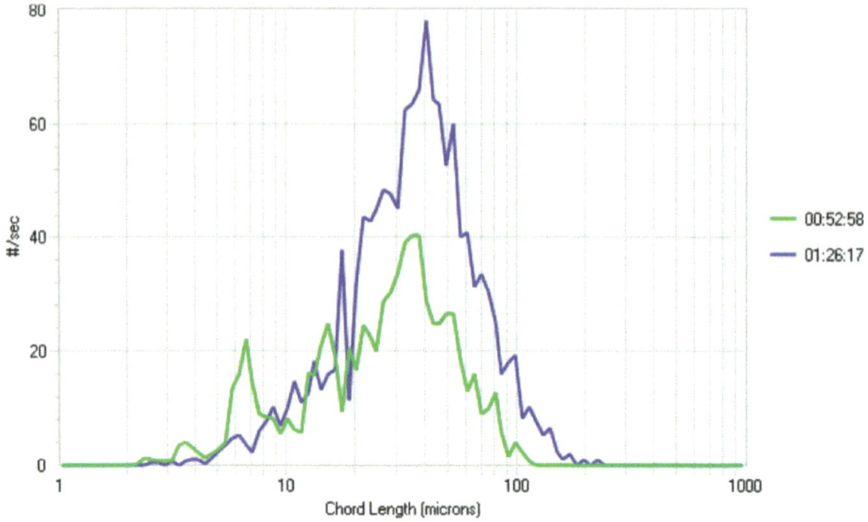

Fig. 16. Chord length distribution obtained in the crystallization onset, green curve and at the end of the experiment.

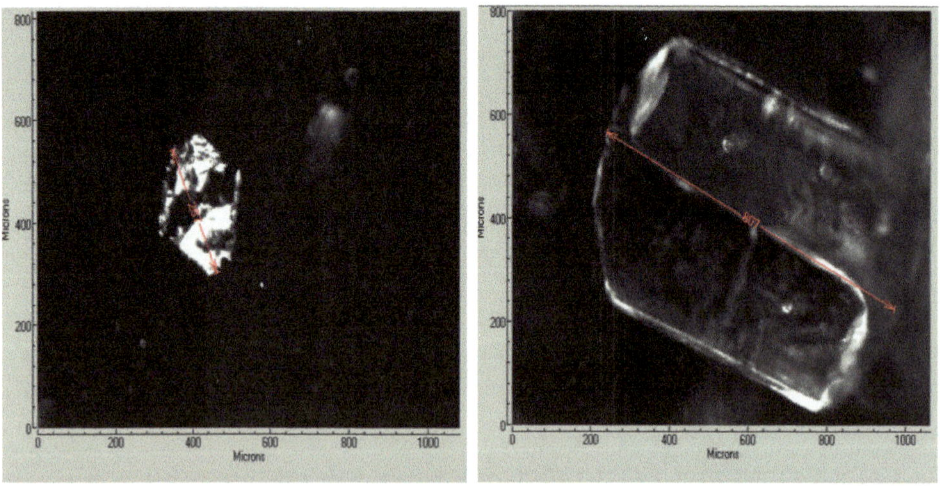

Fig. 17. Real-time particle video microscopy obtained in the crystallization onset (a) and in the end of the experiment (b).

The distributions shown in Figure 16 were displayed on a logarithmic scale on the horizontal axis (length of chords) for easy viewing. The shift to the right side in the chord length distribution indicates crystal growth. The crystal sizes presented in Figure 17 confirm the crystal growth.

The average size of the chord length of adipic acid particles produced in the set of experiments is presented in Table 3.

concentration (% mass)	cooling rate (°C/min)	average size (µm)
13.2	0.2	75.20
13.2	1.0	65.24
15.1	0.2	76.91
15.1	1.0	73.11
17.3	0.2	84.36
17.3	1.0	73.59
19.7	0.2	73.48
19.7	1.0	70.76
22.3	0.2	119.4
22.3	1.0	77.47
25.2	0.2	106.7
25.2	1.0	88.88
28.4	0.2	127.9
28.4	1.0	80.31
31.8	0.2	132.2
31.8	1.0	88.78

Table 3. Average size of the chord length of adipic acid crystals in the different experimental conditions.

The results presented in Table 3 show that the average size of adipic acid crystals increases as the cooling rate decreases. This fact is directly correlated with the metastable zone width, which is directly proportional to the rate of nucleation (Nyvlt et al., 2001). Thus, the largest crystals were observed when the cooling rate was 0.2°C/min, the rate at which the narrowest metastable zone was maintained.

3.6 Determination of metastable zone width of adipic acid in ethanol

The metastable zone width of a system is dependent on the methodology employed for its determination. Depending on the sensitivity of the technique used to detect the onset of crystallization, significant deviations can occur between two different methodologies (Marciniak, 2002). For a satisfactory evaluation of the metastable zone limit, it is necessary that the reactor be cooled at a constant rate. This constant cooling can be seen through the linear behaviour of both the crystallizer temperature and the jacket of the reactor.

The onset crystallization temperatures determined by the different real-time methods are presented in Table 4 and the agreement between these data can be seen in Figure 18.

Adipic acid concentration (%)	T calorimetry (°C)	T infrared (°C)	T FBRM (°C)
13.2	16.45	16.13	16.23
15.1	18.02	17.78	17.90
17.3	24.44	24.34	24.20
19.7	30.91	30.70	30.79
22.3	37.34	37.16	37.28
25.2	42.77	42.70	42.72
28.4	50.30	50.24	50.27
31.8	57.22	57.11	57.19

Table 4. Comparison of the calorimetric method, infrared and total chord counts for the determination of the onset temperature of crystallization of adipic acid in ethanol using a cooling rate of 1°C/min.

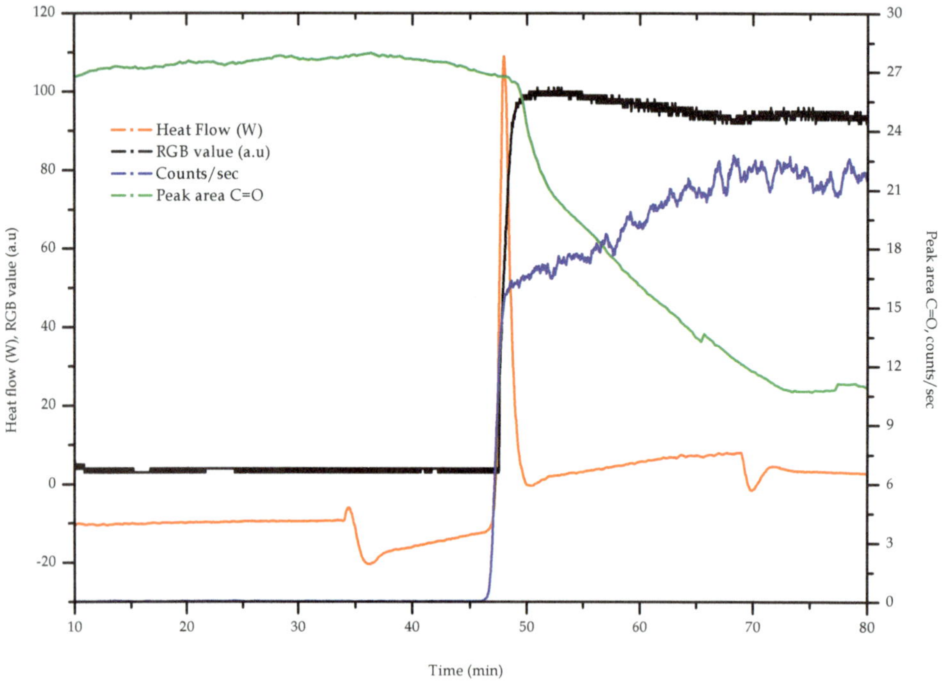

Fig. 18. Curves of the peak area of absorption of carbonyl (C=O), total count (FBRM), image analysis by RGB (**RGB**), calorimetry (heat flow) obtained from the cooling of a solution of 22% adipic acid in ethanol, cooled to 1.0 ° C / min with a stirring rate of 300 rpm.

The results presented in Figure 18, obtained by the different real-time analyses, were similar. In this case, the heat flow measurement presented higher sensibility than the other real-time analyses, but all techniques can be considered equivalent.

3.7 Study of the effect of the cooling rate on the metastable zone width of adipic acid in ethanol

Considering that the calorimetric method presented the highest sensibility, this was the chosen method to present the metastable zone limits for the crystallization of adipic acid by using cooling rates of 1.0 and 0.2ºC/min as presented in Figure 19.

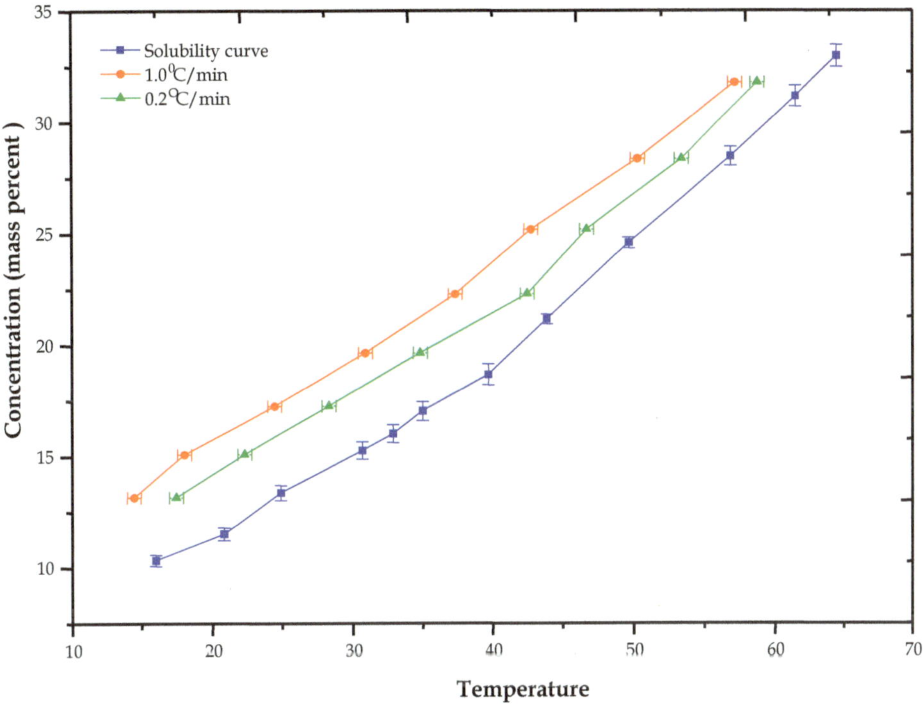

Fig. 19. Solubility curve of adipic acid in ethanol and the metastable zone limits determined at 1.0 and 0.2ºC/min.

The metastable zone width, expressed in terms of maximum cooling achieved, Δt, which is the difference between the saturation temperature, T saturation, and the onset temperature of crystallization, is reduced when the cooling rate is diminished. This statement can be confirmed by the data presented in Figure 19.

4. Conclusion

Considering that the experimental conditions employed during crystallization may affect physical properties of the product, such as its chemical purity, crystal size distribution and polymorphism and that these properties are completely related to the metastable zone width, accurate measurement techniques capable of providing data to construct solubility curves, as well as the onset of crystallization, are of vital importance. Off line analytical methods used to determine solubility curves and to detect the onset of crystallization present much higher errors than those that use real-time analysis for the same purpose. The four real-time analyses used in this work, even though working in different ways, presented equivalent performance, indicating that all of them can be used as process analytical tools to evaluate crystallization processes.

5. References

Alexandre, I.; Hamels, S.; Dufour, S.; Collet, J.; Zammatteo, N.; De Longueville, F.; Gala, JL. & Remacle, J. (2001). Colorimetric Silver Detection of DNA Microarrays. *Analytical Biochemistry*, Vol.295, No.1, (August 2001), pp. 1-8, ISSN 0003-2697

Barton, B.; Gouws, S.; Shaefer, MC. & Zeelie, B. (2003). Evaluation and optimization of the reagent addition sequence during the synthesis of Atrazine (6-chloro-N^2-ethyl-N^4-isopropyl-1,3,5-triazine-2,4-diamine) using reaction calorimetry. *Organic Process Research and Development*, Vol.7, No.6, pp. 1071-1076, ISSN 1083-6160.

Barrett, P. & Glennon, B. (2002). Characterizing the Metastable Zone Width and Solubility Curve Using Lasentec FBRM and PVM Original. *Chemical Engineering Research and Design*, Vol.80, No.7, (October 2002), pp. 799-805 ISSN 0263-8762

Beckmann, W. (2000). Seeding the desired polymorph: Background, possibilities, limitations, and case studies. *Organic Process Research and Development*, Vol.4, No.5, (September 2000), pp. 372-383, ISSN 1083-6160

BenAmor, S.; Colombié, D. & McKenna, T (2002) On line calorimetry: Applications to the monitoring of emulsion polymerization without samples or models of the heat-transfer coefficient. *Industrial and Engineering Chemistry Research*, (May 2002), Vol.41, No.17, pp. 4233-4241, ISSN 0888-5885.

Chen, ZP.; Morris, J.; Borissova, A.; Khan, S.; Mahmud, T.; Penchev, R. & Roberts, KJ. (2009). On-line monitoring of batch cooling crystallization of organic compounds using ATR-FTIR spectroscopy coupled with an advanced calibration method. *Chemometrics and Intelligent Laboratory Systems*, Vol.96, No.1, (March 2009), pp. 49-58, ISSN 0169-7439.

Czapla, F.; Kail, N.; Öncül, A.; Lorenz, H.; Briesen, H. & Seidel-Morgenstern, A. (2010). Application of a recent FBRM-probe model to quantify preferential crystallization of dl-threonine. *Chemical Engineering Research and Design*, Vol.88, No.11, (November 2010), pp. 1494-1504, ISSN 0263-8762

Derdour, L.; Pack, S. K.; Skliar, D.; Lai, C. J. & Kiang, S. (2011). Crystallization from solutions containing multiple conformers: A new modeling approach for solubility and supersaturation. *Chemical Engineering Science*, Vol.66, No.1, (January 2011) pp. 88-102, ISSN 0009-2509.

Dunuwila, D. D.; Carroll II, L. B.; Berglund, K.A. (1994). An investigation of the applicability of attenuated total reflection infrared spectroscopy for measurement of solubility and supersaturation of aqueous citric acid solutions. *Journal of Crystal Growth*, Vol.137 Nos.3-4 (April 1994) pp. 561-568 ISSN 0022-0248

Elizalde, O.; Azpeitia, M.; Reis, M. M.; Asua, J. M.; & Leiza, J. R. (2005). Monitoring emulsion polymerization reactors: Calorimetry versus Raman spectroscopy. *Industrial and Engineering Chemistry Research*, Vol.44 No.18 (July 2005) pp. 7200-7207. ISSN 0888-5885

Fakatselis, T.E. (2002). Residence time optimization in continuous crystallizers. *Crystal Growth and Design*, Vol.2, No.5, (September 2002), pp. 375-379, ISSN 1528-7483

Feng, L. & Berglund, K. A. (2002). ATR-FTIR for determining optimal cooling curves for batch crystallization of succinic acid. *Crystal Growth and Design*, Vol.2, No.5, (September 2002), pp. 449-452, ISSN 1528-7483

Févotte, G. & Klein, J. P. (1995). Crystallization calorimetry for the assessment of batch seeding and cooling police. *The Chemical Engineering Journal*, Vol.59, No.2, (October 1995), pp.143-152, ISSN 1385-8947

Fujiwara, M.; Nagy, Z. K.; Chew, J. W. & Braatz, R. D. (2005). First-principles and direct design approaches for the control of pharmaceutical crystallization. *Journal of Process Control*, Vol.15, No.5, (August 2005), pp. 493-504, ISSN 0959-1524

Gahn, C. & Mersmann, A. (1995). The brittleness of substances crystallized in industrial-processes. *Powder Technology*, Vol.85, No.1, (October 1995), pp. 71-81, ISSN 0032-5910

Gaiao, EN.; Martins, VL.; Lyra, WS.; Almeida, LFA.; Silva, ECS. & Araújo, MCU. (2006). Digital image-based titrations. *Analytical Chimica Acta*. Vol.570, No.2, (Juno 2006), pp. 283-290, ISSN 0003-2670.

Gesthuisen, R.; Krämer, S.; Niggemann, G.; Leiza, J. R.; & Asua, J. M. (2005). Determining the best reaction calorimetry technique: theoretical development. *Computers and Chemical Engineering*, Vol.29, No.2,(January 2005) pp. 349-365 ISSN 0098-1354.

Giulietti, M.; Seckler, M. M.; Derenzo, S.; Ré, M. I. & Cekinski, E. (2001). Industrial crystallization and precipitation from solutions: State of the technique. *Brazilian Journal of Chemical Engineering*, Vol.18, No.4, pp. 423-440, ISSN 0104-6632

Gunawan, R.; Ma, D.L.; Fujiwara, M. & Braatz, R.D. (2002). Identification of kinetics parameters in multidimensional crystallization processes. *International Journal of Modern Physics B*, Vol.16, Nos.1-2, (January 2002), pp. 367-374, ISSN 0217-9792

Hernández, JA.; Heyd, B. & Trystram, G. (2008). On-line assessment of brightness and surface kinetics during coffee roasting. *Journal of Food Engineering*. Vol.87, No.3, (August 2008), pp. 314-322, ISSN 0260-8774.

Jolling, K.; Vandeven, M.; Eynden, JVD.; Amellot, M. & Kerkhove, EV. (2007). A highly reliable and budget-friendly Peltier-cooled camera for biological fluorescence imaging microscopy. *Journal of Microscopy*. Vol.228, No.3, (December 2007), pp. 264-271, ISSN 0022-2720.

Joung, O. J.; Kim, Y. H. & Fukui, K. (2005). Determination of metastable zone width in cooling crystallization with a quartz crystal sensor. *Sensors and Actuators B: Chemical*, Vol.105, No. 2, (March 2005), pp. 464-472, ISSN 0924-4005.

Kadam, S. S.; Mesbah, A.; der Windt, E. V. & Kramer, H. J. M. (2011). Rapid online calibration for ATR-FTIR spectroscopy during batch crystallization of ammonium sulphate in a semi-industrial scale crystallizer. *Chemical Engineering Research and Design*, Vol.89, No.7, (July 2011), pp. 995-1005, ISSN0263-8762.

Kail, N.; Briesen, H.; & Marquardt, W. (2008). Analysis of FBRM measurement by means of a 3D optical model. *Powder Technology* Vol.185 No.3 (July 2008) pp. 211-222, ISSN 0032-5910.

Kougoulos, E.; Jones, A. G.; Jennings, K. H.; & Wood-Kaczmar, M. W. (2005). Use of focused beam reflectance measurement (FBRM) and process video imaging (PVI) in a modified mixed suspension mixed removal (MSMPR) cooling crystallizer. *Journal of Crystal* Growth, Vol.273, Nos.3-4, (January 2005), pp. 529-534, ISSN 0022-0248

Lavanchy, F.; Fortini, S. & Meyer, T. (2004). Reaction calorimetry as a new tool for supercritical fluids. *Organic Process Research and Development*, Vol.8, No.3, pp. 504-510, ISSN 1083-6160.

Lewiner, F.; Klein, J.P.; Puel, F. & Fevotte, G. (2001). On-line ATR FTIR measurement of supersaturation during solution crystallization processes. Calibration and applications on three solute/solvent systems. *Chemical Engineering Science*, Vol.56, No.6, (March 2001), pp. 2069-2084, ISSN 0009-2509

Liang, RQ.; Tan, CY. & Ruan, KC. (2004). Colorimetric detection of protein microarrays based on nanogold probe coupled with silver enhancement. *Journal of Immunological Methods*, (February 2004), Vol.285, No.2, pp. 154-163, ISSN 0022-1759.

Liotta, V. & Sabesan, V. (2004). Monitoring and feedback control of supersaturation using ATR-FTIR to produce an active pharmaceutical ingredient of a desired crystal size. *Organic Process Research and Development*, Vol.8, No. 3, (May 2004), pp. 488-494, ISSN 1083-6160

Lyra, WS.; dos Santos, VB.; Dionízio, AGG.; Martins, VL.; Almeida, LF.; Gaião, EN.; Diniz, PHGD.; Silva, EC. & Araújo, MCU. (2009) Digital image-based flame emission spectrometry. *Talanta*. Vol.77, No.5, (May 2011), pp. 1584-1589, ISSN 0039-9140.

Ma, D. L.; Chung, S.H.; Braatz, R. D. (1999). Worst-case performance analysis of optimal batch control trajectories. *Aiche Journal*, Vol.45, No.7, (July 1999), pp. 1469-1486, ISSN 0001-1541

Maleki, N.; Safavi, A. & Sedaghatpour, F. (2004). Single-step calibration, prediction and real samples data acquisition for artificial neural network using a CCD camera. *Talanta*. Vol.64, No.4, (November 2004), pp. 830-835, ISSN 0039-9140.

Marciniak, B. (2002). Density and ultrasonic velocity of undersaturated and supersaturated solutions of fluoranthene in trichloroethylene, and study of their metastable zone width. *Journal of Crystal Growth*, Vol.236, No.1-3, (March 2002), pp. 347-356, ISSN 0022-0248

Marisol, IW.; Bijou, B. & Von Stockar, U. (1985). Use of a novel heat-flux calorimeter for the measurement of heat generated during growth of *K. fragilis* on lactose. *Thermochimica Acta* Vol.85, No.93, (April 1985), pp. 497-500, ISSN 0040-6031.

Man, Y. B. C.; Syahariza, Z. A. & Rohman, A., Fourier Transform Infrared (FTIR) Spectroscopy: Development, Techniques and Application in the Analyses of Fats and Oils, editor Oliver J. Rees, Nova Science Publishers, Inc. (2010).

Mantelis, CA. & Meyer, T. (2008). Supercritical reaction calorimetry: Versatile tool for measuring heat transfer properties and monitoring chemical reactions in supercritical fluids. *Industrial and Engineering Chemistry Research*, (May 2008), Vol.47, No.10, pp. 3372-3379, ISSN 0888-5885.

Mersmann, A. (2001) *Crystallization Technology Handbook*. Marcel Dekker Inc., New York.

Mullin, J. W. (2001). *Crystallization*. (4thed), Butterwoth-Heinemann, ISBN 075064833.3, Oxford, UK.

Nguyen, TNP. & Kim, KJ. (2008). Kinetic study on hemipenta hydrate risedronate monosodium in batch crystallization by cooling mode Original. *International Journal of Pharmaceutics*, (November 2008), Vol.364, No.1, pp. 1-8, ISSN 0378-5173.

Nýlvt, J.; Hostomsky, J. & Giulietti, M. (2001). *Cristalização*. IPT/UFSCar, ISBN85-85173-63-7, São Paulo, Brazil.

Paulaime, A. M.; Seyssiecq, I. & Veesler, S. (2003). The influence of organic additives on the crystallization and agglomeration of gibbsite. *Powder Technology*, Vol.130, Nos.1-3, (February 2003), pp. 345-351, ISSN 0032-5910

Perlberg, A.; Lorenz, H. & Seidel-Morgenstern, A. (2005). Crystal growth kinetics via isothermal seeded batch crystallization: Evaluation of measurement techniques and application to mandelic acid in water. *Industrial and Engineering Chemistry Research*, Vol.44, No. 4, (February 2005), pp. 1012-1020, ISSN 0888-5885

Podder, J. (2002). The study of impurities effect on the growth and nucleation kinetics of potassium dihydrogen phosphate. *Journal of Crystal Growth*, Vol.237-239 Part 1, (April 2002), pp. 70-75, ISSN 0022-0248.

Pöllänen, K.; Häkkinen, A.; Reinikainen, S. P.; Louhi-Kultanen, M.; & Nyström, L. (2006). A Study on Batch Cooling Crystallization of Sulphathiazole: Process Monitoring Using ATR-FTIR and Product Characterization by Automated Image Analysis. *Chemical Engineering Research and Design*, Vol.84, (January 2006), No.1, pp. 47-59, ISSN 0263-8762.

Qu, H.; Alatalo, H.; Hatakka, H.; Kohonen, J.; Louhi-Kultanen, M.; Reinikainen, S. P.; & Kallas, J. (2009). Raman and ATR FTIR spectroscopy in reactive crystallization: Simultaneous monitoring of solute concentration and polymorphic state of the crystals. *Journal of Crystal Growth* Vol.311, (June 2009), No.13, pp.3466-3475, ISSN 0022-0248

Rauls, M.; Bastosch, K.; Kind, M.; Kuch, St.; Lacmann, R. & Mersmann, A. (2000). The influence of impurities on crystallization kinetics – a case study on ammonium sulfate. *Journal of Crystal Growth*, Vol.213, Nos.1-2, (May 2002), pp.116-128, ISSN 0022-0248.

Safavi, A.; Maleki, N.; Rostamzadeh, A. & Maesum, S. (2007). CCD camera full range pH sensor array. *Talanta*. Vol.71, No.1, (January 2007), pp. 498-501, ISSN 0039-9140.

Seiceira, R.; Higa, CM.; Barreto, AG. & da Silva, JFC. (2005). Determination of kinetic parameters and Hammett ρ from the synthesis of triaryl phosphites using reaction calorimetry. *Thermochimica Acta* Vol.428, No.1-2, pp. 101-104, ISSN 0040-6031.

Sheikhzadeh, M.; Trifkovic, M.; & Rohani, M. (2008). Real-time optimal control of an anti-solvent isothermal semi-batch crystallization process. *Chemical Engineering Science*. Vol.63, (February 2008), No.3, pp.829-839, ISSN 0009-2509.

Shirshov, YM.; Khoruzhenko, VY.; Kostyukevych, KV.; Khristosenko, RV.; Samoylova, IA.; Pavlunchenko, AS.; Samoylov, AV. & Ushenin, YV. Analysis of some alcohols molecules based on the change of RGB components of interferentially colored calixarene films. *Sensors Actuators B: Chemical*. Vol.112, pp. 427-436, ISSN 0925-4005.

Silva, J.F.C. & Silva, A.P.M. (2011). Determination of the Adipic Acid Solubility Curve in Acetone by Using ATR-FTIR and Heat Flow Calorimetry. *Organic Process Research and Development*. Vol.15, No.4, (May 2011), pp. 893-897, ISSN 1083-6160.

Simon, LL.; Nagy, ZK. & Hungerbuhler, K. (2009). Endoscopy-based in situ bulk video imaging of batch crystallization processes. *Organic Process Research and Development*, Vol.13, No.6, pp. 1254-1261, ISSN 1083-6160.

Sun, Y.; Song, X.; Wang, J.; Luo, Y. & Yu, J. (2010). Determination of seeded supersolubility of lithium carbonate using FBRM. *Journal of Crystal Growth*, Vol.312, No.2, (January 2010) pp. 294-300, ISSN 0022-0248.

Teychené, S.; Autret, J. M. & Biscans, B. (2004). Crystallization of efflucimibe drug in a solvent mixture: effects of process conditions on polymorphism. *Crystal Growth and Design*, Vol.4, No.5, (September 2004), pp. 971-977, ISSN 1528-7483

Ulrich, J. & Strege, C. (2002). Some aspects of the importance of metastable zone width and nucleation in industrial crystallizers. *Journal of Crystal Growth*, Vol.237-239 Part 3, (April 2002), pp. 2130-2135, ISSN 0022-0248

Yu, Z. Q.; Tan, R. B. H. & Chow, P. S. (2005). Effects of operating conditions on agglomeration and habit of paracetamol crystals in anti-solvent crystallization. *Journal of Crystal Growth*, Vol.279, No.3-4, (June 2005), pp. 477-488, ISSN 0022-0248

Yu, L. X.; Lionberger, R. A.; Raw, A. S.; D'Costa, R.; Wu, H. & Hussain, A. S. (2004). Applications of process analytical technology to crystallization processes. *Advanced Drug Delivery Reviews*, Vol.56, No. 3, (February 2004), pp. 349-369, ISSN 0169-409X

5

Macromolecular Crystallization Controlled by Colloidal Interactions: The Case of Urate Oxidase

Françoise Bonneté
Institut des Biomolécules Max Mousseron,
Université d'Avignon et des Pays de Vaucluse
France

1. Introduction

Crystallization is a natural or artificial process involving the physical transformation of a fluid or a gas into a regularly organized solid form, the crystal (Fig.1). It occurs in many fields such as health sciences, geosciences, microelectronics, industrial chemical processes.

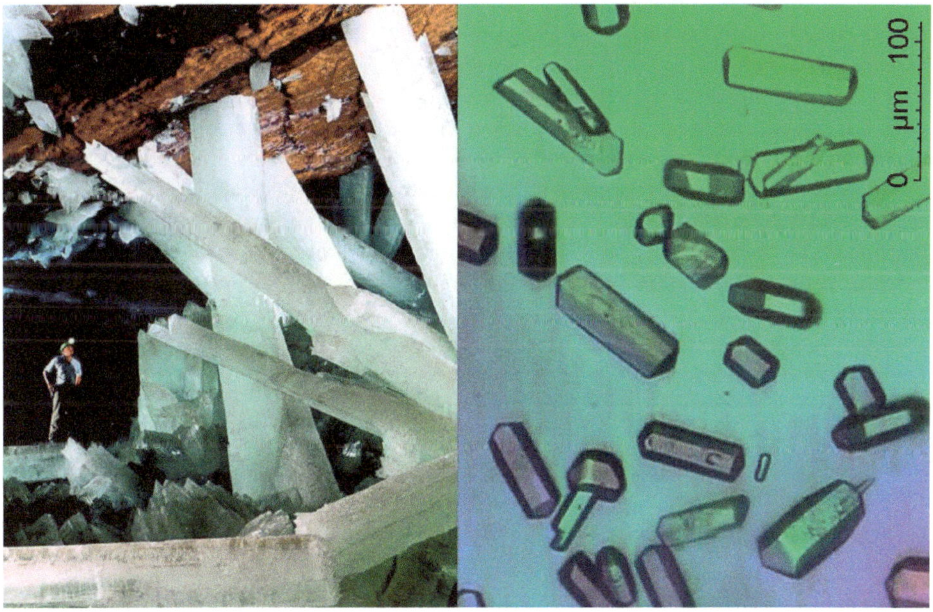

Fig. 1. (Left) Giant Crystal Cave's Mystery Solved (Lovgren, 2007); (Right) Micrometric protein crystals in batch of a therapeutic enzyme, Urate oxidase.

In health sciences, crystals can grow *in vivo* or *in vitro*. *In vivo* this can be due to pathologies (Pande et al., 2001); *in vitro*, crystallization is mainly used to decipher 3D atomic structures of biological macromolecules and understand their structure-function relationship. Originally, crystallization was a method of purification (Sumner, 1926), but now chromatography has replaced protein crystallization in the protein purification process. Today, crystallization is still used for purification and formulation steps in the biotechnology and pharmaceutical industries (Weber et al., 2008). It is a powerful and economical protein purification method, since high-purity proteins can be obtained in a single-step operation. However, crystallization remains an empirical process, still based on trial-and-error methods using commercial crystallization cocktails. Crystallization for protein purification is not always practical, since it requires good knowledge of the phase diagram and a substantial quantity of the protein to be crystallized. Although proteins are composed of amino acids of limited types, their structural diversity makes their crystallization conditions difficult to predict. Crystallization conditions include a wide spectrum of parameters such as pH and buffer type, ionic strength, type and concentration of precipitant, temperature, presence and concentration of surfactant molecules and other additives (e.g. cofactors, inhibitors) and, above all, protein concentration. Finding an efficient method to easily crystallize any protein has been, and can be expected to continue to be, a major challenge. Many methods of protein crystallization have been described (Bergfors, 2009). Since the advent of structural genomic programs, a lot of work has been done on automation and miniaturization of crystallization methods in view of the limited amount of protein material available, particularly membrane proteins. Other methods that limit convection effects on crystal growth have shown their efficacy on diffraction quality: crystallization in microgravity (McPherson et al., 1999), in gels (Ng et al., 2003), in microfluidics (Zheng et al., 2005) and other unconventional methods (Sazaki et al., 2004). All these general crystallization methods have recently been thoroughly reviewed (Sauter et al., 2011). But crystallization is not restricted to certain methodologies. Crystallization is a physical process, which consists of two major distinct but inseparable events, nucleation and growth, governed by both thermodynamic and kinetic factors. A pre-nucleation step can be added, creating a supersaturation state. For a long time it was thought that crystallization obeys no comprehensive theory. In the 80's-90's, rational and physical approaches were developed to understand the fundamentals of nucleation and crystal growth of proteins. The concepts of nucleation and crystal growth were described (Feher & Kam, 1985) and recently reviewed (Chernov, 2003; García-Ruiz, 2003). While growth of macromolecular crystals is well characterized, in particular by direct visualization using atomic force microscopy, there are fewer studies of nucleation, since it requires high supersaturation, that is to say often high concentrations of proteins. Some nucleation studies were performed with a model protein such as lysozyme by dynamic light scattering (Mikol et al., 1989) or small angle X ray (Finet et al., 1998) or neutron scattering (Boué et al., 1993) in order to understand the prenucleation and nucleation steps. This was the beginning of the development of rational approaches based on an understanding of the physical properties of macromolecular solutions. Numerous articles and reviews examined the correlation between crystallization and interactions between macromolecules in solution (Ducruix et al., 1996; Muschol & Rosenberger, 1995), applicable to soluble proteins like membrane proteins (Hitscherich et al., 2000). The measurement of the second virial coefficient, noted indifferently A_2, B_2 or B_{22} depending on the authors, appeared to be a powerful tool for predicting crystallization conditions (George & Wilson, 1994). George & Wilson showed that a restrictive range of values of the second virial coefficient was favorable to crystallization of soluble proteins. This "crystallization slot", which is about [-1x10^{-4}; -8x10^{-4} mol.mL.g^{-2}], corresponds to slight or moderate attractions. Second virial coefficients can be measured using different experimental techniques, including osmotic pressure (Haynes et al., 1992), static light scattering (SLS) (Velev et al., 1998), ultracentrifugation (Behlke & Ristau, 1999), small-angle X-ray scattering (SAXS) (Bonneté et al., 1999), small-angle neutron scattering (SANS) (Gripon et al., 1997), size-exclusion chromatography (Bloustine et al., 2003) and self-interaction chromatography (SIC) (Tessier et al., 2002). Studying different biological macromolecules (protein, virus) of various sizes, molecular masses or net charge (Bonneté & Vivares, 2002), we showed that a dimensionless second virial coefficient normalized to protein excluded volume gave a better representation of the effective pair potential between macromolecules (McQuarrie, 2000) when comparing the effects of physico-chemical parameters leading to crystallization. We found a new dimensionless crystallization slot [-10; 3] in which the Brome mosaic virus (BMV) (Casselyn et al., 2001) was found to crystallize with a slightly positive second virial coefficient a_2 = +2.8. This a_2 value was not inconsistent with attractive interactions since the dimensionless second virial coefficient of a pure hard-sphere potential is equal to 4. Indeed the effective pair potential of biomolecules in solution is the sum of different components including, in particular, hard sphere repulsion, van der Waals attraction, electrostatic effects and depletion attraction. Phase diagrams of biological macromolecules are therefore governed by an appropriate combination of these interaction potentials in solution. Repulsive regimes favor solubility, whereas the presence of attractive potentials may induce a variety of phase transitions, including the desired macromolecular crystallization. These pair potentials are medium range, from a few Å up to tenths of Å, and each of them is under separate control of physico-chemical conditions such as pH, temperature and solvent composition (Hansen & McDonald, 1976; Israelachvili, 1994). The desired fine tuning of the interactions for controlled crystallization requires identifying the individual contributions. In order to do so, theoretical simulations were performed on series of SAXS scattered intensities recorded on hen egg white lysozyme as a function of pH, salt concentration, salt type and temperature, on alpha crystallins (Finet & Tardieu, 2001) and urate oxidase (Vivares et al., 2002) as a function of polyethylene glycol (PEG). In this chapter we will show, using the example of urate oxidase, how to choose physico-chemical conditions, how they control the different interaction forces and how these forces can be modulated to design different phase diagrams for different applications (biocrystallography, or pharmaceutical processes for example).

2. Theoretical and experimental background

It is common practice in the colloid field to calculate phase diagrams from interaction potentials in solution (Asherie et al., 1996). The forces between colloids in solution include hard sphere repulsion plus attraction, whose range may modify the appearance of the phase diagram (Fig.2). In a colloid-polymer mixture, the presence of stable or metastable phase transitions can be determined by the ratio of the attraction range and the colloid size (Lekkerkerker, 1997), the range of the attractive interaction between colloids in solution being controlled by the size of the polymer (Poon, 2002). Predicting the phase diagrams of proteins based on knowledge of interaction forces may therefore be a more effective alternative for controlled crystallization than trial-and-error screenings.

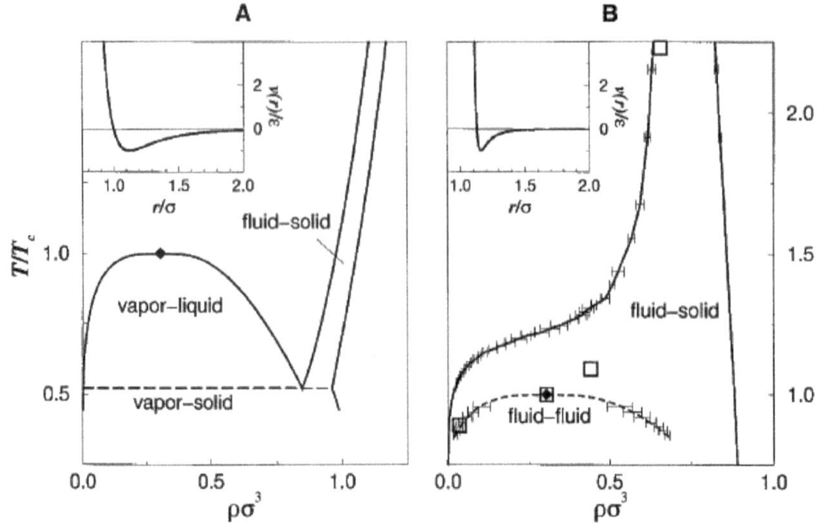

Fig. 2. Typical phase diagram of colloids A) with long-range attraction; B) with short- range attraction (ten Wolde & Frenkel, 1997)

Thus, a thorough understanding of the interaction potentials that govern protein phase diagrams will allow better control of crystal growth. The good news is that this can be applied successfully to biomacromolecule crystallization. The phase diagram of biomacromolecules, usually represented by the solubility curve, is governed by an appropriate combination of interaction potentials (Tardieu et al., 1999). Repulsion favors solubility, whereas attractions induced either by salt or polymer addition favor phase transitions, including crystallization. Ideally, to grow crystals, interaction potentials would be calculated in solution from knowledge of the macromolecule's characteristics, i.e. sequence, molecular mass, charge and isoelectric point as a function of physico-chemical environment (pH, temperature, ionic strength, etc.), leading to its phase diagram. Alas, this is not yet possible. However, using knowledge of the physico-chemical and biochemical characteristics of macromolecules in solution, we can choose relevant parameters to modulate potential forces and control nucleation rate and crystal growth. While it is not yet possible to calculate interaction potentials from macromolecule characteristics, it is possible experimentally to measure the resulting interactions in solution, either repulsive or attractive, through second virial coefficient (A_2) measurements, and then to simulate the underlying pair potentials. A_2 can be measured from the macromolecule concentration dependence of different experimental measurements, such as osmotic pressure, analytical ultracentrifugation, self-interaction chromatography, static light scattering, small angle X-ray or neutron scattering (SAXS or SANS). We used SAXS because it offers additional advantages. By combining SAXS measurements and numerical simulations, we can also analyze forces present in macromolecule solutions, that is to say the different repulsive and attractive components, and study them as a function of common physico-chemical parameters. This type of approach has already been used in the case of colloid-polymer mixtures (Lutterbach et al., 1999a; Lutterbach et al., 1999b; Ye et al., 1996) or biopolymers (Malfois et al., 1996; Vérétout et al., 1989).

2.1 Small Angle X-ray scattering and numerical calculation

To analyze the interaction potentials that control phase diagrams, we combined SAXS and numerical simulations. The total normalized intensity I(c,q), scattered by a solution of monodisperse spherical particles at a scattering angle 2θ, can be expressed as a function of the particle concentration c and the modulus of the scattering vector q, $q = 4\pi\lambda^{-1}\sin\theta$ (note that $q = 2\pi s^{*}$) by:

$$I(c,q) = I(0,q) \times S(c,q) \quad (1)$$

I(0,q) is the intensity scattered by one particle and is usually called the particle form factor. Experimentally, the form factor is generally obtained from curves recorded at low concentrations to avoid interaction effects. The form factor gives information on the particle shape and its oligomeric state. At low angles, the form factor of an ideal solution (c→0, without interaction) can be written:

$$I(c \to 0,q) = I(c \to 0,0).\exp\left(-\frac{R_g^2}{3}q^2\right) \quad (2)$$

Similarly, for non-ideal solutions, the low angle part of the intensity curves recorded as a function of c is written:

$$I(c,q) = I(c,0).\exp\left(-\frac{R_g^2}{3}q^2\right) \quad (3)$$

Therefore a « Guinier plot », i.e. a Ln I(q) plot versus q^2 (Guinier & Fournet, 1955), provides, with intensity at the origin, I(0,0) or I(c,0), and with the structure factor at the origin, S(c,0), since S(c,0)=I(c,0)/I(0,0). Indeed, with interacting spherical particles, departure from ideality can be accounted for by the interference term, S(c,q), usually called the solution structure factor. The value of S(c,q) at zero-q angle gives information on the nature of interactions between particles. With repulsive interactions, the particles are evenly distributed and S(c,0) is lower than 1 (Example Fig. 3A). With attractive interactions, fluctuations in particle distribution are observed and S(c,0) is larger than 1. The nature of the net interactions, either attractive or repulsive, can easily be determined by the plot of the structure factor at the origin, S(c,0), as a function of particle concentration, since it is related to the osmotic with pressure Π by:

$$S(c,0) = \frac{RT}{M}\left(\frac{\partial \Pi}{\partial c}\right)^{-1} \quad (4)$$

With

$$\frac{\Pi}{cRT} = \frac{1}{M} + A_2 c + A_3 c^2 + \ldots \quad (5)$$

the concentration c being expressed in $g.cm^{-3}$.

---

*Note: Depending on SAXS beamlines used for experiments, the scattered intensity was expressed either as a function of $q = 4\pi\lambda^{-1}\sin\theta$ or $s = 2\lambda^{-1}\sin\theta$ in $Å^{-1}$ or nm^{-1}

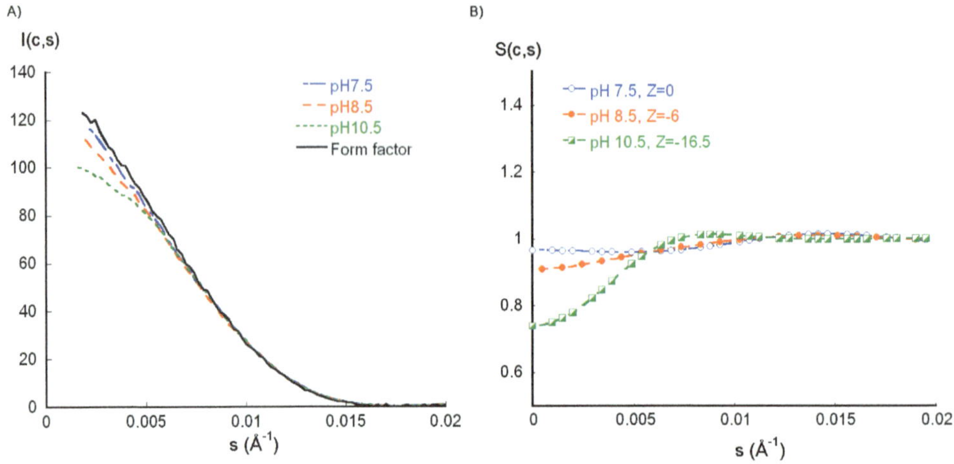

Fig. 3. A) Scattering intensities of urate oxidase as a function of pH. B) Experimental (dots) and fitted (lines) repulsive structure factor of urate oxidase at different pH corresponding to repulsive interactions, S(c,0) < 1.

Therefore, the second virial coefficient can be obtained by the expression:

$$S(c,0) = \frac{I(c,0)}{I(0,0)} = \frac{1}{1 + 2.M.A_2.c + ...} \tag{6}$$

Experimentally the structure factor can be obtained by intensity extrapolation from the expression:

$$S(c,0) = \frac{\lim\limits_{q \to 0} I(c,q)}{\lim\limits_{c \to 0} I(c,0)} \tag{7}$$

The slope of the linear fit gives the coefficient A_2 in mol.ml.g^{-2}.

$$S(c,0) \approx 1 - 2.M.A_2.c \tag{8}$$

If A_2 is positive, the overall interactions are repulsive (Example Fig. 3B); if A_2 is negative, the interactions are attractive.

The solution can be described mathematically as the convolution product of a particle shape and a particle distribution. *S(c,q)* is the Fourier transform of the spherically averaged auto-correlation function *g(r)* of the particle distribution or pair-distribution function:

$$S(c,q) = 1 + \rho \int 4\pi r^2 (g(r) - 1) \frac{\sin rq}{rq} dr \tag{9}$$

where $\rho = cN_a/M$ is the number density of particles, i.e. the number of particles per unit volume and *c* is the particle concentration in g.cm^{-3}. Models and numerical methods based on statistical mechanics are extended to proteins in solution (Lomakin et al., 1996; Malfois, et al., 1996). Calculation of structure factors is based on models of pair particle potential $U(r)$, from which a particle distribution $g(r)$ at equilibrium is inferred. The calculation of $g(r)$ is based on the Ornstein-Zernike (OZ) equation and on the hypernetted chain (HNC) integral equation and uses an iterative method (Belloni, 1985; 1988). For one-component fluids, the OZ relationship between total, $h(r) = g(r) - 1$, and direct, $c(r)$, correlation functions is written in Fourier space (where FT indicates a Fourier transform, normalized by the density ρ):

$$S(c, q) = 1 + \mathrm{FT}h(r) = 1/(1/\mathrm{FT} - c(r)) \qquad (10)$$

If we introduce an auxiliary function $\gamma(r) = h(r) - c(r)$, the integral HNC equation is:

$$g(r) = \exp[-\beta U(r) + \gamma(r)] \qquad (11)$$

where $U(r)$ is the interaction pair potential. To numerically solve the OZ equation with the above closure relation, iterations are used. The structure factor $S(c,q)$, and/or macroscopic properties and thermodynamic variables, are then calculated from the Fourier transform and integrals of pair distribution function $g(r)$ (Hansen & McDonald, 1976). The calculated structure factor is then compared to the experimental structure factor as shown in an example Fig. 3B.

For the simplest "one component" model, only the interaction pair potentials between macromolecules, which interact through solvent and ions, are explicitly considered in the numerical simulations. Since in such calculations the exact potential shape is not critical, we describe the potentials, either attractive or repulsive, in the mathematical form of a Yukawa potential, which is a function of three parameters, hard sphere diameter, σ, depth (strength), J, and range, d, according to:

$$U(r)/k_BT = J\,(\sigma/r)\,\exp[-(r-\sigma)/d] \qquad (12)$$

In the case of binary mixtures of macromolecules (mac) and polymers (pol) i.e. of "two component" systems, the total scattered intensity, $I(c_ic_j,q)$, can be expressed as a function of component concentrations c_i by (Belloni, 1991):

$$\begin{aligned} I(c_ic_j,q) &= \Sigma_i\Sigma_j \sqrt{c_ic_j}\sqrt{I_i(0,q)I_j(0,q)}S_{ij}(q) \\ &= I_{mac}(c_{mac},q) + I_{mac-pol}(c_{mac},c_{pol},q) + I_{pol}(c_{pol},q) \end{aligned} \qquad (13)$$

Because of the presence of a cross term, only SAXS intensities and not structure factors can be compared. The polymer form factor $I_{pol}(0,q)$ can be taken as equal to the Debye form factor, which is valid for a Gaussian coil and has already been used successfully with PEG (Debye, 1946):

$$I_{pol}(0,q) = I_{pol}(0,0)\,(2/x^2)\,(\exp(-x) - 1 + x) \qquad (14)$$

where $x = q^2R_g^2$, and R_g is the polymer radius of gyration. Partial structure factors $S_{ij}(q)$ are related to the Fourier transform of partial pair distribution function $g_{ij}(r)$:

$$S_{ij}(q) = \delta_{ij} + \sqrt{c_ic_j}\int (g_{ij}(r) - 1)\exp(irq)\,dr \qquad (15)$$

where i,j = 1,2, δ_{ij} is the Kronecker symbol (δ_{ij}=0 when i≠j and δ_{ij}=1 when i=j) and r the interparticle distance. To determine the pair distribution function, the usual procedure, as for one component systems, is to use the Ornstein-Zernike equation to link the total, $h_{ij}(r) = g_{ij}(r) - 1$, and the direct, correlation $c_{ij}(r)$ functions. The closure equation is once again the HNC equation. From a set of the three direct potentials, $U_{pol\text{-}pol}(r)$, $U_{mac\text{-}pol}(r)$ and $U_{mac\text{-}mac}(r)$, the theoretical scattered intensity from the binary mixture can be calculated and compared to the experimental scattering curve (Vivares, et al., 2002).

2.2 Direct pair potentials

2.2.1 The DLVO model

The direct protein-protein potential $U_{uox\text{-}uox}(r)$ was chosen equal to the DLVO potential. We had already successfully applied the DLVO model to different proteins at low-salt concentrations (Malfois, et al., 1996; Tardieu, et al., 1999). The DLVO potential is the sum of three potentials: a hard-sphere potential, an electrostatic repulsion and a van der Waals attraction. The hard-sphere potential reflects the fact that proteins cannot interpenetrate, the repulsive coulombic potential is due to the fact that each protein holds the same net charge and the van der Waals attractive potential is the resulting dispersion interaction between proteins. For the sake of simplicity, we chose a Yukawa shape for the coulombic and the van der Waals potentials, a Yukawa shape van der Waals potential having been shown to be sufficient to describe the attractive protein-protein interaction in aqueous solutions.

Respective mathematical expressions of the three potentials are described:

- Hard-sphere potential:

$$U^{HS}_{UOX-UOX}(r) = +\infty \quad r \le \sigma \\ = 0 \quad r > \sigma \tag{16}$$

with σ the protein diameter

- Repulsive coulombic potential:

$$U^{coul}_{UOX-UOX}(r) = Z^2 L_B / [\sigma(1+0.5\sigma/\lambda_D)^2].(\sigma/r).\exp(-(r-\sigma)/\lambda_D) \quad r > \sigma \tag{17}$$

with Z the effective protein charge, L_B the Bjerrum length (equal to $e^2/(4\pi\varepsilon_0\varepsilon_s k_B T) = 7.31Å$ at T = 293.15K with $\varepsilon_s = \varepsilon_{H2O} = 80$) and λ_D the Debye length ($\lambda_D(Å) = 3/\sqrt{I}$ at 293.15 K where I is the ionic strength expressed in mol/l). The potential is expressed in k_BT units.

Repulsive interactions are expected to vary with pH and to be screened with ionic strength (addition of alcohol, which reduces the water dielectric constant, can also reduce the repulsion).

- van der Waals potential:

$$U^{VdW}_{UOX-UOX}(r) = -J_{VdW}.(\sigma/r).\exp(-(r-\sigma)/d) \quad r > \sigma \tag{18}$$

with J_{VdW} (in k_BT units) and d (in Å), respectively, the depth and the range of the potential.

In practice, the van der Waals component is determined at or close to the pI. With small compact proteins, the best fit parameter for depth with a 3Å range was indeed found to be close to the calculated values, 2-3 k_BT (Tardieu, et al., 1999) and the attraction was found to increase with decreasing temperature.

In the DLVO potential, hard-sphere and electrostatic interactions have a repulsive effect which favors solubility. Except possibly at pI, the van der Waals forces are weaker than the coulombic interactions (or even disappear). However, it is clear that the basic interactions considered in the DLVO potential model are generally unable to provide the attraction necessary for macromolecular nucleation and crystal growth. Fortunately, other forces can play a role.

2.2.2 The Hofmeister effect

Salt has long been known to act as a crystallizing agent (Arakawa & Timasheff, 1985). A number of phase diagrams were measured, e.g. by the Ducruix group (Carbonnaux et al., 1995; Guilloteau et al., 1992) and showed that solubility varies with the type of monovalent salt, following the direct/reverse order of the Hofmeister series according to whether the particles are studied at a pH higher/lower than the pI, respectively.

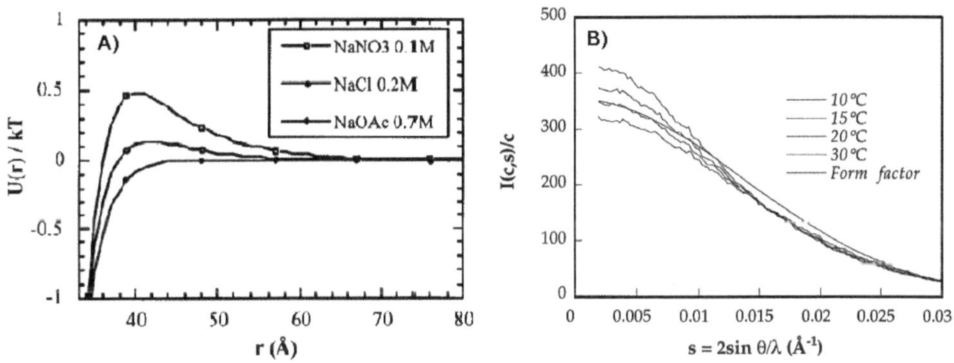

Fig. 4. A) Interaction potential of lysozyme in different salts. B) Normalized scattering intensity of Lysozyme in NaCl 200 mM pH 4.5 at different temperatures (curves in the same order as in legend for both figures).

When the effect of monovalent salts on protein interactions in solution was analyzed (Muschol & Rosenberger 1995, Tardieu *et al.*, 1999), whatever the particle size , at medium ionic strength (> 0.2 M), monovalent anions were observed not only to screen the charges, but to induce an additional attraction, specific to salt type (Finet et al., 2004). This attraction is short-range, about 3Å, and increases with decreasing temperature.

2.2.3 Depletion attraction

The addition of neutral non-interacting polymers to colloidal solutions has long been known to induce a depletion attraction (Asakura & Oosawa, 1954). This depletion interaction can be explained in a simplified manner for an ideal polymer solution.

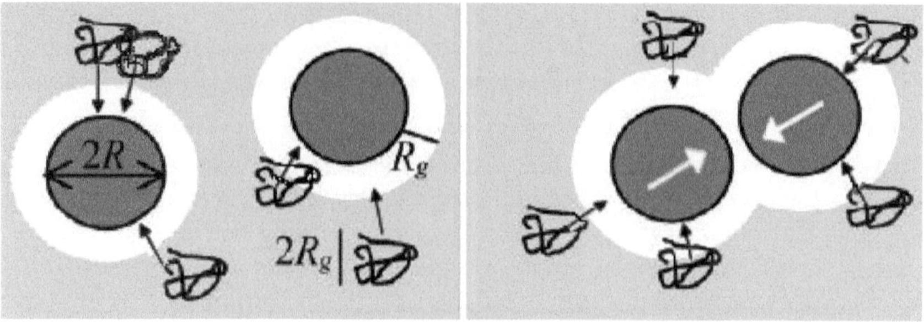

Fig. 5. Schematic illustration of the depletion effect in a colloid-polymer mixture

Molecules of polymer - characterized by their radius of gyration R_g - and colloids cannot mutually interpenetrate, and furthermore, the center of polymers is excluded from a region of thickness R_g around each colloidal particle. This excluded volume is called the depletion zone. When two colloid particles come sufficiently close to each other, their depletion zones overlap and the free volume accessible to the polymer molecules increases, leading to a gain in entropy of the system. Thermodynamically, it is therefore more favorable for the polymer when colloidal particles approach each other, i.e. when there is an attractive interaction between them. This model remains valid as long as polymer molecules do not overlap. To describe the direct polymer-polymer potential $U_{pol-pol}$ (r), we used an approach where the effective potential is finite for all distances between two polymer molecule centers of mass. Molecules of polymer were therefore considered as "soft colloids". The mathematical form chosen for the polymer-polymer potential was a Gaussian form (Bolhuis et al., 2001):

$$U_{pol-pol}(r) = J_{pol}.\exp(-(r / R_{pol})^2) \tag{19}$$

Where J_{pol} (in k_BT units) and R_{pol} (in Å) are the prefactor parameter and the range of the Gaussian potential respectively.

For the direct protein-polymer potential $U_{pol-col}(r)$, we chose a Yukawa form, which depends on only two parameters:

$$\begin{aligned} U_{pol-col}(r) &= +\infty \quad r \le \sigma / 2 \\ &= J_{pol-col} \cdot \frac{\sigma}{2r}.\exp(-(r - \sigma / 2) / d_{pol-col}) \quad r > \sigma / 2 \end{aligned} \tag{20}$$

with $J_{pol-col}$ (in k_BT units) and $d_{pol-col}$ (in Å) the intensity and the range of the potential respectively.

Thus, the numerical simulations provide us with an effective macromolecule-macromolecule potential, $U^{eff}_{col-col}(r)$, which may be written as the sum of the macromolecular interaction potential in the absence of polymer, $U_{uox-uox}(r)$, and of the depletion term, $U_{depletion}$ (r):

$$U^{eff}_{col-col}(r) = U_{col-col}(r) + U_{depletion}(r) \tag{21}$$

The depletion potential is therefore obtained from equation 21. From numerical simulations performed on the binary mixture, urate oxidase-polyethylene glycol (Vivares, et al., 2002), we found that the depth and the range of the effective potential increases with the addition of polymer.

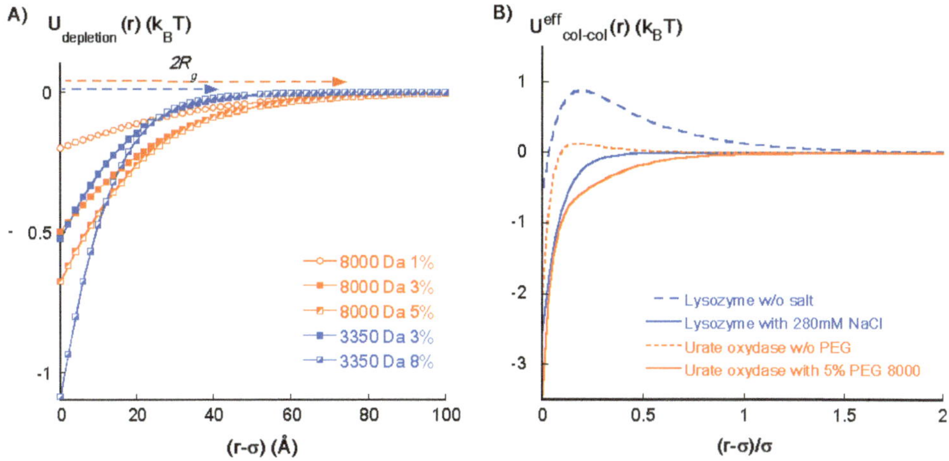

Fig. 6. A) Resulting depletion potential at different PEG concentrations for PEG 8000Da and PEG 3350Da; B) Comparison of the attractive potential induced by salt in lysozyme solutions and by PEG in urate oxidase solutions normalized to the particle diameter.

While depletion potential depth increases with polymer concentration whatever the polymer size, depletion range is approximately equal to $2R_g$ for each polymer, whatever the polymer concentration (Fig. 6A). Crystallization in polyethylene glycol is thus induced by a medium-range attractive potential between macromolecules, in addition to the short-range van der Waals. This result appears in contrast to that observed with small proteins like lysozyme, where crystallization was found to be induced by short-range attraction. The attractive potential range necessary for protein crystallization increases with the size of the macromolecule. In fact, it is interesting to note that, when normalized to the particle diameter, the salt- and PEG-induced interaction potentials that promote the crystallization of lysozyme (14300 Da) and urate oxidase (135000 Da) respectively are quite similar (Fig. 6B). Finally, since salt induces a short-range attraction and since polymer-induced potential range varies both with polymer size and with polymer concentration, by choosing salt in the Hofmeister series as well as polymer concentration and size it is possible to modify the phase diagram and thus control protein crystallization.

Nucleation and crystal growth are inseparable events, both of which depend on supersaturation. The nucleation step controls the structure of the crystalline phase and the number of crystals. Recently, it has been shown that interaction potentials play a key role in the determination of nucleation kinetic parameters (Bhamidi et al., 2005). Depending on the application (biocrystallography, powder diffraction, purification, for example), it appears important to control the nucleation step, supersaturation and, therefore, solubility.

We have seen that, whatever the salt or polymer used, macromolecular crystallization occurs in the presence of an attractive pair potential. We have analyzed how attraction and repulsion can be varied by changes in the environment: pH, ionic strength, type of salt, polymers, and temperature (Tardieu, et al., 1999; Vivares, et al., 2002). With low molecular weight proteins, a coulombic, pH-dependent, repulsive potential plus a short-range, possibly van der Waals, attraction are sufficient to account for the behavior observed at low ionic strength. At higher ionic strengths, salt-specific effects following the (direct or reverse) order of the Hofmeister series correspond to an additional short-range salt-specific attraction. With increasing protein size, the van der Waals contribution becomes negligible. Adding polymers like polyethylene glycol (PEG) induces a depletion attraction. The way forward in growing crystals is therefore a fairly simple extension of these observed effects. While this path is not guaranteed to lead to success, it may well enable us to reduce the time and effort spent on trial-and-error methods.

3. A protein of pharmaceutical interest: Urate oxidase

Lysozyme has long been the model protein most often chosen to study nucleation and crystal growth mechanisms (Drenth, 2005; Gavira & Garcia-Ruiz, 2009; Liu et al., 2010; Vekilov, 2010), since it is easily available at low cost. We used urate oxidase from *Aspergillus flavus* as a new model system to explore the crystallogenesis of proteins in general and the crystallographic structure of urate oxidase in particular. Urate oxidase is used as a protein drug to reduce toxic uric acid accumulation and to treat the hyperuricemic disorders occurring during chemotherapy. Urate oxidase *from Aspergillus flavus* is produced, purified and made commercially available by Sanofi-Aventis (France). Urate oxidase is normally purified using multiple steps of concentration and chromatography (McGrath & Walsh, 2005). Like other proteins (Jacobsen et al., 1998), it could, however, be purified by crystallization (Giffard et al., 2011). Crystallization has the inherent advantages of providing higher final purity yields, not denaturing the protein of interest and often providing some stabilization effects, but it requires a good knowledge of the phase diagram and a substantial amount of the protein to be crystallized. The structure of the urate oxidase from *A. flavus* has been solved in the absence (Retailleau et al., 2004) as well as in the presence of different inhibitors (Retailleau et al., 2005). It is a homotetrameric enzyme of 135kDa with a subunit consisting of 301 amino acids. Although different urate oxidase structures have been determined, its catalytic mechanism is still poorly understood. One possible mechanism for the oxidation of uric acid could be revealed by the precise protonation state of the substrate during the reaction. Neutron crystallography can provide such information and make possible direct determination of the protonation states of the active site residues and substrate analogues (Oksanen et al., 2009). For this purpose, it is necessary to grow large, well ordered, deuterated crystals; here too, therefore, a good knowledge of the phase diagram is required. Finally, whether crystallization is aimed at urate oxidase purification or at growing highly diffracting crystals, a good knowledge of the phase diagram and thus of crystallization conditions are required. We used urate oxidase, then, as a model system both from a fundamental and an applied point of view, in suitable biological and physicochemical conditions, to characterize, simulate and modulate pair potentials present in solution and design appropriate phase diagrams for specific applications. We will show in particular, how macromolecule properties - such as sequence, molecular mass, isoelectric point, stability - can provide the parameters to be screened to obtain adequate phase diagrams.

4. Urate oxidase interactions in solution and implications for crystallization

The stability of urate oxidase from *Aspergillus flavus* was studied by differential scanning calorimetry (DSC) (Bayol et al., 1995) in conjunction with enzymatic activity measurement and size exclusion chromatography. The recombinant urate oxidase is not stable below pH 6 and shows maximum stability between pH 7.25 and 9.5 up to 35 °C. Studies of crystallization were therefore performed at a pH above 7.2 and a temperature below 35 ° C. pH 8 was identified as the value around which the enzyme stabilization activity is optimized (Aleman et al., 1998). The buffer selected for use is a sodium phosphate buffer at pH 8 at a concentration of between 5 mM and 100 mM. It was in this range of urate oxidase stability that different parameters (pH, nature of buffer, temperature, ionic strength, additives) were studied for their effects on interaction forces and their influence on phase diagrams.

4.1 pH and Salt effects

Urate oxidase is a large tetrameric enzyme of 135 kDa stable at pH above 7. Its theoretical isoelectric point (pI), calculated from its primary structure and the pKa of each charged amino acid, is 7.5 (Fig 7), consistent with IEF experiments.

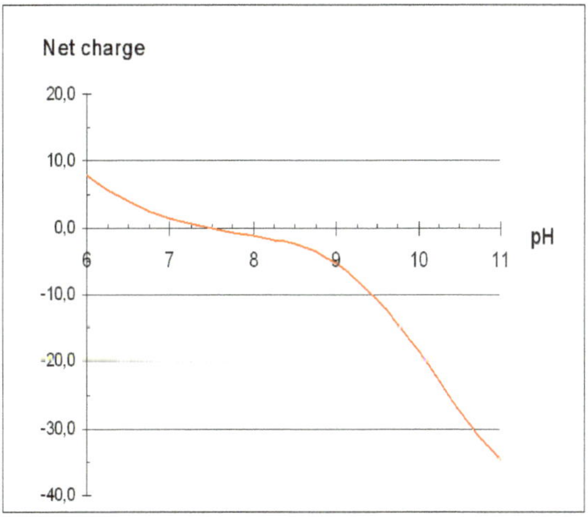

Fig. 7. Urate oxidase net charge as a function of pH

By definition, at pI without any other crystallizing agents, the protein net charge is zero. Thus, there is no electrostatic repulsion. The van der Waals attraction may therefore induce instability and aggregation of proteins. Usually protein solubility is minimum at pI (Riès-Kautt & Ducruix, 1999). At pH above pI, the urate oxidase negative net charge increases as pH increases. We therefore expect urate oxidase electrostatic repulsions to increase as the net charge increases.

A compilation of second virial coefficient (A_2) values of urate oxidase as a function of pH, determined from previous SAXS experiments (Giffard et al., 2008; Vivares & Bonneté, 2002) (Fig. 8A), shows that A_2 is positive whatever the pH above pI, as expected for repulsive interactions, and increases as both pH and protein net charge increase. However, as pH approaches pI, A_2 remains positive, which is not consistent with pure van der Waals attraction at pI. This may be due to residual charges on the protein. In contrast, for a negative second virial coefficient measured in Tris buffer pH 8 close to pI without addition of salt (Fig. 8B), the second virial coefficient increases, becoming positive as salt concentration increases.

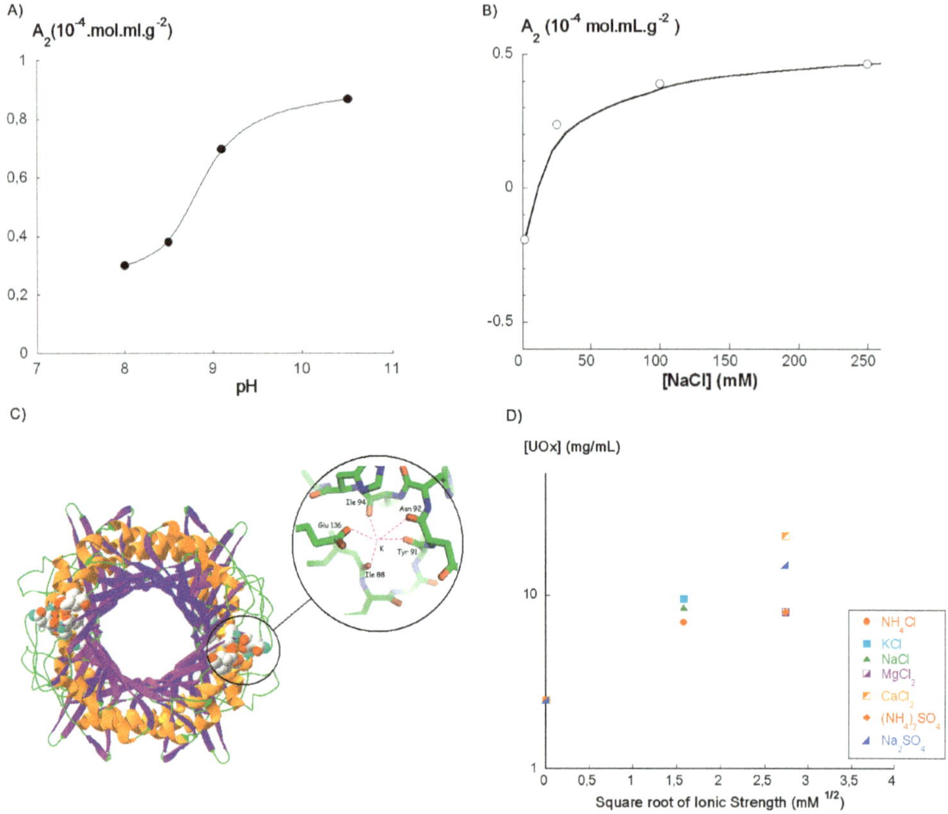

Fig. 8. A) Variation of second virial coefficient (A_2) as a function of pH at 20 °C: pH 8.0 with sodium phosphate buffer; pH 9.1 sodium borate buffer; pH 10.5 sodium carbonate buffer. B) Variation of A_2 in Tris buffer pH with addition of sodium chloride. C) Cationic binding site on the urate oxidase surface (courtesy of G. Marassio). D) Solubility of urate oxidase as a function of different salt concentrations.

This surprising result suggests that the addition of sodium salt, in particular the addition of sodium cation, induces an increase in repulsion ($A_2 > 0$). Most articles report that addition of salt decreases solubility, which is correlated with the fact that salt induces attractive interactions through charge screening. This effect, known as salting-out (Arakawa & Timasheff, 1985), is generally observed at medium and high salt concentrations. At low ionic strength, the opposite effect is expected, i.e. salting-in, where solubility increases with addition of salt. However, salting-in has only sporadically been reported (Faber et al., 2007). Our result is explained by X-ray crystallography (Fig. 8C), which shows that the salting-in effect observed with urate oxidase at pH 8 in Tris buffer results from the direct binding of cations to specific sites on the surface of the protein. A similar effect was recently reported (Gibb & Gibb, 2011), which suggests that salting-in is induced by ion binding. This salting-in effect is not specific to sodium cation, since it was observed with different cations such as K^+, NH_4^+, Mg^{++}, Ca^{++} (Fig. 8D). In all cases, the addition of salt induces positive A_2 values, an indicator of repulsive interactions with rasburicase, leading to an increase in solubility. Finally, the positive A_2 values of urate oxidase in sodium phosphate pH 8 or Tris buffers with 10mM salt are consistent with the high solubility of the protein in these buffers, and the negative A_2 value found in tris buffer pH 8 without salt is consistent with attraction inducing crystallization of urate oxidase (Fig. 9A). Obtaining crystals thus made solubility measurements possible (Fig. 9B).

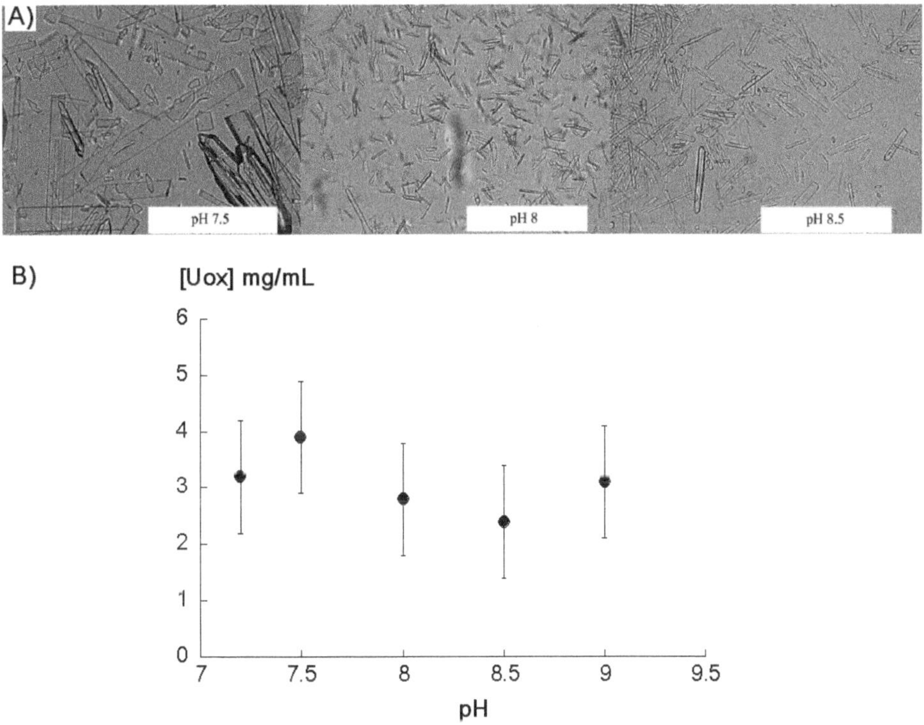

Fig. 9. A) Optical microscopy images of microcrystalline urate oxidase in Tris buffer at different pH values (from Collings et al., 2010). B) Solubility of urate oxidase as a function of pH in Tris buffer w/o salt added.

Even though these crystals were not of sufficient quality for high resolution crystallography, they were perfectly suitable for powder diffraction for space group determination (Collings et al., 2010). Attractive interactions with second virial coefficients in the "crystallization slot" defined by George & Wilson are undoubtedly necessary conditions to obtain crystals; however these interactions must be controlled upstream of the crystallization process to design adequate phase diagrams for appropriate crystallography.

4.2 Buffer type and temperature effects

pH and salt effects observed on second virial coefficient variations are easily explained by a change in the coulombic repulsive interactions, either 1/ due to a decrease in the protein net charge, either by shifting pH towards pI or by screening the repulsion with salt at pH far from pI, or 2/ due to an increase in the protein net charge, by ion binding near the pI. The first assumption was verified by characterizing the different underlying interaction potentials (Vivares & Bonneté, 2002) and will not be discussed here. The ion binding effect underlines the importance of choice of solubilization buffer, especially since the objective is to test temperature effect. In previous work, a slight temperature effect was observed on the scattered intensity at small angles at pH 10.5 between 283 and 293 K, and this was attributed to a weak van der Waals attraction. In order to characterize the temperature effect on interaction forces and its implications for crystallization, SAXS experiments were performed over a larger range of temperature from 278 to 303 K at a constant pH closer to the isoelectric point, where the repulsive component is weaker. Since Tris buffer is very sensitive to temperature variations (dpKa/dT = -0.028), we used a borate buffer whose pH is known to be less temperature-dependent (dpKa/dT = -0.008). The scattered intensities of urate oxidase in Tris buffer and in borate buffer at pH 8 as a function of temperature are shown in Fig. 10.

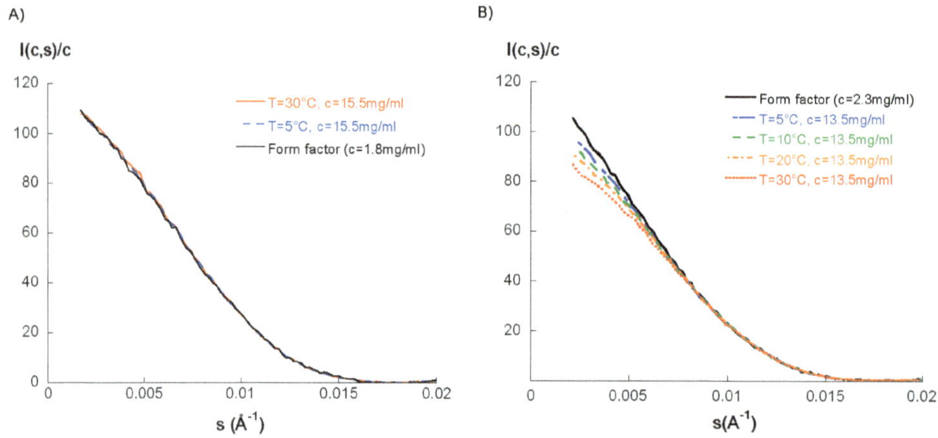

Fig. 10. Variation of scattered intensity of urate oxidase as a function of temperature: A) in tris buffer pH 8; B) in borate buffer pH 8.

It is clear that buffer type influences interactions in urate oxidase in solution. Whereas no variation in intensity is observed in Tris buffer, in borate buffer intensity increases as temperature decreases, probably due to an increase in van der Waals attraction, as previously observed and modelled with lysozyme (Tardieu, et al., 1999).

The second virial coefficient remains positive, probably due to the presence of sodium cations in borate buffer, but decreases as pH decreases (Fig.11A). Since pH in borate buffer does not vary with temperature, the net charge and the repulsive interactions of urate oxidase does not vary. The overall decrease in A_2 can only be induced by an increase in attractive interactions, the van der Waals attraction. In contrast, in Tris buffer, pH increases as the temperature decreases, inducing an increase in the net charge at pH above pI and thus an increase in repulsive interactions, matching the increase in attraction with temperature. This effect can induce the dissolution of crystals obtained at room temperature when stored in cold rooms, for example (data not shown but observed with urate oxidase), even though solubility is direct (Fig. 11B), as in the case of urate oxidase. When care is taken to keep pH constant whatever the temperature, and not to add cations to the buffer, it is therefore possible to obtain crystals and measure solubility (Fig. 11B). Temperature is an important parameter to modify attractive interactions and induce and control crystallization. Neththertheless, the appropriate buffer still needs to be chosen and a significant change in solubility needs to be measured if large crystals are to be grown (Budayova-Spano et al., 2007).

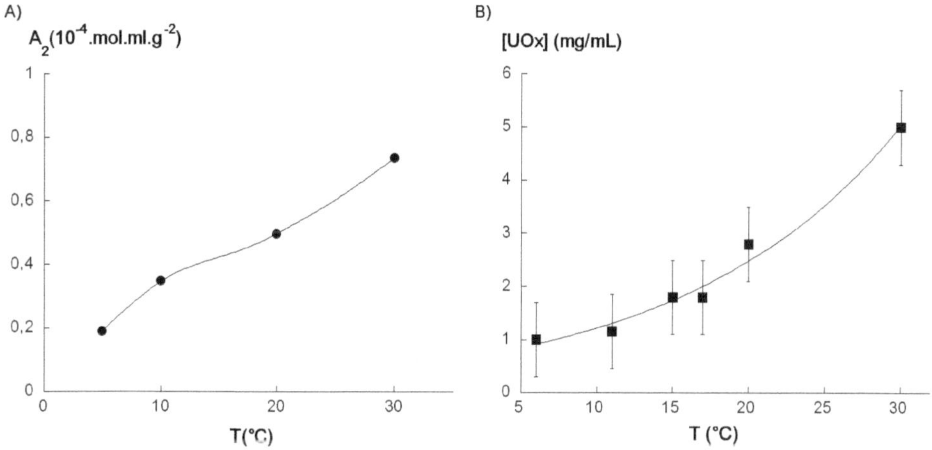

Fig. 11. A) Variation of urate oxidase A_2 with temperature in Na borate buffer pH 8; B) Solubility of urate oxidase at pH 8 in tris buffer w/o cations as a function of temperature

4.3 Usual and unusual polymers

Salts are known to act as crystallizing agents for many proteins, often small compact proteins, by inducing a short-range attraction. Ions, and in particular anions, were observed, at medium ionic strength, not only to screen charges but to induce an attraction, specific to the salt type (Ducruix, et al., 1996). We have seen here with urate oxidase, a large protein, that specific ion binding can induce solubilization of the protein by inducing a repulsive interaction, preventing its crystallization with salts. When salts are not effective in crystallizing proteins, neutral polymers can do the job (Bonneté et al., 2001; Budayova et al., 1999; Hitscherich, et al., 2000). Polyethylene glycol (PEG), a non-adsorbing neutral polymer, has long been used for protein crystallization (McPherson, 1976). The solubility (actually the precipitation) of proteins in solution containing PEG was studied (Atha & Ingham, 1981). A theoretical model was proposed by Asakura-Oosawa to explain the attraction between colloids due to addition of these polymers (§2.2.3). The effectiveness of polymer in precipitating proteins increases with polymer size and concentration, since it increases the depletion zone. By studying the crystallization of urate oxidase, our group greatly contributed to a better understanding of the mechanism of depletion attraction in the presence of different polyethylene glycols. We characterized the attractive interactions between urate oxidase induced by addition of three different PEGs (Fig. 12A), obtaining crystals (Fig. 12B) in conditions where A_2 was found negative in the "crystallization slot". Compared to salt, a major advantage of PEG in crystallization is that the depth and range of the attraction can be varied almost at will, simply by changing polymer size and concentration. However, it may be appropriate to couple the effects of salts (screening of charges or ion binding) and/or of pH with the effects of polymers to modulate the overall interactions and control the phase diagram for specific design.

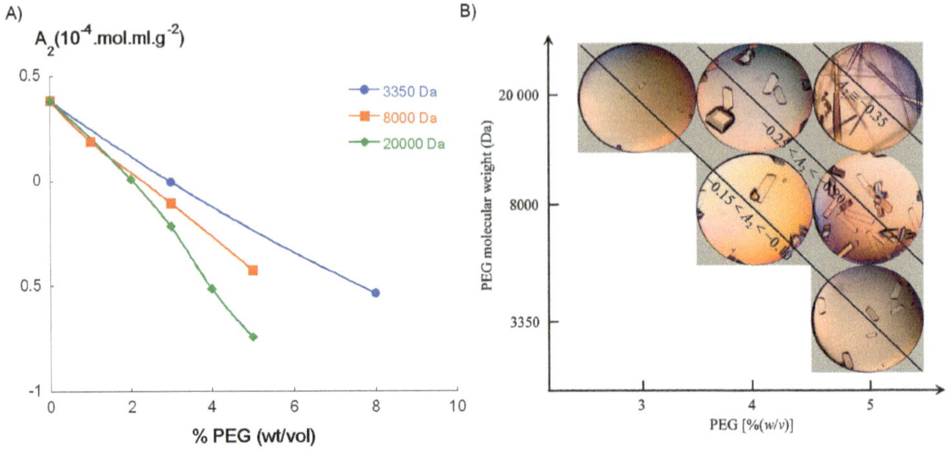

Fig. 12. A) Second virial coefficient urate oxidase in different PEG solutions in Tris pH8. B) Crystals of urate oxidase obtained in attractive ($A_2 < 0$) conditions.

However, as a function of |pH-pI|, the addition of PEG may be insufficient to induce an overall attraction, as shown in figure 13A. Thus, at pH 10, urate oxidase presents repulsive

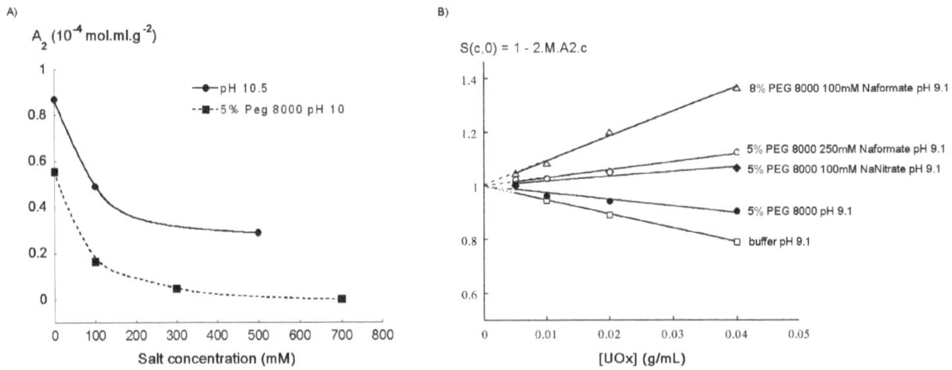

Fig. 13. A) Second virial coefficient of urate oxidase at pH 10 as a function of salt addition far from pI with and w/o PEG. B) Zero-angle structure factor in different physico-chemical conditions as a function of urate oxidase concentration; the slope measures the second virial coefficient (the slope is positive, A_2 is negative, attraction and vice versa)

interactions due to the protein negative net charge (see Fig 7). The depletion attraction induced by 5% PEG 8000 is not sufficient to change repulsive interactions into attractive interactions. To induce an overall attraction by addition of PEG, it is necessary to screen charges by addition of salt (Fig13B) or to be close to pI (Fig. 12A), where electrostatic repulsion is lowest.

At pI, regardless of whether salt is present in the buffer (Tris pH8 in our case), the addition of PEG to urate oxidase solutions induces attraction and leads to a decrease in solubility (Fig. 14). However, at pH 8 the solubility of urate oxidase without salt in the buffer is lower than the solubility of urate oxidase with salt. Since nucleation and crystal growth are driven by the supersaturation β, with β the ratio [initial concentration] / [solubility], a limited variation in solubility and therefore in supersaturation will be unfavorable to the growth of large crystals, and will rather favor the nucleation of small crystals.

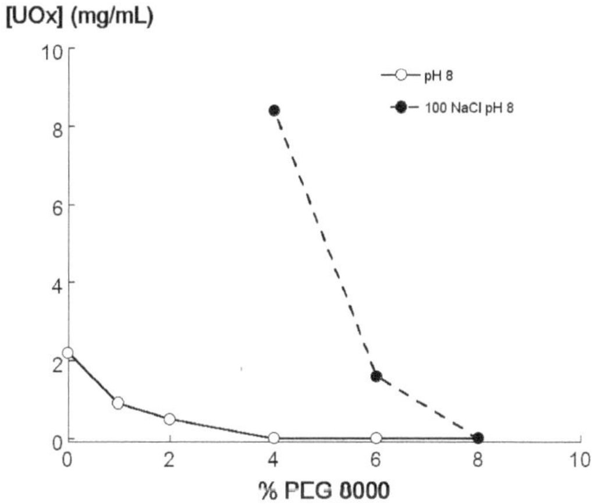

Fig. 14. Solubility of urate oxidase as a function of % of PEG with and w/o salt

As soon as crystals can be obtained, it is therefore possible to characterize the phase diagram with one or several solubility curves depending on the presence of polymorphs and the metastable liquid-liquid phase separation (Fig. 15).

In some cases, salts and neutral polymers are not effective for crystallization or cannot be used in pharmaceutical processes. Other crystallizing agents that can be used include organic compounds such as ethanol (Boyer et al., 1999), isopropanol, methanol, 2-methyl 2,4-pentanediol (Costenaro et al., 2001). Polymers such as amphiphilic multi-block polymers are used in the cosmetic and pharmaceutical industries, more often as emulsifiers, solubilizers, dispersing and wetting agents in the preparation of solid dispersions than as crystallizing agents in biocrystallography or in crystallization processes. We characterized a new class of crystallizing agent for soluble protein crystallization, compatible with both pharmaceutical processes and high-resolution structure determination in bio-crystallography. Poloxamer P188 is a nonionic co-polymer surfactant with a tri-block structure, composed of two hydrophilic segments, poly(oxyethylene) (PEO), and a central hydrophobic segment, poly(oxypropylene) (PPO), linked by ether bonds. The resulting construct can be represented as $HO(C_2H_4O)a(C_3H_6O)b(C_2H_4O)aH$, where a is about 75 and b is about 31 (Takáts et al., 2001) (Fig. 16). Its average molecular weight is around 8400 $g.mol^{-1}$

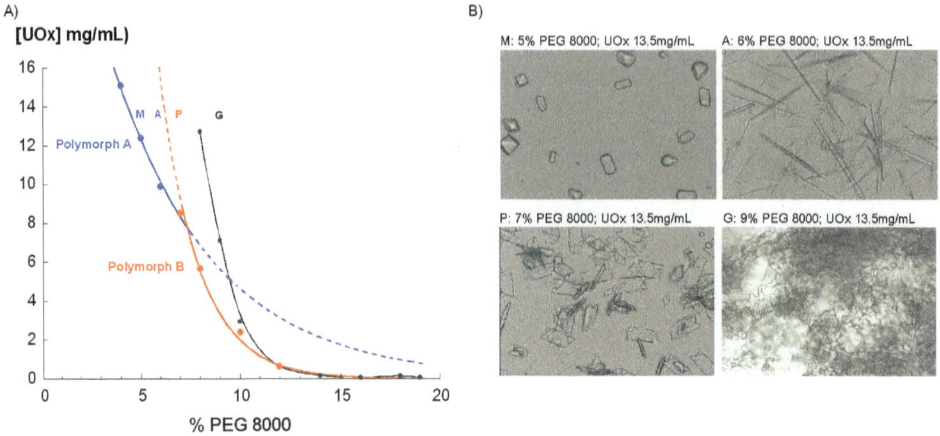

Fig. 15. Phase diagram of urate oxidase in presence of polyethylene glycol 8000 and optical microscopy images of different conditions in this phase diagram (M, A are polymorph A, P polymorph B and G a liquid-liquid phase transition) (Vivares et al., 2006).

and its cmc (critical micellar concentration) is about 0.1 % w/v (Schmolka, 1977). Poloxamer is similar to the usual crystallizing agent for urate oxidase, PEG 8000, which is a linear hydrophilic polymer consisting of approximately n = 180 poly(oxyethylene) units and has a molecular weight of about 8000 g.mol^{-1}. Because of their amphiphilic structure, poloxamers have surfactant properties that make them useful in pharmaceutical applications. They can be used to increase water solubility of hydrophobic, oily substances as well as to increase the miscibility of two substances with different hydrophobicities. They are also used as model systems for drug delivery (Adams et al., 2003) applications. Recently, they have been shown to function as artificial chaperones to facilitate refolding of denatured proteins in solution or to suppress aggregation. In general, all these applications require low concentrations of poloxamer, typically below its cmc and involving monomeric poloxamer in solution. Below the cmc, the hydrophobic segment of polymer can non-specifically interact with exposed hydrophobic domains, preventing aggregation and aiding in the refolding of proteins (Lee et al., 2006).

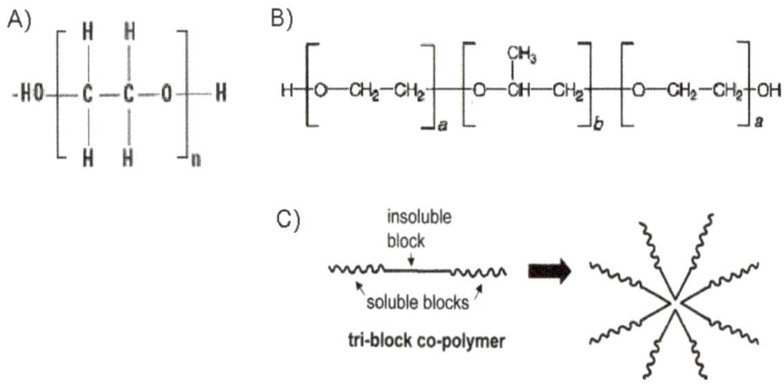

Fig. 16. Chemical structure of poloxamer P188 (A), PEG (B); C) Scheme of micellization of poloxamer at c > cmc (~ 0.1%)

However, high concentrations of poloxamer have also been reported to induce protein aggregation (Garcia, 1975). The addition of 12 to 16 % of the block copolymer to plasma or serum induces the precipitation of high molecular weight proteins including antibodies. This suggests that it would be possible to use high concentrations of poloxamer, thus probably in its micelle form, to induce protein crystallization by depletion, as observed with PEG. To this end, we explored the interactions and crystallization of urate oxidase by addition of poloxamer P188. The second virial coefficient is positive without poloxamer in Tris with 30mM KCl, as expected from previous studies, and increases as the concentration of poloxamer increases up to its cmc (about 0.1%). Above cmc, the A_2 decreases and becomes negative at concentrations higher than 1%. Solubility of urate oxidase was therefore measured in the same conditions, as a function of concentration of poloxamer P188. As with PEG 8000, solubility decreases as the concentration of poloxamer increases and is higher with poloxamer P188 than with PEG 8000 for concentrations lower than 6 % (Fig. 17). Thus, the surfacting or crystallizing nature of poloxamer P188 depends on its concentration. Below 0.1 %, i.e. its cmc, poloxamer is monomeric. In this range of concentrations, it is possible that the hydrophobic central PPO block non-specifically interacts with solvent-exposed hydrophobic patches at the protein's surface, while the hydrophilic surfactant chains remain exposed to the aqueous phase, increasing the solubility and the repulsion of urate oxidase. Above 0.1 % poloxamer, micelles can form. The attractive effect of poloxamer at concentrations above 1 % can be explained by a depletion effect driven by poloxamer micelles. The same effect has been suggested with nanoemulsions whose micelles, after surface saturation by surfactant, do not adsorb on the surface but rather cause attraction by a depletion mechanism (Wulff-Pérez et al., 2009).

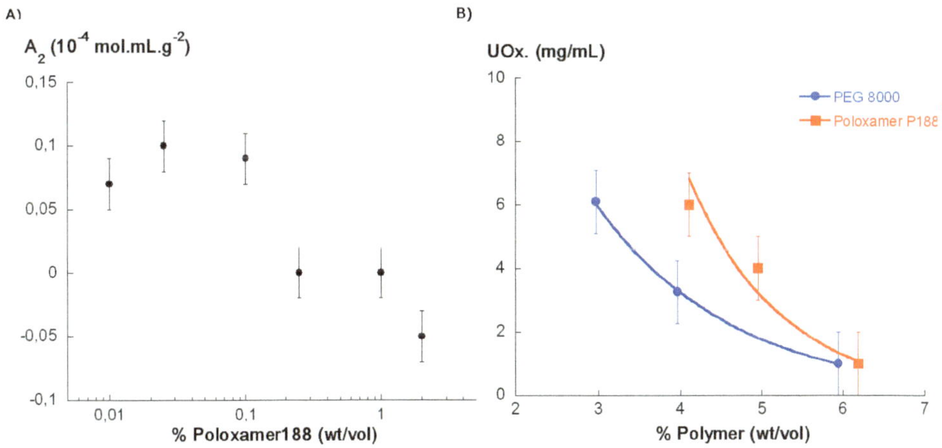

Fig. 17. A) Second virial coefficient of urate oxidase in solution as a function of % Poloxamer P188. B) Comparison between solubility of urate oxidase in PEG 8000 and Poloxamer P188 in 30 mM KCl, 50 mM Tris pH 7.5 at 20°C.

Poloxamer P188 can thus be used as a crystallizing agent for urate oxidase. When other amphiphilic surfactants such as poloxamer P407, polysorbate 20 and polysorbate 80 were studied, poloxamer P407 was found to have a similar structure and molecular formula to poloxamer P188, where a and b are 100 and 60 respectively. The molecular weight of the hydrophobic core is 4000 g.mol-1, which represents 30 % of the total mass of the polymer. P407 average molecular weight is about 13300 $g.mol^{-1}$. As shown in Fig. 18A, urate oxidase solubility in poloxamer P407 is higher than in either poloxamer P188 or PEG 8000, when the remaining solution components are kept fixed (i.e. 50 mM Tris pH 7.5, 30 mM KCl). Unlike PEG, urate oxidase solubility increases with the mass of the poloxamer used. This result suggests that poloxamer has a solubilizing effect proportional to its hydrophobic core content, and this compensates for its precipitating by depletion. The effect of polysorbates, another class of emulsifiers used in the preparation of pharmaceuticals and food, was also tested. Polysorbates are oily liquids derived from PEGylated sorbitan (a sorbitol derivative) esterified with fatty acids. They all have a hydrophilic moiety characterized by twenty oxyethylene-(CH_2CH_2O)- groups, while the hydrophobic segments vary according to the polysorbate compound. Polysorbates 20 and 80 effectively induce crystallization of urate oxidase (Fig. 18B). Urate oxidase solubility is higher with polysorbates 20 and 80 than with PEG 8000 and poloxamers P188 and P407. As with poloxamers, it appears that the longer the polysorbate hydrophobic chain, the higher the solubility of urate oxidase. Nevertheless, the modification observed in the habit of the urate oxidase crystals with 16% polysorbate 80 suggests that this concentration of surfactant could affect crystal growth, and possibly protein structure.

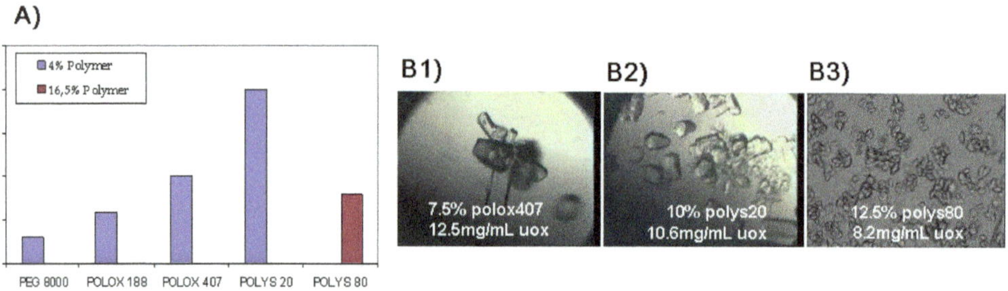

Fig. 18. A) Urate oxidase solubility with various polymers (in 30 mM KCl, 50 mM Tris pH7.5, 20°C). B) Crystals of urate oxidase in 30 mM KCl, 50 mM Tris pH7.5, 20°C: B1) 12.5 $mg.mL^{-1}$ uox with 7.5 % poloxamer P407; B2) 10.6 $mg.mL^{-1}$ uox with 10 % polysorbate 20; B3) 8.2 $mg.mL^{-1}$ uox with 12.5 % polysorbate 80.

4.4 Urate oxidase crystal design for high pressure crystallography

Macromolecules were long supposed to have their own crystallization conditions, since they have a unique sequence. However, the fact is that macromolecules in solution interact through different weak interaction forces. They can be considered as charged colloids under the influence of attraction forces (van der Waals, depletion, Hofmeister) and repulsion forces (hard sphere, electrostatic). These forces are medium-range and each of them is under the separate control of physico-chemical conditions such as pH, temperature and solvent composition. These forces govern macromolecule properties in solution such as solubility, phase separation (liquid-liquid or liquid-solid). The fine tuning of these interaction forces makes the design of phase diagrams for specific applications possible. The overall attractive potential leading to crystallization can be obtained in different ways. We used urate oxidase as a model system, since it was available in large quantities, to explore a wide spectrum of physico-chemical conditions and to see how these parameters influence pair potentials and phase diagrams. For example, the salting-in effect by increasing repulsion allows us to modulate the position of the solubility curve in the urate oxidase phase diagram, leading to a more or less large metastable zone, in order to grow either large single crystals or numerous small crystals. Another example is the use of amphiphilic polymer below or above its cmc (critical micellar concentration), allowing us either to favor the solubility of proteins or to induce their crystallization. In the following example, we describe a strategy for growing protein crystals suitable for macromolecular crystallography under high pressure (HPMX). Studying proteins under high pressure encompasses a wide range of objectives, from understanding the physical chemistry of protein interactions with water to practical applications in food processing. Many proteins are studied using a variety of techniques applied under high pressure, such as spectroscopic techniques, NMR spectroscopy, as well as a wide range of scattering techniques including static and dynamic light scattering, neutron scattering and X-ray crystallography. Such high-pressure approaches are generally used to solve problems involving macromolecule structural changes. For High Pressure Macromolecular Crystallography (HPMX), obtaining the right protein crystal is a challenge. First, as for any crystallographic study, it is necessary to obtain a reasonably large crystal preferentially in a high symmetry space group. Secondly, the crystal has to fit the pressure cell as well as possible, and be grown by a method which allows the crystal to be picked out easily, for example the hanging drop technique. Finally, the crystal has to remain stable in its mother liquor under pressure. Indeed, the effect of pressure on protein nucleation and crystallization can vary widely. In some cases, increased pressure increases the nucleation rate (Visuri et al., 1990), while in other cases, the protein solubility increases, decreasing the nucleation rate (Lorber et al., 1996). Pressure can affect solubility and therefore lead either to dissolution of a crystal or to secondary nucleation. To circumvent the pressure effect on secondary crystal nucleation in the cell, the crystal must be placed in a mother liquor containing no protein and a high concentration of crystallizing agent, since variations in solubility caused by pressure are known to be reduced by increasing the crystallizing agent concentration.

To grow a suitable crystal for HPMX, we first need a low supersaturation β which will lead to a small number of large crystals rather than a large number of small crystals, finally yielding a crystal in equilibrium with a very low protein solution, i.e. at high crystallizing agent concentration. To optimize urate oxidase crystallization conditions, we first characterize the phase diagram without salt, i.e. the conditions where solubility is lowest, as a function of PEG percentage (Fig. 19A) and explore different crystal growth conditions by using the Microbatch technique (Chayen et al., 1992).

The rasburicase crystals obtained as a function of PEG 8000 are of two different shapes: tabular crystals when the percentage of PEG is below 4%, and plate-like crystals when PEG percentage is above 4% (Fig. 19B). Unfortunately, crystals obtained at the lowest solubility, i.e. when percentages of PEG are greatest, are not suitable for crystal diffraction studies, being too numerous and poorly faceted. A suitable tabular crystal grown at a lower PEG percentage is picked out and placed in a glass cell. According to the solubility curve in 50mM Tris pH8, rasburicase is totally insoluble in a solution of more than 10% PEG 8000. To control the stability of the crystal, protein-free mother liquor (15% PEG 8000, 50mM tris pH 8.0) is added to the crystal. The solubility, i.e. the concentration of the protein solution Δc remaining around the picked-out crystal, being too high, this leads to a secondary nucleation around the crystal (Fig. 19C). Indeed, the remaining protein solution around the crystal diluted in the protein-free mother liquor is supersatured and induces a new nucleation. The crystallization conditions must therefore be a compromise between solubility which is sufficient to grow large crystals but not too high, so as to avoid secondary nucleation when the mother liquor is added.

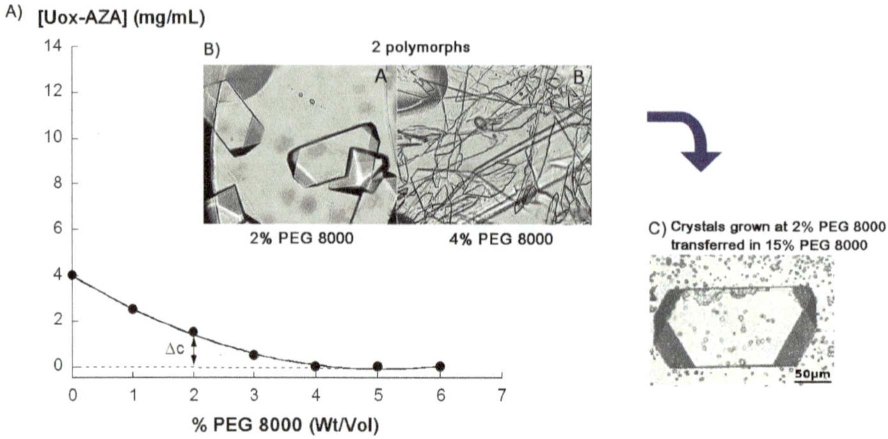

Fig. 19. A) Solubility of urate oxidase complexed with 8-azaxanthine, in Tris buffer pH 8, without salt, as a function of PEG 8000; B) Micrographs of urate oxidase crystals grown in microbatch at two PEG percentages; C) Micrograph of urate oxidase crystals transferred in 15% PEG 8000.

Therefore, to obtain a massive-habit crystal at high PEG percentage, we added 100mM NaCl, despite the salting-in effect which increases solubility, and determined the new solubility curve as a function of PEG 8000 (Fig. 20A).

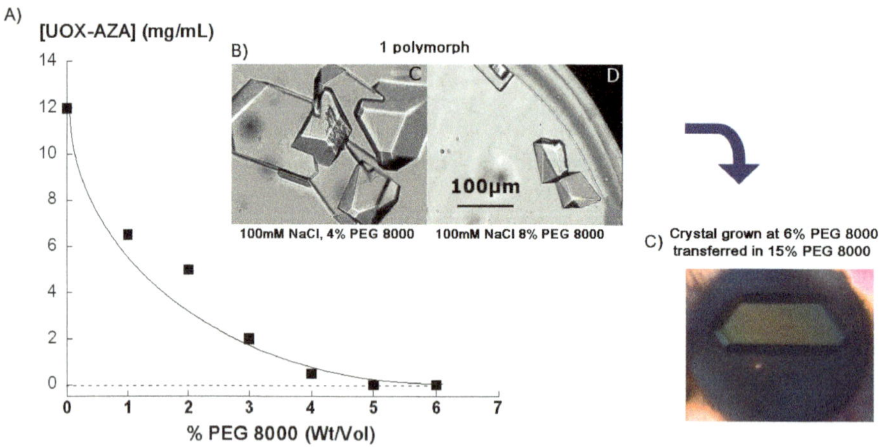

Fig. 20. A) Solubility of urate oxidase complexed with 8-azaxanthine, in Tris buffer pH 8, with 100mM NaCl, as a function of PEG 8000; B) Micrographs of urate oxidase crystals grown in microbatch at two PEG percentages; C) Micrograph of urate oxidase crystal transferred in 15% PEG 8000.

Crystal habits become more suitable for crystal diffraction studies (Fig. 20B) with sizes compatible with the pressure cell dimensions. We successfully transfer a crystal grown in 3.2mg/mL of urate oxidase 6% PEG 100mM NaCl Tris 50mM pH8 into a solution of 15% PEG 8000, 100mM NaCl, 50mM tris pH 8.0. It remains stable for more than one week without secondary nucleation appearing (Fig. 20C). At this concentration of protein, no micro-crystals are generated from the solution surrounding the crystal, whereas at lower PEG and higher protein concentrations in the surrounding solution, many micro-crystals grow.

5. Conclusion

Distribution of macromolecules in solution, phase diagrams and the crystallization process are governed by an appropriate combination of interaction potentials in solution. Three types of parameters or additives appear to play a crucial role: pH, salt and polymers. Whatever the diversity of protein sequences, it is possible, via second virial coefficient measurements, to limit the number of trials for a first screening of crystallization conditions and to rationalize crystallization. The design of small crystals or large diffracting crystals then results from a subtle mix of strong or weak repulsions and attractions, which control the position of the solubility curve and the metastable zone in the phase diagram.

6. Acknowledgment

This work was carried out by Denis Vivarès (1998-2003) and Marion Giffard (2006-2009) during their PhD theses. We gratefully acknowledge the generous support of Bertrand Castro, Mohamed El Hajji and François Ragot from Sanofi-Aventis (France) for this project on fundamental crystallization of urate oxidase since 1998, and their subsequent PhD financial support (2006-2009) enabling us to study the crystallization of urate oxidase for industrial applications.

7. References

Adams, M. L.; Lavasanifar, A. & Kwon, G. S. (2003). Amphiphilic block copolymers for drug delivery. *Journal of Pharmaceutical Sciences*, Vol. 92, No. 7, pp. 1343-1355,

Aleman, C.; Bayol, A.; Breul, T. & Dupin, P. (1998). Stable liquid composition containing urate oxidase and lyophilized composition for its preparation. Sanofi, Paris. United States

Arakawa, T. & Timasheff, S. (1985). Theory of protein solubility. *Methods in Enzymology*, Vol. 114, No. pp. 49-77, ISSN: 0076-6879

Asakura, S. & Oosawa, F. (1954). On the interaction between two bodies immersed in a solution of macromolecules. *J. Chem. Phys.*, Vol. 22, No. pp. 1255-1256,

Asherie, N.; Lomakin, A. & Benedek, G. B. (1996). Phase diagram of colloidal solutions. *Physical Review Letters*, Vol. 77, No. 23, pp. 4832-4835, ISSN: 0031-9007

Atha, D. H. & Ingham, K. C. (1981). Mechanism of precipitation of proteins by polyethylene glycols. Analysis in terms of excluded volume. *The Journal of biological chemistry*, Vol. 256, No. 23, pp. ISSN: 0021-9258

Bayol, A.; Dupin, P.; Boe, J. F.; Claudy, P. & Létoffé, J. M. (1995). Study of pH and temperature-induced transitions in urate oxidase (Uox-EC1.7.3.3) by microcalorimetry (DSC), size exclusion chromatography (SEC) and enzymatic activity experiments. *Biophysical Chemistry*, Vol. 54, No. 3, pp. 229-235, ISSN: 0301-4622

Behlke, J. & Ristau, O. (1999). Analysis of the thermodynamic non-ideality of proteins by sedimentation equilibrium experiments. *Biophysical Chemistry*, Vol. 76, No. 1, pp. 13-23, ISBN 0301-4622

Belloni, L. (1985). A hypernetted chain study of highly asymmetrical polyelectrolytes. *Chemical Physics*, Vol. 99, No. 1, pp. 43-54,

Belloni, L. (1988). Self-consistent integral equation applied to the highly charged primitive model. *J. Chem. Phys.*, Vol. 88, No. 8, pp. 5143-5148,

Belloni, L. (1991). Interacting monodisperse and polydisperse spheres, In: *Neutron, X-ray and light scattering,* Series, N.-H.D., pp. 135-155, Elsevier Science Publishers B.V., Amsterdam-Oxford-NewYork-Tokyo

Bergfors, T. (2009). Protein Crystallization ISBN-13: 978-0972077446,

Bhamidi, V.; Varanasi, S. & Schall, C. A. (2005). Protein Crystal Nucleation: Is the Pair Interaction Potential the Primary Determinant of Kinetics? *Langmuir*, Vol. 21, No. 20, pp. 9044-9050, ISBN 0743-7463

Bloustine, J.; Berejnov, V. & Fraden, S. (2003). Measurements of Protein-Protein Interactions by Size Exclusion Chromatography. *Biophysical Journal*, Vol. 85, No. 4, pp. 2619-2623, ISSN: 0006-3495

Bolhuis, P. G.; Louis, A. A.; Hansen, J. P. & Meijer, E. J. (2001). Accurate effective pair potential for polymer solutions. *J.Chem.Phys.*, Vol. 114, No. 9, pp. 4296-4311,

Bonneté, F. & Vivares, D. (2002). Interest of the normalized second virial coefficient and interaction potentials for crystallizing large macromolecules. *Acta Cryst*, Vol. D58, No. pp. 1571-1575, ISSN: 0907-4449

Bonneté, F.; Finet, S. & Tardieu, A. (1999). Second virial coefficient: variations with lysozyme crystallization conditions. *J. Crystal Growth*, Vol. 196, No. pp. 403-414, ISSN: 0022-0248

Bonneté, F.; Vivares, D.; Robert, C. & Colloc'h, N. (2001). Interactions in solution and crystallization of *Aspergillus flavus* urate oxidase. *J. Crystal Growth*, Vol. 232, No. pp. 330-339, ISSN: 0022-0248

Boué, F.; Lefaucheux, F.; Robert, M. C. & Rosenman, I. (1993). Small angle neutron scattering study of lysozyme solutions. *Journal of Crystal Growth*, Vol. 133, No. 3-4, pp. 246-254, ISBN 0022-0248

Boyer, M.; Roy, M.-O.; Jullien, M.; Bonnete, F. & Tardieu, A. (1999). Protein interactions in concentrated ribonuclease solutions. *J. Crystal Growth*, Vol. 196, No. 2-4, pp. 185-192,

Budayova, M.; Bonneté, F.; Tardieu, A. & Vachette, P. (1999). Interactions in solution of large oligomeric protein. *J. Crystal Growth*, Vol. 196, No. pp. 210-219, ISSN: 0022-0248

Budayova-Spano, M.; Dauvergne, F.; Audiffren, M.; Bactivelane, T. & Cusack, S. (2007). A methodology and an instrument for the temperature-controlled optimization of crystal growth. *Acta Crystallographica Section D-Biological Crystallography*, Vol. 63, No. pp. 339-347, ISSN: 0907-4449

Carbonnaux, C.; Riès-Kautt, M. & Ducruix, A. (1995). Relative effectiveness of various anions on the solubility of acidic Hypoderma lineatum collagenase at pH 7.2. *Protein Science*, Vol. 4, No. 10, pp. 2123-2128, ISSN: 0961-8368

Casselyn, M.; Perez, J.; Tardieu, A.; Vachette, P.; Witz, J. & Delacroix, H. (2001). Spherical plant viruses: interactions in solution, phase diagrams and crystallization of brome mosaic virus. *Acta Crystallographica Section D-Biological Crystallography*, Vol. 57, No. pp. 1799-1812, ISSN: 0907-4449

Chayen, N. E.; Stewart, P. D. S. & Blow, D. M. (1992). Microbatch Crystallization under Oil - a New Technique Allowing Many Small-Volume Crystallization Trials. *Journal of Crystal Growth*, Vol. 122, No. 1-4, pp. 176-180, ISSN: 0022-0248

Chernov, A. A. (2003). Protein crystals and their growth. *Journal of Structural Biology, Macromolecular crystallization in the structural genomics era*, Vol. 142, No. 1, pp. 3-21, ISSN: 1047-8477

Collings, I.; Watier, Y.; Giffard, M.; Dagogo, S.; Kahn, R.; Bonnete, F.; Wright, J. P.; Fitch, A. N. & Margiolaki, I. (2010). Polymorphism of microcrystalline urate oxidase from Aspergillus flavus. *Acta Crystallographica Section D-Biological Crystallography*, Vol. 66, No. pp. 539-548, ISSN: 0907-4449

Costenaro, L.; Zaccai, G. & Ebel, C. (2001). Understanding the crystallisation of an acidic protein by dilution in the ternary NaCl-2-methyl-2,4-pentanediol-H2O system. *J. Crystal Growth*, Vol. 232, No. pp. 102-113,

Debye, P. (1946). The Intrinsic Viscosity of Polymer Solutions. *J.Chem.Phys.*, Vol. 14, No. pp. 636-646,

Drenth, J. (2005). The nucleation of lysozyme from a fluctuation point of view. *Crystal Growth & Design*, Vol. 5, No. 3, pp. 1125-1127, ISSN: 1528-7483

Ducruix, A.; Guilloteau, J. P.; Riès-Kautt, M. & Tardieu, A. (1996). Protein interactions as seen by solution X-ray scattering prior to crystallogenesis. *J. Crystal Growth*, Vol. 168, No. pp. 28-39, ISSN 0022-0248

Faber, C.; Hobley, T.; Mollerup, J.; Thomas, O. R. T. & Kaasgaard, S. (2007). Study of the Solubility of a Modified Bacillus licheniformis r-Amylase around the Isoelectric Point. *Journal of Chemical and Engineering Data*, Vol. 52, No. pp. 707-713, ISSN: 0021-9568

Feher, G. & Kam, Z. (1985). Nucleation and growth of protein crystals: general principles and assays. *Method Enzymol.*, Vol. 114, No. pp. 77–112, ISSN: 00766879

Finet, S. & Tardieu, A. (2001). alpha-cristallin interaction forces studied by small angle X-ray scattering and numerical simulations. *J. Crystal Growth*, Vol. 232, No. 1-4, pp. 40-49, ISSN: 0022-0248

Finet, S.; Bonneté, F.; Frouin, J.; Provost, K. & Tardieu, A. (1998). Lysozyme crystal growth, as observed by small angle X-ray scattering, proceeds without crystallization intermediates. *European Biophysics Journal*, Vol. 27, No. 3, pp. 263-271, ISBN 0175-7571

Finet, S.; Skouri-Panet, F.; Casselyn, M.; Bonneté, F. & Tardieu, A. (2004). The Hofmeister effect as seen by SAXS in protein solutions. *Current Opinion in Colloid & Interface Science*, Vol. 9, No. 1-2, pp. 112-116, ISSN: 1359-0294

Garcia, L. A. (1975). Production of antisera comprising fractionating plasma or serum with an ethylene oxide-polyoxypropylene block copolymer. Baxter Laboratories, Inc. United States of America

García-Ruiz, J.-M. (2003). Nucleation of protein crystals. *Journal of Structural Biology, Macromolecular crystallization in the structural genomics era*, Vol. 142, No. 1, pp. 22-31, ISSN: 1047-8477

Gavira, J. A. & Garcia-Ruiz, J. M. (2009). Effects of a Magnetic Field on Lysozyme Crystal Nucleation and Growth in a Diffusive Environment. *Crystal Growth & Design*, Vol. 9, No. 6, pp. 2610-2615, ISSN: 1528-7483

George, A. & Wilson, W. W. (1994). Predicting protein crystallization from a dilute solution property. *Acta Crystallographica*, Vol. D50, No. 4, pp. 361-365, ISSN: 0907-4449

Gibb, C. L. D. & Gibb, B. C. (2011). Anion Binding to Hydrophobic Concavity Is Central to the Salting-in Effects of Hofmeister Chaotropes. *Journal of the American Chemical Society J. Am. Chem. Soc.*, Vol. 133, No. 19, pp. 7344-7347, ISSN: 0002-7863

Giffard, M.; Colloc'h, N.; Ferté, N.; Castro, C. & Bonneté, F. (2008). Salting-in effect on urate oxidase crystal design. *Crystal Growth and Design*, Vol. 8, No. pp. 4220-4226, ISSN: 1932-6203

Giffard, M.; Ferté, N.; Ragot, F.; El Hajji, M.; Castro, B. & Bonneté, F. (2011). Urate Oxidase Purification by Salting-in Crystallization: Towards an Alternative to Chromatography. *Plos One*, Vol. 6, No. 5, pp. WOS:000290483600005

Gripon, C.; Legrand, L.; Rosenman, I.; Vidal, O.; Robert, M. C. & Boué, F. (1997). Lysozyme-lysozyme interactions in under- and super-saturated solutions: a simple relation between the second virial coefficients in H2O and D2O. *Journal of Crystal Growth*, Vol. 178, No. 4, pp. 575-584, ISBN 0022-0248

Guilloteau, J.-P.; Rie`s-Kautt, M. M. & Ducruix, A. F. (1992). Variation of lysozyme solubility as a function of temperature in the presence of organic and inorganic salts. *Journal of Crystal Growth*, Vol. 122, No. 1-4, pp. 223-230, ISSN: 0022-0248

Guinier, A. & Fournet, G. (1955). Small angle scattering of X-rays Wiley, New York

Hansen, J. P. & McDonald, I. R. (1976). Theory of simple liquids Academic Press, London - New York - San Francisco

Haynes, C. A.; Tamura, K.; Korfer, H. R.; Blanch, H. W. & Prausnitz, J. M. (1992). Thermodynamic properties of aqueous .alpha.-chymotrypsin solution from membrane osmometry measurements. *The Journal of Physical Chemistry*, Vol. 96, No. 2, pp. 905-912, ISSN: 0022-3654

Hitscherich, C. J.; Kaplan, J.; Allaman, M.; Wiencek, J. & Loll, P. J. (2000). Static light scattering studies of OmpF porin : implications for integral membrane protein crystallization. *Protein Sci.*, Vol. 9, No. pp. 1559-1566, ISSN: 0961-8368

Israelachvili, J. (1994). Intermolecular and surface forces New York

Jacobsen, C.; Garside, J. & Hoare, M. (1998). Nucleation and growth of microbial lipase crystals from clarified concentrated fermentation broths. *Biotechnology and Bioengineering*, Vol. 57, No. 6, pp. 666-675, ISSN: 0006-3592

Lee, R. C.; Despa, F.; Guo, L.; Betala, P.; Kuo, A. & Thiyagarajan, P. (2006). Surfactant copolymers prevent aggregation of heat denatured lysozyme. *Annals of Biomedical Engineering*, Vol. 34, No. 7, pp. 1190-1200, ISI:000239085800012

Lekkerkerker, H. N. W. (1997). Strong, weak and metastable liquids. *Physica A*, Vol. 244, No. pp. 227-237, ISSN: 0378-4371

Liu, Y. X.; Wang, X. J. & Ching, C. B. (2010). Toward Further Understanding of Lysozyme Crystallization: Phase Diagram, Protein-Protein Interaction, Nucleation Kinetics, and Growth Kinetics. *Crystal Growth & Design*, Vol. 10, No. 2, pp. 548-558, ISSN: 1528-7483

Lomakin, A.; Asherie, N. & Benedek, G. B. (1996). Monte Carlo study of phase separation in aqueous protein solutions. *Journal of Chemical Physics*, Vol. 104, No. 4, pp. 1646-1656, ISSN: 0021-9606

Lorber, B.; Jenner, G. & Giégé, R. (1996). Effect of high hydrostatic pressure on nucleation and growth of protein crystals. *J. Crystal Growth*, Vol. 158, No. 1-2, pp. 103-117, ISSN: 0022-0248

Lovgren, S. (2007). Giant Crystal Cave's Mystery Solved. *National Geographic News*

Lutterbach, N.; Versmold, H.; Reus, V.; Belloni, L. & Zemb, T. (1999a). Charge-stabilized liquidlike ordered binary colloidal suspensions. 1. Ultra-Small-Angle X-ray Scattering characterization. *Langmuir*, Vol. 15, No. pp. 337-344, ISSN: 0743-7463

Lutterbach, N.; Versmold, H.; Reus, V.; Belloni, L.; Zemb, T. & Lindner, P. (1999b). Charge-stabilized liquidlike ordered binary colloidal suspensions. 2. Partial structure factors determined by Small-Angle Neutron Scattering. *Langmuir*, Vol. 15, No. pp. 345-352, ISSN: 0743-7463

Malfois, M.; Bonneté, F.; Belloni, L. & Tardieu, A. (1996). A model of attractive interactions to account for liquid-liquid phase separation of protein solutions. *J. Chem. Phys.*, Vol. 105, No. B, pp. 3290-3300, ISSN: 0021-9606

McGrath, B. & Walsh, G. (2005). Directory of Therapeutic Enzymes CRC Press Inc, ISBN-10: 0849327148,

McPherson, A. (1976). Crystallization of proteins from polyethylene glycol. *The Journal of biological chemistry*, Vol. 251, No. 20, pp. 6300-6303, ISSN: 0021-9258

McPherson, A.; Malkin, J. A.; Kuznetsov, Y. G.; Koszelak, S.; Wells, M.; Jenkins, G.; Howard, J. & Lawson, G. (1999). The effects of microgravity on protein crystallization: evidence for concentration gradients around growing crystals. *Journal of Crystal Growth*, Vol. 196, No. 2-4, pp. 572-586, IBSN 0022-0248

McQuarrie, D. A. (2000). Statistical mechanics Sausalito, CA

Mikol, V.; Hirsch, E. & Giegé, R. (1989). Monitoring protein crystallization by dynamic light scattering. *FEBS Letters*, Vol. 258, No. 1, pp. 63-66, ISBN 0014-5793

Muschol, M. & Rosenberger, F. (1995). Interactions in under- and supersaturated lysozyme solutions. Static and dynamic light scattering results. *J. Chem. Phys.*, Vol. 103, No. 24, pp. 10424-10432, ISSN: 0021-9606

Ng, J. D.; Gavira, J. A. & García-Ruíz, J. M. (2003). Protein crystallization by capillary counterdiffusion for applied crystallographic structure determination. *Journal of Structural Biology, Macromolecular crystallization in the structural genomics era*, Vol. 142, No. 1, pp. 218-231, ISBN 1047-8477

Oksanen, E.; Blakeley, M. P.; Bonnete, F.; Dauvergne, M. T.; Dauvergne, F. & Budayova-Spano, M. (2009). Large crystal growth by thermal control allows combined X-ray and neutron crystallographic studies to elucidate the protonation states in Aspergillus flavus urate oxidase. *Journal of the Royal Society Interface*, Vol. 6, No. pp. S599-S610, ISSN: 1742-5689

Pande, A.; Pande, J.; Asherie, N.; Lomakin, A.; Ogun, O.; King, J. & Benedek, G. B. (2001). Crystal cataracts: human genetic cataract caused by protein crystallization. *PNAS*, Vol. 98, No. 11, pp. 6116-6120, ISBN 0027-8424

Poon, W. C. K. (2002). The physics of a model colloid-polymer mixture. *Journal of Physics-Condensed Matter*, Vol. 14, No. 33, pp. R859-R880, ISSN: 0953-8984

Retailleau, P.; Colloc'h, N.; Vivares, D.; Bonneté, F.; Castro, B.; El Hajji, M.; Mornon, J. P.; Monard, G. & Prangé, T. (2004). Complexed and ligand-free high resolution structures of Urate oxidase (Uox) from *Aspergillus flavus*: a re-assignation of the active site binding mode. *Acta Cryst*, Vol. D60, No. pp. 453-462, ISSN: 0907-4449

Retailleau, P.; Colloc'h, N.; Vivares, D.; Bonneté, F.; Castro, B.; El Hajji, M. & Prangé, T. (2005). Urate oxidase from Aspergillus flavus: new crystal-packing contacts in relation to the content of the active site. *Acta Cryst.*, Vol. D61, No. pp. 218-229, ISSN: 0907-4449

Riès-Kautt, M. & Ducruix, A. (1999). From solution to crystals with a physico-chemical aspect, In: *Crystallization of nucleic acids and proteins - A practical approach*, Giegé, R., pp., Oxford University Press, Oxford, New York, Tokyo

Sauter, C.; Lorber, B.; McPherson, A. & Giegé, R. (2011). Crystallization, In: *General methods In International Tables of Crystallography; Crystallography of Biological Macromolecules*, pp. In press., Chichester

Sazaki, G.; Moreno, A. & Nakajima, K. (2004). Novel coupling effects of the magnetic and electric fields on protein crystallization. *Journal of Crystal Growth*, Vol. 262, No. 1-4, pp. 499-502, ISBN 0022-0248

Schmolka, I. (1977). A review of block polymer surfactants. *Journal of the American Oil Chemists' Society*, Vol. 54, No. 3, pp. 110-116,

Sumner, J. (1926). The isolation and crystallization of the enzyme urease. Preliminary paper. *Journal of Biological Chemistry*, Vol. 69, No. 2, pp. 435-441,

Takáts, Z.; Vékey, K. & Hegedüs, L. (2001). Qualitative and quantitative determination of poloxamer surfactants by mass spectrometry. *Rapid Communications in Mass Spectrometry*, Vol. 15, No. 10, pp. 805-810,

Tardieu, A.; Le Verge, A.; Riès-Kautt, M.; Malfois, M.; Bonneté, F.; Finet, S. & Belloni, L. (1999). Proteins in solution : from X-ray scattering intensities to interaction potentials. *J. Crystal Growth*, Vol. 196, No. pp. 193-203, ISSN: 0022-0248

ten Wolde, P. R. & Frenkel, D. (1997). Enhancement of protein crystal nucleation by critical density fluctuations. *Science*, Vol. 277, No. pp. 1975-1978,

Tessier, P. M.; Vandrey, S. D.; Berger, B. W.; Pazhianur, R.; Sandler, S. I. & Lenhoff, A. M. (2002). Self-interaction chromatography: a novel screening method for rational protein crystallization. *Acta Cryst.*, Vol. D58, No. 10 Part 1, pp. 1531-1535, ISSN: 0907-4449

Vekilov, P. G. (2010). Nucleation. *Crystal Growth & Design*, Vol. 10, No. 12, pp. 5007-5019, WOS:000284675100001

Velev, O. D.; Kaler, E. W. & Lenhoff, A. M. (1998). Protein interactions in solution characterized by light and neutron scattering: comparison of lysozyme and chymotrypsinogen. *Biophys. J.*, Vol. 75, No. 6, pp. 2682-2697, ISSN: 0006-3495

Vérétout, F.; Delaye, M. & Tardieu, A. (1989). Molecular basis of eye lens transparency. Osmotic pressure and X-ray analysis of alpha-crystallin solutions. *J. Mol. Biol.*, Vol. 205, No. 4, pp. 713-728, ISSN: 0022-2836

Visuri, K.; Kaipainen, E.; Kivimaki, J.; Niemi, H.; Leisola, M. & Palosaari, S. (1990). A New Method for Protein Crystallization Using High Pressure. *Nature Biotechnology*, Vol. 8, No. 6, pp. 547-549,

Vivares, D. & Bonneté, F. (2002). X-ray scattering studies of Aspergillus flavus urate oxidase: towards a better understanding of PEG effects on the crystallization of large proteins. *Acta Cryst.*, Vol. D58, No. pp. 472-479, ISSN: 0907-4449

Vivares, D.; Astier, J.-P.; Veesler, S. & Bonnete, F. (2006). Polymorphism of urate oxidase in PEG solutions. *Crystal Growth & Design*, Vol. 6, No. pp. 287-292, ISSN: 1528-7483

Vivares, D.; Belloni, L.; Tardieu, A. & Bonneté, F. (2002). Catching the PEG-induced attractive interaction between proteins. *Eur. Phys. J. E.*, Vol. 9, No. pp. 15-25,

Weber, M.; Jones, M. J. & Ulrich, J. (2008). Crystallization as a Purification Method for Jack Bean Urease: On the Suitability of Poly(Ethylene Glycol), Li2SO4, and NaCl as Precipitants. *Cryst. Growth Des.*, Vol. 8, No. 2, pp. 711-716, ISBN 1528-7483

Wulff-Pérez, M.; Torcello-Gómez, A.; Gálvez-Ruíz, M. J. & Martín-Rodríguez, A. (2009). Stability of emulsions for parenteral feeding: Preparation and characterization of o/w nanoemulsions with natural oils and Pluronic f68 as surfactant. *Food Hydrocolloids*, Vol. 23, No. 4, pp. 1096-1102,

Ye, X.; Narayanan, T.; Tong, P.; Huang, J. S.; Lin, M. Y.; Carvalho, B. L. & Fetters, L. J. (1996). Depletion interactions in colloid-polymer mixtures. *Phys. Rev. E*, Vol. 54, No. 6, pp. 6500-6510, ISSN: 1063-651X

Zheng, B.; Gerdts, C. J. & Ismagilov, R. F. (2005). Using nanoliter plugs in microfluidics to facilitate and understand protein crystallization. *Current Opinion in Structural Biology*, Vol. 15, No. pp. 548–555, ISSN: 0959-440X

6

Phenacetin Crystallization: Cooling Regimes and Crystal Morphology

Humphrey A. Moynihan and Dawn M. Kelly
Dept. of Chemistry / Analytical and Biological Chemistry Research Facility, University College Cork, Cork Ireland

1. Introduction

Phenacetin crystallizations were undertaken to gain an insight into how to better control the various parameters of a crystallization process. Good control of the crystallization process allows the design of experiments to control certain characteristics of the final crystal product. These include crystal morphology, phase, particle size and crystal size distribution (CSD). These physical attributes of a crystal population are thought to be achieved through the control of the nucleation and growth of crystals during the crystallization process. One of the first attempts to control crystallization processes was proposed by Griffiths (Griffiths, 1925), who suggested the idea of "controlled cooling" in batch crystallizations by maintaining the supersaturation in the metastable region in order to improve the product CSD. Garside et al. performed experimental studies employing this concept of controlled cooling in crystallization processes, which included simple strategies such as isothermal operation and linear cooling (Garside et al., 1972). Since these early crystallization control approaches, there have been many strategies that indirectly affect the CSD of a crystallization process, often involving a change in temperature or the addition of an antisolvent during a pre-defined timescale so as to follow a given supersaturation profile in the phase diagram. These profiles are obtained using simple trial-and-error experimentation or more complex model-based or direct-design approaches (Abu Bakar et al., 2009a). Woo et al. proposed an adaptive concentration control strategy that employs the measurement of the number of particle counts per unit time provided by *in situ* laser backscattering in order to detect the onset of nucleation and adapt the operating curve accordingly (Woo et al., 2009). In a move towards new strategies for directly affecting the CSD of a crystallisation process, Abu Baker et al. (Abu Bakar et al., 2009a) used a model-free approach to crystallization control known as 'direct nucleation control' (DNC), in which the number of counts measured by an *in situ* laser probe is directly controlled using a feedback control strategy. An alternative strategy is to chart an oscillating cooling regime within the metastable zone of the crystallization, in which the overall cooling trend drives crystal growth while the temperature fluctuations provided by the oscillating cool allows partial dissolution (Abu Bakar et al., 2009b). The result should be an equalizing of crystal size and shape. Both of these approaches were utilized in the present study. Phenacetin was selected as the subject for this study as it is an Active Pharmaceutical Ingredient - now withdrawn (Prescott, 1980) - and is not known to be crystal polymorphic (Hansen, 2006).

This chapter will present work on phenacetin crystallization using two forms of modified cooling regimes in an attempt to directly affect the product CSD of the crystallization process. We begin with a cooling regime using turbidity counts as a termination clause for each cooling and heating step. We then move to a more in-depth study in Section 3.2, where the cyclic cooling paths that had predefined temperature parameters were used, ensuring that a cyclic temperature profile was maintained within the limits of the metastable zone width (MSZW). This temperature profile was coupled with a seeding regime where each experiment was seeded with crystals that were sieved to give specific size ranges of less than 150 μm (fine particles), between 150-300 μm (medium particles) and over 500 μm (coarse particles). The objective of both of these types of modified cooling regimes is to provide a narrower range of crystal sizes and shapes than would otherwise be obtained by a simple linear cooling regime.

2. Experimental methods

Phenacetin was purchased from Sigma-Aldrich. GPR grade absolute ethanol was purchased from Carbon Group Chemical Ltd. (Cork, Ireland) and was used in all of the experiments. Ethanol was chosen as a solvent for these experiments because the solubility of phenacetin in ethanol varies significantly over the temperature range 10 °C to 30 °C, allowing for the development of a practical crystallization method. The determination of phenacetin solubility and metastable zone width is described below. Phenacetin crystallizes cleanly from ethanol to give well-formed needles which are largely free from the inclusions of other defects.

A HEL Autolab jacketed reactor vessel equipped with a HEL Lasertrack *in situ* particle sizing laser probe was used to monitor the presence and size of phenacetin crystals on a 1 L scale. The laser wavelength was 795 nm. It is possible to get readings of particles as small as 0.5 μm. A PTFE PT100 thermocouple gave *in situ* temperature measurements of the crystallization medium. The temperature of the jacket fluid (Huber DW-Therm thermal fluid, operating range −90 °C to 200 °C) was controlled by a Huber Unistat 815 circulation thermostat. The system was entirely controlled from one PC using HEL WinISO software, allowing the control of stirring rates, the addition of material (the pumping rate and mass pumped), and the heating parameters (heating or cooling rates, reactor vessel temperature set point, jacket fluid temperature set point). The software allowed the pre-programming of crystallization regimes containing as many separate steps as required, e.g., multiple heat/cool steps, solvent addition steps, etc. The steps could be programmed to terminate and move to a next step upon reaching a specified set point, such as the reactor temperature or the number of particles observed.

Crystal habits were observed using a Nikon Polarizing Microscope Eclipse 50i POL and photomicrographs were taken on a Nikon Digital Sight DS-Fi1 digital camera. Figure 1 shows a schematic representation of the experimental setup for phenacetin crystallization experiments.

In the experiments described below, in which seeding was used, crystals obtained from water were used as seed crystals because they typically have a plate-like habit. The plate-like habit provides a greater surface area for growth and so is more suitable for seeds than the needle-like habit obtained from ethanol. The crystals formed from water also have rougher surfaces than the ethanol-derived needles, again assisting crystal growth. In addition, any differences in habit after the crystallization of phenacetin from ethanol using these seed crystals would assist in the determination of whether growth or nucleation dominated these processes. The seed crystals were sieved to give the size ranges described below using a set of U.S.A. standard testing sieves (Avantech Manufacturing) in mesh sizes of 150 μm, 300 μm and 500 μm. The seeds were added to crystallisations in an ethanol slurry using seed crystals with a weight in grams equal to 20% of the total batch concentration.

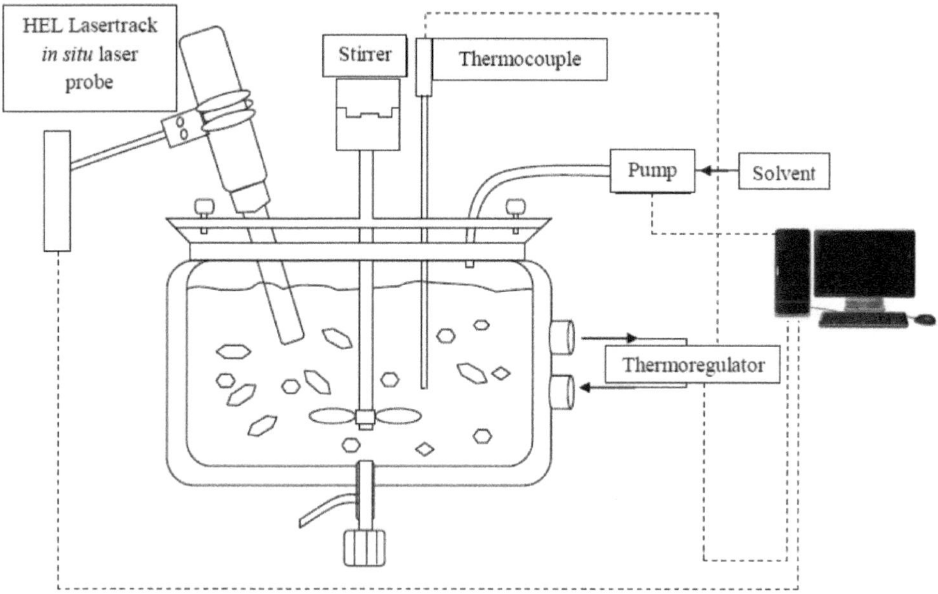

Fig. 1. A schematic representation of the experimental setup for phenacetin crystallization experiments.

Only one crystalline form of phenacetin has been reported to date (Hansen, 2006). To check that no new polymorphs or solvates of phenacetin were obtained during these experiments, powder X-ray diffraction (PXRD) patterns were recorded for all of the samples and compared with the theoretical pattern generated from the structure reported by Hansen et al. (Hansen, 2006). PXRD was performed at an ambient temperature using a Stoe Stadi MP diffractometer operating in transmission mode with a linear PSD detector, with an anode current of 40 mA, an accelerating voltage of 40 kV and Cu Kα_1 X-radiation (λ = 1.5406 Å), typically over a scan range of 3.5° to 60° 2θ, scanning in steps of 2° for 90 s per step. The samples were held between acetate foils. The theoretical patterns were generated from the crystallographic information file downloaded from the Cambridge Structural Database (CSD) - CSD reference code PYRAZB21 - using the THEO function on the Stoe WinXPOW software package. Figure 2 shows a typical example of a PXRD pattern of a phenacetin batch with the theoretical pattern overlaid, with excellent correspondence between the observed and theoretical patterns.

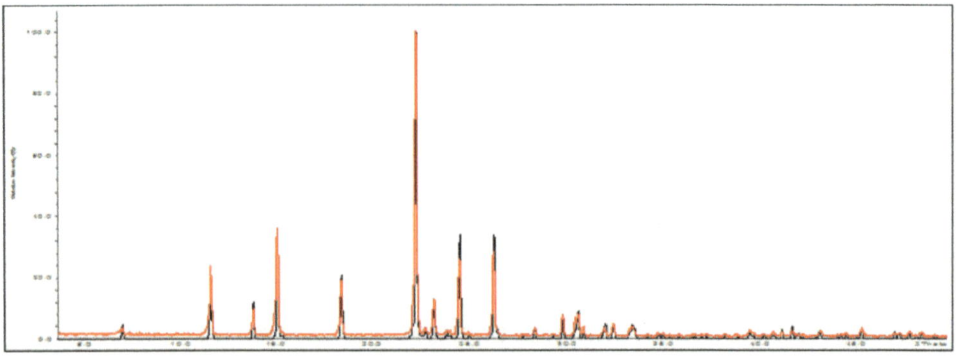

Fig. 2. Comparison of the theoretical PXRD pattern of phenacetin (black) generated from single crystal data (Hansen, 2006; CSD reference code PYRAZB21) overlaid with the experimental pattern (red) of a batch of phenacetin crystallized from ethanol.

3. Results and discussion

3.1 Cooling with feedback control

The metastable zone width (MSZW) of phenacetin from ethanol was first found by a heat / cool experiment in which the concentration was diluted after each heat /cool step until enough data points were collected to plot a solubility/super-solubility diagram. Figure 3 is the experiment progression diagram, illustrating the alternate heating and cooling cyclic programme from which the data points were collected, and Figure 4 is the solubility/super-solubility diagram of phenacetin in ethanol.

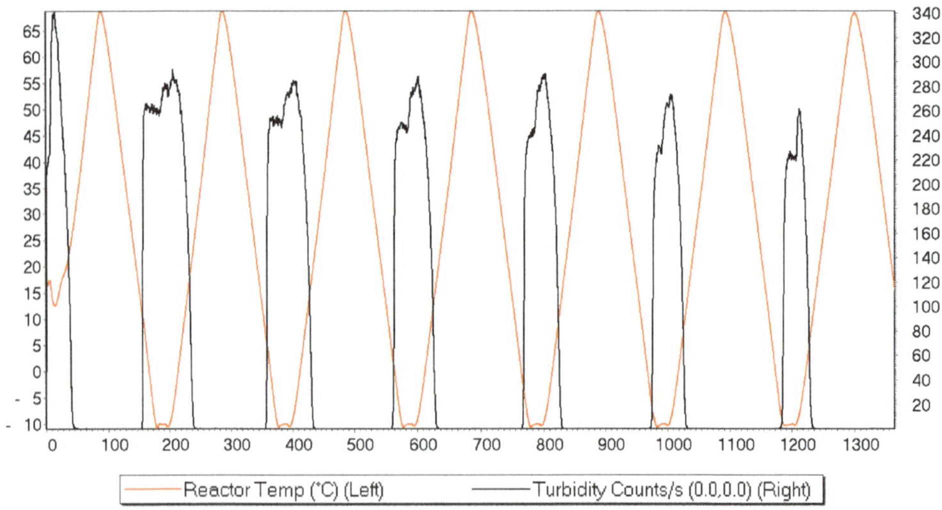

Fig. 3. The experiment progression diagram for measuring the metastable zone width (MSZW) of phenacetin in ethanol.

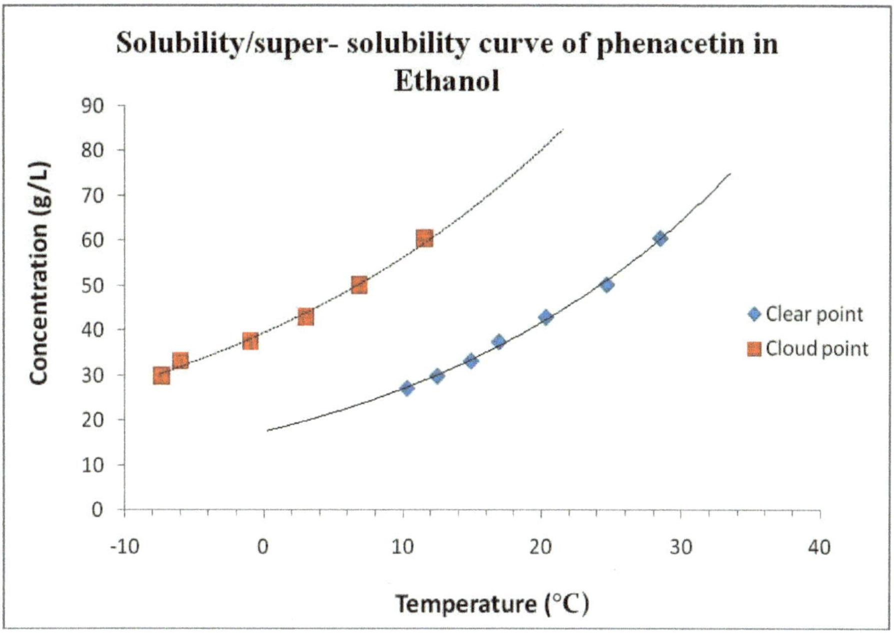

Fig. 4. The solubility/super-solubility diagram of phenacetin from ethanol.

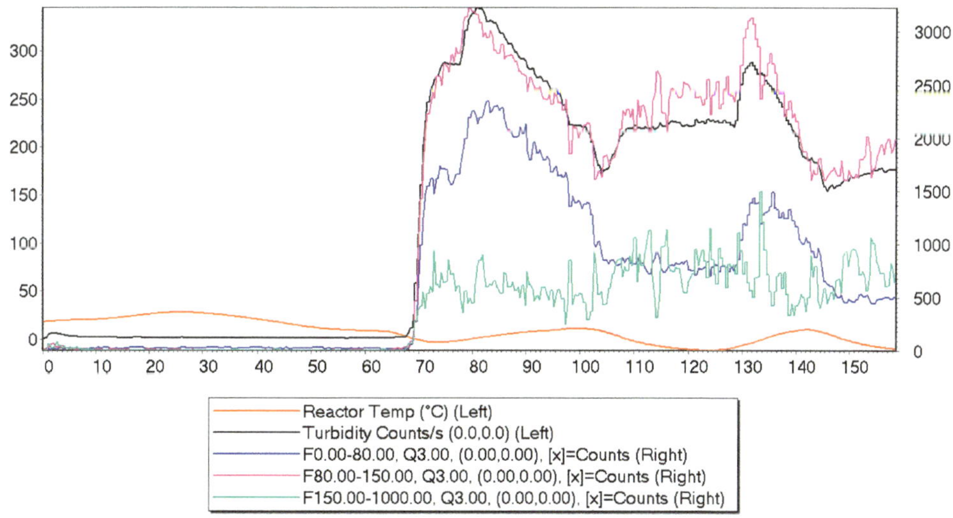

Fig. 5. The progression diagram of a heat/cool experiment of phenacetin in ethanol using turbidity counts as a termination clause (F is the number of particles inside the given size range - e.g., the size range 0.00-80.00 is the content between 0 - 80 µm – and Qz(x) is the less-than-size for parameter x and weight z, where x is counts and z = 3 is interpreted as the volume equivalent distribution).

Our first experiments were undertaken using the phenacetin crystallizations in ethanol involved controlling the crystallization using turbidity counts as a termination clause for each heating and cooling step. The first experiment involved a batch without seeding. Phenacetin, at a concentration of 50 g L^{-1}, was dissolved in ethanol, stirred at 180 rpm and followed by a slow cool at -0.5 °C / min until the phenacetin crystallized out of solution. The termination clause was set on the cooling step so that the cooling was terminated when the turbidity counts exceeded a set number. In this case, the programme was set to cool until the crystal population reached a turbidity of 300 counts / s. Once the turbidity counts reached the pre-set level, the experiment began to heat again until the turbidity counts dropped below another pre-set number. In this experiment, the lower limit was set at 290 counts / s, thereby terminating the step and beginning the cooling process once again. Ideally, this cyclic heating and cooling should be continued until a steady state is reached, which should result in a crystallization process where the turbidity counts are maintained at the pre-set number. Figure 5 gives this experiment's progression diagram of phenacetin in ethanol under this regime.

Using turbidity counts as the terminating clause for controlling the temperature in crystallization batches that were not seeded proved difficult. The temperature response to this terminating clause was slow, causing large fluctuations in the turbidity counts and the fine and medium total particle counts. The coarse particle counts remained relatively stable. The crystals collected at the end of the experiment were needles of varying size. PXRD analysis proved that no new forms of phenacetin were present. Since the experiment was carried out in ethanol, the crystallization of needle shaped crystals was expected. Figure 6 is a photomicrograph of the crystals collected at the end of the experiment.

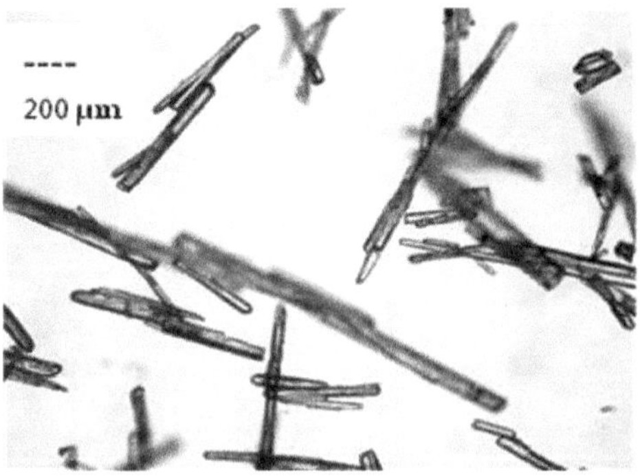

Fig. 6. Photomicrograph of crystals at the end of the experiment from a batch of a phenacetin crystallization without seeding.

It was then proposed to seed a phenacetin crystallization experiment, using the same conditions as before and with seeds of varying sizes, in order to see if better control could be achieved. It was decided to use phenacetin crystals obtained from water as the seed crystals because they typically have a plate-like habit. Any differences in habit after the crystallization of phenacetin from ethanol using these seed crystals would assist the determination of whether growth or nucleation was dominating these processes. Figure 7 is an experimental design showing the proposed heating and cooling path in the seeded experiment and Figure 8 is the experiment progression diagram of a seeded phenacetin crystallization in ethanol using turbidity counts as the termination clause.

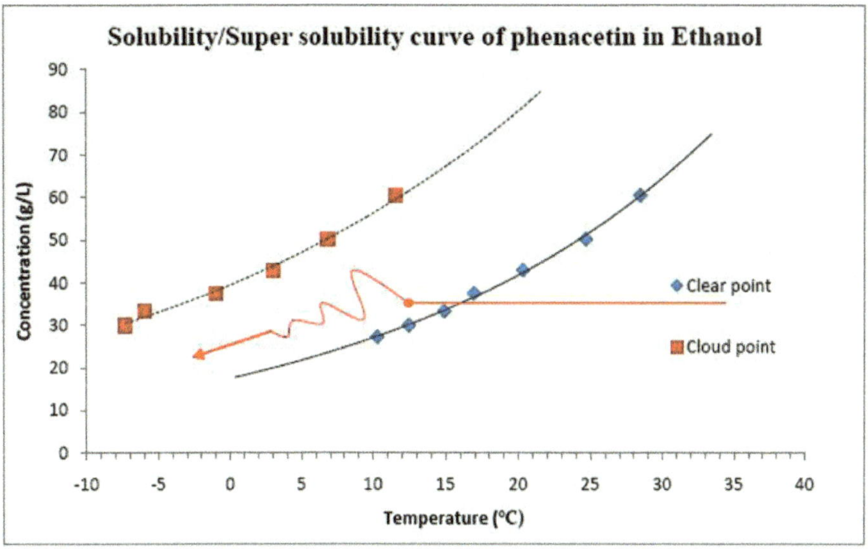

Fig. 7. Design of an experiment showing the proposed heating and cooling path using turbidity counts as the termination clause for each step (the red dot indicates the point of seeding).

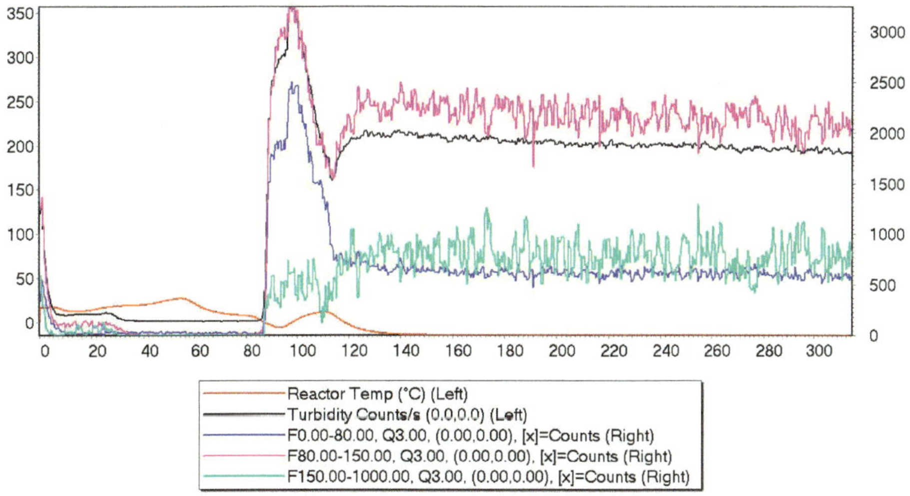

Fig. 8. The experiment progression diagram of a seeded phenacetin crystallization in ethanol using turbidity counts as a termination clause (F is the number of particles inside the given size range - e.g., the size range 0.00-80.00 is the content between 0 - 80 µm - and Qz(x) is the less-than-size for parameter x and weight z, where x is counts and z = 3 is interpreted as the volume equivalent distribution)

The use of turbidity counts as the terminating step to control the temperature in seeded phenacetin crystallizations resulted in a more controlled experiment. The crystallization was slowly cooled to within the solubility limit at a cooling rate of -0.5 °C / min. The batch was seeded with phenacetin crystals of various sizes and the crystallization was rapidly cooled at a rate of -5 °C / min. When the turbidity counts exceeded the stated level - which in this experiment was a turbidity of 200 counts / s - the process began to heat, causing the dissolution of fines and medium particles. Once the turbidity counts had reached the desired level (190 counts / s), cooling began again. The coarse and medium particles began to rise (indicating growth) while the fines remained steady. Figure 9 shows photomicrographs of the seed crystals, which are of varying size, and the end of experiment crystals, which show a more uniform crystal size and are no longer needle-like. The final crystals have a regular prismatic habit, unlike than the needles obtained from ethanol without seeding or the plates obtained from water. This reflects the operation of crystal growth from ethanol on plate-like seeds obtained from water, generating a new and more desirable prismatic habit.

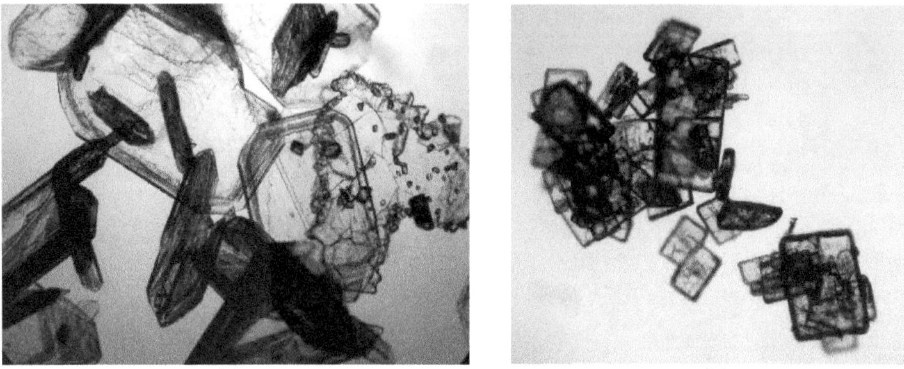

(a) (b)

Fig. 9. Photomicrographs of (a) the seed crystals of phenacetin of varying sizes and (b) end of experiment crystals.

3.2 Cyclic cooling

The next series of experiments used cyclic cooling paths that had predefined temperature parameters, ensuring that a cyclic temperature profile was maintained within the MSZW limits. Rather than using the turbidity counts as a termination clause, these experiments followed plans that were programmed before the experiment began, using the data from the MSZW diagram to find the temperature limits. Figure 10 shows the design of the experiment for this crystallization.

In order to achieve the best control, it was decided to carry out a series of phenacetin crystallizations from ethanol, keeping the crystallization conditions constant throughout but varying the seed crystals. Each experiment was seeded with crystals that were sieved to give a specific size range. The three size ranges used were seed crystals under 150 μm, between 150-300 μm and over 500 μm. All of the crystallization experiments were seeded with crystals grown from water and added to an ethanol slurry using seed crystals with a weight in grams equal to 20% of the total batch concentration. Figure 11 gives the experiment progression diagram of the phenacetin crystallization from ethanol that was seeded with crystals grown from water within the size range 150-300 μm.

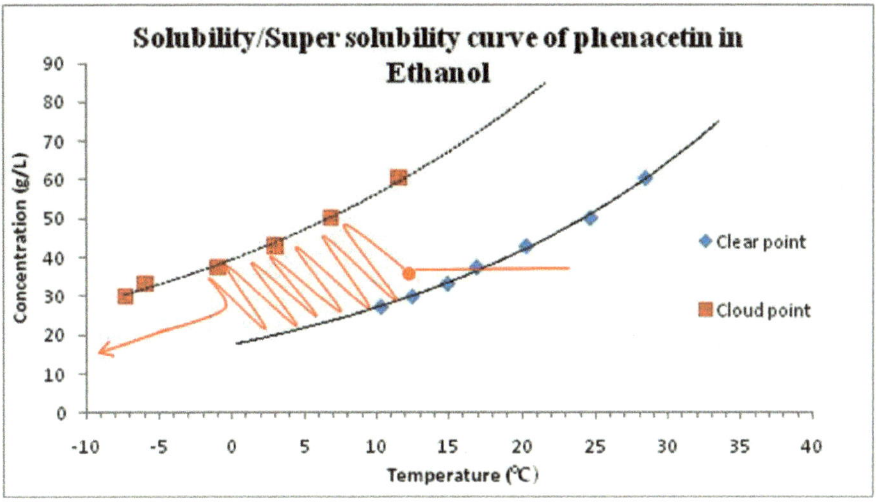

Fig. 10. An experiment design illustrating the pre-set path of the cyclic heating and cooling programme for phenacetin crystallizations in ethanol.

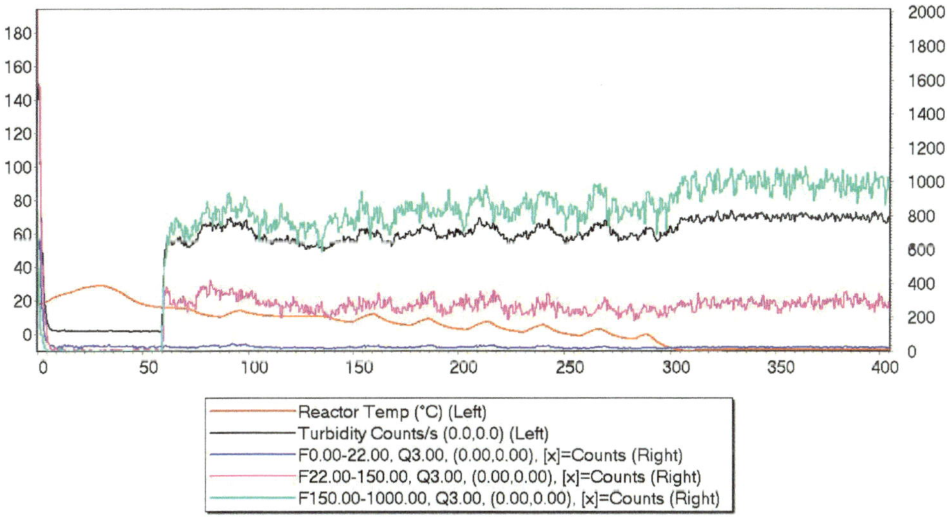

Fig. 11. Experiment progression diagram of the seeded (crystals between 150-300 µm) phenacetin crystallization from ethanol using a pre-set cyclic plan (F is the number of particles inside the given size range - e.g., the size range 0.00-80.00 is the content between 0 - 80 µm - and Qz(x) is the less-than-size for parameter x and weight z, where x is counts and z = 3 is interpreted as the volume equivalent distribution).

The experiment progression diagram shows the point of seeding as the point where there is a simultaneous increase in the turbidity counts (the black line), the medium-sized particles (the pink line indicating particles between 80-150 µm) and the coarse particles (light blue line indicating particles between 150-1000 µm). It should be noted that at this point there is no increase in the fine particles (the dark blue line indicating particles between 0-80 µm). This is expected as the crystallization is seeded when the temperature just reaches the limit of the solubility curve, indicating a saturated solution and that spontaneous nucleation has not occurred. Therefore, the only particles in the solution are the seed crystals. Since these crystals are within the size range of 150-300 µm, no fine particles are present at this stage. As the solution is cooled in a cyclic heat/cool manner, any nucleation will register a rise in the counts for the fraction, indicating fine particles (the dark blue line). It can be seen from the above diagram that the dark blue line stays close to the baseline throughout the experiment, with some small intermittent increases, indicating that a small amount of nucleation occurs when the crystallization is cooling and that these fine particles are being dissolved during heating. This illustrates that this cyclic heating and cooling program is effective in controlling the nucleation process and riding the sample of fine particles, which tend to be problematic when filtering large scale crystallizations. This experiment progression diagram shows that throughout the crystallization, the medium-sized particles (the pink line) are maintained at a steady level with a slight increase in the coarse particles (the light blue) indicating growth. Figure 12 presents photomicrographs of the seed crystals grown from water and the end of experiment crystals.

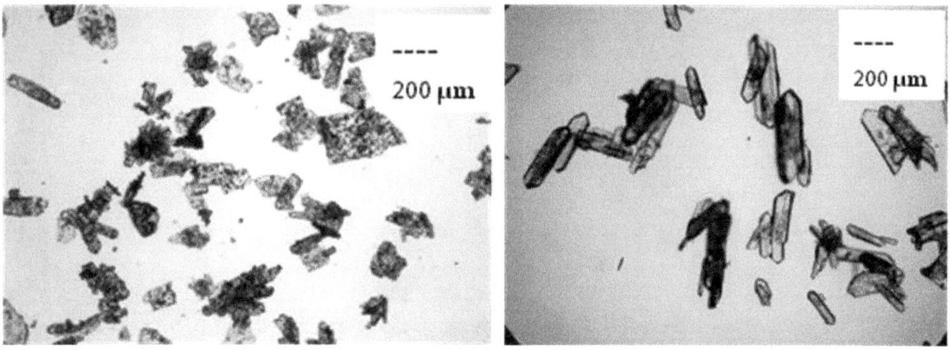

(a) (b)

Fig. 12. Photomicrographs of (a) seed crystals between 150-300 µm and (b) end of experiment crystals.

The experiment was seeded with crystals grown from water which were plate-like in shape, and the end of experiment crystals were elongated plate-like needles. If nucleation occurs after the experiment was seeded, then these crystals will be very needle-like in shape. If the crystallization predominantly favours growth, then those seed crystals that are plate-like to begin with would have a distorted shape due to the solvent effects on the phenacetin crystals. In theory, if very fine particles precipitate during the cooling process, indicating nucleation, then these would be dissolved on the heating step. This continual heating and cooling would ensure that enough time is given to the growth of the seed crystals, while at the same time ridding the crystallization process of fine particles. The heating cycle would also ensure that the very large crystals are dissolved slightly - thereby reducing their size - the result of which would likely be a crystal batch where there is a good crystal size distribution, with the majority of the crystals falling under a mean crystal size. Figure 13 gives the particle size distributions of the crystals at the beginning and at the end of the experiment.

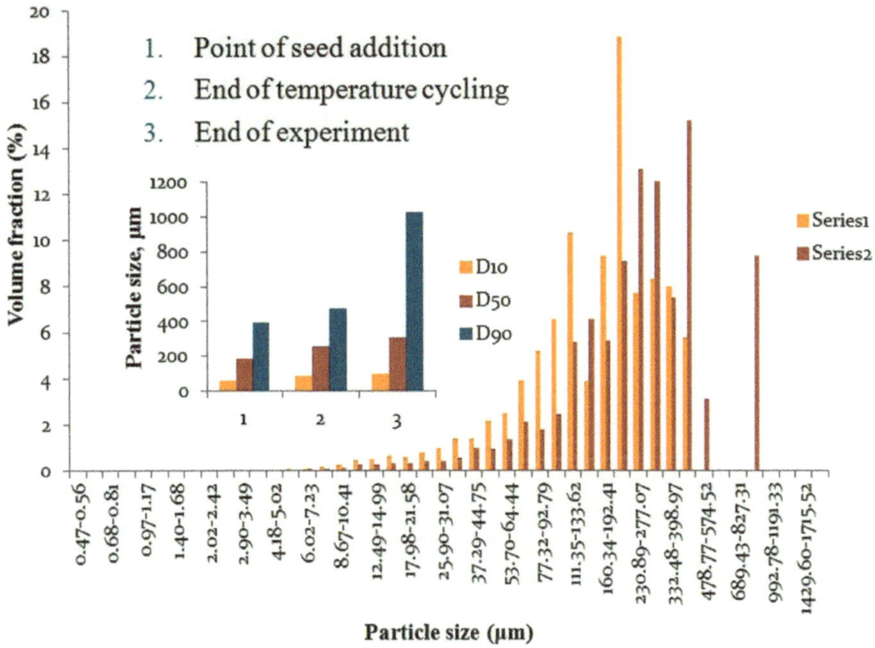

Fig. 13. Crystal size distributions of crystals shortly after the experiment is seeded (series 1) and at the end of the experiment (series 2), showing an increase in the particle size distribution and indicating growth (inset is a bar chart of the D10, D50 and D90 distributions at three points throughout the experiment).

The crystal size distributions in Figure 13 show an upward shift in the particle size from the beginning of the experiment (series 1) to the end of the experiment (series 2), which is indicative of growth. Also shown in Figure 13 are the distributions D10 (meaning the size under which 10% of the crystal population falls), D50 (the size under which 50% of the crystal population falls) and D90 (the size under which 90% of the crystal population falls). The inset of Figure 13 gives these size distributions for the crystal population at the beginning of the experiment, after the cyclic heat/cool program and at the end of the experiment. It agrees with the particle size distributions and also indicates growth. Figure 14 displays photomicrographs of a seed crystal and an end of experiment crystal.

The photographs of the crystals at the beginning and the end of the experiment show the growth of those seed crystals that began plate-like and then grew to elongated plate-like needles. The difference in size also indicates growth. When we take the seed crystal into consideration, we can see that the crystals are flat and plate-like in shape, with round edges and a rough surface. In contrast, the end of experiment crystals tend to be elongated plate-like needles with a rigid, defined shape and a smooth surface.

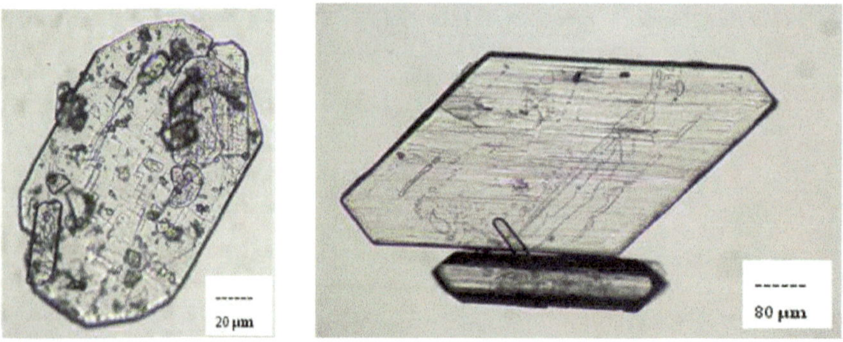

Fig. 14. Photomicrographs illustrating the difference between the seed crystals and the end of experiment crystals.

This experiment was repeated using the same experimental conditions and seeding with crystals (grown from water) under 150 µm. Figure 15 gives the experiment progression diagram of this crystallization. The experiment progression diagram in Figure 15 shows the point of seeding at the point where the turbidity counts (the black line) and the fine (the dark blue line), medium (the pink line) and coarse (the light blue line) particles increase simultaneously. During the cyclic heating and cooling, there is a steady decrease in the fine particle counts (the dark blue line) and a steady increase in the coarse particle counts (the light blue line), both trends indicating growth. The medium particle counts (pink line) remained steady. Figure 16 displays photomicrographs of the seed crystals under 150 µm and the end of experiment crystals. Once again, the crystals grow to form plate-like needles at the end of the experiment.

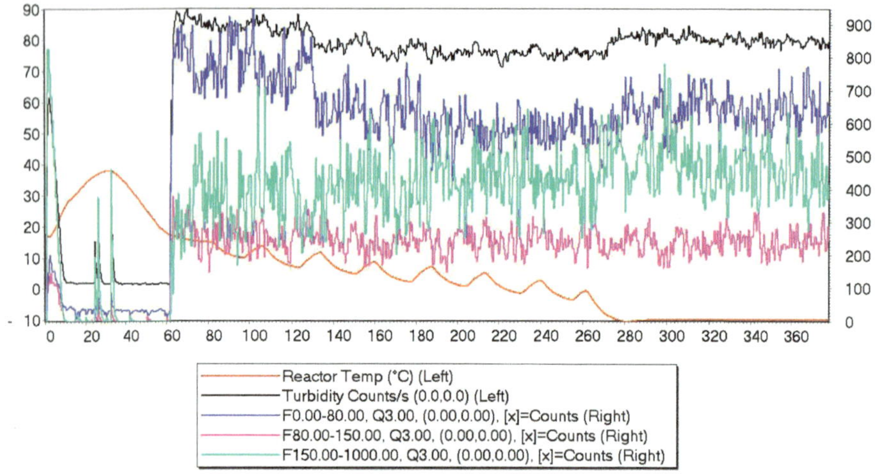

Fig. 15. The experiment progression diagram of a pre-set cyclic heating and cooling crystallization that was seeded with crystals under 150 µm, grown from water (F is the number of particles inside the given size range - e.g., the size range 0.00-80.00 is the content between 0 - 80 µm - and Qz(x) is the less-than-size for parameter x and weight z, where x is counts and z = 3 is interpreted as the volume equivalent distribution).

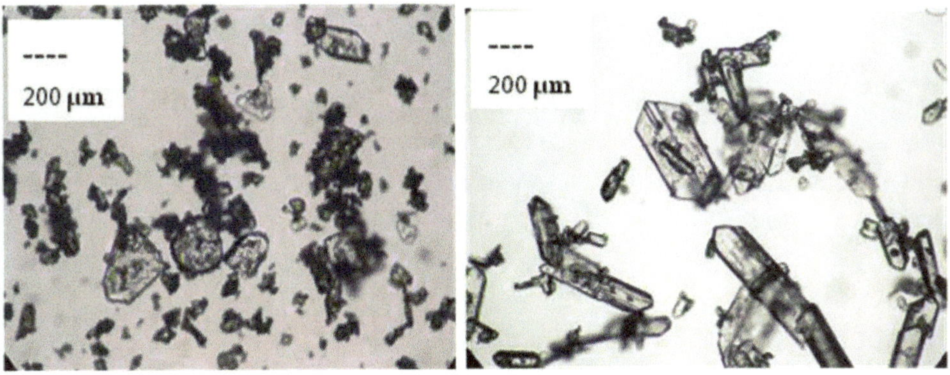

Fig. 16. Photomicrographs of (left) seed crystals grown from water under 150 µm and (right) end of experiment crystals, which are plate-like needles.

Figure 17 is a bar chart of the D10, D50 and D90 particle sizes at various points in the crystallization process, which indicates the growth of the crystal. Figure 18 shows particle size distributions of the crystal population at various points in the phenacetin crystallizations seeded with crystals under 150 µm, with the cumulative distribution in the inset. Both of these distributions show an increase in particle size throughout the experiment, indicating growth.

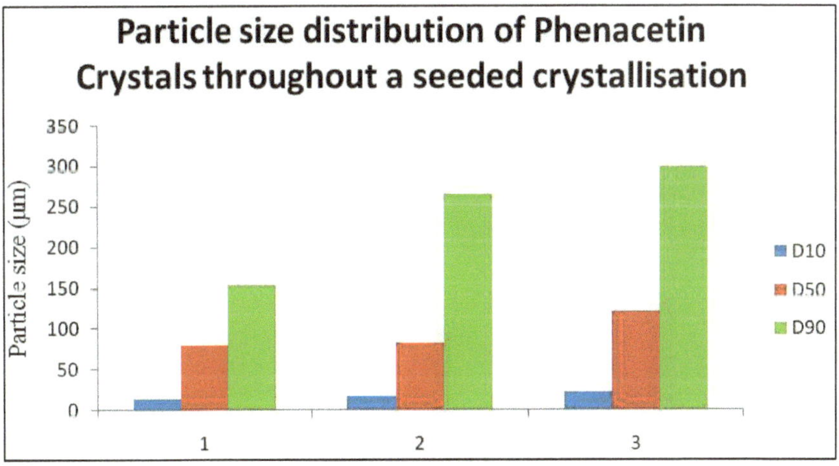

Fig. 17. A bar chart of the D10, D50 and D90 particle sizes of phenacetin crystals throughout the seeded experiments, with the seed crystals grown from water and under 150 µm.

The final experiment in this series used phenacetin crystallizations seeded with particles over 500 µm. Figure 19 is the experiment progression diagram of this crystallization. In this experiment progression diagram, the point of seeding is when the turbidity counts (the black line) and the coarse particle counts (the light blue line) increase simultaneously. Throughout the crystallization, the fine particle counts (the dark blue line) and the medium-sized particle counts (the pink line) pepper the base line, showing control over nucleation

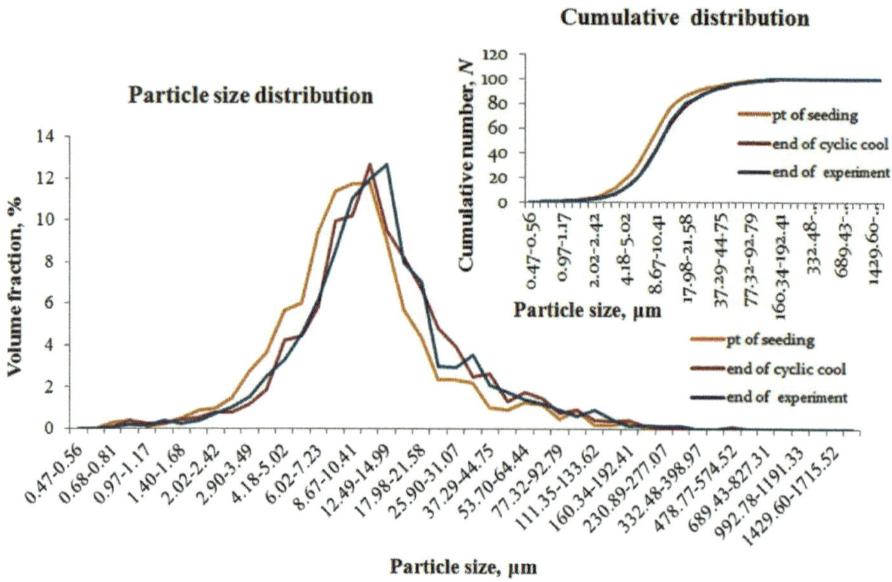

Fig. 18. Particle size distributions taken throughout a seed crystallization seeded with crystals under 150 µm and grown from water.

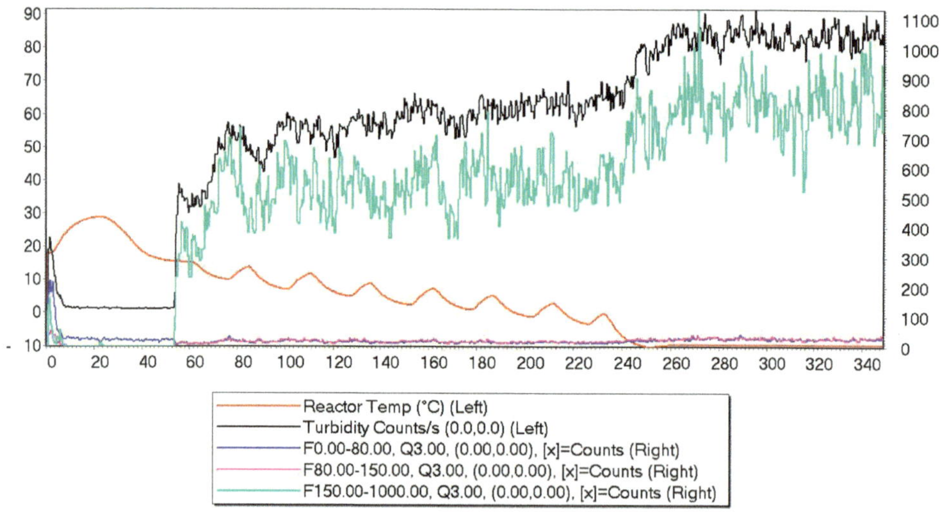

Fig. 19. Experiment progression diagram of a cyclic heating and cooling phenacetin crystallization seeded with crystals over 500 µm, grown from water (F is the number of particles inside the given size range - e.g. the size range 0.00-80.00 is the content between 0 - 80 µm - and Qz(x) is the less-than-size for parameter x and weight z, where x is counts and z = 3 is interpreted as the volume equivalent distribution).

and which is maintained at a very low level. Figure 20 shows photomicrographs of seed crystals and end of experiment crystals showing that little growth has occurred. The plate-like crystals from the water become sharper around the edges, showing the change in shape from the plates to plate-like needles due to the solvent interaction.

Fig. 20. Photomicrographs of (a) seed crystals over 500 µm, grown from water and (b) plate-like needles at the end of experiment.

Figure 21 displays particle size distributions of phenacetin crystallizations seeded with crystals over 500 µm, indicating that little growth occurs throughout the crystallization with relatively little change in particle size distribution. The cumulative distribution in the inset also indicates this lack of growth.

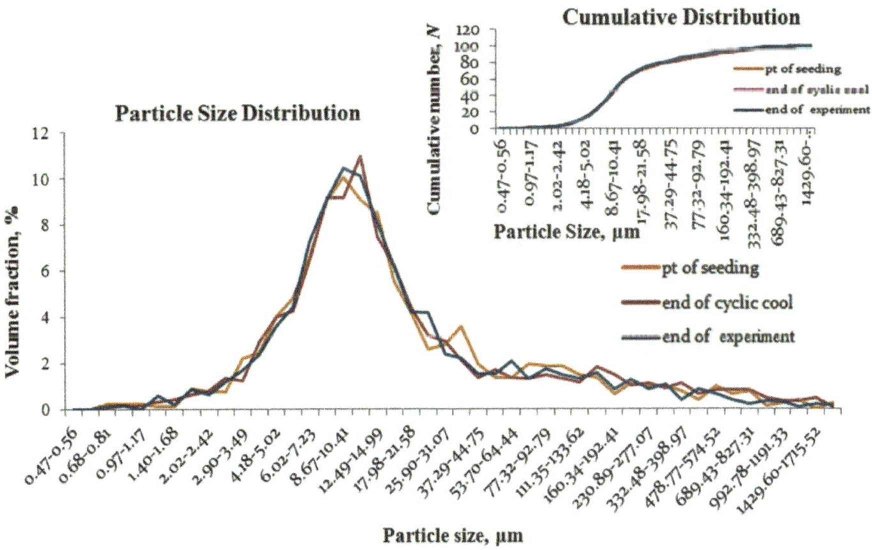

Fig. 21. Particle size distribution of the crystal population at the beginning, after cyclic heat/cool and at the end of the experiment, seeded with crystals over 500 µm (inset is the cumulative distribution at the same points, indicating that little growth is occurring).

The crystal size distributions indicate that little or no growth is taking place in this crystallization, which is not too surprising as the crystals were quite large initially and may have been at their optimum size.

The series of experiments of phenacetin crystallizations using seed crystals grown from water resulted in particle size distributions that were easy to interpret. We now wanted to see how these particle size distributions looked when they were of a crystal population which contained crystals with different dimensions, such as needles. This experiment involved phenacetin crystallizations in ethanol, using seed crystals grown from ethanol. The experiment progression diagram is given in Figure 22.

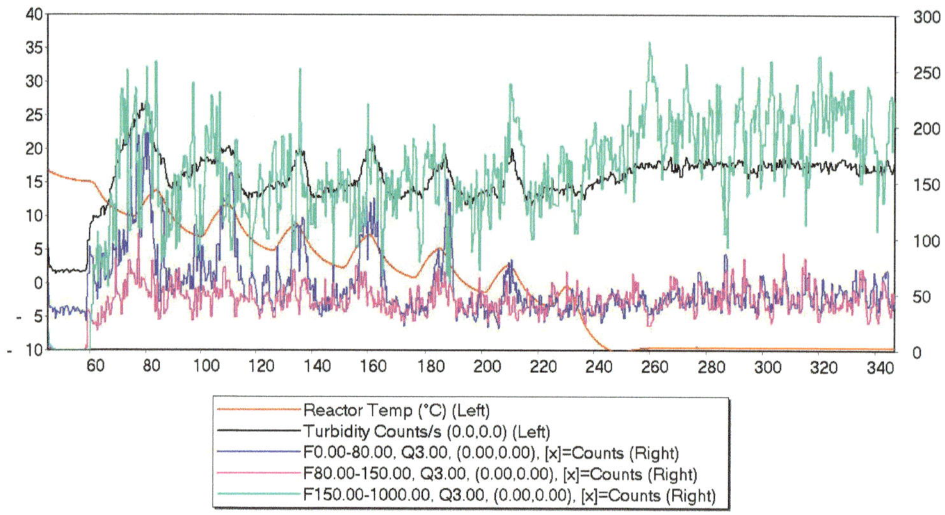

Fig. 22. Experiment progression diagram of phenacetin crystallization seeded with crystals grown from ethanol (F is the number of particles inside the given size range - e.g., the size range 0.00-80.00 is the content between 0 - 80 µm - and Qz(x) is the less-than-size for parameter x and weight z, where x is counts and z = 3 is interpreted as the volume equivalent distribution).

The experiment progression diagram shows a general trend where the fine particle counts (the dark blue line) and medium-sized particle counts (the pink line) maintain a low level and where there is a rise in the coarse particle counts (the light blue line). In Figure 22, the reason for the significant noise in the measurement is likely to be due to the shape of the needles and how they are interpreted as they pass the window of the laser probe. The longest face along the length of a needle may be interpreted as a large particle, while the narrow face at either end of the needle may be interpreted as a fine particle. Thus, every time this needle moves past the window of the laser probe, depending on its orientation, it may be assigned a different particle size leading to significant noise in the measurement. Figure 23 displays photomicrographs of the seed crystals and the end of experiment crystals.

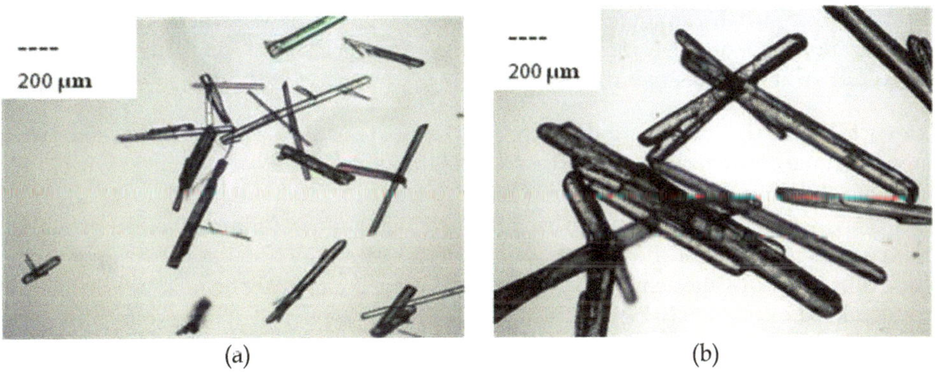

Fig. 23. Photomicrographs of (a) seed crystals grown from ethanol and (b) end of experiment crystals.

These photomicrographs indicate growth with the crystals maintaining their needle-like shape, as would be expected. Figure 24 shows particle size distributions of the crystal population at various points throughout the crystallization.

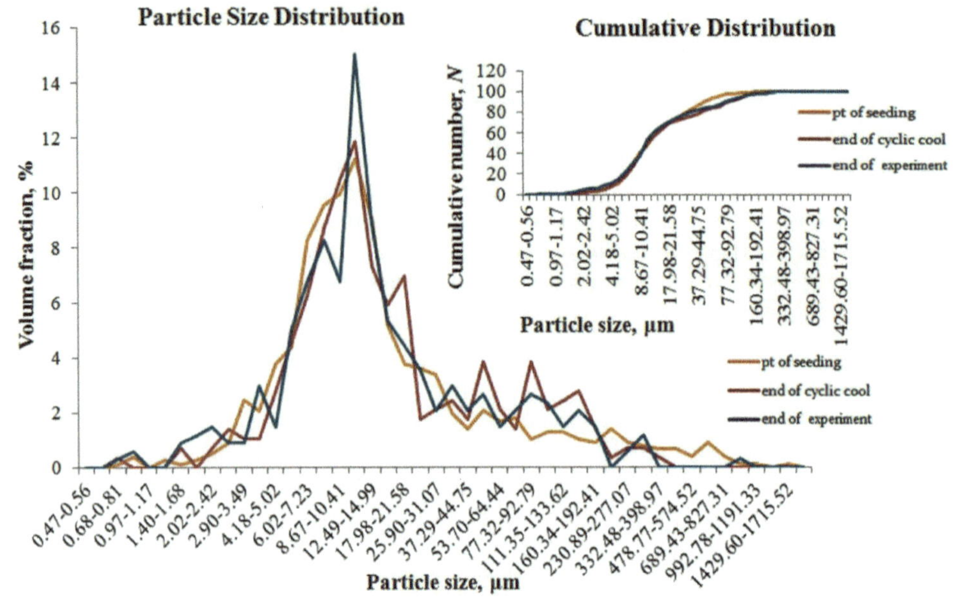

Fig. 24. Particle size distributions of phenacetin crystal populations at various points in a crystallization seeded with crystals grown from ethanol (inset is the cumulative crystallization at the same points).

The particle size distributions indicate growth, as there is an upward shift in particle size throughout the crystallization. These distributions also indicate an environment where there are many differing crystals sizes, as can be seen from the multimodal nature of the trend lines. This is most likely explained by differing orientations of the needle-like crystals as they pass the window of the laser probe. The laser sees the crystal as a two dimensional shape and measures the longest chord length. When the crystal shape is needle-like, the population of crystals passing the sensor can appear as one of the crystals with many varying sizes, depending upon the orientation of the crystals (Barthe, 2008). This is a negative aspect of *in situ* laser probe technology and makes the particle size distributions more difficult to interpret.

The use of seeded batch cooling crystallizations with temperature cycling may be preferable for the control of nucleation and growth, enhancing the size uniformity of the crystals. Cyclic temperature cooling programmes may be preferable to linear temperature cooling programs for the size uniformity of crystals (Lindenberg et al., 2009). Single crystal analysis and microscopy can be used to follow the direction of the growth of seed crystals during a crystallization process.

4. Conclusions

Crystallizations of phenacetin from ethanol were carried out in a 1 L HEL Autolab vessel fitted with a HEL Lasertrack *in situ* particle sizing laser probe. Two forms of modified cooling regimes were examined, each of which was intended to provide a narrower range of crystal sizes and shapes rather than a straightforward linear cool. The first cooling regime involved using turbidity counts as a termination clause for each heating and cooling step. The termination clause was set on the cooling step, so that the cooling was terminated when the turbidity counts exceeded a set number. Once the turbidity counts reached the pre-set level, the experiment began to heat again until the turbidity counts dropped below another pre-set number. Using this approach for crystallization batches that were not seeded proved difficult. The temperature response to this terminating clause was slow, causing large fluctuations in the turbidity counts and the fine and medium total particle counts. The use of turbidity counts as the terminating step for controlling temperature in seeded phenacetin crystallizations resulted in a more controlled experiment, giving more uniform crystals. It was also possible to generate much more prismatic crystals in this way, in contrast to the highly needle-like crystals obtained without seeding.

The second approach used cyclic cooling paths that had predefined temperature parameters, ensuring that a cyclic temperature profile was maintained within the MSZW limits. Each experiment was seeded with crystals that were sieved to give specific size ranges of less than 150 µm, between 150-300 µm and over 500 µm. The use of the seeds in the 150-300 µm range gave relatively uniform elongated plates of good quality. The use of seeds under 150 µm gave rise to elongated plates of reasonable uniformity. Those seeds over 500 µm grew the least during the cooling process and were the least effective in generating uniformity of size and shape, possibly due to these seed crystals' being so large that they had potentially already reached their optimum size. Overall, the pre-programmed cyclic cooling regime was found to be easier to operate and provided good quality crystal batches when used in conjunction with sieved seeds.

5. Acknowledgements

The authors are grateful for financial support from the Solid State Pharmaceutical Cluster (Science Foundation Ireland grant 07/SRC/B1158).

6. References

Abu Baker, M. R.; Nagy, Z. K.; Tan, R. B. H.; Saleemi, A. N. & Rielly, C. D. (2009) The Impact of Direct Nucleation Control on Crystal Size Distribution in Pharmaceutical Crystallization Processes. *Crystal Growth and Design*, 9, 1378-1384

Abu Bakar, M. R.; Nagy, Z. K. & Rielly, C. D. (2009) Seeded Batch Cooling Crystallization with Temperature Cycling for the Control of Size Uniformity and Polymorphic Purity of Sulfathiazole Crystals. *Organic Process Research and Development*, 13, 1343-1356

Barthe, S. C. (2008) Investigation and modeling of the mechanisms involved in batch cooling crystallization and polymorphism through efficient use of the FBRM. PhD Thesis, Georgia Institute of Technology, 119-138

Garside, J.; Gaska, C. & Mullin, J. W. (1972) Crystal Growth Rate Studies with Potassium Sulphate in a Fluidized Bed Crystallizer. *Journal of Crystal Growth*, 13/14, 510-516

Griffiths, H. (1925) Mechanical Crystallization. *J. Soc. Chem. Ind. Trans.*, 44, 7T-18T

Hansen, L. K.; Perlovich, G. L. & Bauer-Brandl, A. (2006) Redetermination of *p*-ethoxyacetanilide (phenacetin). *Acta Crystallographica*, E62, o2712-o2713

Lindenberg, C.; Krattli, M.; Cornel, J.; Mazzotti, M. & Brozio, J. (2009) Design and Optimization of a Combined Cooling/Antisolvent Crystallization Process. *Crystal Growth and Design*, 9, 1124-1136

Prescott, L. F. (1980) Kinetics and metabolism of paracetamol and phenacetin. *British Journal of Clinical Pharmacology*, 10, 291-298

Woo, Xing Yi; Nagy, Z. K.; Tan, R. B. H. & Braatz, R. D. (2009) Adaptive Concentration Control of Cooling and Antisolvent Crystallization with Laser Backscattering Measurement. *Crystal Growth and Design*, 9, 182-191

7

Thin Film Growth Through Sputtering Technique and Its Applications

Edgar Alfonso, Jairo Olaya and Gloria Cubillos
Universidad Nacional de Colombia
Colombia

1. Introduction

During the last decade the dc and rf sputtering techniques have been used extensively in their two configurations – balanced and unbalanced magnetron. The main applications have been in the fields of industry and research. Examples of industrial applications are: decorative thin films (Raymond & Baham, 1999), hard wear-resistant thin films (Rodil & Olaya, 2006), low-friction thin films (Heimberg *et al.*, 2001) corrosion-resistant thin films (Flores *et al.*, 2006), and thin films used as a protective optical system (Stefan *et al.*, 2008), as well as maybe the most interesting applications, thin films used in the electronic industry (Monroy *et al.*, 2011). In the research field, the investigation has been oriented toward understanding the main physical mechanisms, such as: interaction between charged particles and the surface of the target material, adherence between the substrate and the deposited material, and chemical reactions near the substrate, as well as the influence of the deposit parameters (substrate temperature, working pressure, density power applied to the target). This research has produced thin films with a high degree of crystallinity and with the possibility of various industrial applications.

Moreover, researchers have made an effort to improve the system of operation. These efforts have been initiated through the so-called conventional or balanced magnetron sputtering in the early 1970s (Waits R, 1978), followed by the development of unbalanced systems in the late 1980s (Window, 1986) and its incorporation into multi-source "closed-field" systems in the early 1990s (Teer, 1989). Finally, the sputtering technique can increase the rate of deposition and ion energy by applying a unipolar high power pulse of low frequency and low duty cycle to the cathode target, referred to as high-power impulse magnetron sputtering (HiPIMS) or high-power pulsed magnetron sputtering (HPPMS). Common to all highly ionized techniques is very high density plasma. Implementing these discharges in sputter deposition technology modifies the surface of components, bringing improvements in mechanical, chemical, optical, electronic, and many other properties of the material. High-current glows are transient discharges operating at simultaneously high voltage (> 300 V) and high current density (> $100 mAcm^{-2}$). They have recently proven successful for the deposition of thin-film materials. These developments have made it possible to have an exceptionally versatile technique, suitable for the deposition of high-quality, well-adhered films of a wide range of materials with high rates of deposition. Table 1 show the main applications obtained in the last decade with the magnetron sputtering (balanced and unbalanced) rf and dc versions.

Technique	Coating	Reference
Balanced magnetron	Nano-composites of NC-TiC wear applications	(Pavel, *et al.*, 2011)
Balanced magnetron	Optical properties of AlSiN nano-composites	(Liu, *et al.*, 2009)
Balanced magnetron	Hard coatings to decorative applications	(Raymond & Baham, 1999)
Balanced magnetron	Nd-Fe-B Film for magnetic applications	(Liu, *et al.*, 20007)
Balanced magnetron	Optical applications.	(Stefan, *et al.*, 2008)
Unbalanced magnetron	Hard films for corrosion and wear applications	(Rodil & Olaya, 2006).
Unbalanced magnetron	Nb Films for biological applications	(Olivares, *et al.*,2011)
Unbalanced magnetron	NbO films for biological applications	(Ramirez, *et al.*, 2011)
Unbalanced magnetron	Electrical applications of NbN films.	(Olaya, *et al.*, 2008,)
Unbalanced magnetron	Diamond-like carbon films for infrared transmission enhancement	(Lingxia & Xu, 2000).
High power pulsed magnetron sputtering	Thin films for automotive engineering	(Bewilogua, *et al.*, 2009)
High power pulsed magnetron sputtering	TiAlCN/VCN films for tribological applications	(Kamath, *et al.*, 2011)

Table 1. Technological applications of thin films obtained with magnetron sputtering

In sputtering there are two means of operation: dc (diode and triode) and ac (radiofrequency), which also function in two configurations: magnetron dc (balanced and unbalanced) and magnetron ac (balanced and unbalanced).

In dc (diode) discharge, the cathode electrode is the sputtering target and the substrate is placed on the anode, which is often at ground potential (Vossen &Cuomo, 1978). The applied potential appears across a region very near the cathode, and the plasma generation region is near the cathode surface. The cathode in dc discharge must be an electrical conductor, since an insulating surface will develop a surface charge that will prevent ion bombardment of the surface. This condition implies that dc sputtering must be used to sputter simple electrically conductive materials such as metals, although the process is rather slow and expensive compared to vacuum deposition. An advantage of dc sputtering is that the plasma can be established uniformly over a large area, so that a solid large-area vaporization source can be established.

On the other hand, in dc sputtering the electrons that are ejected from the cathode are accelerated away from the cathode and are not efficiently used for sustaining the discharge. To avoid this effect, a magnetic field is added to the dc sputtering system that can deflect the electrons to near the target surface, and with appropriate arrangement of the magnets, the electrons can be made to circulate on a closed path on the target surface. This high current of electrons creates high-density plasma, from which ions can be extracted to sputter the target material, producing a magnetron sputter configuration (Penfold, 1995). A disadvantage of the magnetron sputtering configuration is that the plasma is confined near the cathode and is not available to active reactive gases in the plasma near the substrate for reactive sputter deposition. This difficulty can be overcome using an unbalanced magnetron configuration (see Fig. 1), where the magnetic field is such that some electrons can escape from the cathode region (Windows & Savvides, 1986). A disadvantage of the unbalanced magnetron is that the current of escaping electrons is not uniform, and the plasma generated is not uniform.

In ac sputtering, working at frequencies below about 50 kHz, the potential on the target is periodically reversed, and the ions have enough mobility so that a dc diode-like discharge, where the total potential drop is near the cathode, can be formed alternately on each electrode. The substrate chamber walls can be used as the counterelectrode. At frequencies above 50 kHz, the ions do not have enough mobility to allow establishing a dc-diode-like discharge and the applied potential is felt throughout the space between electrodes. The electrons acquire sufficient energy to cause ionizing collisions in the space between the electrodes. When an rf potential with a large peak-to-peak voltage is capacitively coupled to an electrode, an alternating positive-negative potential appears on the surface. During part of each half-cycle, the potential is such that ions are accelerated to the surface with enough energy to cause sputtering, while in alternate half-cycles, electrons reach the surface and prevent any charge buildup. Rf sputtering can be used to sputter insulating material, although the sputtering rate is low. A major disadvantage of rf sputtering of dielectric targets is that most insulating materials have poor thermal conductivity and high coefficients of thermal expansion, and are usually brittle materials. Since most of the bombarding energy produces heat, this means that large thermal gradients can be generated that result in fracturing the target if high power levels are used.

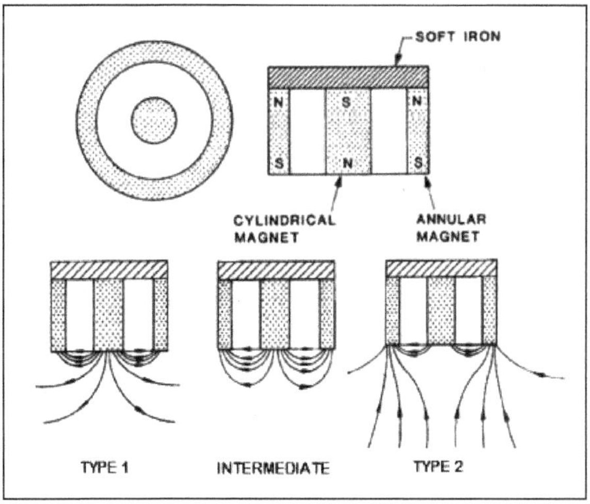

Fig. 1. Schematic configuration of the a- balanced magnetron (intermediate) and b- unbalanced magnetron (type 1and type 2 (B. Window & N. Savvides, 1986)

Rf sputtering can be used with a magnetic field in balanced and unbalanced configurations to obtain a result similar to dc-like diode discharge. In fig 2 the different configurations of dc and rf sputtering are shown.

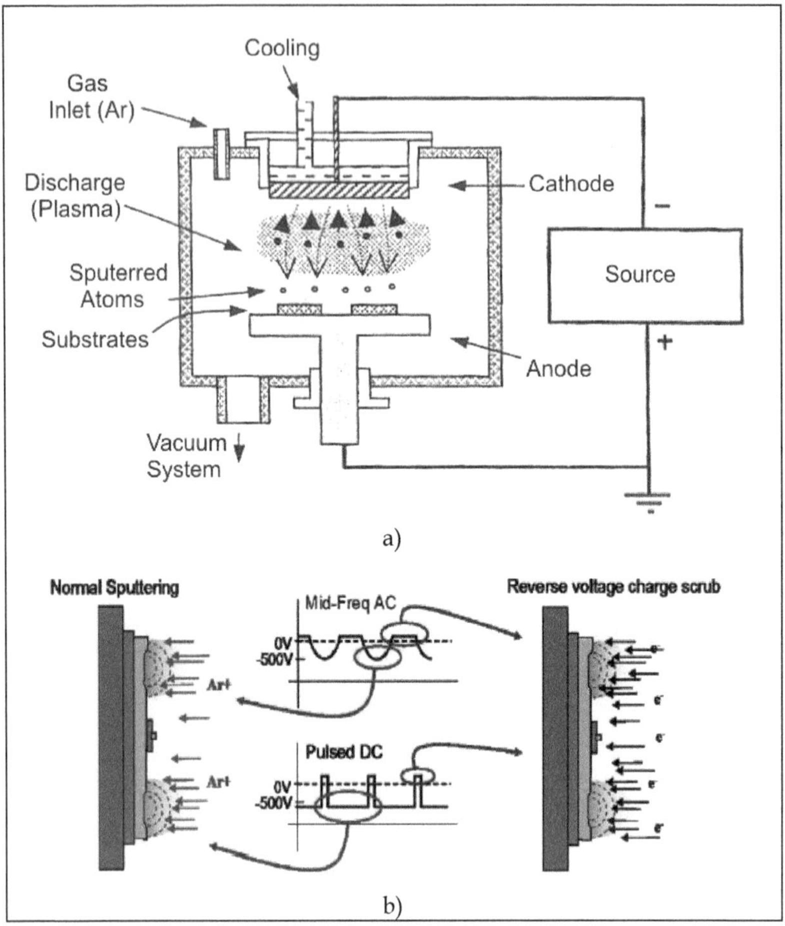

Fig. 2. Sputtering configurations a) dc sputtering, b) rf sputtering

It is important to state that in all the cases discussed above, the target and the substrate were facing (on- axis sputtering). In this configuration, the highly energetic electrons irradiate the substrates and/or the growing surface of the thin films during deposition. Off-axis sputtering reduces the effects of the irradiation of the high-energy particles. In off-axis sputtering, the substrates are settled at the outside of the discharge plasma. The thickness distribution of thin films deposited by off-axis sputtering will be larger than that for on-axis sputtering. A rotating substrate holder with a metal shadow mask is used for the reduction of the thickness distribution of the off-axis sputtering. Under a suitable design, the thickness distribution is less than 2% for substrates of 100 × 100 mm in an rf sputtering system using a 5-inch target (Shibahara *et al.*, 1987).

In this chapter we will present the physical parameters involved in the growth of thin films; also discussed will be the influence that the growth parameters have on the degree crystallinity of the films, the chemical characterization, and the optical characterization of the films; and finally, we will discuss the residual stress, hardness, and corrosion and wear resistance of thin films.

2. Physical sputtering

The main physical phenomenon involved in the sputtering technique is the momentum transfer between energetic atomic-sized particles (usually ions of noble gases) and the atoms of the surface of the material (target). During the interchange of momentum, many effects can be produced on the elastic and inelastic collisions; in the first kind of collision, mainly reflected particles can be found (neutrals, ions of the target and the gas). In the second kind, the collisions can present secondary electrons, UV/visible photons, X-ray and implanted particles; schematically, Fig. 3 shows different processes that may occur during the interaction between charged particles and the surface of the material.

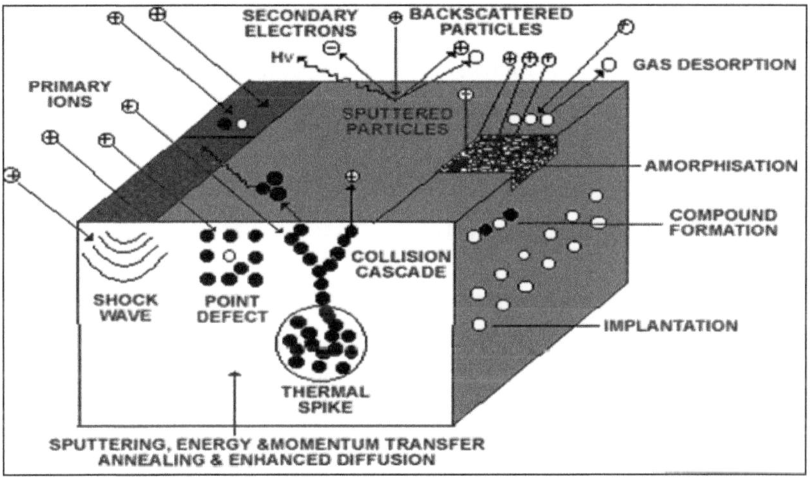

Fig. 3. The main physical process produced in sputtering technique (Weissmantel, 1983).

The momentum-transfer theory for physical sputtering was proposed early on, but was replaced by the "hot-spot" theory, in which the process of thermal vaporization is involved. The confusion about the physical process present in sputtering has only been overcome thanks to the work of Guntherschulze in the 1920´s and 30´s and Wehner *et al.* in the 1950´s and 60´s, who demonstrated that the effects produced in sputtering could only be explained by the momentum transfer theory. These effects can be summarized as:

1. The sputtering yield (ratio of atoms sputtered to the number of high-energy incident particles) depends on the mass of the bombarding particle as well as their energy. For ion energies from 100 eV to 1000 eV the sputtering yield can be calculated as:

$$Y = \frac{3\alpha}{4\pi^2} \frac{4m_1 m_t}{(m_1 + m_t)^2} \frac{E}{U} \tag{1}$$

where m_i is the atomic mass of the bombarding incident ion, m_t the atomic mass of the target, U the binding energy of the surface atom of the target, E the energy of the incident ion, and α depends on the ratio of the masses of the target atom and the incident ion (monotonically increased with mi/m_t; for a ratio 0.1 α= 0.17 for a ratio 10 α=1.4) (Ochiati, 1986).

The sputtering yield is sensitive to the angle-of-incidence of the bombarding particle.There is a "threshold energy" below which sputtering does not occur no matter how high the bombarding flow.

2. Many sputtering atoms have kinetic energies much higher than those of thermally evaporated atoms.
3. Atoms ejected from single crystals tend to be ejected along the directions of the close-packed planes in the crystal.
4. In a polycrystalline material, some crystallographic planes are sputtered faster than are others.
5. Atoms sputtered from the alloy's surface are deposited in the ratio of the bulk composition, not their relative vapor pressure, as in the case of thermal vaporization.
6. Sputtering yields decrease at very high energies because the ions lose much of their energy far below the surface.
7. The sputtering yield is rather intensive to the temperature of the sputtering target.
8. There is no sputtering by electrons even at very high temperature.
9. The secondary electron emission by ion bombardment is low, whereas high rates from thermo electron emission would be expected if high temperatures were present.

Effects 1 through 7 above are important for the growth of films by sputter deposition. This is particularly true for low pressure (<5 m Torr).

3. Physical models that explain the microstructure of thin film growth through sputtering

The microstructure of thin films is related to the mobility of the adatoms during growth. The energy supply to the atoms is provided by the following mechanism: a- thermal effect, b- ionic bombarding and c- chemical reactions at the substrate. The effects that are produced by these mechanisms in the growth of thin films can be explained by the structure zone model (SZM). The SZM model can determine the morphology and microstructure of the films as a function of the adatoms, regardless of the kind of material. The parameters that the SZM model includes for determining the microstructure of the films are basically the substrate temperature, the final working pressure, the bias voltage applied to the substrate, and the thermal characteristics of the target. For example, in the research of Movchan and Demchishin (Movchan & Demchisshim, 1969), it has been established that the microstructure of the thin films of Ti, Ni, ZrO_2 y Al_2O_3 is related to the normalized temperature, *i. e.* Ts/Tm (Ts is the temperature of the substrate and Tm is the melting temperature of the target). Movchan *et al.* have shown that in metallic films there are three well-defined zones.

The first zone is Ts/Tm <0.3. This zone is formed by small and elongated grains that form a columnar structure with porous morphology and weakly binding grains. The columnar structure is produced by a low diffusion, a low mobility of the atoms adsorbed by the substrate surface, and the atomic shadow effects, which are produced by varying velocity in the growth the columns and the various incidence angles at which the atoms arrive at the surface of the substrate. In the second zone, $0.3 \leq Ts/Tm \leq 0.45$, the substrate temperature increasing homogeneous which leads to a higher diffusion of the adatoms, which produce a dense structure with a higher degree of binding among the columns and the borders between columns, with borders of the grain beginning to form. In this zone, the size of the grain can be increased and the grains extended in equiaxed form, from interface substrate-film to film surface. In the third zone, $Ts/Tm > 0.45$, the volumetric diffusion size has a great influence on the morphology of the film, due to the increase in the diffusion into the grains, which produces growth of the grains, formation of the equiaxed grain and re-crystallization. These effects produce a greater crystalline structure.

Thorton (Thorton, 1974) elaborated the zone classification, considering the final working pressure, because this growth parameter can change both the kinetic energy of the ions that arrive at the substrate and the mean free path of the particles, which allows an increase or decrease in the bombardment of the surface of the substrate, which in turn determines the mobility the adatoms in that surface. In the Thorton model, the T zone as a transitional zone between first and second zone discussed above was added. The T zone is formed by grains defined by the limits of the low porosity. The surfaces of the T zone are denser and less rough than the two surfaces around them (see fig 4).

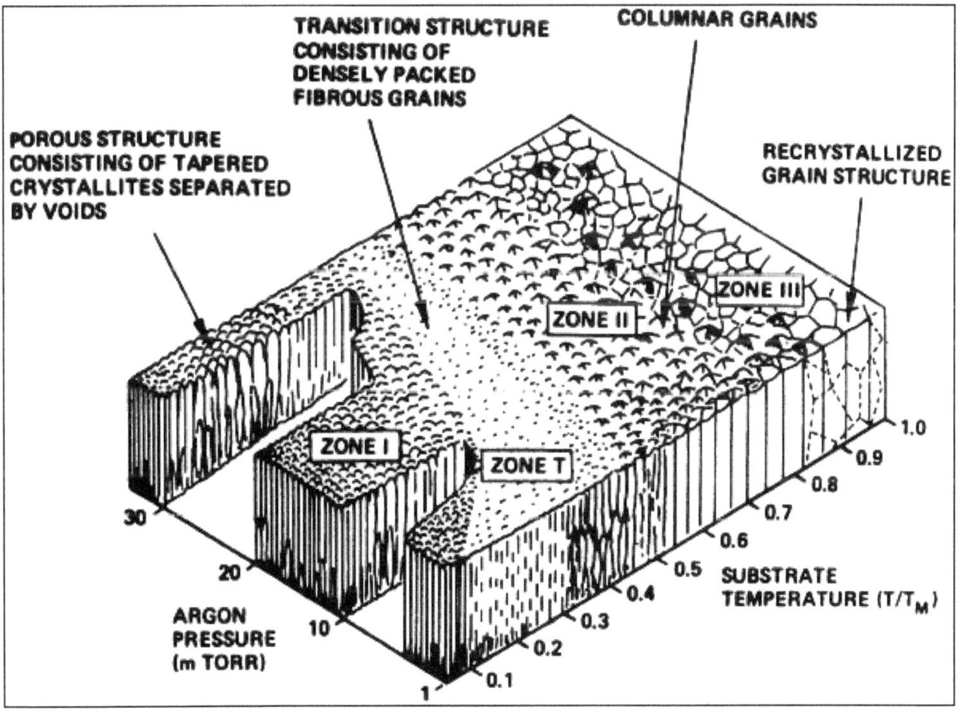

Fig. 4. Thorton zone model (Thonton, 1974)

Moreover, Messier (Messier & Giri, 1984) found that in thin films of TiB_2, BN and SiC there is a non-linear limit between the first zone and T zone, which is a function of the bias voltage applied to the substrate. The bias voltage of the substrate has the same effect on the mobility and adsorption of the atoms as an increase in the temperature, so when it is increased, the T zone increases and the first zone decreases, resulting in denser thin films and with a high degree of crystallinity.

The bias voltage also influences the mechanical properties of thin films; for example, with a voltage (≥100V) applied to the substrate, the deformation of the lattice increases, causing high residual efforts and low adherence between the substrate and the film.

A new three-dimensional SZM model has been produced in order to explain the simultaneous influence of the normalized temperature, the bias voltage, and the relation between ion density current (J_i) and density deposited atoms (J_a). In Fig. 5, it can observed that it is possible to obtain the third zone with high density using a combination high-density ionic current, intermediate values of normalized temperature, and low bias voltage (Kelly & Arnell, 1998).

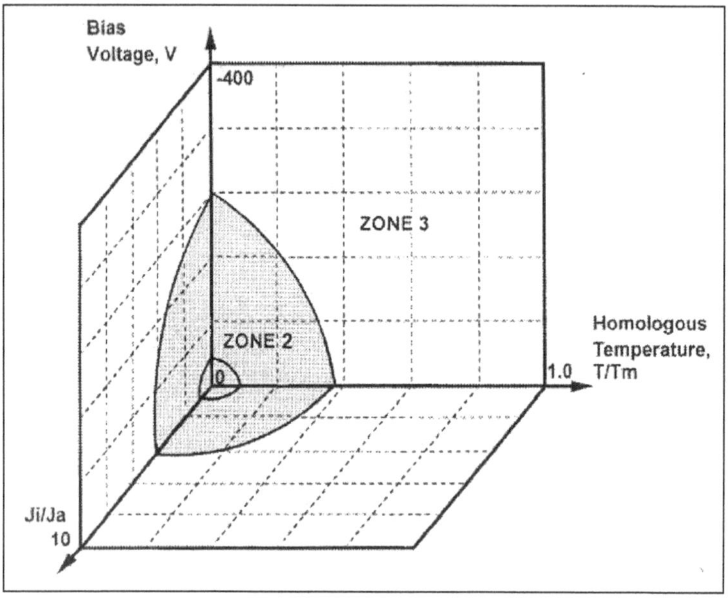

Fig. 5. The zone model including the relation between current density of ions and current density of atoms (Kelly & Arnell, 1999).

Other parameters of growth that affect the microstructure are the power supply to the target and gas flows. For example, changing the nitrogen flow during the growth of TiN and NbN films changes the preferred orientation of the films from the [111] direction to the [200] direction (Alfonso *et al.*, 2010). Explaining the change in the direction of growth of metallic nitride films is a very complex procedure, as is shown in table 2, where it can be observed that any parameter change can influence the direction of growth of the film. There are three models that explain the preferential growth direction: a) the thermodynamic model (Pelleg *et. al.*, 1991); b) the kinetic model (Greene *et al.*, 1995), and c) the atomic model (Petrov *et al.*, 2003).

Material	Parameter	Change of orientation	References
TiN	Ji/Ja	(111) → (200)	(Green *et al.*, 1995)
	N2/Ar	(111) → (200)	(Noda *et al.*, 2002)
	Th	Mixed → (200) → (111)	(Pelleg *et al.*, 1991)
	Vb	(200) → (111) → (220)	(Kobayashi & Doy, 1984)
	W	(200) → (111)	(Je *et al.*, 1997).
	T	(111) → (200)	(Cheng *et al.*, 2002)
AlN	P	axis c → axis a	(Lee & Y. Lee, 1994)
	ds-t	axis c → axis a	(Ishihara *et al.*, 1998)
	λ/ds-t	axis a → axis c	(Ishihara *et al.*, 1998)
	N2/Ar	axis a → axis c	(Lee & Y. Lee, 1994)
	Vb	axis a→ axis c→ axis a	(Wang & Zhao, 1997).
	W	axis a → axis c	(Lee & Y. Lee, 1994)
TaN	N2/Ar	(111) → (200) → (111)	(Nie *et al.*, 2001)
	Vb	(111) → (220)	(Lin & Lee, 2000)
	Ji/Ja	(111) → (200)	(Shi *et al.*, 2002)

Table 2. Relationship between the deposition parameters and film preferential orientation. Ji/Ja: ion-atom flow ratio, N_2/Ar: flow ratios, t h: thickness, Vb: bias voltage, W: power supply, Ts: substrate temperature, P: pressure, ds-t: target-substrate distance y λ: free mean path.

In the thermodynamic model, it has been established that the growth orientation in thin films is produced at the thermodynamic equilibrium, which it reaches when the total energy (W_{hkl}) of the system substrate film is at a minimum. In this case, W_{hkl} is formed by the addition of surface energy (S_{hkl}) and deformation energy (U_{hkl}), S_{hk} passivation; energy is produced by unsaturated bonding at the surface. S_{hkl} can be calculated by:

$$S_{hkl} = \frac{\Delta H N_{hkl}}{Z} \left[\frac{J}{m^2} \right] \quad (2)$$

where ΔH is the sublimation energy, N_{hkl} is the number of unsaturated bondings per atom at the plane (hkl), and Z is the amount of coordination among neighbors. Equation 2 does not include impurities.

On the other hand, deformation energy is related to intrinsic efforts in the film. Considering only two dimensions, U_{hkl} can be calculated by:

$$U_{hkl} = \epsilon_{hkl}^2 E_{hkl} (1-\mu) \quad (3)$$

where ε_{hkl} is the deformation along the plane (hkl) of the film, E_{hkl} is the elastic module of the plane (hkl), and μ is the Poisson ratio. Research has established that the values of ε_{hkl} and E_{hkl} are different for (111), (220) and (200) planes (Ma *et al.*, 2004), and using equations (2) and (3) an order relation has been obtained: $S_{111}>S_{220}>S_{200}$ and $U_{200}>U_{220}>U_{111}$. These relations imply that the preferential orientation is determined by a configuration of minimum total energy, which results in competition between the (111), with minimum deformation energy, and (200) planes, with minimum surface energy.

Moreover, Pelleg has researched the variation of the total energy (W_{hkl}) for the (111), (220) and (200) planes as a function of the thickness of the TiN film deposited at 373K through the rf sputtering technique. These results have shown that the surface energy does not change with the thickness, the deformation energy increases proportionally with the thickness, and the slope of the straight line represents the deformation per volume.

In the kinetic model, it is proposed that the kinetic process of the ions is involved in the orientation of the growth, which influences the surface substrate, the energy and flow of the ions being the main parameters. A thin film grows in the [111] direction because the adatoms have limited mobility, caused by the low temperature of the substrate and the restricted flow of the ions. Increasing the mobility, either through increasing the substrate temperature or raising the relation J_i/J_a above five, a preferential orientation in the [200] direction is obtained.

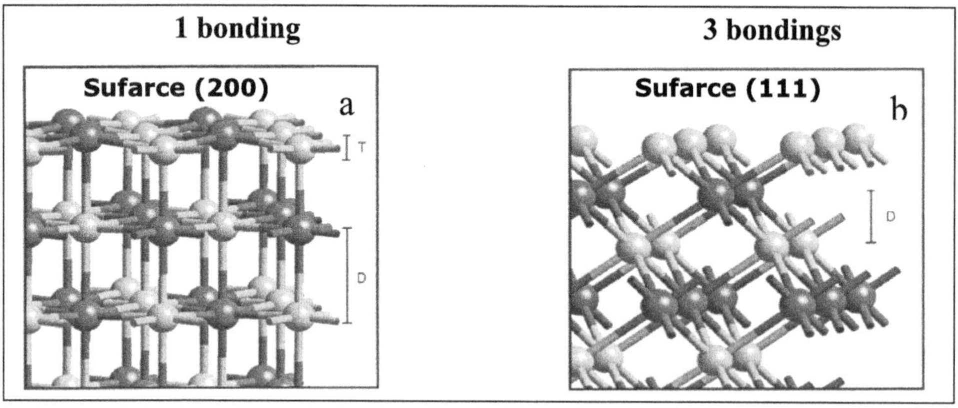

Fig. 6. Surface growth on a- (200) planes and b- (111) planes. The growth on the (200) plane has 1 binding, while the surfaces grown on (111) planes have 3 bandings (M. Marlo & V. Milman, 2000).

Finally, the atomistic model considers that in thin films deposited at high temperatures the thermodynamic parameters control the orientation of the growth, favoring planes with low energy, *i.e.* the (200) planes. This behavior can be understood if adatoms with low diffusion, as in the case of the cation Cl in NaCl, which have three bonds in the (111) surface, and have high diffusion in surface (200), are considered to have only one bond (see Fig. 6), leading to the conclusion that if NaCl thin films are grown at high temperatures, they probably will grow along the [200] direction.

On the other hand, the same model establishes that the degree of ion bombardment is influenced by the flow ratio between the density current of the ions and the density current of the atoms, Ji/Ja, and the ion energy, Ei, and both depend on the pressure, substrate target distance, and substrate bias (Ensinger, 1998; Losbichle & Mitterer, 1997). One parameter that combines both the ion energy and the flow has been suggested by Musil and Kadlec (Musil

& Kadlec, 1990), Ziemann and Kay for Pd coatings (Ziemann and Kay, 1993) and passivation for CBN films (Kulisch, *et al.*,1999). This represents the energy deposited per dense particle. This bombardment-induced mobility parameter, known as the energy parameter Ep, is defined as:

$$E_p = \frac{J_i}{J_a} \times E_i \tag{4}$$

where E_i is incident energy of the ions and is obtained from the plasma potential (V_p) and substrate bias (V_s) and the elementary charge by the expression $E_i = e(V_s - V_p)$.

3.1 Physical mechanisms of thin film growth

So far, we have studied the influence of the growth parameters on the structure and microstructure of thin films; now we discuss the main physical mechanisms involved in the nucleation and crystallization of the films over the substrate. Studies done through X-ray diffraction, optical diffraction and mainly through electron diffraction have allowed establishing that there are three mechanisms of the nucleation and growth of thin films, which depend on the thermodynamic parameters of the deposit and the substrate surface interaction between the adatoms and the substrate material (Green, 1994). The three basic modes are: (i) Volmer–Weber model, (ii) Frank–Van der Merwe model and (iii) Stranski–Krastanov model. A schematic illustration of each of these modes of growth is shown in Fig. 7.

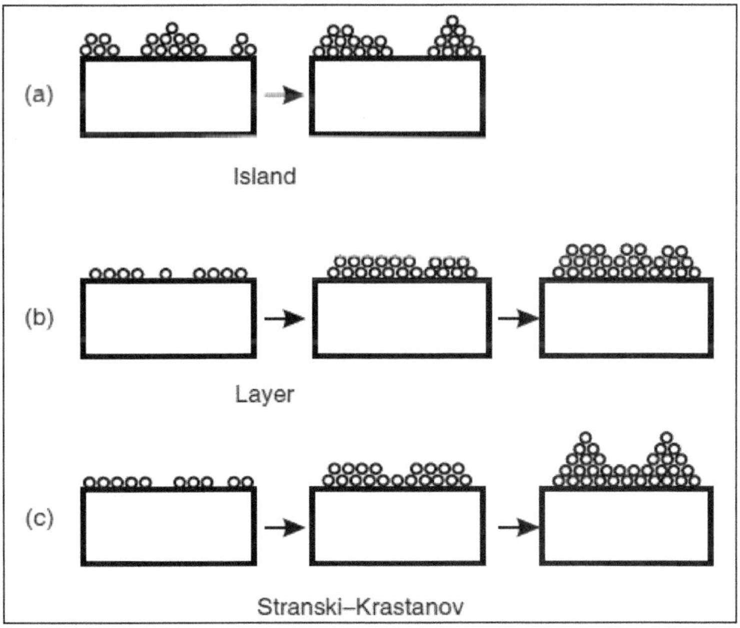

Fig. 7. Modes of growth of film: a- Volmer–Weber island growth b- Frank–vander Merwe layer growth and c- Stranski-Krastanov layer plus island growth (Harsha, 2005).

In the Volmer–Weber model, equilibrium exists in a three-dimensional crystal of the film in contact with the substrate, while the rest of the substrate is devoid of any condensed phase. Nucleation of film occurs in the form of discrete three-dimensional nuclei on the surface of the substrate, for example lead on graphite. Both the number of nuclei and the size of a given nucleus increase. Finally, the nuclei grow in size until they intersperse with each other to form a continuous film.

In the Frank and Vander Merwe model, nucleation occurs in the form of a monolayer island of the deposit. Eventually the monolayer's grow together to form a complete continuous monolayer of the deposit. The process repeats itself so that the deposit grows in a layer-by-layer manner, for example rare gases on graphite. In this growth, the interaction between the substrate and the layer atoms is stronger than that between neighboring layer atoms. Layer-by-layer growth is hindered by elastic constraints at the solid–solid interface.

The Stransky and Krastanov (S-K) model combines the features of layer-by-layer growth and discrete three-dimensional nucleation The S-K nucleation is common with metal-on-metal deposition and at low temperatures where the surface mobility is low (Greene, 1987). The conditions for these types of growth are generally described in terms of thermodynamics and surface energy considerations. In this model, nucleation and growth occurs as in the layer-by-layer mode, so that a finite number of monolayer's is produced. Subsequent formation of film occurs by formation of discrete nuclei. The lattice mismatch between the substrate and the deposit cannot be accommodated when the layer thickness increases, so the three-dimensional growth follows the layer-by-layer growth. Alternatively, symmetry or orientation of the overlayers with respect to the substrate might be responsible for the production of this growth mode.

The foregoing models were summarized by Barna and Adamik (Barna & Adamimik, 1998), who established that the growth of the films has the following evolution: nucleation, island growth, coalescence of islands, formation of polycrystalline islands and channels, development of continuous structure, and thickness growth.

According to Barna, the nucleation starting the growth of individual islands takes place on the substrate surface at the very first stage of the condensation (primary nucleation) or later on the bare substrate surface area developing upon liquid-like coalescence (secondary nucleation). A peculiar case of nucleation shows up on the surface of a growing crystal when its growth is blocked by a surface covering layer (SCL) of an impurity phase. This is the repeated nucleation. The primary nucleation starts the condensation and the film growth on the whole substrate surface simultaneously, while the secondary and the repeated nucleation initiates the start of the growth locally in later stages of film formation. It is important to note that on amorphous substrates the nuclei are randomly oriented.

Crystal growth is the fundamental structure-forming phenomenon which incorporates the depositing material into the condensed phase. Two main cases of crystal growth should be considered in the case of polycrystalline thin films: a- the growth of discrete crystals dispersed on the substrate surface, (Fig. 8a and b), and b- the growth of crystals which are parts of a polycrystalline structure (Fig. 8c and d). Crystals growing from the nuclei are randomly oriented due to the random orientation of the nuclei. The complete coalescence of the crystals touching each other produces a grain coarsening, resulting also in the development of discrete single crystals and is connected to some changes in the orientation controlled mainly by the minimization of the substrate–crystal interface energy. The intersection lines of the crystal side faces and the substrate present a specific structural precondition for the growth of these crystals. The intersection lines can be active or passive in the monolayer nucleation on the side crystal faces. In the first case, the movement of the monolayer growth steps proceeds from the intersection line to the top of the crystal, while in the second case, the movement of the growth steps proceeds in the direction of the intersection line. In the presence of impurities, the direction of the movement of the growth steps will be important in determining the location of the developing second phase, *e.g.*, SCL.

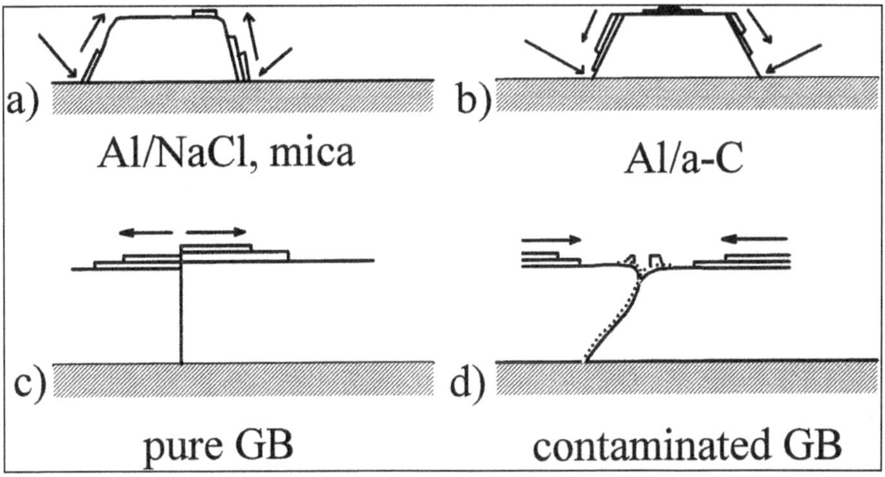

Fig. 8. Types of crystal growth in polycrystalline thin films: a- and b- growth of dispersed individual crystals on the substrate surface, the intersection line between the substrate and the side crystal faces are active (a) and passive (b) in the monolayer nucleation; c- and d- role of grain boundaries in the growth of crystals as parts of a polycrystalline matrix, pure grain boundary active in the monolayer nucleation (a), contaminated grain boundary passive in the monolayer nucleation (b). (Barna & Adamimik)

4. Influence of the normalized temperature (T_s/ T_m) on the structure and microstructure of thin films

The structural behavior, as a function of the normalized temperature of $Ts/T_m<0.3$ is shown in Fig. 9. In general, the patterns recorded from the various films grown present the same reflections as the target (δ-NbN cubic phase), but show important differences in the relative intensities, in particular those corresponding to planes (111) and (200), which increase with an increase of the absolute temperature. Regarding this, it is important to point out that contrary to what occurs in the target and the rest of the film patterns, the XRD pattern of the film grown at 553 K shows a greater intensity for the reflection from plane (111) than from that corresponding to the (200) plane. While the (111) plane in the target pattern has a relative intensity of 100% compared to 86% for the (200) plane, in the 553 K sample the relative intensity of the (200) plane is approximately 6% of the corresponding (111) plane. This result indicates that the supply of additional 22.4 meV to the substrate (which is the energy difference between room temperature and 553 K) does not bring about changes in the polycrystalline character of the materials, but gives place to the growth of films with preferred orientation along the (111) plane.

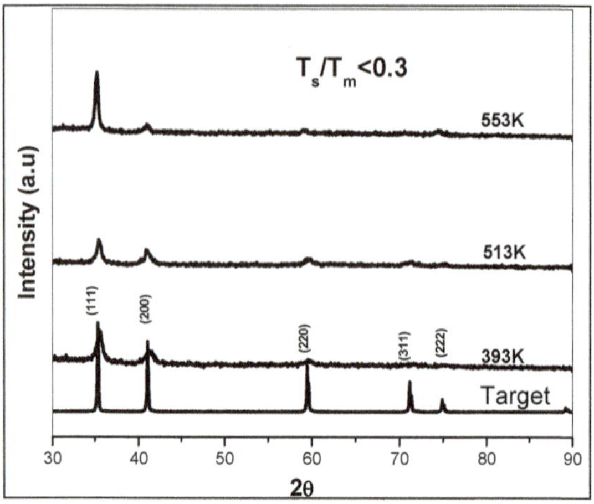

Fig. 9. XRD pattern of NbN thin films deposited at a normalized temperature (T_s/T_m<0.3) at 300W of power.

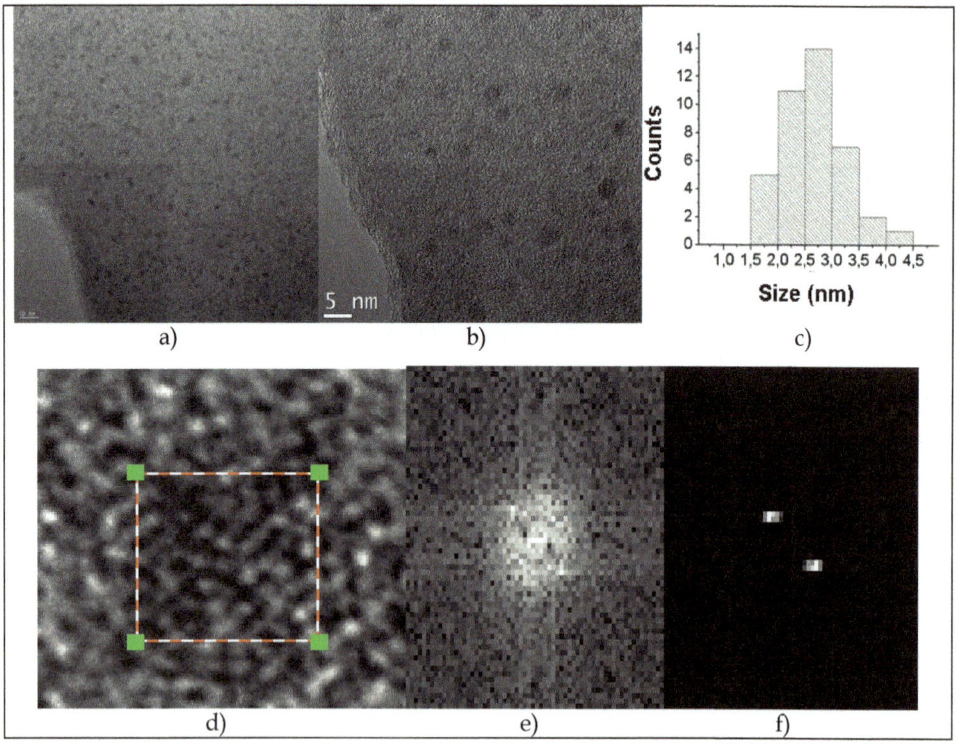

Fig. 10.a) and 10.b) TEM micrographs using multibeam configuration of NbN nanoparticles distributed homogeneously; c) average grain size of NbN nanoparticles; d) HRTEM micrographs shows atomic planes; e) reciprocal space of a NbN nanoparticle; and f) distance between (200) planes.

Fig. 10 shows the micro-structural analysis carried out on the NbN films through transmission ion electron microcopy (TEM). Micrographs 10a and 10b correspond to the image obtained in multibeam configuration, and in them one can observe an amorphous matrix with crystalline nanoparticles with spherical morphology that are homogenously distributed along the growth plane of the film. Using Gatan software, it was determined that the average grain size of the nanoparticles was 2.7±0.6 nm with normal distribution (fig. 10c). When doing magnification of nanoparticles of Fig. 10a, we can observe an interferential pattern produced by the atoms belonging to the NbN film (fig. 9d). Higher magnifications of the nanoparticles visualized in Fig. 10a let us observe atomic planes as a product of the interferential pattern produced by the arrangement of atoms belonging to the NbN film (Fig.10d). Image processing of Fig. 10d, using the Fourier transform, lets us visualize the reciprocal space of one NbN nanoparticle (Fig. 10e). Applying a mask (Fig. 10f) over the corresponding diffracting spots and directly measuring the distance between them (0.224 nm), we can confirm that they correspond to distances between (200) planes of δ-NbN. These results confirm the analysis of XRD, which established that NbN grows preferentially along the (200) plane.

To produce thin films with normalized temperature larger than 0.3 it is necessary to have ultrahigh-vacuum equipment. This condition limits the production the thin films through magnetron sputtering. However, there are studies such as that of Frederick (Frederic & Gall, 2005), who grew CrN thin films on MgO substrate through magnetron sputtering with Ts/tm between 0 .43 to 0.48. These authors found that the films grown at 993 K present complete single-crystal structure with smooth surfaces. The root-mean-square surface roughness for 230-nm-thick layers decreases from 18.8 to 9.3 to 1.1 nm as Ts is raised from 873 to 973 to 1073 K.

4.1 Influence of the gas flow on the structure and microstructure of thin films

In a different set of experiments, nitrogen gas was introduced into the deposition chamber (maintaining the final working pressure constant) in order to study the influence that the addition of this gas during deposition exerts on the structural and micro structural properties of the NbN films. Fig. 11 shows the diffraction patterns recorded from the films grown at 300 W, 553 K and different nitrogen flows. The results obtained make clear that in all cases a preferential growth appears along the (200) plane (Fig.11a). The relative intensity of this diffraction peak is so high that it makes it impossible to distinguish the diffraction peaks corresponding to other planes. To determine the polycrystalline character of the film, we carried out X-ray diffraction experiments at grazing incidence. Fig. 11b shows a representative example. It is clear from Fig. 11 that the films show the same diffraction peaks as the target, confirming their polycrystalline character, although having a preferential orientation along the (200) plane (texture index, 0.65). The grain sizes deduced by Scherer's equation (Cullity, 2001) for the different films grown along the (200) plane vary from 35 nm (Φ= 2 sscm) to 42 nm (Φ = 6 sscm). These results indicate that incorporation of nitrogen during the fabrication process favors the preferential growth of the d-NbN phase along the (200) plane. This effect has also been produced during the growth of TiN films, due to the change of the preferred orientation of the films from the [111] direction to the [200] direction when the gas flow is increased.

The incorporation of nitrogen during the deposition process implies changes in the dynamics of the plasma, since the increase in the number of nitrogen molecules increases the probability of collisions, promoting a larger number of chemical reactions on the substrate surface. These reactions can be explained using a model of low energy (<20 eV) ion bombardment during film growth at a Ts/Tm ratio ranging between 0.1 and 0.3 (a condition which is satisfied in this case) (Petrov & Barna, 2003). According to this model, 25 eV are sufficient to cause collision dissociation of the N_2 ions, providing a continuous source of atomic nitrogen. The nitrogen readily chemisorbs on the (200) planes but not on the N-terminated (111) planes. This in turn reduces the mean free path of the metal cation on the (200) plane due to capture by the nitrogen atoms, and promotes the formation of a NbNi

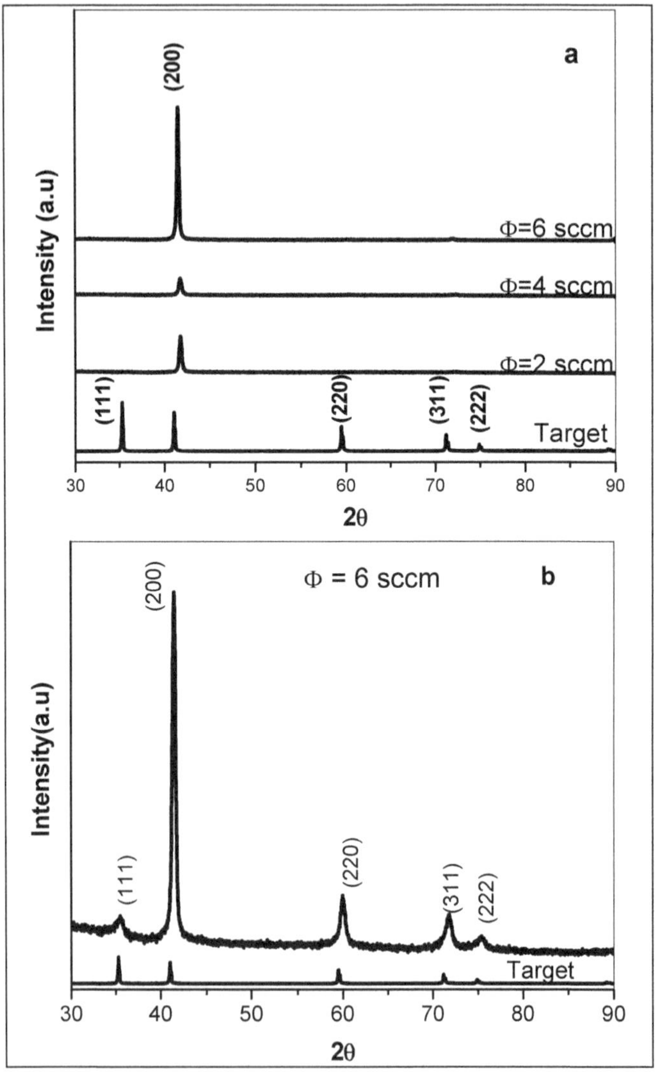

Fig. 11. a) XRD patterns θ-2θ configuration and b) Grazing angle (2°) XRD pattern recorder from the NbN film, grown at 300 W, 553 K, 20 sccm of Ar and 6 sccm of N_2.

(i = 1–4) admolecule or islands of adatoms (Petrov & Barna, 2003). This can be considered as causing an additional decrease in the (200) surface energy relative to that of the (111) plane. Consequently, the presence of the nitrogen atoms reduces the flow of cations from the (200) to the (111) planes, resulting in the orientation of the growth along the [200] direction. In this example, the growth of the films depends on the final working pressure as well as the bombardment energy of the ions.

The SEM study (Fig. 12a and 12b) of film growth as a function of the gas flow indicates that the NbN films present a compact granular structure, with a columnar growth of the type described by Movchan and Demchisshim (Movchan & Demchisshim, 1969), having an average thickness of 0.7 μm, which implies a deposition rate of 20 nm/min.

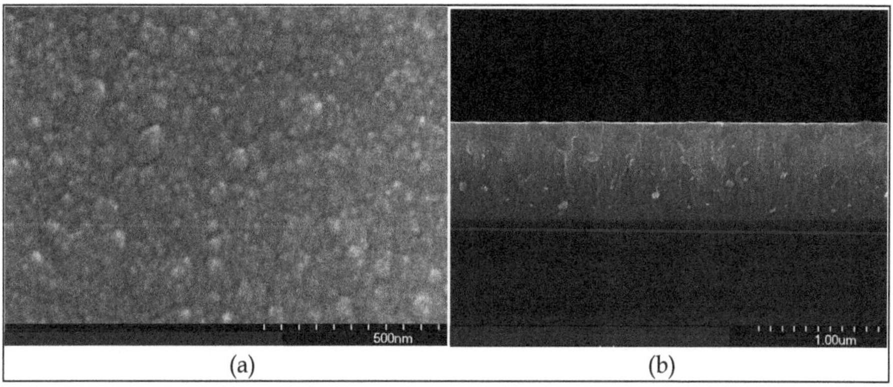

Fig. 12. a) Micrograph of the morphology of NbN thin film. b) Micrograph of the cross-section of NbN film, grown at 300 W and 553 K, 20 sccm of Ar and 6 sccm of N_2.

An interesting example of the growth of thin films in a reactive phase through the rf sputtering technique is ZrN_xO_y thin films, which grow at different flow ratios (N_2/O_2), but with the final working pressure constant (7.4 X10^{-1} Pa). The results of the XRD studies are shown in figures 13a and 13b, where it is possible to observe the influence that the flows have on the crystallographic of the films. Fig. 13a shows the XRD pattern of the films where the nitrogen flow was maintained constant (2.5 sccm) and the oxygen flow was varied, and Fig. 13b shows the XRD pattern of the films where the oxygen flow was maintained constant (2.0 sccm) and the nitrogen flow was varied.

The results obtained allowed establishing that there is an optimum flow ratio of 1.25 in which growth of a film with a high degree of crystallinity is reached. This behavior is very similar to the NbN films discussed above; therefore, the physicochemical mechanisms involved in the growth of thin films are the same for nitrogen molecules as for oxygen molecules, due to the higher reactivity of oxygen as compared to nitrogen (Martin, *et al.*, 2001).

The SEM study (Fig. 14a) of ZrN_xO_y films that were grown at a flow ratio of 1.25 present a highly homogenous and maybe very compact surface in which it is not possible to find contrast, and therefore the growth mechanism is not well defined. The microstructure of the films was evaluated through AFM studies (Fig. 14b). The micrograph reveals that the average size grain was 150 nm with an average rugosity of 5.9 nm.

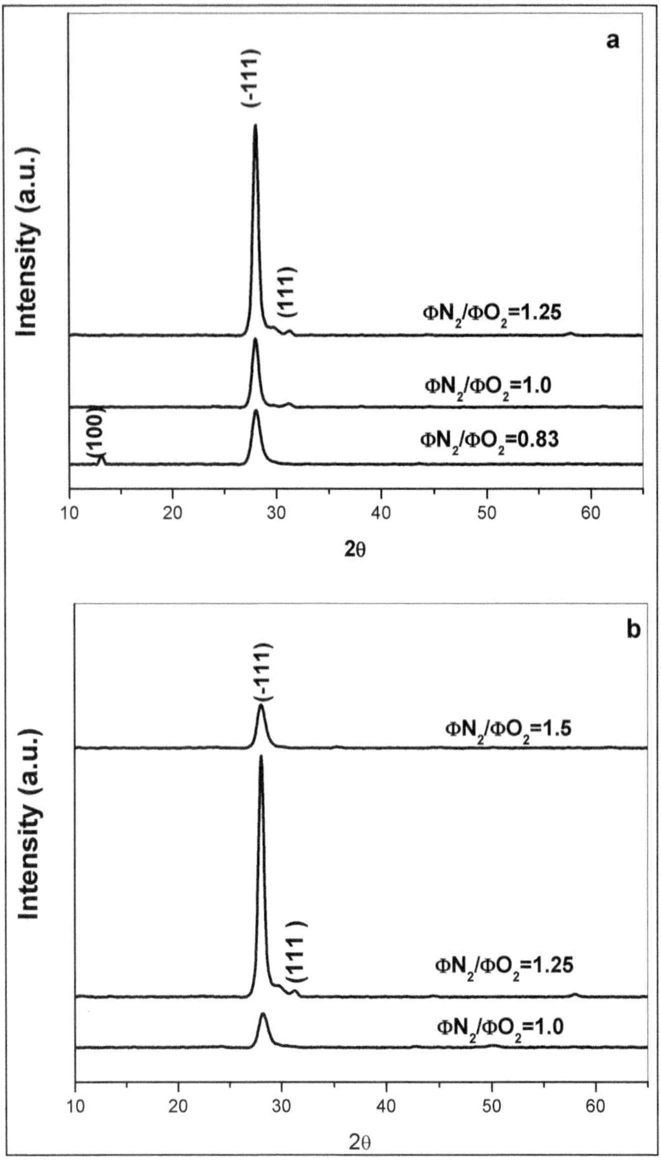

Fig. 13 . XRD Patterns recorded from ZrN_xO_y thin films deposited on common glass at 623K and different ratios flow ($\Phi N_2/\Phi O_2$). a- ΦN_2 constant at 2.5 sccm and b- ΦO_2 constant at 2.0 sccm.

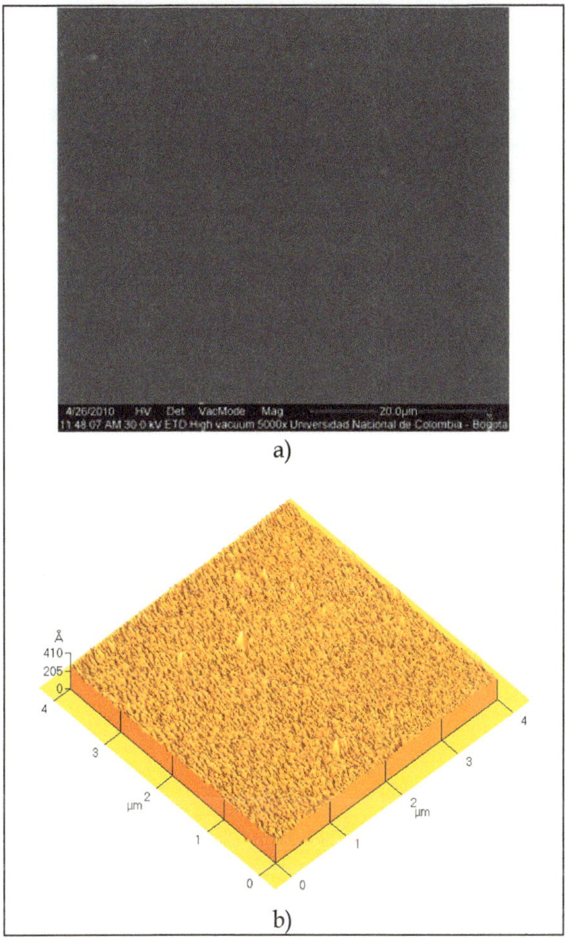

Fig. 14. a- Micrograph of the morphology of ZrN_xO_y thin film. b- AFM micrograph of the ZrN_xO_y film, deposited at 350 W, 623 K and $\Phi N_2/\Phi O_2$= 1.25

4.2 Influence of the power on the structure and microstructure of thin films

The XRD patterns of MgO films (Fig. 15a and 15b) recorded from films grown at room temperature shows the influence which the power supply applied to the target has on the structural behavior of thin films. The films grown within the range from 150 to 200W showed amorphous behavior (not shown in the XRD pattern). On the other hand, starting at 250W, all films showed the (200) plane corresponding to magnesium oxide in the FCC phase. In the XRD pattern it can also be observed that the intensity of the (200) plane is so high that it makes it impossible to distinguish the diffraction peaks corresponding to other planes. To determine the polycrystalline character of the film, we carried out X-ray diffraction experiments at grazing incidence. From Fig. 15b it is clear that the films present two main diffraction peaks, (200) and (220), which belong to the target, confirming their polycrystalline character, although showing a preferential orientation along the (200) plane (texture index, 0.85).

Another example that shows the influence of the power supply applied to the target on the crystallization of the films is the growth of ZrN_xO_y thin films; the films were grown from the Zr target in the reactive phase in an atmosphere of nitrogen and oxygen (Φ_N= 2.5 sccm, Φ_O=2.0 sccm). In the XRD pattern of Fig. 16, it is possible to determine that there is a threshold power (250W) for producing the crystallization of the film on the substrate. At this power, the film grows with a high degree of crystallization along the (111) plane. This behavior can be explained by the energy model, since increasing the power on the target implies increasing the energy of the ions that are bombarding the substrate and therefore improving the mobility of the adatoms, which produce chemical reactions and atomic grouping along the planes with the higher surface energy, which generates films with a high degree of crystallinity.

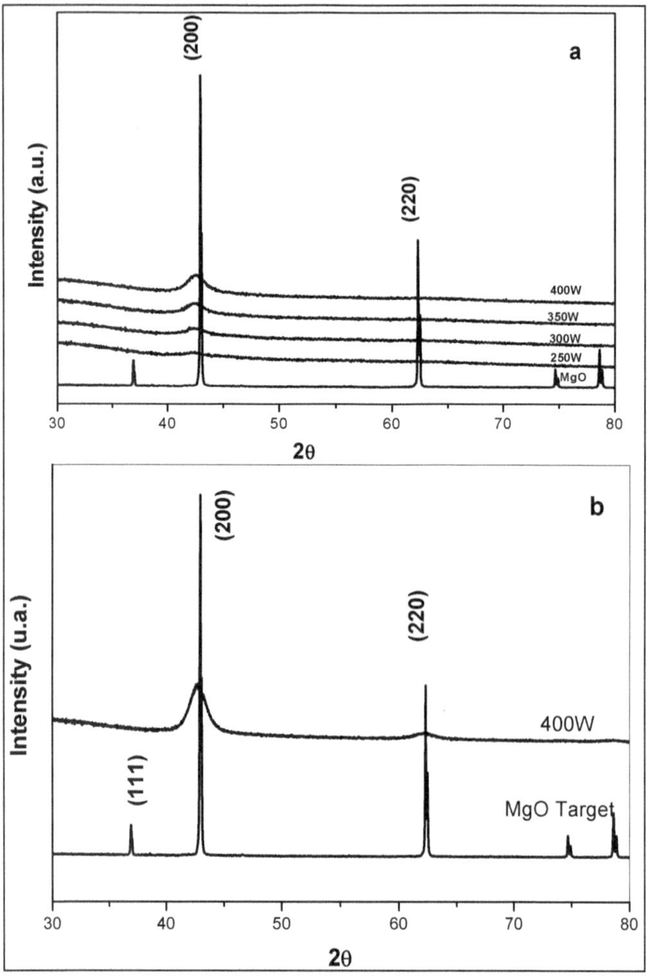

Fig. 15. a- XRD patterns recorded from MgO films deposited at 293 K and different power supplies applied to the target b- Grazing angle (3°) XRD pattern recorded from a MgO film grown at 400 W. The target XRD pattern is included as reference.

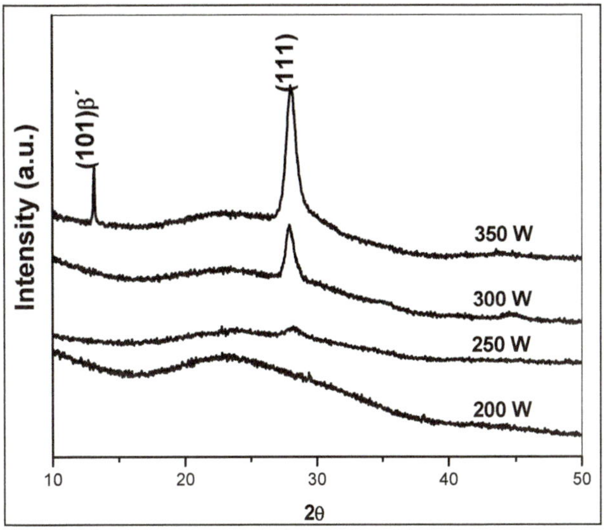

Fig. 16. XRD Patterns recorded from ZrN_xO_y thin films deposited on common glass at 623K at different powers applied to the target.

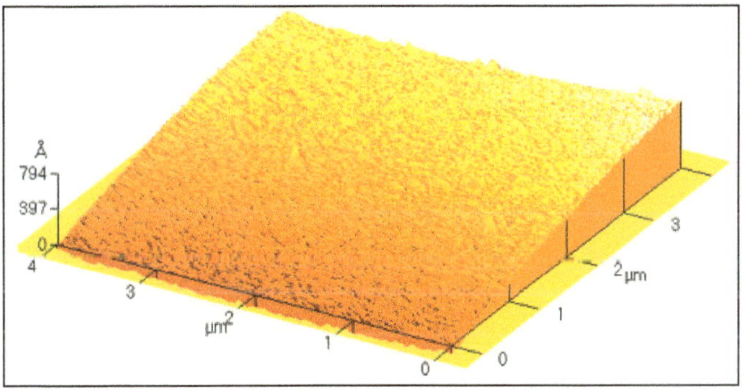

Fig. 17. Atomic force micrographs the MgO thin films deposited at 400 W.

The study of the microstructure of thin films was carried out using atomic force microscopy (AFM). The micrograph of Fig. 17 shows that in the area swept by the cantilever point (16µm^2) the film has an average roughness of 20 nm and an average grain size on the order of 110 nm. These results established that there is a threshold of power for the growth of nano-structured MgO thin films with high texture without intentional heating of the substrate.

4.3 Influence of the substrate bias voltage (V_s) on the structure and microstructure of thin films

Fig. 18 shows the influence that the substrate bias voltage has on the growth of the Ti alloy thin films deposited on steel and glass substrates. The films grown on steel at -100V showed, in addition to the diffraction peaks of the substrate material (note that in all the XRD patterns recorded from the films deposited onto steel diffraction peaks from the substrate are still visible), a quite broad peak, which corresponds to the (002) plane of the α- phase of Ti. The XRD pattern corresponding to the films deposited at –120 V present two well-defined peaks at 35.1° (100) and 38.4° (110), which we associate with the (α/β)-alloy phase. The film deposited at a bias voltage of -160 V shows a peak at 38.4°, which can be associated with the (110) plane of the β- phase of Ti. When the bias voltage increases to -200 V, the XRD pattern shows only one high-intensity peak at 38.4°(110), which can be associated with the β- phase of Ti. It can also be observed that the width of the diffraction peaks is narrower for the films deposited at higher bias voltages, especially for that deposited at -200 V, which suggests that increasing the bias voltage increases the grain size of the deposited

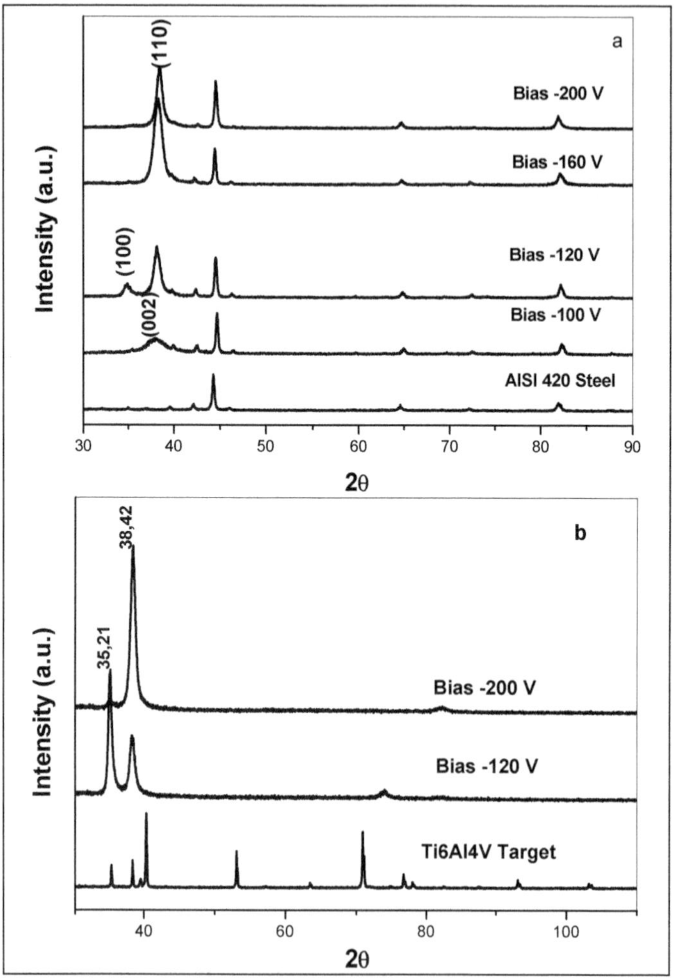

Fig. 18. (a). XRD pattern of the Ti6A14V films deposited by rf magnetron on steel substrates. The substrate appears as a reference. (b) XRD pattern of the films growth on glass and XRD pattern of theTi6A14V in bulk.

films. The films grown on glass at 120V and -200V show a similar behavior, although there are two differences: the two peaks appearing in the XRD pattern of the film grown at -120 V show a different intensity ratio from those of the film grown on steel, and the XRD pattern of the film grown at –200 V shows an additional low-intensity peak at 82.3°, which we associate with the (220) reflection of the β-Ti phase.

In summary, it follows from the XRD data that the substrate bias voltage (Vs) has a great influence on the structure of the deposited films: an increase of the bias voltage promotes the growth of different phases of titanium having well-defined different preferential crystallographic orientations. The increase of V_s is also reflected in a better crystallinity of the deposited films. The data also show that the crystalline phases formed at different V_s are the same in both substrates (see fig 18b). These results are in accord with that established in the foregoing section, where it was indicated that the increasing Vs improves the density of the film, which in turn allows obtaining films with a high degree of crystallinity.

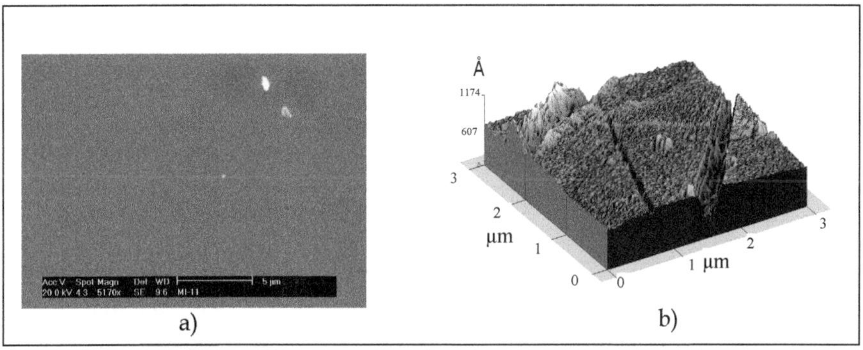

Fig. 19. a) Micrograph of a Ti6A14V film grown on steel with bias voltage of – 200 V. (b) Atomic force Micrograph recorded from Ti6Al4V grown on steel at bias voltage of – 200V.

The analysis of the microstructure of the Ti6Al4V was made through scenic electron microscopy (SEM) and an atomic force microscope (AFM). Fig. 19a shows the SEM results; the recorded micrograph of the film grown at –200V reveals a film with excellent texture, high homogeneity and denser microstructure with grain refinement produced under the enhanced plasma bombardment, which is induced by the substrate bias voltage. Through Scherer's equation (Cullity, 2001) and using the broadening of the (110) and (110) planes of the Ti β-phase, it was found that the average grain size was 13 nm for the (100) plane and 16.5 nm for the (110) plane (Alfonso *et al.*, 2005). Fig. 19b shows the results of the AFM, which indicated that the films grown at -200 V possess a roughness of 20 nm for a scan area of 3μmx3μm.

4.4 Influence of the energy parameter (E_p) on the structure and microstructure of thin films

The X-ray diffraction patterns of the CrNx films deposited at different Ep values are shown in Fig. 20. It was necessary to use two different scales, because as the Ep increased, the intensity of the CrN (200) peak increased by two orders of magnitude and became narrower. At Ep values lower than 30 eV/atom, we observed that the films mainly showed a [200] orientation with traces of the CrN[111] and CrN[220] orientations, while at higher Ep values only the CrN[200] orientation was detected. An estimation of the ⟨200⟩ texture was obtained comparing the (200) peak intensity (I200) relative to the intensity of all the orientations (Ii) appearing in the XRD pattern. Table 3 summarizes some experimental conditions of the deposition of the CrN films.

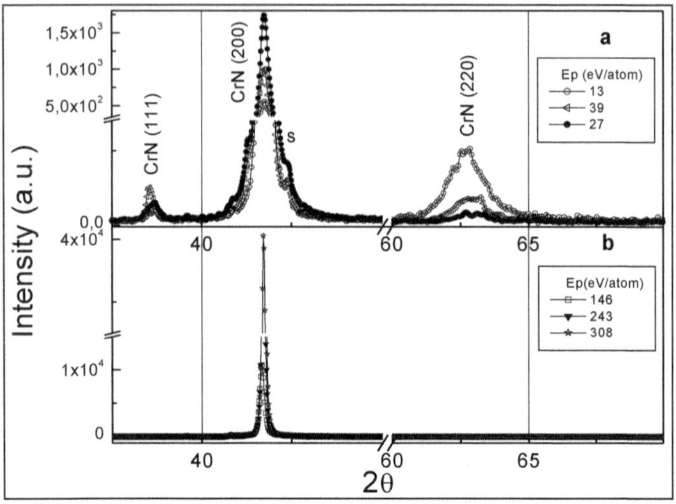

Fig. 20. X-ray diffraction patterns of CrNx coatings: (a) low Ep values, (b) high Ep values.

Sample	d s-t (cm)	Vs (V)	R (μm/min)	Ei (eV)	Ji/Ja	Ep (eV/atom)	$F_{<200>}$
CrN-3	3	0	0.10	10	1.3	13	0.80
CrN-4	4	0	0.064	16	2	32	0.89
CrN-5	5	0	0.051	12	2.2	26	0.90
CrN-6	6	0	0.035	13.5	2.6	35	0.92
CrN-7	7	0	0.026	14.5	2.7	39	0.93
CrN-8	8	0	0.019	17	2.7	46	0.92
CrN-3-100	3	-100	0.10	110	1.3	143	0.99
CrN-4-100	4	-100	0.064	115.6	2.0	231	0.99
CrN-5-100	5	-100	0.05	112.3	2.2	247	1.0
CrN-6-100	6	-100	0.034	113.5	2.6	295	1.0
CrN-7-100	7	-100	0.026	114.5	2.7	310	1.0
CrN-8-100	8	-100	0.019	117	2.7	316	1.0

Table 3. Summary of experimental growth conditions of CrN thin films and some of the results: deposition distance ds–t, r.f. bias voltage Vs, deposition rate R, ion energy Ei, ion-atom flow Ji/Ja, energy parameter Ep and texture factor $f_{<200>}$.

The results of the microstructure behavior of the CrN thin films are in good agreement with that of the theory of Petrov *et al.*, which was explained above. In the case of CrN thin films, increasing the ion energy by applying a higher substrate bias and keeping low ion–atom flows resulted in a film densification and a change in the preferred orientation from [111] to [002] with a subsequent increase in the levels of stress and the incorporation of argon ions. Moreover, by increasing the energy to around 20 eV and working at higher ion–atom ratios, the residual stress remained low, but the texture of the films changed abruptly and completely from the [111] to the [002] direction. According to the Petrov Model, the key factor was that the energy of the N_2 ions should be near 20 eV to promote the dissociation of the ions through collision with the film surface, in this way providing a supply of atomic nitrogen which can chemisorb on [002] oriented grains and later capture metal atoms, resulting in the development of the ⟨002⟩ texture. In CrN films, it was observed that at the lowest Ep values (Ep<50 eV/atom) with $f_{(200)}$ ~80%, the film texture was equivalent to the transition regime in Petrov *et al.*'s research. The ion energy and ion-atom flows were not sufficiently low to produce pure ⟨111⟩ texture, as explained above, but were also not sufficiently high to induce 100% ⟨002⟩. In this transition regime, a competitive growth between the high-trapping (111) plane and the low surface energy (002) plane was established, since as the ds-t increased, the ion energy of the N_2 ions approached 20 eV (see Fig. 20). As the Ep value increased, the conditions favored the formation of the ⟨002⟩ texture. Thus we considered Petrov's theory to be a rather good model to explain the microstructure evolution of the texture in metal nitride films. The competition of the different orientations during growth also affected the microstructure, as observed in Fig. 21a-d. As the ion bombardment increased, the microstructure changed from a non-ordered columnar growth to well-oriented grains that look like fibers coming out from the substrate, *i.e.* equiaxed grains.

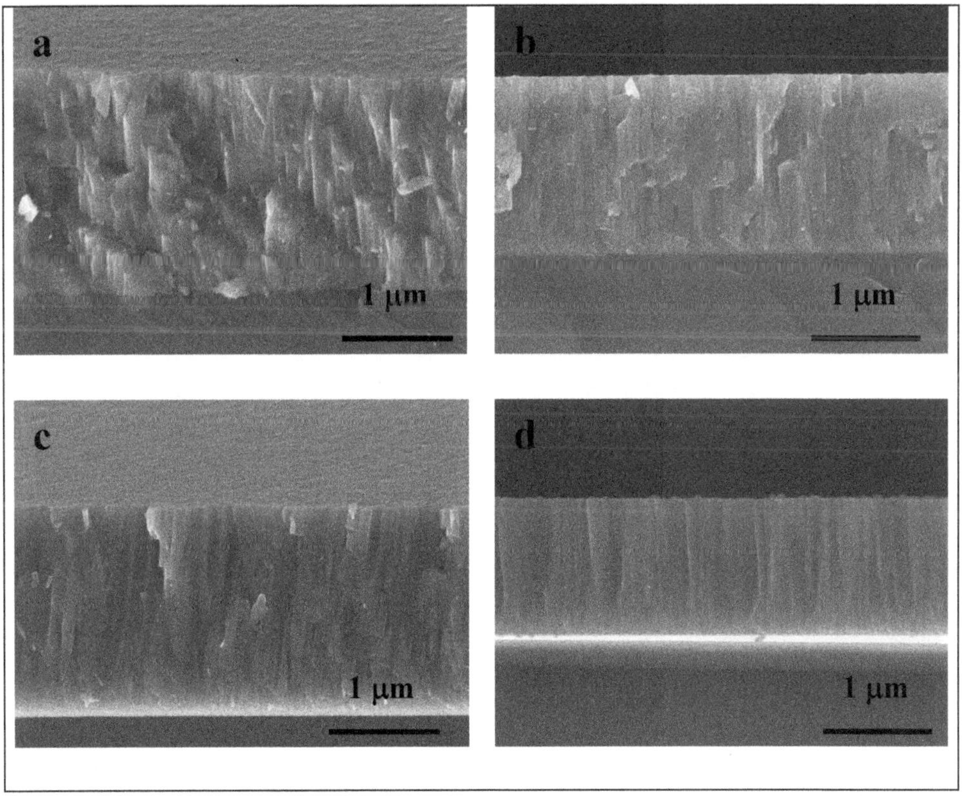

Fig. 21. Cross-section SEM images of samples deposited at increasing Ep values (a) 13, (b) 46, (c) 143 (d) 316 eV/atom, respectively.

5. Chemical characterization of thin films

5.1 X-ray photoelectron spectroscopy (XPS) analysis

The last 10 nm of the surface of thin films can be analyzed using X-ray photoelectron spectroscopy (XPS). As an example, we show the results of XPS on Ti6Al4V thin films. The spontaneously passive film formed on the deposited Ti6Al4V films upon exposure to the air was studied through XPS. Fig. 4 shows high resolution narrow-scan spectra recorded from the Ti 2p, Al 2p, V 2p and O 1s spectra recorded from one of the samples. All the narrow-scan spectra recorded from the rest of the samples are almost identical to those presented in Fig. 22.

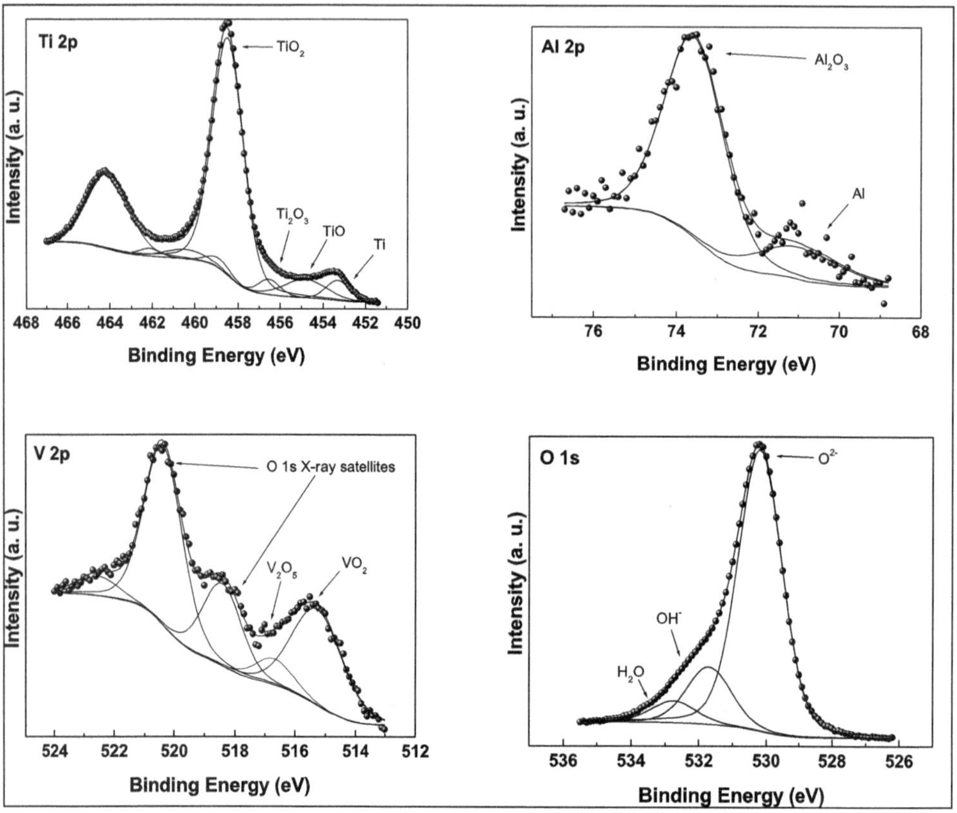

Fig. 22. 2p, Al 2p, V 2p and O 1s XPS narrow-scan spectra recorded from a film deposited at a bias voltage of -140 V.

The Ti 2p spectrum shows several contributions, and is similar to that reported previously by other authors for the spontaneously-formed passive oxide layer on bulk Ti6Al4V alloy. The spectrum is dominated by a major TiO_2 contribution (BE Ti $2p_{3/2}$=458.4 eV, 82%), and shows smaller Ti (BE Ti $2p_{3/2}$=453.3 eV, 5%), TiO (BE Ti $2p_{3/2}$=454.8 eV, 9%) and Ti_2O_3 (BE Ti $2p_{3/2}$=456.5 eV, 4%) contributions. The Al 2p spectrum shows a major Al_2O_3 contribution at 73.6 eV and a minor Al^0 contribution at 71.0 eV. Again this spectrum is similar to that observed for the passive oxide layer on a Ti6Al4V alloy. The O 1s peak shows three different contributions: a main one at 530.2 eV that can be associated with metal-oxygen bonds, and two much less intense contributions at 531.7 eV and 532.8 eV, which correspond to the presence of OH^- groups and chemisorbed water, respectively. The fit of the V 2p spectrum is complicated by the fact that the V peaks overlap strongly with the O 1s $K\alpha_{3,4}$ X-ray satellite peaks. Therefore, the whole spectrum was refined considering the presence of these satellites and the presence of vanadium peaks. The results show the presence of an intense peak at 515.4 eV, which we associate with the presence of VO_2, and a less intense vanadium contribution at 516.7 eV, which corresponds to V_2O_5. This finding contrasts with those of Milosev (Milosev *et al.*, 2000), who did not find oxidized vanadium in the passive layer formed on bulk Ti6Al4V (Alfonso *et al.*, 2006).

These results show that the films that have been grown through rf magnetron sputtering reproduced the stiochoimetry of the target, although on the surface of the thin films chemical reactions will be produced, forming hydroxides and passivation layers.

5.2 Energy dispersive X-ray (EDX) analysis

The elemental composition of thin films is determined through energy dispersive X-ray analysis (EDX), although is important to indicate that elements such as oxygen and nitrogen are not possible to assess because the binding energies are very near to the binding energy of the carbon. As an example, the EDX results for NbN nanoparticle are shown. Fig. 23 shows the elements that constitute the nanoparticles of NbN (N and Nb) that formed the NbN studied in section 4, and the elements of the substrate (common glass) on which the thin films were grown. The Cu belongs to the sampling.

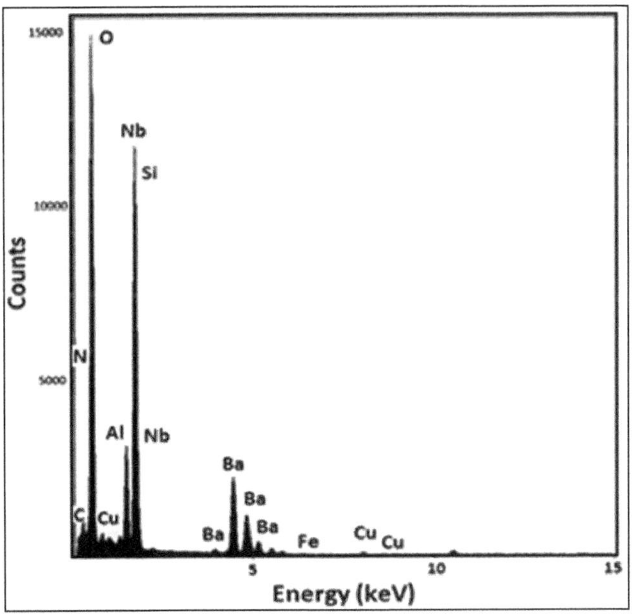

Fig. 23. EDX spectrum of NbN nanoparticle that integrates the NbN thin films.

6. Optical characterization of thin films

An important characteristic of thin films is the optical behavior, since this determines possible industrial applications, which range from transparent coatings to optical filters; optical studies are carried out through measurements of transmittance, absorbance and reflectance. Fig. 24 shows the transmittance percent as a function of the wavelength on MgO thin films that have been grown through rf sputtering with different power supplies applied to the target.

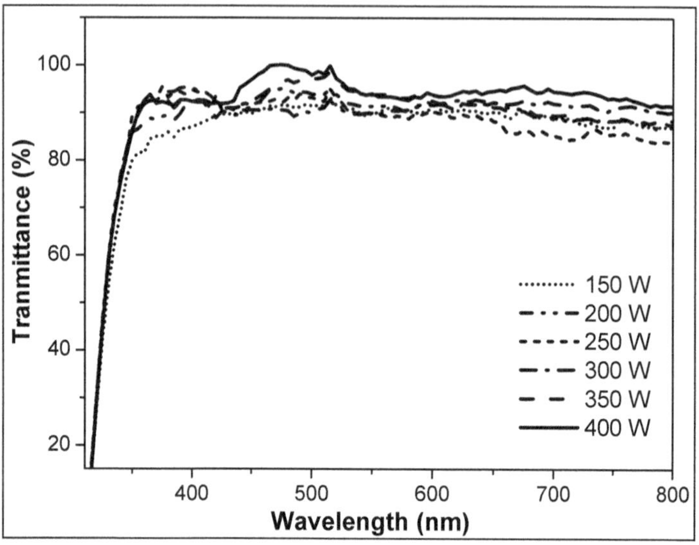

Fig. 24. Transmission spectrum vs wavelength the films the MgO with various strengths of power supply to the target.

The results show (see Fig. 24) that the films have percentages of transparency that range from 84% to 95% in films grown at 150 and 400 W, respectively. These results indicate that films that were grown at a higher power have a high value of transmittance. This behavior is due to the fact that the films are denser and possibly present lower diffraction.

7. Applications of thin films

Surface modification by means of thin film deposition is an important industrial process used to protect basic materials against wear, fatigue, corrosion and many other surface-related damage phenomena (Vyas *et al.*, 2010; Yang *et al.*, 2008; Wesley, 2001; Li *et al.*, 2009; Kumar, A. & Kaur, 2009). The modern methods of plasma-assisted physical vapor deposition techniques provide great flexibility for designing films with specific chemistry and microstructure, leading to coatings with unique properties. Among these, ceramic coatings deposited on metallic substrates have shown excellent improvement of the surface properties, such as a low friction coefficient and a high degree of hardness with associated good wear resistance and also corrosion resistance to aggressive environments; (Yang, *et al.*, 2008; Wesley, 2012).

Films of CrN, TiN, ZrN, TaN and NbN were deposited using an unbalanced magnetron sputtering system with different energy parameters to investigate its effect on some film properties. Fig. 23 shows the main results obtained in these films. In general, it may be observed that the residual compressive stresses, determined using the curvature method and Stoney's equation (Stoney, 1909), and increased with the energy parameter, although for some films the variation was very small.

Fig. 25a shows the microhardness of the substrate–film system. Hardness measurements were made on samples deposited on AISI M2 tool steel. The hardness of the substrate was 800 kg mm^{-2}. The hardness of the films was 2–3 times higher than the substrate hardness. Group IV nitrides generally are harder than those of group V (Hofmann, 1990). Moreover, the data showed that the hardness increased as the energy parameter increased. This could be attributed to the increase in ion bombardment on the substrate surface, increasing adatom mobility and producing denser films, and the effect on hardness of compressive stresses present in these films (Olaya *et al.*, 2009).

Fig. 25b presents the wear coefficient, Kwear, which was evaluated using a ball cratering system, which is a micro scale abrasion test. Two-body grooving abrasion was the wear mechanism observed for all coatings (not shown), probably due to the micro-cutting action of abrasive particles that were dragged across the ball, basically remaining fixed to the ball surface during the test (Adachi, K. & Hutchings, 2003). The wear coefficients were all in the 10^{-6} mm^3 $N^{-1}m^{-1}$ range, and the lowest values were obtained for NbN and TaN films. It may be seen that there was a slight decrease in k wear as the energy parameter increased, but the variation was too small to be considered important. In addition, no clear trend was observed as the unbalance coefficient was increased. This might be a consequence of variations in other film properties, such as the coefficient of friction or the roughness, parameters known to exert a strong influence on the wear performance of the surfaces.

Ceramic films like metallic nitrides on a metal substrate are commonly believed to be immune to corrosion. Fig. 25 c-d shows the potentiodynamic polarization curves for the films deposited on AISI 304 substrate at the two energy parameters and the AISI 304 substrate in an electrolyte of 0.5 M H_2SO_4 + 0.05 M KSCN. The corrosion resistance of a material in the polarization curve is determined by its ability to retain low current densities as the electric potential increases. Quantitatively, Tafel analysis was used to determine the corrosion potential, Ecorr, while the critical passivation current density, Icrit, was estimated from the maximum anodic dissolution current before passivation. The corrosion behavior of PVD ceramic-coated steels in aqueous solutions has been increasingly investigated in recent years (Vyas *et al.*, 2010; Yang *et al.*, 2008; Wesley, 2001; Li *et al.*, 2009; Kumar, A. & Kaur, 2009). One of the major drawbacks is the presence of defects, which are associated with the growth process in PVD ceramic coatings. These coating defects (*e.g.* pores) are particularly deleterious for corrosion protection, since they provide direct paths through which the electrolyte can reach the coating/substrate interface, where localized galvanic corrosion occurs due to the difference in the corrosion potential between the coating and the steel. A lot of research has been undertaken to deal with this problem of coating defects (Stansbury& Buchnan, 2000; Kaciulis *et al.*, 2000; Lang & Yu, 2001; Lee *et al.*, 2009; Chou & Huang, 2003) particularly for transition metal nitride coatings, which in general have excellent wear and corrosion resistance and therefore are widely used in industry.

The critical passivation current density shows that in contrast to the other properties of the films, there was a decrease in the corrosion resistance of the films as the energy parameter increased with Ep. The critical passivation current density is proportional to the exposed area of the substrate due to the existence of pores or pinholes in the film (Olaya *et al.*, 2005). The value of I_{crit} depends on the grain limits that join the columns due to the fact that they can contain vacancies, micropores, pinholes, and possibly microcracks, allowing the diffusion of electrolytes of the corrosive solution toward the substrate and increasing degradation of the coatings.

Therefore, the results reported in Fig. 25 c-d suggest that as the energy parameter increased there was more substrate area exposed at the bottom of the pinholes. These pinholes or defects are usually localized at the grain boundaries, which are defined by the crystal growth process, which consequently models the final film structure. The rise in the corrosion current with Ep was very small for the NbN and TiN films, but significantly higher for TaN, ZrN and CrN. However, the higher ion bombardment induced by the degree of magnetic field created more defects, increasing the residual stresses and therefore deteriorating the film-substrate adhesion and consequently the response of the coatings to the corrosion products, as was visually observed at the end of the corrosion test for the TaN film. This phenomenon can be explained by galvanic coupling produced by a difference in the corrosion potential between coated and uncoated specimens. The potential difference is characterized by anodic dissolution of the substrate material with a high anodic current density at the defect site, leading to an adhesion failure of the coating.

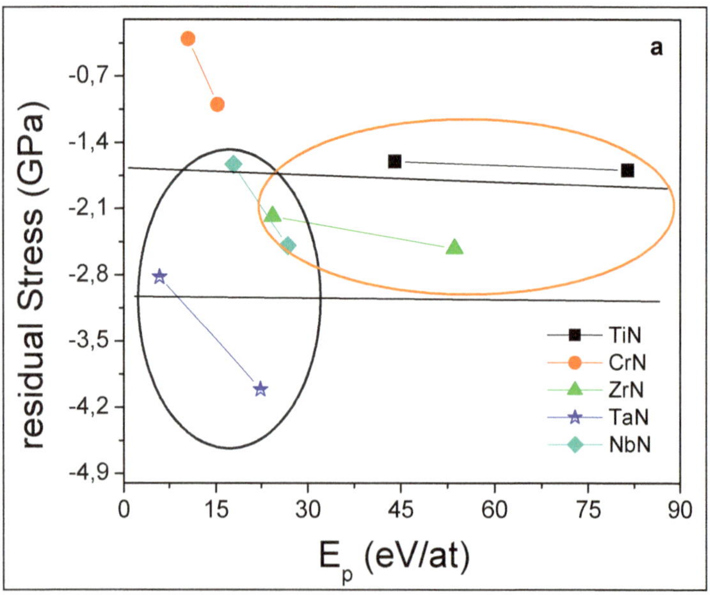

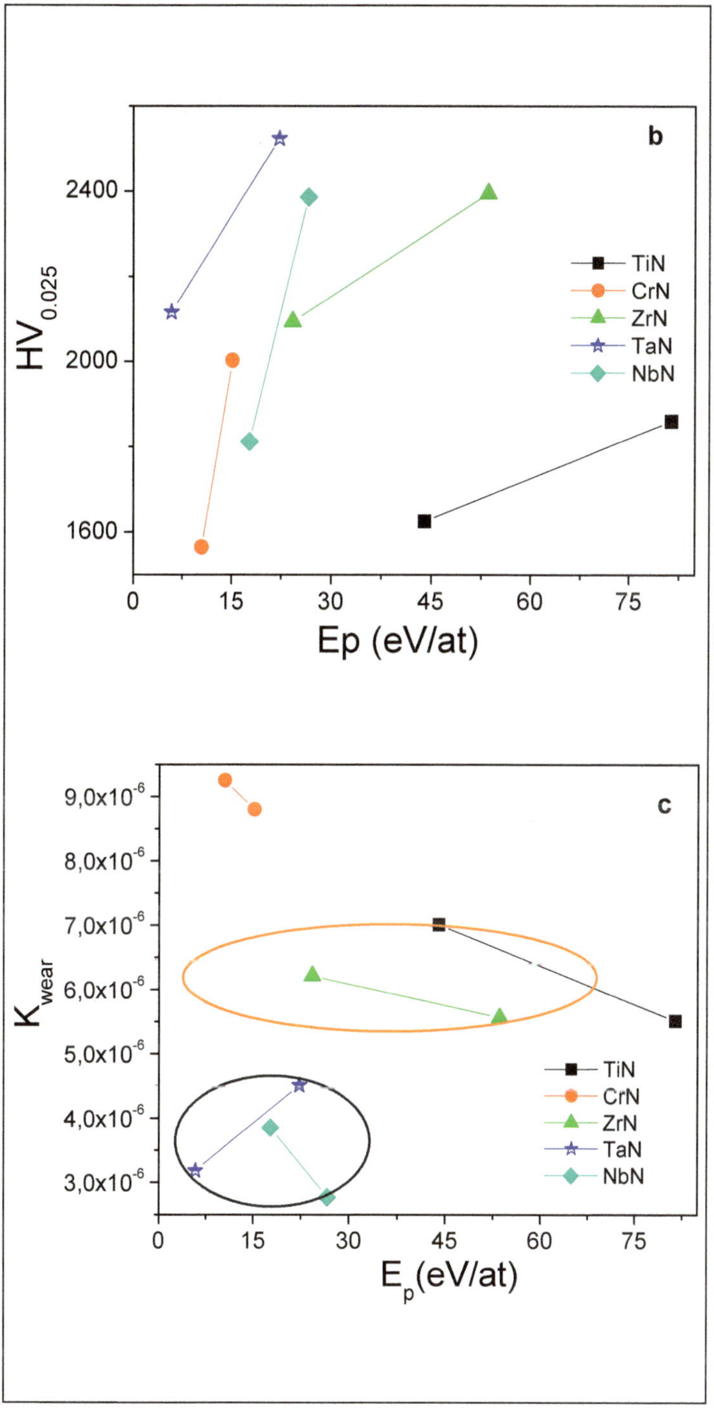
b
2400
2000
1600
$HV_{0.025}$
0
15
30
45
60
75
Ep (eV/at)
TiN
CrN
ZrN
TaN
NbN
c
9,0x10^{-6}
8,0x10^{-6}
7,0x10^{-6}
6,0x10^{-6}
5,0x10^{-6}
4,0x10^{-6}
3,0x10^{-6}
K_{wear}
0
15
30
45
60
75
E_p(eV/at)
TiN
CrN
ZrN
TaN
NbN

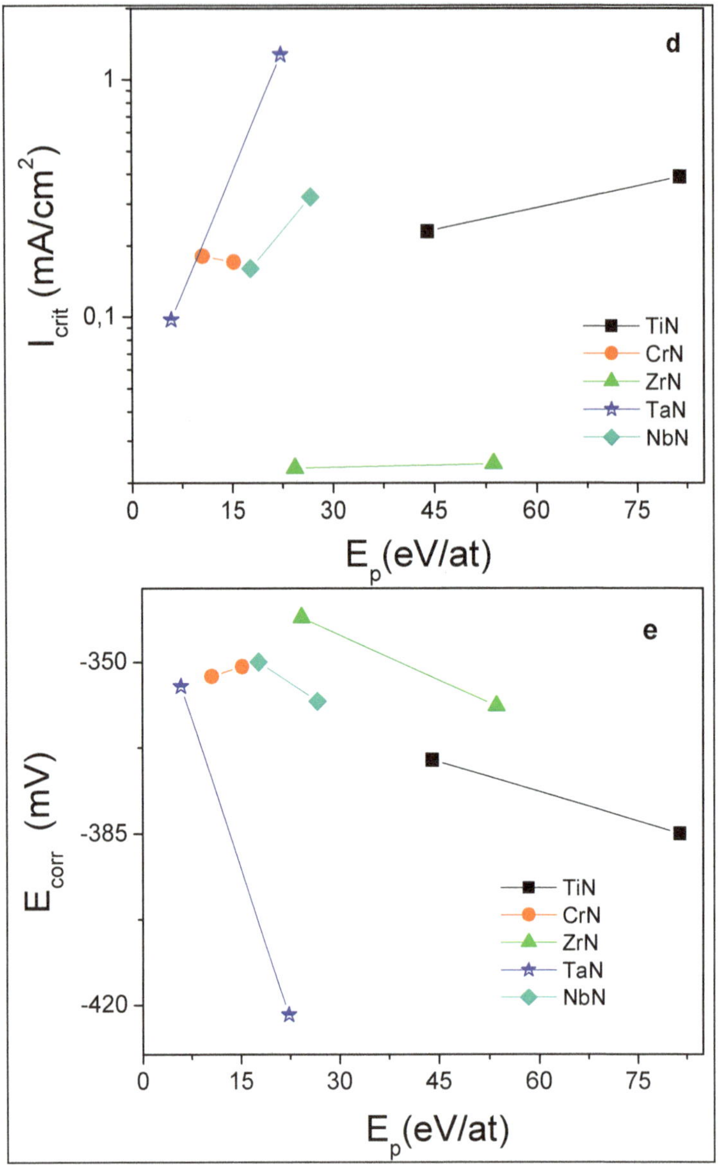

Fig. 25. Summary of mechanical and chemical properties of films deposited through sputtering. a) residual stress b) hardness c) wear coefficient d) critical passivation current and d) corrosion potential

8. Conclusion

In this chapter, the main physical-chemical models utilized to explain the growth of thin films through the rf sputtering technique have been presented, and the influence that the growth parameters have on the crystallinity and the micro-structure of the thin films has been discussed, and based on the described models, examples have been provided of the thin films' growth under each of the growth parameters. The fundamental idea of presenting these basic theories is to introduce the science of materials to young researchers in the world, helping them to understand by means of examples the basic concepts that apply to the growth of thin films, with the hope that a better understanding of these theories will produce new and better applications.

9. References

Adachi, K. & Hutchings, I., Wear-mode mapping for the micro-scale abrasion test. *Wear*, Vol 255, issues 1-6, (august-september 2003) 23-29.

B. A. Movchan and A. V. Demchishin, Study of the Structure and Properties of Dioxide thin, Fiz. Met. Metalloved. Vol. 28, (1969) 83-90.

B. D. Cullity, s: R. Stock, Elements of X-ray diffraction, Prentice Hall, 2001.

B. Rauschenbach, J. W. Gerlach, Texture Development in Titanium Nitride Films Grown by Low-Energy Ion Assisted Deposition, Cryst. Res. Technol. Vol. 35, (2000) 6-7.

B. Wang, Y.N. Zhao, Z. He, The effects of deposition parameters on the crystallographic orientation of AlN films prepared by RF reactive sputtering, Vacuum. Vol. 48, (1997) 427-429.

B. Window & N. Savvides, Charged particle fluxes from planar magnetron sputtering sources, J. Vac. Sci. Technol. A4 (1986) 196–202

Barna, P. , Adamik, M., Fundamental structure forming phenomena of polycrystalline films and the structure zone models. Thin Solid Films, Vol. 317, (April 1998) 27-33.

C. Lin, C. Lee, Growth of Tantalum Nitride Films on Si by Radio Frequency Reactive Sputtering of Ta in N2/Ar Gas Mixtures: Effect of Bias, J. Electrochem. Soc. Vol. 147, (2000) 713-718.

C. S. Shin, D. Gall, Y. W. Kim, N. Hellgren, I. Petrov, J.E. Greene, Development of preferred orientation in polycrystalline NaCl-structure δ-TaN layers grown by reactive magnetron sputtering: Role of low-energy ion surface interactions , J. Appl. Phys. 92, (2002) 5084.

C.R. Weissmantel, Deposition of Metastable Films by Ion Beam and Plasma Techniques, *Proc. 9th Int. Vacuum Congress and 5th Int. Conf. on Solid Surfaces* (Madrid), 26 Sept-1 Oct 1983, p 229-308

Chou, W., Yu, G. & Huang, J. Corrosion resistance of ZrN films on AISI 304 stainless steel substrate. *Surface and Coatings Technology*. Vol. 167, issue 1, (april 2003) 59-67.

Emery, C., Chourasia,A., Yashar, P., A study of CrNx thin films by X-ray photoelectron spectroscopy J. Elect. Spectros. Relat. Phenom. Vol. 104, issue 1-3, (July 1999) 91-97.

Flores, J. F.; Olaya, J. J.; Colás, R.; Rodil, S. E.; Valdez, B. S.; Fuente, I. G. Corrosion behaviour of TaN thin PVD films on steels. Corrosion Engineering, Science and Tecjnology, Vol. 4, issue 2, (June 2006) 168-176.

G. G. Stoney, The measurement of stress in layered structures, in: Proc. Roy. Soc. London Vol. A82, (1909) 172.

G. Kamath , A.P. Ehiasarian, Y. Purandare, P.Eh. Hovsepian, Tribological and oxidation behaviour of TiAlCN/VCN nanoscale multilayer coating deposited by the combined HIPIMS/(HIPIMS-UBM) technique, Surface & Coatings Technology 205 (2011) 2823–2829

Goldfarb, I., Pelleg, J., Zevin, L. and Croitoru, N. (1991). Lattice distortion in thin films of IVB metal (Ti, Zr, Hr) Nitrides, *Thin Solid Film*, Vol. 200, issue 1, (May 1991) 117-127., (May 1991), 117-127.

Greene, J. E., Low Energy Ion Bombardment during Film Deposition from the Vapor Phase: Effects on Microstructure and Microchemistry, *Solid State Technol.*, 30,(4), (1987)115.

Greene, J. E., Nucleation, Film Growth, and Microstructural Evolution, *Deposition Processes for Films and Coating*, 2nd edition, (R. Bunshah,ed.), Ch. 13, Noyes Publications (1994).

H. Liu, W. Tang, D. Hui, L. Hei, F. Lu. Characterization of (Al, Si)N films deposited by balanced magnetron sputtering, Thin Solid Films, Vol. 517, issue 21, 1 (September 2009) 5988-5993.

H.B. Nie, S.Y. Xu, S.J. Wang, L.P. You, Z. Yang, C.K. Ong, J. Li and T.Y.F. Liew, Structural and electrical properties of tantalum nitride thin films fabricated by using reactive radio-frequency magnetron sputtering, Appl. Phys. A. Mater. Sci. process, Vol. 73 (2001) 229- 236.

Hofmann S., Target and substrate surface reaction kinetics in magnetron sputtering of nitride coatings, Thin Solid Films, Vol. 191, (1990) 335-348.

Hwan-Chul Leea, Guen-Hong Kimb, Soon-Ku Hongc, Ki-Young Leea, Yoon-Joong Yonga, Chang-Hwan Chunb, Jai-Young Leea, Influence of sputtering pressure on the microstructure evolution of AlN thin films prepared by reactive sputtering, Thin Solid Films, Vol. 261, issue 1-2, (1 June 1995) 148-153.

I. Milosev, M. Metikos-Hukovic and H.-H Strehblow, Passive film on orthopaedic TiAlV by X-ray photoelectron, Biomaterials, Vol. 21, (2000), 2103.

J. E. Alfonso, Fernando Pacheco, Alvaro Castro P, J. Torres, Influence of the substrate bias voltaje on the crystallographic structure and mechanical properties of Ti6Al4V coatings, deposited by rf magnetro, Phys. Stat. sol. Vol. 10, (2005) 3786-3789.

J. E. Alfonso, J. Torres, and J. F. Marco,. Influence of the Substrate Bias Voltage on the Crystallographic Structure and Surface Composition of Ti6A14V Thin Films Deposited by rf Magnetron Sputtering, Brazilian Journal of Physics, Vol. 36, issue 3, (September 2006) 994-996.

J. E. Greene, J.-E. Sundgren, L. Hultman, I. Petrov, and D. B. Bergstrom, Development of Preferred Orientation in Polycrystalline TiN Layers Grown by Ultra-High-Vacuum Reactive Magnetron Sputtering, Appl. Phys. Letters, Vol. 67, (1995) 2928.

J. H. Je, D. Y. Noh, H.K. Kim, K.S.Liang, Preferred orientation of TiN films studied by a real time synchrotron x-ray scatteringJ. Appl. Phys. 81, (1997) 6126.

J. Musil, S. Kadlec, Reactive sputtering of TiN films at large substrate to target distances, Vacuum, Vol. 40 (1990) 435.

J. Pelleg, L. Z. Zevin, S. Lungo, Lattice distortion in thin films of IV B metal (Ti, Zr, Hr) nitrides Thin Solid Film, Vol. 197, (1991) 117.

J. R. Frederick, D. Gall, Surface morphological evolution of epitaxial CrN(001) layers, J. Appl. Phys., Vol. 98, (2005) 054906.

J.A. Heimberg, K.J. Wahl, & I. L. Singer, Superlow friction behavior of diamond like carbon coatings: Time and speed effects, Applied Physics Letters, Vol. 79, issue 17, (april 2001) 2449-2451.

J.J. Olaya, L. Huerta, S.E. Rodil, R. Escamilla, Superconducting niobium nitride films deposited by unbalanced magnetron sputtering , Thin Solid Films, Volume Vol. 516, issue 23, (1 october 2008) 8768-8773

K. Aso, M. Hayakawa, H. Matsuda, K. Hayashi, W. Ishikawa, On specimen preparation by radio frequency planar magnetron sputtering , J. Vac. Sci. technol. Vol. A4, issue 1, (1986) 19-25.

K. Bewilogua a, G. Bra¨uer a, A. Dietz a, J. Ga¨bler a, G. Goch (1)b, B. Karpuschewski (1)c, B. Szyszka, Surface technology for automotive engineering, Manufacturing Technology, Vol. 58, (2009) 608–627

K. S. Harsha, Principles of Physical Vapor Deposition of Thin Films, Elsevier, 2005.

K. Shibahara, S. Nishino, and H. Matsunam, Antiphase-domain-free growth of cubic SiC on Si(100), Appl. Phys. Lett., 50, (1987) 1888.

Kaciulis, S., Mezzi, A., Montesperelli, G. Lamastra, F., Rapone, M. Casadei, F., Valente, T. & Gusmano, G. Multi-technique study of corrosion resistant CrN/Cr/CrN and CrN : C coatings. *Surface and Coatings Technology*, Vol. 201, issue 1-2, 12, (6 september 2006) 313-319.

Kelly,P., Arnell, R., Magnetron sputtering: a review of recent developments and applications. *Vacuum*, Vol. 56, (September 1999) 159-152.

Kelly,P., Arnell, R., The influence of magnetron configuration on ion current density and deposition rate in a dual unbalanced magnetron sputtering system. *Surface and Coatings Technology*, Vol. 108–109 (1998) 317–322

Kumar,A. & Kaur, D., Nanoindentation and corrosion studies of TiN/NiTi thin films for biomedical applications. *Surface and Coatings Technology*, Vol. 204, issue 6-7, (14 december 2009), 1132-1136.

L. Monroy, J. Olaya, M. Rivera2, A. Ortiz3, G. Santana. Growth study of Y-Ba-Cu-O on buffer layers and different substrates made by ultrasonic spray pyrolysis. Rev. LatinAm. Metal. Mat., Vol. 32 (October 2011) 21-29

Lang, F. & Yu, Z. Surf. Coat. Technolology, The corrosion resistance and wear resistance of thick TiN coatings deposited by arc ion plating. *Surface and Coatings Technology*, Vol. 145, issue 1-3, (august 2001) 80-87.

Lee, K., Kim, W., Young, J., Won, S. & CHOE, H. Effects of TiN film coating on electrochemical behaviors of nanotube formed Ti-xHf alloys. *Transactions of Nonferrous Metals Society of China*, Vol. 19, issue 4, (august 2009) 857-861.

Lingxia Hang , Y. Yin, Junqi Xu, Optimisation of diamond-like carbon films by unbalanced magnetron sputtering for infrared transmission enhancement , Thin Solid Films, Volume 515, Issue 1, (25 September 2006) 357-361.

M. Ishihara, S. J. Li, H. Yumoto, K. Akashi, Y. Ide, Control of preferential orientation of AlN films prepared by the reactive sputtering method Thin Solid Films, Vol. 316 (1998) 152.

M. Marlo, V. Milman, Density- functional study of bulk and surface properties of titanium nitride using different exchange-correlation functional, Physical Review B, Vol. 63, Num. 4 (January 2000), pp. 2899-2907

Martin, O. Banakh, A.M.E. Santo, S. Springer, R. Sanjines, J. Takadoum, F. Levy, Appl. Surf. Sci. Vol. 185 (2001) 123–133.

Mayrhofer,P., Tischler, G., Mitterer, C., Microstructure and mechanical/thermal properties of Cr-N coatings deposited by reactive unbalanced magnetron sputtering, Surf. Coat. Technol. Vol.142-144, (July 2001), 78-84.

Messier, R., Giri, A.P., Roy, R.A., Revised structure zone model for thin film physical structure. *Journal of Vacuum Science and Technology, A: Vacuum, Surfaces, and Films*, 2(2, Pt. 1), 500-3 .

Mitsunori Kobayashi, Yoshihiko Doi, TiN and TiC coating on cemented carbides by ion plating, Thin Solid Films, Vol. 54, Issue 1, 2, (October 1978) 67-74

Olaya, J., Rodil, S., Muhl, S. & Sánchez, E. Comparative study of chromium nitride coatings deposited by unbalanced and balanced magnetron sputtering. Thin Solid Films, Vol. 474, issues 1-2, (march 2005), 119-126.

Olaya, J., Rodil,S., Marulanda, D & B. Bhushan, Propiedades mecánicas de nitruros metálicos depositados con UBM: tecnología eficiente y ambientalmente limpia. *Revista Mexicana de física.* , Vol.55, issue 6 (november 2009), 425-431.

P. Losbichler, C. Mitterer, Non-reactively sputtered TiN and TiB2 films: influence of activation energy on film growth, Surf. Coat. Technol. Vol. 97, (1997) 567.

P. Ziemann, E. Kay, Correlation between the ion bombardment during film growth of films and their structure and electrical properties J. Vac. Sci. Technol., A, Vac. Surf. Films, 1 (1983) 512.

P.B. Barna, M. Adamik, in: F.C. Matacotta, G. Ottaviani Eds., Science and Technology of Thin Films World Scientific Publ., Singapore, 1995.

P.B. Barna, M. Adamik,Fundamental structure forming phenomena of polycrystalline films and the structure zone models., Thin Solid Films, 317, (208) 27-33.

Pavel Souček, Tereza Schmidtová, Evaluation of composition, mechanical properties and structure of NC-TiC/a-C:H coatings prepared by balanced magnetron sputtering, Surface and Coatings Technology, In Press, Accepted Manuscript, Available online 16 September 2011.

Petrov,I, Barna, P, Hultman,L, Greene,J., Microstructural evolution during film growth. *J. Vac. Sci. Technol. A,* Vol 21, No 5, (September 2003), 117-128.

Raymond Constantin, Performance of hard coatings, made by balanced and unbalanced magnetron sputtering, for decorative applications Baham Miremad, Surface and Coatings Technology, Vol. 120-121, (November 1999) 728-733.

René Olivares-Navarrete , Jhon Jairo Olaya, Claudia Ramirez and Sandra Elizabeth Rodil, Coatings 1 (2011) 72-87 G. Ramírez, S.E. Rodil, H. Arzate, S. Muhl, J.J. Olaya Biocompatibility of Niobium coatings, Applied Surface Science, Vol. 257, (2011) 2555-2559

S.E. Rodil and J. J Olaya, Unbalanced magnetic field configuration: plasma and film properties, Journal of Physics: Condensed Matter, Vol. 18,. 1703-1720, 2006.

Siegwart, R. (2001), Indirect Manipulation of a Sphere on a Flat Disk Using Force Information. *International Journal of Advanced Robotic Systems,* Vol.6, No.4, (December 2009),12-16.

Stansbury E. and Buchnan R. *Fundamentals of Electrochemical Corrosion*. First edition. ASM International). ISBN 0-87170-676-8, United States of America(2000).

Stefan Wrehde, Marion Quaas, Robert Bogdanowicz, Hartmut Steffen, Harm Wulff, Rainer Hippler, Optical and chemical characterization of thin TiNx films deposited by DC-magnetron sputtering, Vacuum, Vol. 82, (2008) 1115–1119

T. Q. Li, S. Noda, Y. Tsuji, T. Osawa, and H. Komiyama, J. Vac. Sci. Technol. Vol. A 20 (3), (2002) 583-588.

Teer D.G. A , magnetron sputter ion-plating system, Surf Coat Technol, Vol. 39-40, (1989) 565.

Thornton, J., Influence of apparatus geometry and deposition conditions on the structure and topography of thick sputtered coating, J. Vac. Sci. Technol., Vol. 11, (1974) 666-670.

W. Ensinger, Tribological behavior of duplex coating improved by ion implantation, Surf. Coat. Technol. Vol. 99, (1998) 1

W.F. Liu, S. Suzuki, K. Machida, Magnetic properties of Nd–Fe–B film magnets prepared by RF sputtering, Journal of Magnetism and Magnetic Materials, Vol. 308, Issue 1, (January 2007) 126-130.

Waits R.K. J Planar magnetron sputtering, Vac. Sci. Technol. Vol. 15, Issue 2, (1978) 179-87.

Window B, Savvides N, Unbalanced dc magnetron as source high ion flux, J Vac. Sci. Technol. Vol. A 4, Issue 2 (1986) 453-456.

Y. Ochiai, K. Aso, M. Hayakawa, H. Matsuda, K. Hayashi, W. Ishikawa, On specimen preparation by radio frequency planar magnetron sputtering , J. Vac. Sci. technol. Vol. A4, Issue 1, (1986) 19-25

Y.H. Cheng, B.K. Tay, S.P. Lan, Influence of deposition temperature on the structure and internal stress of TiN films deposited by filtered cathodic vacuum arc, J. Vac. Sci. Technol. Vol. A 20, (2002) 1270.

8

Crystallization by Antisolvent Addition and Cooling

Marco Giulietti* and André Bernardo
Chemical Engineering Department Federal University of São Carlos UFSCar
Brasil

1. Introduction

Crystallization is the second most important separation process in chemical industry after distillation. Crystallization consists of a solid disperse phase formation into a continuous medium, which usually is a liquid solution in industrial systems. This solid phase formation occurs in two main steps: the appearance of transition structures between solid and fluid phase, or nucleation; and the growth of these structures into solid particles, crystals. The solution concentration must be higher than the equilibrium concentration at that temperature (solubility) in order to nucleation and crystal growth occur. The difference between actual concentration and equilibrium concentration is called supersaturation and is the driving force of crystallization. Supersaturation can be generated in the system by cooling, solvent evaporation, or changing of medium - addition of an antisolvent which reduces the solute solubility in the resultant system, or changing the solute by chemical reaction producing another substance with much lower solubility. Frequently other secondary processes occur, as agglomeration and breakage of those particles, which affect the final product (crystal) size distribution.

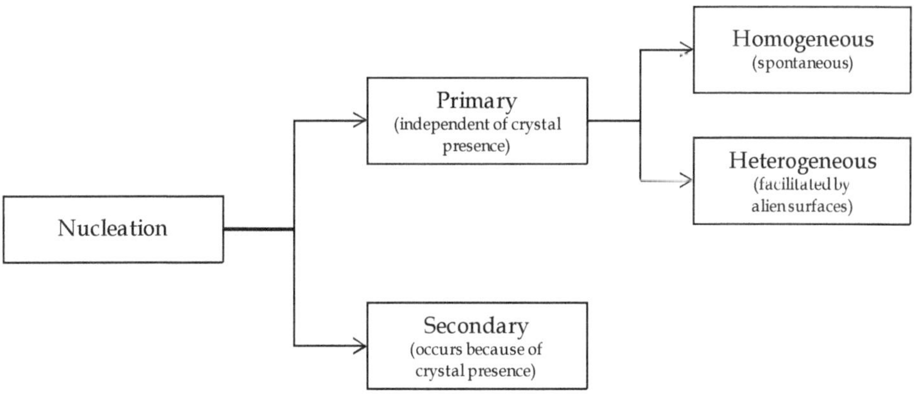

Fig. 1. Mechanisms of nucleation

* Corresponding Author

Nucleation may occur by different mechanisms. It may be primary, when is independent of crystal presence in solution, or secondary, when occurs as result of crystal presence in solution. When nucleation is primary, it may occur as a homogeneous mechanism, i.e. spontaneous depending only of supersaturation degree, or heterogeneous when is facilitated by alien surfaces (reactor or agitator wall, for instance). These different mechanisms are shown in Figure 1.

Primary homogeneous nucleation is the result of successive equilibrium of solute clustering (Figure 2). When a stable cluster is formed, it remains in solution as a transition structure from which the solid dispersed phase may appear. This stable cluster is called nucleus (Mullin, 2001).

$$\begin{aligned} &A + A \leftrightarrow A_2 \\ &A_2 + A \leftrightarrow A_3 \\ &\;\vdots \\ &A_{n-1} + A \leftrightarrow A_n \end{aligned}$$

Fig. 2. Solute clustering

The excess Gibbs energy is result of a resistance for the formation of a new surface, or the product of the surface tension and the cluster area, and a tendency to reach equilibrium, product of chemical potential and cluster volume:

$$\Delta G = \Delta G_S + \Delta G_V = \gamma \cdot A_{cluster} - \Delta\mu \cdot V_{cluster} = \gamma \cdot 4\pi r^2 - \Delta\mu \cdot \frac{4}{3}\pi r^3 \qquad (1)$$

where γ is the surface energy of the system solute-solvent, $\Delta\mu$ is the chemical potential, $A_{cluster}$ is the cluster surface area, $V_{cluster}$ is the cluster volume, and r is the cluster radius (therefore, there is an implicit hypothesis considering cluster spherical)

Because the clusters start to grow from very small sizes (theoretically from the solute volume), it is possible to infer that initially the surface term of the excess Gibbs energy is larger than the volume term. As clusters continue to grow, they reach a critical size which corresponds to the maximum value of the excess Gibbs energy (the critical Gibbs energy). From this point on, Gibbs energy starts to decrease, and crystals grow from the existent stables clusters (nuclei). Mathematically, the critical size is calculated deriving equation (1) and equating to zero the value:

$$r_{crit} = \frac{2\gamma}{\Delta\mu} \qquad (2)$$

The above equations as well as Gibbs-Thomson equation for non-electrolytes (Mullin, 2001) may be manipulated to obtain an expression for the chemical potential:

$$\ln\frac{c(r)}{c^*} = \ln S = \frac{2M\gamma}{RT\rho r} \qquad (3)$$

where M is the molecular weight of solute, S is the supersaturation ratio defined as c/c^*, R is the gas constant, T the absolute temperature, ρ the crystal density and r the particle radius.

Manipulating equation (3) and considering equation (2):

$$\ln S = \frac{2M\gamma}{RT\rho r} = \frac{2\gamma\upsilon}{\upsilon \frac{R}{V} T \frac{m}{M} r} = \frac{2\gamma\upsilon}{\frac{Rn}{N_A} Tr} = \frac{2\gamma\upsilon}{kTr} \Rightarrow \Delta\mu = \frac{kT\ln S}{\upsilon} \tag{4}$$

where υ is the molecular volume, N_A is the Avogadro number, n is the number of mols, and k is the constant of Boltzmann.

Considering the homogeneous nucleus radius definition (2) and the obtained expression for the chemical potential (4), the critical Gibbs energy for homogeneous nucleation is:

$$\Delta G_{crit} = \frac{16\pi\gamma^3\upsilon^2}{3(kT\ln S)^2} \tag{5}$$

The critical Gibbs energy may be interpreted as an energy barrier to be transposed to the appearance of the nucleus, as primary nucleation depends on ordered clustering of solute until a critical size (or critical radius if cluster and nucleus are considered spherical). The practical consequences of this barrier are the metastability and the induction time. Metaestability is the state of a clear supersaturated solution, i.e. there is driving force for crystallization but there is no crystal in the system yet. The metastable state may persist for different induction times - time interval between the supersaturation imposition and the appearance of first nucleus - depending on supersaturation degree. As this supersaturation degree increases, smaller induction times are observed until the limit of instantaneous nucleation. At this point, the metastable zone width (abbreviated as MZW), i.e. a region between equilibrium concentration and actual solution concentration, was reached.

In primary heterogeneous nucleation, an alien surface (dust particle, reactor wall etc.) facilitates nucleation (Figure 3):

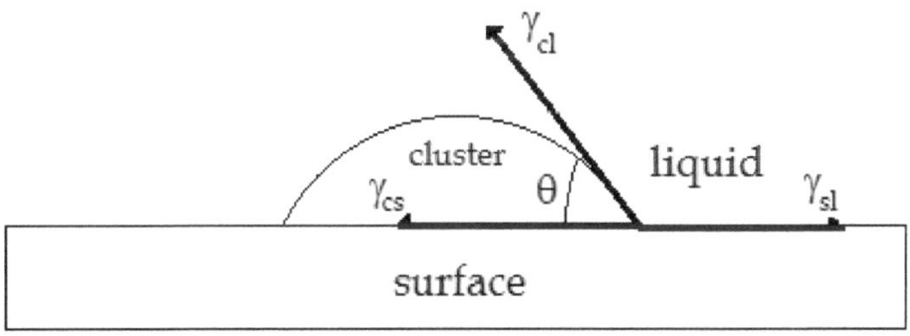

Fig. 3. Surface energies in heterogeneous nucleation. γ_{cs}, γ_{sl}, γ_{cl} are, respectively surface energies between cluster and surface, surface and liquid, and cluster and liquid.

Therefore, in heterogeneous nucleation there is a contact angle θ and a factor Φ which express the affinity between the cluster and the surface (Mullin, 2001):

$$\cos\theta = \frac{\gamma_{sl} - \gamma_{cs}}{\gamma_{cl}} \tag{6}$$

$$\Phi = \frac{(2+\cos\theta)(1-\cos\theta)^2}{4} \tag{7}$$

The factor Φ varies from 0 (total affinity between cluster and surface) to 1 (no affinity). The primary nucleation rate, J, is the rate of appearance of nucleus in a given volume:

$$J = \frac{dN}{Vdt} \tag{8}$$

where N is the number of nucleus, t is time and V is the system volume.

For its characteristics, primary nucleation may be modeled as an Arrhenius thermally activated rate:

$$J = A\exp\left(-\frac{\Delta G}{kT}\right) \tag{9}$$

Equations 5, 7 and 9 allow determining a primary nucleation rate:

$$J = A\exp\left(-\frac{16f\gamma^3\upsilon^2}{3k^3T^3(\ln S)^2}\right) \tag{10}$$

Where f is a factor that equals 1 for homogeneous nucleation or is lower than 1 for heterogeneous nucleation (Bernardo et al, 2004). Consequently, it should be expected that in industrial systems heterogeneous nucleation occurs before homogeneous, and the metastability or the induction times be smaller than in homogeneous nucleation.

It is reported that melts frequently demonstrate abnormal nucleation characteristics - nucleation rate follows the expected exponential curve as supersaturation is imposed to the system, but reaches a maximum and decreases for higher supersaturation values. It was suggested that the viscosity of melts increase starkly with cooling and restricts molecular movement inhibiting the formation of ordered structures. It was modeled by a modification in equation 10, including a 'viscosity' term (Mullin, 2001):

$$J = A\exp\left(-\frac{16f\gamma^3\upsilon^2}{3k^3T^3(\ln S)^2} + \frac{\Delta G'}{kT}\right) \tag{11}$$

where ΔG' is the activation energy for molecular motion across the system, and is very large for viscous liquids and glasses.

When there are crystals suspended in solution, they may be shorn by relative movement of liquid or broken by collision with other crystals or with crystallizer or impellers surfaces. The consequence of these mechanical processes is the appearance in suspension of small embryos which allow the growth of new crystals, i.e. secondary nucleation. As it exists because of physical interactions between crystals and the system, it cannot be modeled by thermodynamic equations, as primary nucleation. The wide-spread solution is to relate secondary nucleation rate B_0 with process variables which may cause secondary nucleation, as supersaturation ΔC (the difference between C and C*), impeller rotation W, and concentration of solids M_T in a power law equation (Myerson and Ginde, 2002):

$$B^0 = k_N W^i M_T^j \Delta C^n \tag{12}$$

Seeding is a common practice in industry. It consists of adding a small quantity of crystal in the supersaturated solution that will facilitate the crystal growth by the existence of a surface. If the process of crystallization is seeded, it is expected that secondary nucleation be the dominant mechanism of nucleation.

The crystal growth may be defined as the variation in time of the characteristic size of the crystal:

$$G = \frac{dL}{dt} \tag{13}$$

where G is the crystal growth rate and L is the characteristic size of the crystal.

There are many theories to explain and model crystal growth. Briefly explaining, these theories may be grouped in three sets: surface energy theories, that postulate that shape of growing crystals search the minimum energy condition, are limited to molecular modeling studies; diffusion theories, which states that crystal growth is limited by the diffusion of the solute to the crystal surface; adsorption theories, which states that the integration of the solute molecule to the crystal surface is the rate-determining step. For the crystallization of melts, the crystal growth may be limited by heat release. It seems quite obvious that crystal growth may be limited by diffusion of solute to the surface, integration of the solute in the surface, or by the heat release depending on the system composition or the process conditions (Mulin, 2001). In the engineering practice, it is common to describe crystal growth rate as a power law (Myerson and Ginde, 2002):

$$G = k_G \Delta C^g \tag{14}$$

where k_G is a constant which may vary with temperature according to an Arrhenius-like equation, and g is the crystal growth order, generally a number between 1 and 2.

1.1 Antisolvent crystallization

As already mentioned, supersaturation may be generated by changing the solubility of the system by the addition of an antisolvent – a liquid miscible with the solvent which reduces solute solubility in this new mixed solvent. An advantage of the antisolvent crystallization is that the process can be carried out at temperatures near the ambient temperature. It is quite convenient for heat-sensitive substances. Also, the process would demand less energy than a solvent evaporation process. However, the solvent-antisolvent mixture must be separated in order to recover and recycle one or both solvents. Another advantage of antisolvent crystallization is that the change in solvent composition may favor one crystalline structure in those cases where the solute may crystallize in two or more crystalline phases (what is called polymorphism), and only one of them is desired for product application. Because of these characteristics, antisolvent crystallization has been widely used to crystallize pharmaceutical products, which are generally sensitive to degradation by heating and frequently have polymorphism occurrence.

Takiyama et al. (2010) utilized the antisolvent crystallization to control the formation of two possible polymorphs of indomethacin (alpha and gamma forms).To obtain only the desired stable polymorph, it is required to avoid the precipitation of meta-stable polymorph crystals; they postulate that in the antisolvent crystallization, the solubility profiles are essential data for crystallization operation design to selectively isolate the target polymorph. In antisolvent crystallization, indomethacin was dissolved in acetone and heptane as used as antisolvent. Agitation speed and rate of addition control, as well as seeding gamma crystals were done to obtain the desired gamma form. The experiments were done at 288 and 313 K.

Granberg et al. (1999) investigated the influence of solvent composition on the antisolvent crystallization of paracetamol in acetone-water mixtures where extra water was added as antisolvent and concluded that supersaturation degree and not solvent composition defines induction time. They noticed increasing nucleation and agglomeration rate with increasing initial supersaturation, but at a given initial supersaturation, the solvent composition has no clear influence on the crystalline product characteristics. Crystal growth rate showed good relationship with solubility. Their work concluded that antisolvent modifies solubility and crystal shape, but has low influence in crystallization kinetics, governed by supersaturation degree.

Analyzing data of benzoic acid crystallization in water-ethanol solution by the water addition, Kubota (2008) concluded that solvent composition has no effect on induction time or primary nucleation rate, which could be modeled considering only the supersaturation imposed to the system as water as added.

Antisolvent crystallization may be combined with cooling strategies to enhance crystallization. Sheikhzadeh et al. (2008) implemented an adaptive MIMO neuro-fuzzy logic control for crystallization of paracetamol in isopropanol-water system in which water was added as antisolvent and temperature was varied from 40 to 10°C. When seeds were added, product yield reached 99%, while unseeded experiments reached 95% product yield. Seeding allowed to significantly reduce batch time without reduction in crystal mean size.

In combined cooling antisolvent crystallization, it seems that when antisolvent is added before cooling, the results are better than the opposite. Studying crystallization of paracetamol in isopropanol-water system in which water was added as antisolvent, Knox et al. (2009) increased the yield from 78.4% to 93.5% when antisolvent was added before cooling.

Nagy et al. (2006, 2008) utilized the method of moments to model the combined cooling and antisolvent crystallization in order to obtain the optimum recipe for crystallizing lovastatin (a hypolipidemic agent in drugs) in acetone/water mixture and achieve a maximized crystal size. Compared to cooling-only strategy, antisolvent-only strategy improved the product mean size in 15%, while combined strategy improved the mean size in 22%. The width of particle size distribution was lowered in 17 and 23% when only antisolvent and combined cooling antisolvent was used, respectively, compared to cooling-only strategy.

2. Crystallization of sugars

The major amount of crystallization studies for sugars, sucrose is almost an exception, were made concerning the undesirable crystallization or the need to impose the crystallization of these substances in food formulation (Hartel, 2001; Hartel et al., 2011). Very little research concerns industrial crystallization processes, and even for sucrose, industrial processes remain based on practical operational practice. Fundamental studies on sugars solutions behavior have been being done, almost always concerning the behavior of food formulation.

In fact, the most used sugars in industrial formulations - sucrose, glucose, fructose, lactose - seem to have specific characteristics that become the study of their crystallization quite complex. These substances are all very soluble in water, which implies that their industrial processes of crystallization will have very high initial (feed) concentration. Figure 4 shows the solubility of sucrose, glucose, fructose, lactose in water.

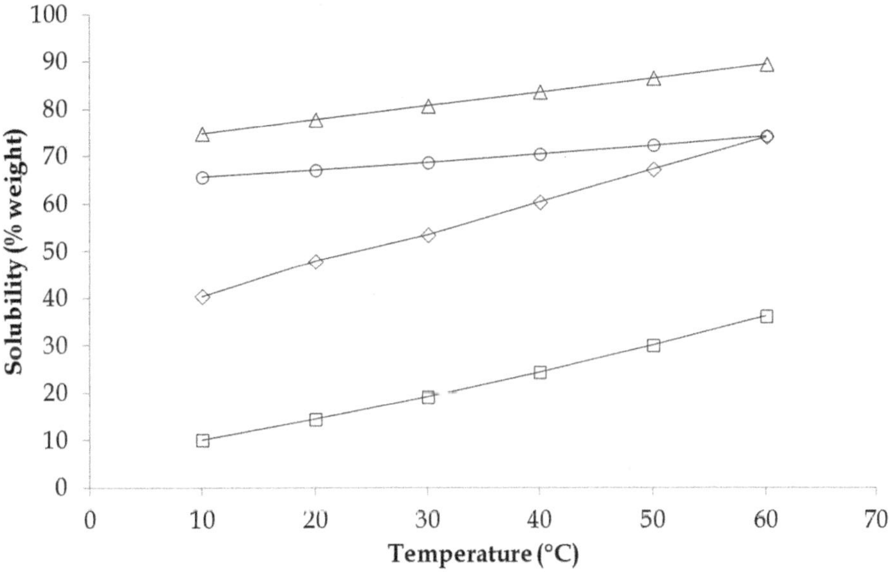

Fig. 4. Solubility in water of fructose △ (Silva, 2010), sucrose ○ (Ouiazzane et al., 2008), glucose ◇ (Alves et al., 2007), lactose □ (Brito, 2007).

It may be observed in Figure 1 that despite of lactose, all the other sugar have solubilities larger than 40 % weight. Even lactose have its solubility larger than 30 % weight at 60°C, where the others haves solubilities larger than 60 % weight.

Other important physical properties to be considered are solution viscosity and glass transition temperature T_g. Chirife and Buera (1997) presented a simple equation to predict the viscosity of sugar solutions at 20°C. Table 1 utilized this equation to calculate viscosities for selected sugars in a 60% weight solution. Hartel et al. (2011) compiled T_g values for different sugars (Table 2). For temperature values around or below T_g, the solution will be expected to exhibit a very low mobility.

Sugar	Viscosity (cP)
Glucose	36,4
Fructose	32,9
Sucrose	59,4
Lactose	65,7

Table 1. Viscosity at 20°C for 60% weight solution of selected sugars

Sugar	T_g (°C)
Glucose	31
Fructose	5-10
Sucrose	62-70
Lactose	101

Table 2. Glass transition temperatures of selected sugars

Considering a sugar crystallization process, if no seed is added, primary nucleation rate will occurs as predict equation (11), i. e. nucleation rate will grow exponentially until a maximum and decrease for higher supersaturation. Crystal growth rate will be limited by diffusion of solute to crystal face, which implies that even seeding policies has limited efficiency.

Further, metastable zone width (MZW) may have reach 30°C as reported in several references (Gharsallaoui et al. (2008); Brito (2007); Silva (2010)). The combination of large MZW, flat solubility curves (Figure 1) and high viscosity makes simple cooling crystallization practically unfeasible. Mathlouthi and Genotelle (1998) compare sucrose crystallization to a 'hurdle race', where viscosity seems to be a minor hurdle and the disassociation of hydration water a major one.

2.1 The role of water affinity in sugar crystallization

Mathlouthi and Genotelle (1998) considered two steps in sucrose crystallization: diffusion of sucrose molecules from the bulk solution to the interface crystal/solution and the incorporation of these molecules to the crystal after releasing their hydration water. Utilizing X-ray diffraction and laser-Raman spectroscopy, they concluded that hydrogen bonds between sucrose molecules in concentrated solutions is so strong that it hinders completely the free diffusion of molecules, and that diffusion in concentrated solutions is not due to viscous flow, but to the transfer of water molecules from one sucrose molecule to another by rotation of these sugar molecules. Consequently, water would diffuse in the concentrated solution and sucrose molecules would remain immobile, becoming the migration of hydration water from the crystal surface to the bulk solution very likely to be the controlling step in sucrose crystal growth.

Gharsallaoui et al. (2008) studied the interactions between water and disaccharides (sucrose, maltitol, and trehalose) in saturated solution and in crystallization conditions. According to them, narrowest metastable zone width was observed for maltitol and the largest for trehalose, because of the higher affinity of trehalose for water. They conclude that the crystallization of anhydrous disaccharides in aqueous solution necessitates that hydration water be removed and evacuated from crystal integration surface to the bulk solution to allow the growth of crystals. It seems to occur because for disaccharides in dilute or concentrated aqueous solutions, folding around the glycosidic linkage and hydrogen bonding influences very much the solution behavior. Specifically in the case of properties such as solubility, viscosity and molecular arrangements that take place before crystallization. As a general rule, high T_g sugars exhibit a greater degree of freedom to rearrange hydrogen bonds during changes in temperature than low T_g sugars.

Molinero et al. (2004) utilizing atomistic simulations investigated the nature of combination of water and glucose in supercooled solutions and concluded that there is a concentration limit not to have water freezing and keep a glassy state for all system.

Bensoussi et al. (2010) compared the metastable zone width of aqueous solutions of sucrose, maltitol, mannitol and xylitol, and attributed the observed differences to the interactions between water and solute molecules, as well as the conformation of solute molecule in solution. Further, they concluded that these factors are at the origin of solution properties like viscosity, diffusivity and surface tension, which interfere in nucleation and crystal growth. They concluded that nucleation of sugars is affected by their solubility, as it affects viscosity and the consequent solution diffusivity. Besides, the stability of bonds established with water may also affect nucleation. Sucrose and xylitol, for instance, have high potential of forming stable hydrogen bonds with water, as well as more favourable water-sugar interactions than to sugar-sugar interaction, which implies in large MZW, low capacity to form spontaneous nuclei, and high hydrophilic behavior. Mannitol and maltitol have similar MZW despite the difference in their solubility in water, because of, in the case of maltitol, its high viscosity at saturation and flexibility of glucitol moiety which decreases the stability of water-maltitol interactions. On the other hand, mannitol has low affinity for water and low viscosity, which favors the conditions to form spontaneous nuclei; its rigid conformation explains the ease of nucleation and the narrow metastable zone.

Shortly, as general rule, sugars have high affinity with water, which frequently implies in highly viscous, highly soluble solutions with large MZW, low ease of nucleation and small crystal growth rates.

2.2 Antisolvent cooling crystallization of sugars

Sugars are very polar compounds, which explain their affinity with water. The dielectric constant of water is 78.54 at 25°C. A 50 weight percent of sucrose aqueous solution has its dielectric constant equals to 60.19; a similar dextrose solution has a dielectric constant of 63.39 (Malmberg and Maryott, 1950). Dielectric constants of ethanol and acetone at 25°C are 24.55 and 20.7, respectively. As dielectric constant provides a good measure of a system polarity, it is obvious that aqueous solutions of sugars are much more polar that the common organic solvents. The solubility of a solute in aqueous solution should be decreased by the addition of an organic solvent with a dielectric constant lower than that of water. Another factor which contributes to precipitation by organic solvents is the redistribution of water and the organic solvent around solute molecule (Arakawa and Timasheff, 1985). In fact, water-organic solvent mixture cannot be regarded as a continuous medium in the vicinity of a sugar molecule, since the sugar surface may be a mosaic of regions with different polarities and different affinities for the solvent components. Furthermore, large organic molecules as sugars may also be excluded by steric hindrance.

If an organic solvent is mixed to aqueous solution of sugar as antisolvent, it is possible to suppose that it would surround the hydrophobic moieties of sugar molecule surface, dehydrating sugar molecule by a steric hindrance mechanism. This dehydration of sugar molecule, exposed to a less polar medium in which it has little affinity to solvent, would decrease viscosity (increasing solute mobility) and facilitate sugar-sugar interaction. The result would be higher ease of nucleation and higher crystal growth rate.

Following, it is presented results of antisolvent cooling crystallization of fructose utilizing ethanol as antisolvent, and for lactose utilizing acetone, ethanol, and iso-propyl alcohol. Crystallization was evaluated utilizing Nýlvlt's method to calculate crystallization kinetics (Nývlt et al., 2001).

Nýlvlt's method utilizes a set of at least nine experiments of no-seeded batch cooling crystallization, with three different cooling rates and three initial concentrations, to determine the apparent average crystal growth rate, expressed as a power-law equation similar to equation (13), and the apparent average nucleation rate $\dot{N}$, expressed as a power-law equation:

$$\dot{N} = k_N M_T^c \Delta C^n \tag{15}$$

where c value allows to comprehend the nucleation mechanism - c = 0 means true primary or secondary nucleation; c = 1 means that crystal-crystal interaction provokes nucleation; c = 2 means that friction between crystals provokes nucleation.

For all experiments, the metastable zone width ΔT_{max} is measured, and from equation:

$$\log \Delta T_{max} = \frac{1-m}{m} \log \frac{dC^*}{dT} - \frac{1}{m} \log k_N + \frac{1}{m} \log \frac{dT}{dt} \tag{16}$$

where m is the apparent nucleation order and C* is the solubility, it is possible to obtain the values of k_N and m by a multiple linear regression.

The cumulative mass distributions of crystals M(L) may be described as the function:

$$M(L) = 100\left(1 + z + \frac{z^2}{2} + \frac{z^3}{6}\right)\exp(-z) \tag{17}$$

where $z = L/(G \cdot t_{batch})$, is the crystal dimensionless size. It is possible to calculate the z values iteratively, and the average crystal growth rate for that experiment may be calculated from the relation between z and L. Crystal mean size L_m is the L value for z = 3, and

$$G = \frac{L_m}{3t_{batch}} \tag{18}$$

The linear coefficient of z-L relation is called z_n and

$$f(z_n) = 100\left(1 + z_n + \frac{{z_n}^2}{2} + \frac{{z_n}^3}{6}\right)\exp(-z_n) \tag{19}$$

The nucleation rate be calculated by the following equation

$$\dot{N} = \frac{27M_T G}{2k_v\rho_C f(z_n)(L_m - L_n)^4} \tag{20}$$

A multi-linear regression of $\dot{N}$ as function of G and M_T, gives:

$$\ln\dot{N} = \ln\left(\frac{k_N}{k_g^{n/g}}\right) + c\ln(M_T) + \frac{n}{g}\ln G \tag{21}$$

Nucleation order may be calculated:

$$n = \frac{4(m-1)}{3\frac{g}{n}+1} \tag{22}$$

2.3 Antisolvent cooling crystallization of fructose

Silva (2010) studied the antisolvent cooling crystallization of fructose utilizing ethanol as antisolvent. She varied the initial concentration of the aqueous solution of fructose, the quantity of added ethanol expressed as ratio ethanol/water (E/S) and the cooling rate. The agitation rate was 500 rpm and the final temperature was 30°C for all experiments. The results of MZW are shown in table 3.

As it was expected, MZW decreases with added ethanol quantity. The crystals yield, not shown in table, was more than 93% of available fructose quantity for all experiments. The obtained crystals had cubic habit, and agglomeration occurred in all experiments. The crystal mean size, and the crystallization kinetics, calculated by Nývlt's method had no significant difference with the quantity of ethanol added (Table 4).

Initial concentration (% weight)	Ethanol (E/S)	Saturation temperature (°C)	Nucleation temperature (°C)	Cooling rate (°C/min)
86,88	1.5	50.5	30 after 40 min	0.60
86,88	4.0	50.5	30 after 30 min	0.55
86,88	6.0	50.5	42	0.55
86,88	9.0	50.5	47	0.58
88,10	1.5	55	30 after 130 min	0.50
88,10	4.0	55	30 after 20 min	0.55
88,10	6.0	55	38.5	0.60
88,10	9.0	55	42	0.55
89,36	4.0	60	38	0.22
89,36	6.0	60	40.5	0.65
89,36	9.0	60	46.5	0.65

Table 3. MZW of antisolvent cooling crystallization of fructose with ethanol as antisolvent

Flood et al. (2000) studying the same system also concluded that ethanol quantity and temperature did not affect significantly the crystal growth rate. They cited other studies that concluded the same.

Initial concentration (% weight)	Ethanol (E/S)	Product mean size (μm)	G (10^6·m/s)	$\dot{N}$ (10^{-11}·#/m³s)
86,88	1.5	42.50	1.28[1]	202.1[1]
86,88	4.0	109.7	3.14[1]	12.36[1]
86,88	6.0	44.07	1.01	133.5
86,88	9.0	56.75	1.27	49.54
88,10	1.5	51.64	2.82[1]	192.5[1]
88,10	4.0	42.06	1.15[1]	224.1[1]
88,10	6.0	44.02	1.04	156.1
88,10	9.0	63.82	1.46	35.43
89,36	4.0	55.87	1.22	62.32
89,36	6.0	43.95	1.03	177.0
89,36	9.0	60.07	1.36	56.62

[1]As in these experiments nucleation occurred after cooling cessation, calculated kinetic parameters G and $\dot{N}$ must be considered cautiously.

Table 4. Crystallization kinetics of fructose calculated by Nývlt's method

2.4 Antisolvent cooling crystallization of lactose

Brito (2007) studied the antisolvent cooling crystallization of lactose utilizing ethanol (at different pH), isopropanol, and acetone as antisolvents. She varied the initial concentration of the aqueous solution of lactose, the final temperature and the cooling rate; the quantity of added antisolvent was always the same quantity of water in solution (E/S = 1), and the agitation rate was 350 rpm for all experiments. The results of MZW are shown in tables 5, 6 and 7.

C_i (% weight)	pH	Saturation (°C)	Final (°C)	Nucleation (°C)	Cooling rate (°C/min)
25.54	4.00	60	25	52	0.35
36.95	7.00	60	25	55	0.52
33.24	12.41	60	25	53	0.58

Table 5. MZW of antisolvent cooling crystallization of lactose with ethanol as antisolvent

Tables 5, 6 and 7 show that adding the same quantity of antisolvent of water in solution MZW almost disappear for all studied conditions. Tables 8, 9 and 10 present the calculated kinetic parameters (Nývlt's method).

Results presented in tables 8, 9 and 10 allow to conclude that pH has an important role in ethanol cooling crystallization of lactose - crystal growth rate and yield increase with pH, and nucleation rate decreases. For isopropanol, an increase in batch time almost always implies in higher yields and larger mean sizes due to the increase in growth rate. Also, for the same temperature variation, higher initial concentration (that means higher average supersaturation) implies in larger crystal sizes.

C_i (% weight)	Saturation (°C)	Final (°C)	Nucleation (°C)	Cooling rate (°C/min)
42.86	70	20	68	1.00
42.86	70	30	69	0.34
42.86	70	40	68.5	0.53
35.48	60	20	57	1.10
35.48	60	30	57	0.59
35.48	60	40	58	0.33
31.03	52	10	50	0.33
31.03	52	20	49	0.30
31.03	52	30	51.5	0.42

Table 6. MZW of antisolvent cooling crystallization of lactose with isopropanol as antisolvent

C_i (% weight)	Saturation (°C)	Final (°C)	Nucleation (°C)	Cooling rate (°C/min)
35.06	50	20	50	0.15
35.00	50	20	50	0.17
35.07	50	20	48	0.21
35.07	50	20	50	0.27
35.00	50	20	50	0.51
35.07	50	25	50	0.52
35.07	50	35	50	0.45
33.32	45	20	43	0.61
33.32	45	25	43	1.00
33.32	45	30	43.5	0.44
29.88	40	10	39	0.66
30.00	40	20	40	1.19
29.98	40	25	40	0.50

Table 7. MZW of antisolvent cooling crystallization of lactose with acetone as antisolvent

For acetone, larger concentration experiments had a yield of about 90%, increasing cooling rate (which means to increase average supersaturation) caused nucleation rate to increase and mean crystal size to decrease. For lower initial concentration, increasing cooling rate seems to decrease yield and mean size. Intermediate concentration experiments had an increase of mean size with cooling rate.

The presented results corroborated the expectation that the addition of an organic antisolvent eases the crystallization of sugar, as the antisolvent would decrease sugar-water interaction increasing solute mobility (Miranda et al., 2009). An indirect measurement of this effect is the sugar solubility in the solvent mixture. Figure 5 presents lactose solubility in different pH values and in a mixture of 50 percent weight of water and ethanol. Figure 6 presents solubility of lactose in water and in mixture of water with different solvents.

C_i (%weight)	pH	Cooling rate (°C/min)	L_m (μm)	G (10^5·m/s)	$\dot{N}$ (10^{-11}·#/m³s)	Yield (%)
25.54	4.00	0.35	121.18	0.734	5.683	62.4
36.95	7.00	0.52	115.77	3.473	5.308	46.04
33.24	12.41	0.58	168.37	5.051	2.313	82.46

Table 8. Crystallization kinetics of lactose (antisolvent ethanol) calculated by Nývlt's method

C_i (% weight)	Cooling rate (°C/min)	L_m (μm)	G (10^5·m/s)	$\dot{N}$ (10^{-11}·#/m³s)	Yield (%)
42.86	1	103.5	0.6029	3.6751	86.23
42.86	0.34	91.74	5.5047	7.4673	85.72
42.86	0.53	68.09	7.1675	13.769	80.42
35.48	1.1	81.44	6.1077	11.027	49.60
35.48	0.59	53.27	0.2948	18.195	75.65
35.48	0.33	89.98	6.8336	6.2648	49.20
31.03	0.33	47.00	3.1688	28.797	61.01
31.03	0.3	56.47	5.2936	21.703	73.74
31.03	0.42	80.70	5.6968	8.3657	43.85

Table 9. Crystallization kinetics of lactose (antisolvent isopropanol) calculated by Nývlt's method

C_i (% weight)	Cooling rate (°C/min)	L_m (μm)	G (10^5·m/s)	$\dot{N}$ (10^{-11}·#/m³s)	Yield (%)
35.06	0.15	118.71	0.3458	0.7797	88.74
35.00	0.17	99.04	0.3396	1.6071	92.63
35.07	0.21	53.98	0.2722	21.347	89.81
35.07	0.27	77.31	0.4179	4.0916	91.55
35.00	0.51	70.11	0.7130	11.599	86.60
35.07	0.52	93.24	1.1655	6.2562	47.45
35.07	0.45	106.43	1.9351	5.2757	41.55
33.32	0.61	113.84	1.3940	3.3583	73.74
33.32	1.00	72.96	1.7510	17.515	43.85
33.32	0.44	73.94	0.9859	10.476	82.40
29.88	0.66	114.01	1.2906	2.0553	55.72
30.00	1.19	82.07	2.3449	1.2720	40.91
29.98	0.50	149.77	2.2466	1.5606	58.08

Table 10. Crystallization kinetics of lactose (antisolvent acetone) calculated by Nývlt's method

Figure 5 explains the variation in the obtained yields in experiments of table 8, as solubility of lactose has a maximum value in neutral pH and also decreases with ethanol addition. However solubility curves do not explain the yields themselves. In systems where lactose has very low solubility, it would be expected that lactose has low interaction with solvent system and, therefore, high mobility of lactose molecule. A probable reason why lactose crystal yields are about 40% even in system with low solubility could be mutarotation (Miranda et al., 2009). Lactose molecule has two conformations (anomers), α and β forms, and below 93.5°C α-form is the constituent of stable crystals. It is known that mutarotation is affected by temperature, pH and solution impurities. Further, α-form crystallization rate may be faster than mutarotation rate, causing mutarotation to be the rate-determining step for crystallization (McLeod, 2007).

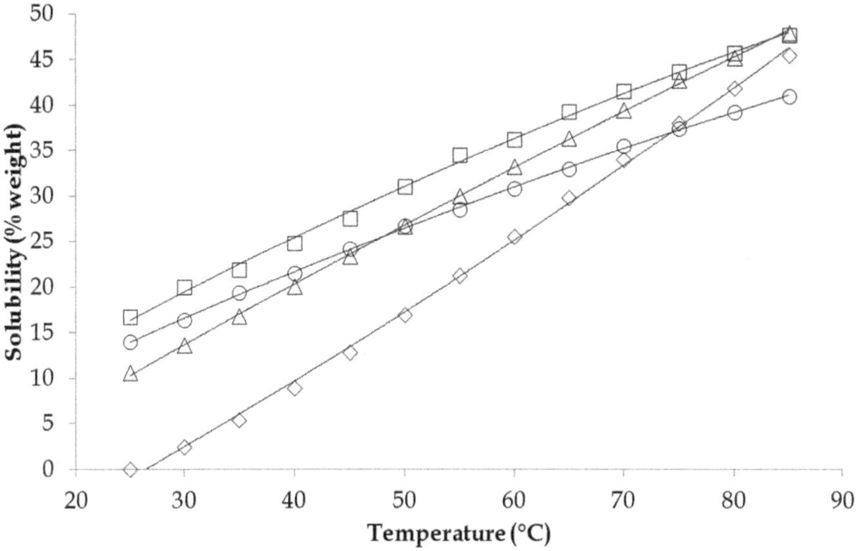

Fig. 5. Solubility of lactose in water at pH 4.0 (◇), water pH 7.0 (□), water pH 12.41 (△), and a mixture 50% weight ethanol water (○).

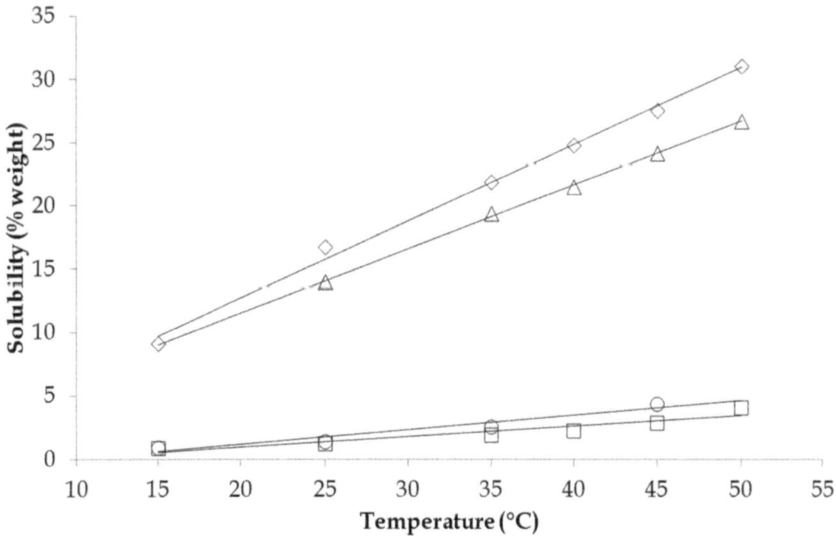

Fig. 6. Solubility of lactose in different solvents: water (◇), solution 50% weight acetone-water (□), solution 50% weight ethanol-water (△), solution 50% weight isopropanol-water (○).

Flood at al. (2000) describes fructose mutarotation issue. According to them, fructose interconverts naturally in solution in five tautomeric forms by mutarotation, but only the β-D-fructopyranose form crystallizes. In aqueous solutions the mutarotation rates would be higher than the crystallization kinetics, but in aqueous ethanolic solutions mutarotation would be sufficiently slow to move the tautomeric equilibrium away from the equilibrium β-D-fructopyranose. So, it would be important express supersaturation in terms of the tautomer, β-D-fructopyranose. In experiments driven by Silva (2010), crystal yield was always higher than 93%, but lower than 100%. In lactose crystallization experiments of Brito (2007), higher yield were about 90%. Therefore, it seems that mutarotation may reduce crystallization yields for lactose and fructose.

From an industrial perspective, in which the maximum crystal yield is an aim, it is important to emphasize that cooling and antisolvent addition must be combined. Figure 7 shows fructose solubility in weight percentage as function of water content in solvent (a mixture of ethanol and water) for the temperatures of 20°C and 60°C.

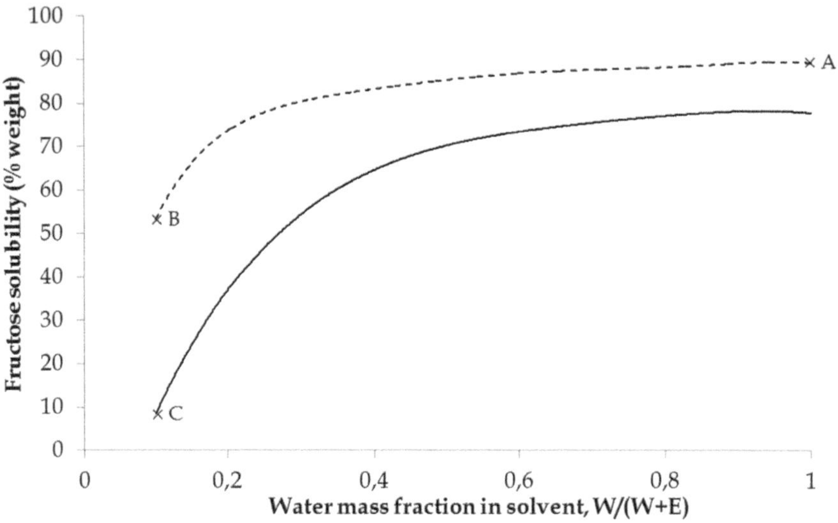

Fig. 7. Fructose solubility in a mixture of ethanol and water at 20°C (continuous line) and at 60°C (dashed line).

The data showed in Figure 7 is based on Silva (2010) work. Figure 7 allows to understand that despite the addition of antisolvent eases crystallization, crystal yield is strongly dependent on final temperature. For 10% of water in solvent (thus 90% ethanol), fructose solubility varies from 53,13% at 60°C to 9,75% at 20°C. For instance, 1000 kg of fructose aqueous solution at 60°C (point A in Figure 7) would contain 106,9 kg of water and 893,1 kg of fructose. Adding 962,1 kg of ethanol to this system, total solvent content would be 1069 kg, with 10% weight of water. In this solvent system, more than 1200 kg of fructose could be dissolved at 60°C (point B in Figure 7) – more than initial quantity with no crystal produced – but only 115,5 kg of fructose would be soluble at 20°C (point C in Figure 7) – giving a theoretical crystal yield of 87% of total fructose, desconsidering possible mutarotation effects, or 81% of total fructose, considering that in Silva's experiments, yield was always more than 93% of available fructose.

3. Conclusion

Crystallization of sugars may be improved by adding an organic liquid antisolvent (as alcohol or ketone) and cooling the system. This addition shuts nucleation hindrance off, as it decreases system viscosity. Simultaneously, the antisolvent competes with solute for water of hydration, throwing solute out of the solution: promoting crystallization. As the solubility of sugars in the mixture water-organic solvent is much lower than in water only, antisolvent addition increases the crystallization rate. Cooling the system maximizes the drowning-out effect.

Sugars may present complex structures which interconvert in solution by mutarotation. As generally only one anomer crystallizes, mutarotation may decrease crystals yield. Mutarotaton may be affected by pH, temperature, and by solvent composition. However, for the studied cases of fructose and lactose, antisolvent cooling crystallization showed to be advantageous even considering mutarotation occurrence.

Fructose was studied utilizing ethanol, and lactose was studied utilizing ethanol, acetone and isopropanol. It is possible to vary antisolvent addition rate and cooling rate simultaneously (Nagy et al., 2008), allowing to optimize crystal quality in industrial operations. The presented kinetic data as well as shape and size distribution measurement for antisolvent cooling crystallization corroborate its utilization in industrial operation.

Combined antisolvent and cooling crystallization is an important technique for obtaining products that are difficult to crystalize due to inherent solution properties like high viscosities, large metastable zone width, low kinetic of nucleation and growth, like sugars and others materials. The good choice of antisolvent must be done carefully with preliminary experiments that can allow to obtain high yields and easiness of solvent recovery. Optimal path to combine the two techniques, antisolvent and cooling, to obtain good crystal size distribution must be evaluated for each particular system, taking into account the couples solvent-antisolvent, solute-mixed solvent. The phase diagram of this ternary system is very important to evaluate that path and possible yield.

4. References

Alves, L. A., Almeida e Silva, J. B., Giulietti, M. (2007). Solubility of D-Glucose in Water and Ethanol/Water Mixtures, *J. Chem. Eng. Data*, 52, 2166-2170.

Arakawa, T., Timasheff, S. N. (1985). Theory of Protein Solubility, in: *Methods of Enzymology 114*, edited by: Wyckoff, H. W., Hirs, C. H. W., Timasheff, S. N., Academic Press.

Bensouissi, A., Roge, B., Mathlouthi, M. (2010). Effect of conformation and water interactions of sucrose, maltitol, mannitol and xylitol on their metastable zone width and ease of nucleation, *Food Chemistry* 122 443–446

Bernardo, A., Calmanovici, C. E., Miranda, E. A. (2004). Induction Time as an Instrument to Enhance Comprehension of Protein Crystallization, *Crystal Growth and Design*, 4 799-805.

Brito, A. B. N. (2007). *Study of lactose crystallization in different solvents*, PhD Thesis (in portuguese), Federal University of São Carlos, São Carlos.

Chirife, J., Buera, M. P. (1997). A Simple Model for Predicting The Viscosity of Sugar and Oligosaccharide Solutions, *Journal of Food Engineering* 33 221-226.

Flood, A. E., Johns, M. R., White, E. T., Crystal Growth Rates and Dispersion for D-Fructose from Aqueous Ethanol, *AIChE Journal*, Vol. 46, No. 2, 239-246 (2000).

Gharsallaoui, A., Roge, B., Mathlouthi, M. (2008). Water-disaccharides interactions in saturated solution and the crystallisation conditions, *Food Chemistry* 106 1329–1339

Granberg, R. A., Bloch, D. G., Rasmuson, A. C. (1999). Crystallization of paracetamol in acetone-water mixtures, *Journal of Crystal Growth* 198/199 1287-1293

Hartel, R. W., Crystallization in foods, in: *Handbook of Industrial Crystallization*, edited by Myerson, A. S., Elsevier, 2001.

Hartel, R. W., Ergun, R., Vogel, S., Phase/State Transitions of Confectionery Sweeteners: Thermodynamic and Kinetic Aspects, *Comprehensive Reviews in Food Science and Food Safety*, 10, 17-32 (2011)

Knox, M., Trifkovic, M., Rohani, S., Combining antisolvent and cooling crystallization: Effect of solvent composition on yield and metastable zone width, *Chemical Engineering Science* 64 (2009) 3555 - 3563

Kubota, N., An interpretation of the metastable zone width concerning primary nucleation in antisolvent crystallization, *Journal of Crystal Growth* 310 (2008) 4647–4651.

Malmberg, C. G., Maryott, A. A., Dielectric Constants of Aqueous Solutions of Dextrose and Sucrose, *Journal of Research of the National Bureau of Standards* Vol. 45, No. 4, October 1950, 299-303.

Mathlouthi, M., Genotelle, J., Role of water in sucrose crystallization, *Carbohydrate Polymers* 37 (1998) 335–342.

McLeod, J., *Nucleation and Growth of Alpha Lactose Monohydrate*, PhD Thesis, Massey University, 2007.

Miranda, E. A., Bernardo, A., Hirata, G.A.M., Giulietti, M,. Crystallization of lactose and whey protein. In: Jane Célia dos Reis Coimbra, Jane S. R.; Teixeira, J. A.. (Org.). *Engineering aspects of milk and dairy products*. 1 ed. Boca Raton, FL, EUA: CRC Press Taylor & Francis Group, 2009, v. 1, p. 121-154.

Molinero, V., Çagın,T., Goddard, III, W.A., Mechanisms of Nonexponential Relaxation in Supercooled Glucose Solutions: the Role of Water Facilitation, *J. Phys. Chem. A* 2004, 108, 3699-3712

Mullin, J. W., Crystallization, 4th Edition, Butterworth-Heineman, London, 2001.

Myerson, A. S.; Ginde, R., Crystals, crystal growth and nucleation, in: *Handbook of Industrial Crystallization*, edited by Allan S. Myerson, 2nd edition, Butterworth, Woburn, 2002.

Nagy, Z.K., Fujiwara, M., Braatz, R.D. Optimal control of combined cooling and antisolvent pharmaceutical crystallization, *Proceedings of BIWIC 2006 13th International Workshop in Industrial Crystallization*, Delft, The Netherlands, 2006.

Nagy, Z.K., Fujiwara, M., Braatz, R.D., Modelling and control of combined cooling and antisolvent crystallization processes, *Journal of Process Control*, 18, 856–864, 2008.

Nývlt, J., Hostomský, J., Giulietti, M., *Cristalização*, EdUFSCar, São Carlos, 2001 (in portuguese).

Ouiazzane, S., Messnaoui, B. Abderafi, S. Wouters, J., Bounahmidi, T., Estimation of sucrose crystallization kinetics from batch crystallizer data, *Journal of Crystal Growth* 310 (2008) 798–803.

Sheikhzadeh, M., Trifkovic, M., Rohani, S. (2008). *Chemical Engineering Science* 63 1261 - 1272

Silva, A. T. C. R. (2010). *Study of fructose crystallization in different media*, MSc. Dissertassion (in portuguese), Federal University of São Carlos, São Carlos.

Takiyama, H., Minamisono, T., Osada, Y., Matsuoka, M. (2010). Operation design for controlling polymorphism in the antisolvent crystallization by using ternary phase diagram, *Chemical Engineering Research and Design*, 88, 1242–1247.

9

Metal Induced Crystallization

Ahamad Mohiddon Mahamad and Ghanashyam Krishna Mamidipudi
School of Physics, University of Hyderabad, Hyderabad, India

1. Introduction

Polycrystalline silicon thin films have attracted the attention of semiconductor industries in the past few decades due to their wide applications in thin film transistors, solar cells, display units and sensors (Schropp & Zeman, 1998; Choi et al., 2005; Mahan et al., 2008). Polycrystalline Si thin films are generally fabricated by crystallizing amorphous Si (a-Si) thin films, because these can render larger grains compared to the conventional poly Si film deposition. As a consequence, a variety of methods for lowering the crystallization temperature of a-Si have been developed. Excimer laser annealing is one of the promising ways to achieve large grain size poly Si films at lower substrate temperatures. Its high costs and nonuniform grain size, however, are significant obstacles that prevent its wide use (Parr et al., 2002). The other promising technique is the solid phase crystallization method. But this technique is essentially a high-temperature process and many substrates, including most forms of glass, cannot withstand the thermal processing. In order to achieve lower costs and have a wider range of application, inexpensive materials such as glass and special polymers have to substitute quartz or Pyrex™ substrates. In the case of glass substrates, all of the processing steps need to be limited to temperatures below 550 °C. The other known technique is rapid thermal annealing (RTA). In RTA infrared radiation is used as a heating source, and has the advantage of the high heating speed (up to 60 °C/s) that reduces the crystallization time. In RTA radiation is applied in pulses to heat the sample without heating the glass substrate (which is transparent to the infrared radiation). However, the grain size obtained in the crystallization of a-Si is also in the range of a few micrometers.

In an effort to reduce the crystallization temperature and crystallization time, and to increase the grain size, metal-induced crystallization (MIC) has been investigated as an alternative crystallization process for thin-film device fabrication. The MIC process involves the deposition of a-Si films on top of which a layer of suitable metal is deposited. This bilayer of metal and Si is then annealed in a furnace at temperatures ranging from 150 to 700 °C for durations between one minute to several hours leading to crystallization of the a-Si. As a result the Si nanocrystals grow in the metal/metal silicide/metal oxide matrix and their growth rate depends on the annealing conditions of the bilayer. Thus, to grow the Si nanocrystals, both thermodynamics and kinetics of the process have to be understood in detail. In the MIC process, metals like aluminum, nickel, gold, silver etc are used to decrease the crystallization temperature below 700 °C. Among the metals employed for the study of the MIC (Jin et al.,1999; Kumar et al.,2007; Kumar et al.,2008; Mohiddon & Krishna, 2011), the preferred metal up to date has been nickel (Ni) due to its low residual metal contamination in the poly-Si region (Wang & Wong, 2001). The mobility and the transfer characteristics of the poly Si TFT are improved by reducing the leakage current caused by the metal agents. To achieve this, metal-induced lateral crystallization (MILC) has been introduced. Ni induced MILC TFTs' are now widely used in large display applications. Large number of metals have been investigated to crystallize the a-Si at possible low temperature. These metals are classified in to two groups basing on the mechanism they follow in crystallizing the a-Si matrix. One group of metals forms eutectic with a-Si (follows Layer exchange mechanism) and the other reacts with a-Si and forms silicide. The eutectic phase forming metals can produce extended continuous poly-Si films with large grain size of 10 μm (Oliver & Hartmann, 2000). However, these metals contaminate the so produced *c*-Si and degrade its properties. Thus, the search for metals that would lead to crystallization of *a*-Si thin film at low temperatures and also reduce the contamination is a subject of recent research.

The aim of this review is to present an overview of the current state of understanding of the mechanisms involved in MIC and to discuss some of the recent work of the current authors in this area. The motivation is to crystallize a-Si films using metals that have not been investigated in detail so far, in addition to Ni which is a well studied metal used in the MIC process. The two other metals that were selected for this study are Tin (Sn) and Chromium (Cr). The former is known to form a eutectic alloy with the Si whereas the latter is a silicide/oxide forming metal, similar to Nickel. Furthermore, results on two different types of geometries are presented; the first involves the deposition of metal on top of the a-Si film, while in the second geometry the a-Si film is deposited on top of the metal film. This provides very interesting insights in to the mechanisms that cause crystallization of a-Si.

2. Mechanisms of metal induced crystallization

The main mechanisms of metal induced crystallization are discussed in this section.

2.1 Layer exchange mechanism

The proposed formation of poly-Si films on foreign substrates by eutectic forming metals relies on the overall layer exchange of adjacent Si and metal films during the transformation of amorphous to polycrystalline Si. The phenomenon is named as "layer exchange mechanism". The general driving force behind metal-induced crystallization is the reduction of the free energy of the silicon material during the transformation of the amorphous to the crystalline phase. When *a*-Si is in contact with certain metals, electronic screening of the covalent bonding in the Si material occurs, which weakens the Si bonds, and therefore, facilitates the inter diffusion of the metal and silicon atoms (Hiraki, 1980). The representatives of the eutectic forming metals are Au, Ag and Al. Aluminum-induced crystallization [AIC] is extensively studied metal to form extended continuous poly-Si films of large-grained material on glass, which is desirable for thin film solar cells.

The AIC comprises the diffusion of Si atoms into the Al layer, which occurs due to the fact that adjacent Al and Si layers are not in thermal equilibrium at elevated temperatures. According to the Al/Si phase diagram, up to 1.5 at.% of Si can be dissolved in Al at temperatures below the eutectic temperature of 577 °C (Murray & McAlister, 1984). The initial interaction of the Al and *a*-Si layers involves the partial dissolution of the Al or Si oxide interface layer, depending on the layer sequence, to allow the inter diffusion of the materials. The Si oxide layer is transformed by the Al into a mixture of Al oxide and an Al_xSi phase. This newly formed Al_xSi phase provides a diffusion channel for the Si and Al atoms (Nast & Wenham, 2000). The evaporated Al films are of polycrystalline nature, there are four different diffusion paths available for the diffusion of the Si atoms. 1) Diffusion inside the Al grains, 2) Diffusion along Al grain boundaries, 3) Diffusion along the Al/*a*-Si interfaces, and 4) Diffusion along the glass/Al interface.

The diffused Si atoms start crystallizing near the Al/*a*-Si interface. The Al grain boundaries are preferential nucleation sites for the dissolved Si atoms, since they are the sites of low critical free energy for nucleus formation. According to the random successive nucleation model, at early stage of crystallization, Si nuclei are formed at preferred sites located at the Al/Si interface. The released heat (crystallization energy and strain energy) leads to rise the local temperature in ambient regions. The temperature field propagates fast, since the thermal diffusion is faster than the atomic diffusion, so that the heat flow can stimulate new nuclei appearing randomly in nearby region. These nuclei of the next generation also produce a local temperature rise. This process repeats many time during annealing, resulting in a fractal formation.

Silicon atoms, dissolved in the Al layer, have a high mobility and, hence, diffuse quickly within the film and/or along the interfaces towards the growing Si grains (McCaldin & Sankur, 1971). It is this fast growth within the Al layer compared to any other growth normal to the layer structure that leads to the formation of a continuous poly-Si film. At an early stage of the crystallization process the newly formed Si grains within the Al matrix are far apart and do not influence each other. The nucleation rate strongly depends on the defects and grain boundaries in the Al layer as well as the temperature and corresponding silicon concentration. Si solute depletion occurs up to the effective diffusion distance in the vicinity of the growing grain (Zener, 1949). In this region the possibility of new nucleation decreases with increasing depletion towards the advancing Al/Si grain interface. When the effective diffusion distances of adjacent grains begin to overlap, competition for the available Si atoms dissolved in the Al layer occurs. At this point, the possibility of new nuclei formation decreases. This type of crystallization pattern, where isolated grains start to interfere at an early stage of the process, is different to solid-phase crystallization of amorphous silicon (Spinella et al., 1998).

The grain-size distribution is dependent on the ratio of the grain growth to the nucleation rate. This ratio increases with decreasing annealing temperature (Nast & Wenham, 2000). The decrease in grain growth due to a lower annealing temperature is less than the decrease in nucleation rate. The effective diffusion distance is longer at lower annealing temperatures leading to larger depletion areas around the growing grains, which prevents further nucleation. Therefore, the grains grow to a larger size before impingement occurs. The nucleations as well as the diffusion-controlled growth are thermally activated processes. Once the size of the Si grains is equivalent to the thickness of the Al layer, they solely advance laterally since they are constrained by the substrate and the *a*-Si/Al interface. Throughout the poly-Si growth process the Al layer is gradually displaced. The continuous supply of Si atoms from a-Si layer and gradual displacement of Al atoms leads to the growth of poly Si film with the large grain size. As an alternative member of this group of metals, we have investigated the tin induced a-Si crystallization.

2.2 Mechanism of diffusion assisted crystallization

Two main mechanisms of diffusion assisted crystallization are proposed in literature. One is based on the diffusion and movement of Ni in the *a*-Si matrix causing crystallization of Si around dispersed nanostructured $NiSi_2$ seeds (Ni diffusion assisted MIC); the other is based on the formation of the silicides at the interface between Ni and Si layers, which then moves into the *a*-Si matrix, leading to crystallization of Si (MILC)

According to the first mechanism, the metal (Ni) at first diffuses through *a*- Si matrix and then forms $NiSi_2$ at a sufficient high temperature. This silicide will then act as seed for the crystallization of *a*-Si. The concept of metal diffusion is proposed on the basis of the experimental observations reported in the literature. Park et al. [Park et al., 2001] reported secondary ion mass spectroscopy (SIMS) and transmission electron microscope studies on excimer laser annealed Ni induced crystallization of poly Si. They measured the melting temperatures as a function of Ni content and reported that *a*-Si, just below the Ni layer, melted at lower temperature and the melting point increases with movement away from the Ni rich layer. The change in melting point was attributed to change in Ni concentration. Their explanation for the observed behaviour was that, the Ni diffuses through *a*-Si matrix only when a sufficiently high temperature is reached. Ferri et al. [Ferri et al., 2001] studied diluted metal contaminated *a*-Si system by detailed Raman spectroscopy measurement and came to similar conclusions.

The knowledge of formation, structure and electrical behaviour of a metal-semiconductor interfaces plays an important role in semiconductor technology. The nickel monosilicide (NiSi) is a low resistance silicide in Ni-Si binary system, and is a key material to reduce the contact resistance of gate and source-drain regions. I. H. Hong et al. (Hong et al., 2006) carried out scanning photoelectron spectromicroscopy on the a-Si/Ni films, which were deposited by chemical vapor deposition at ultra high vacuum pressure. They state that when the Si/Ni interface is heated, Ni_2Si, which is an unstable phase with high resistivity, is first formed at 200-300 °C. At around 300 °C or above, Ni_2Si starts transformation into a low resistivity phase NiSi. The Ni_2Si phase disappears at above 400°C. Another high resistivity phase, $NiSi_2$, nucleates at above 750 °C. Another method of enhancing metal diffusion is electric field assisted MIC (FAMIC). In FAMIC, the crystallization of a-Si is enhanced by applying an electric field to the metal/Si binary system.

The two major driving forces for the MIC process are (1) the free-energy gradient between silicon and metal silicides, (2) the concentration gradient caused by atomic diffusion in metal/silicon stacked films. In addition to these driving forces, the applied electrical field introduces an additional driving force to lower the activation energy for crystallization. Lee (Lee et al., 2000) suggested in the study of a Cu/Si FAMIC system that the electrons can charge up at the a-Si surface of the Cu silicide/a-Si interface under an applied voltage. The Cu ions with positive charges could then migrate at a faster speed, resulting in a directional and rapid lateral movement of Cu silicide/a-Si interface and so the crystallized poly-Si can be formed at the enhanced speed. The similar enhanced growth rate is reported in Ni/Si system. It has been postulated that the enhanced growth rate is a result of field-enhanced diffusion of nickel atoms (Park et al., 1999; Yoon et al., 2001). Having an effective charge of -0.3299e in the Si lattice, Ni atoms move toward the positive electrode under an electric field

(Yoon et al., 2001). The other mechanism of FAMIC considers the bombardment of Ni atoms by electrons traveling in an *a*-Si layer. Nickel atoms are pushed toward the positive electrode under a large number of low-momentum collisions (electron wind). Also, energetic electrons are capable of breaking bonds by exciting the bonding electrons through the collision. In fact both field-enhanced diffusion under electric force and electron bombardment effects are responsible for the enhanced crystallization in FAMIC. In both mechanisms, the diffusion of metal is a common argument. Grisenti et al. (Grisenti et al., 2008) reported the Extended x-ray absorption fine structure (EXAFS) analysis on Ni induced *a*-Si produced by co-sputtering. They came to conclusion that, Ni segregates as $NiSi_2$ even at temperatures as low as 200 °C and this $NiSi_2$ enhances the MIC of *a*-Si. The transformation from one kind of silicide to other is well explained by Ni diffusion through Ni/Si interface and this mechanism is widely accepted in the semiconductor technology.

2.3 Metal Induced Lateral Crystallization (MILC)

In case of MILC first metal silicides are formed at the interface between metal and Si layers at the lower annealing temperatures. On further annealing, nucleation of c-Si on metal silicide precipitate starts, which is then followed by migration of the metal silicide precipitates throughout the a-Si thin film, resulted in crystallization of the entire a-Si film. This mechanism is supported by the following experimental observations. Jin et al. (Jin et al, 1998) studied the XPS depth profile of Ni (10 nm) covered a-Si films (70 nm) which was annealed at 500 C for 1 hr. Ar ions with energy of 4 keV were used for sputtering during depth profiling. They reported the concentration variation of selected elements (Ni, O and Si) across the 100 nm stack. From their results it is found that the Ni concentration first decreased with Ar sputtering then stabilized to about 4% in the bulk of the film and finally increased to about 10% near the interface of the film. No Si was detected on the surface of the stack before the Ar sputtering, indicating the diffusion of Si in Ni and NiO_x was slow at the heat treatment temperature of 500 °C. The presence of excess metallic Ni at the bottom of the stack was attributed to the diffusion of the $NiSi_2$ nodules. Hayzelden et al. (Hayzelden & Batstone, 1993) reported an in-situ transmission electron microscopy study on Ni induced *a*-Si crystallization and suggested a similar mechanism for the diffusion of $NiSi_2$ in *a*-Si matrix. According to the mechanism proposed by these authors, Si first crystallizes on one of the eight faces of $NiSi_2$ crystalline seeds and then it dissociates as Ni and Si atoms at the *c*-Si/ $NiSi_2$ interface. The Nickel atoms then diffuse and react again with *a*-Si to form $NiSi_2$ at *a*-Si/ $NiSi_2$ interface. The cause of such dissociation, diffusion and reformation of $NiSi_2$ was explained by chemical potential free energy of the reactions at different interfaces. The chemical potential of Ni is lower at $NiSi_2$/a-Si interface, while the chemical potential of Si is lower at the $NiSi_2$/c-Si interface. Thus Ni moves to the $NiSi_2$/a-Si interface and Si atoms are froced to diffuse to the $NiSi_2$/c-Si interface. The consumption of a-Si at the $NiSi_2$/a-Si interface and diffusion of the Ni atoms result in the growth of needle like Si crystallites. The diffusion of Ni in crystalline Si is much faster than that in a-Si. This supports the fast diffusion of Ni from the c-Si interface to the a-Si interface through a $NiSi_2$ crystallite. Hwang et al. (Hwang & Li, 2005) studied the Auger depth profile on Ni induced MIC and came to a similar conclusion. Our recent work on the transmission electron microscopy study of Ni induced MIC of a-Si films also supports the $NiSi_2$ diffusion mechanism. The details of the experiment and result are presented in the section 4.2 of the chapter.

3. Tin induced a-Si crystallization

Tin is a representative of the eutectic forming group along with Al and Au. The Sn-Si alloy has a relatively low eutectic temperature of 232 °C (Jeon et al., 2010) compared other members of the group. Our recent work on Tin induced a-Si revealed that it follows the eutectic forming metal mechanism well below the Al-Si system (Mohiddon & Krishna, 2011). Jeon et al. (Jeon et al., 2010) studied the Sn-Si system for growing Si nanowires and quoted that Sn is a favorable catalyst for low temperature synthesis of Si nanowires.

The experimental details are as follows. Sn films of 500 nm thickness were deposited onto Borosilicate Glass (BSG) substrates by resistive thermal evaporation. Si films of 500 nm thickness were deposited by electron beam evaporation over the Sn layers. Fig. 1 shows the block diagram of experimental deposition setup of electron beam evaporator and block diagram of film annealing setup. The starting materials were granular pure silicon powder (99.999%) and Sn (99.99% pure). A pressure 5×10^{-6} Torr was maintained throughout the depositions. The depositions were carried out at ambient temperature and in all cases the substrate to source distance was kept constant at 10 cm. The thickness of the films was measured after deposition using a surface profilometer (model XP-1 of Ambios Technology, USA). The films were annealed in a furnace atmosphere at different temperatures for 1 hr. X-ray diffraction patterns were recorded on a powder x-ray diffractometer (CPS120 of Inel, France) machine equipped with a Co K*a* x-ray source (wavelength = 0.178896 nm) and gas phase position sensitive detector.

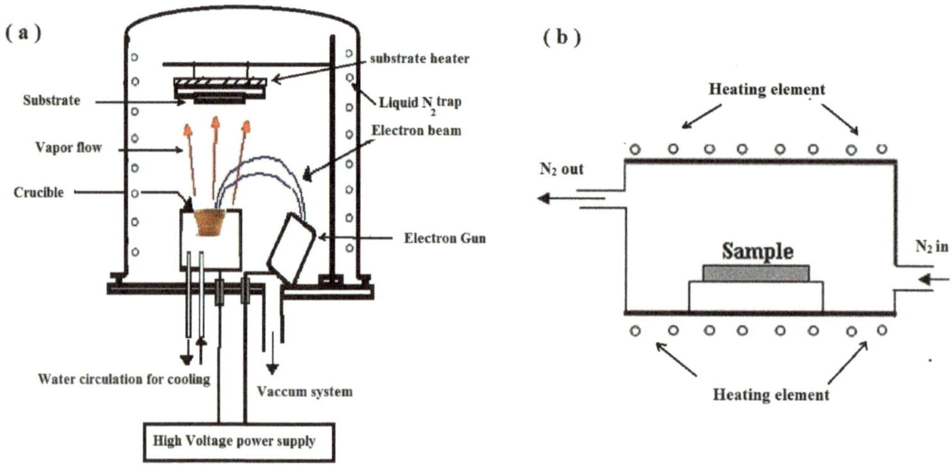

Fig. 1. Schematic diagram of (a) film deposition by Electron beam evaporator (b) annealing of the film in muffle furnace

X-ray diffraction patterns of as deposited BSG/Sn/*a*-Si stacks and stacks post deposition annealed at different temperatures from 300 °C to 500 °C each for 1 hour are presented in Fig. 2. XRD pattern of Sn film deposited on BSG and annealed at 500 °C for 15 hrs is also presented in the same figure. It is evident from the XRD pattern of the annealed Sn film that, it is well crystallized and the pattern can be identified as belonging to the orthorhombic phase of SnO_2 (the peaks were indexed according to the PCPDF file no-781063). The as deposited BSG/Sn/*a*-Si stack, shows sharp diffraction peaks along with a broad diffuse hump around 2θ = 27°, due to the amorphous substrate and *a*-Si film. The peaks can be assigned to the (200), (101), (220), (211), (112) and (312) planes of the tetragonal phase of metallic Sn (the peaks are indexed according to the PCPDF file no- 23456). There is no evidence for the presence of crystalline Si.

On annealing this stack at 300 °C for 1 hr, there is a very interesting transformation in the crystalline behavior of the films. Significantly, all the peaks due to metallic Sn are completely suppressed and the x-ray diffraction pattern resembles that of an amorphous film except for a very diffuse peak at 2θ = 33.58° assigned to the (111) plane of the diamond cubic form of Si. The intensity of this peak increases with increasing annealing temperature up to 500 °C as shown in Fig. 2 indicating that the extent of crystallization of Si is improved by increasing the annealing temperature. However, there is no evidence for crystalline Sn in the films. The reason for this is, most probably, partial oxidation of Sn. The peak at 2θ = 33.58° assigned to the (111) plane of diamond cubic silicon (PCPDF file no- 23345) is shifted by 0.45° which is attributable to strain in the film. Evidently, the onset of crystallization of *a*-Si occurs at 300 °C by contacting with the Sn metal. M Joen et al. have also reported Sn induced crystallization of *a*-Si at 300 °C (Jeon et al., 2010). Significantly, there is no evidence for the formation of silicides in the Sn-Si system even after annealing at 500 °C. Deconvolution of the peak at 2θ = 33.58° provides more insight into the mechanism of crystallization of *a*-Si. The peak was deconvoluted by smoothening and best fitting with a Gaussian function. The data is shown in the inset of Fig.2 for the samples annealed at 300, 400 and 500 °C. Closer observation of the peak corresponding to the sample annealed at 500 °C shows that the peak can be resolved into two peaks centered at 33.7 and 34°. The deconvoluted peaks fitted using Gaussian function is shown in the inset of the Fig 2. The peak centered at 2θ=34° is assigned to the (113) plane of orthorhombic SnO_2 phase (PCPDF file no- 781063). By eliminating the effect of SnO_2 phase in 500 °C annealed samples, the crystallite size is estimated as 18 nm. The average nanocrystal size calculated by Scherrer's equation (Mohiddon & Yadav, 2008) is found to increase from 5 nm to 14 nm with increasing the annealing temperature from 300 to 400 °C. The solid solubility of Sn in Si is known to be nil, *i.e.* they are immiscible in the solid state (Leonard & Koch, 1992). Hence it is not expected to form a silicide. X-ray diffraction data presented earlier indicates that this is the case even in the present samples. The crystallographic evolution of the c-Si phase in the current case is traced by studying the dependence of the crystallite size of Sn and Si on the annealing temperature. This is plotted in Fig. 3, from which it is evident that the crystallite size of Si is zero in the as-deposited film while that of the Sn phase is high. Interestingly, there is a cross-over region below 300 °C when the crystallite sizes of both phases are low. This indicates that at this temperature, the film is essentially amorphous. On increasing the annealing temperature to 400 and 500 °C, there is further increase in the crystallite size of Si while crystallite size of the Sn phase is invariant at zero. This behavior is very similar to that observed in the case of Al induced crystallization (AIC) of a-Si (McCaldin & Sanku, 1971; Zener,1949;Spinella et al.,1998) with one very crucial difference. No peaks corresponding to crystalline Sn are observed on annealing the films unlike the AIC case. This is evidently due to oxidation of Sn during the annealing process as revealed by the deconvolution of the (111) peak of Si in the inset of Fig.2. Thus we conclude from this work that *a*-Si crystallizes when it is brought in contact with Sn metal layer without involving the formation of a silicide, exactly as followed by eutectic forming metals at 300 °C.

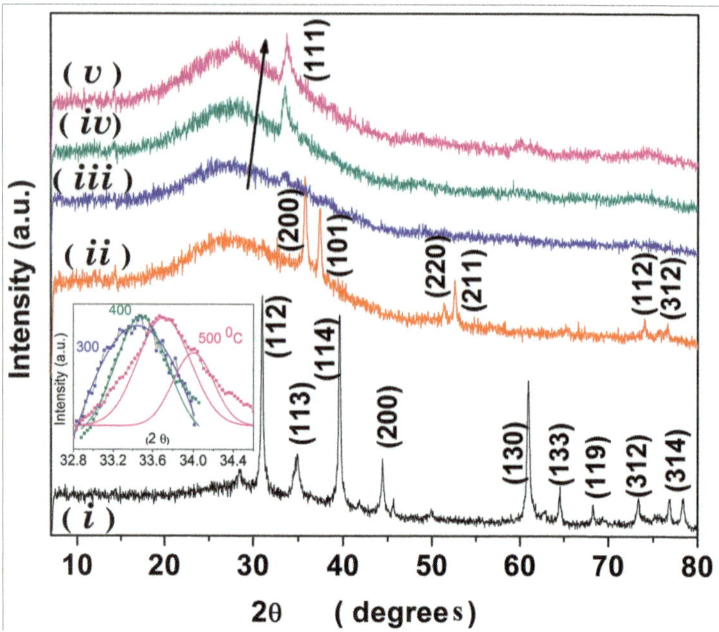

Fig. 2. X-ray diffraction pattern of (*i*) BSG/Sn film annealed at 500 °C for 15 hours. (*ii*) as-deposited BSG/Sn/a-Si stack and BSG/Sn/a-Si stack annealed at (*iii*) 300, (*iv*) 400 and (*v*) 500 °C along with the expanded and smoothened (111) peak of *c*-Si for BSG/Sn/a-Si stack, deconvoluted to show the presence of SnO_2 (Mohiddon et al., 2012).

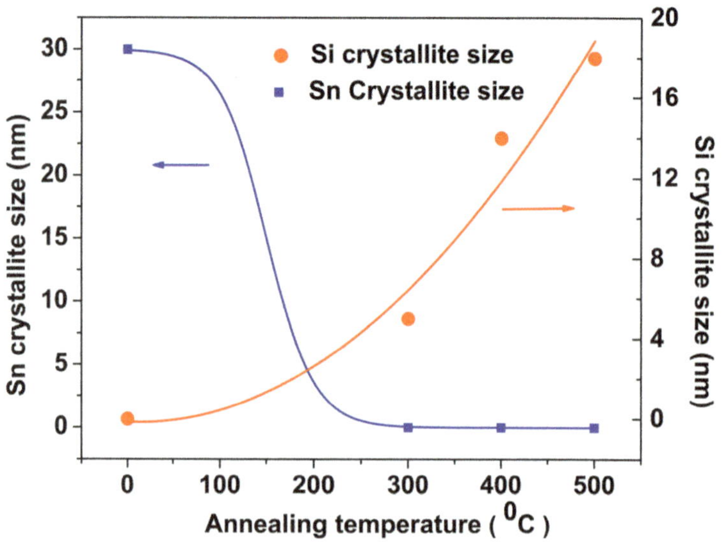

Fig. 3. Variation in crystallite size of Si and Sn phases as a function of annealing temperature (Mohiddon et al., 2012).

4. Nickel and Chromium induced crystallization

The second group of metals used for MIC, reacts with silicon and forms silicides. e.g.: Ni, Cr, Pd, Mo etc. A very small concentration of metals contamination crystallizes large part of *a*-Si and hence they are highly useful in electronic applications. In this section we review our work on Nickel and Chromium induced crystallization of amorphous Si films.

4.1 Nickel induced crystallization

Nickel induced MIC is one of best representatives of this group. To investigate Nickel based MIC, two types of experiments were carried out. In the first case, a 400 nm *a*-Si film is deposited on a fused silica (FS) substrate followed by 200 nm Ni film forming a FS/*a*-Si(400nm)/Ni(200nm) (FSN) stack. The depositions were carried out at different substrate temperatures from 200 to 400 °C. After the deposition is completed, the films were annealed in the vacuum, in the same chamber without removing the vacuum for 30 min at the deposition temperature. The second growth process involves the deposition of 200 nm Ni films on fused silica (FS) substrates followed by deposition of 400 nm amorphous Si films at 200 °C substrate temperature, forming a FS/Ni(200nm)/a-Si(400nm) (FNS) stack.

All films were deposited by the electron beam evaporation technique. The starting materials were granular pure silicon powder (99.999%) and nickel powder (99.99% pure). The vacuum chamber was evacuated using a diffusion–rotary pump combination equipped with a liquid nitrogen trap. A pressure 5×10^{-6} Torr was maintained throughout the depositions. In all cases the substrate to source distance was kept constant at 10 cm. The thickness of the film is measured after deposition using a surface profilometer (model XP-1 Ambios Tech., USA). X-ray diffraction patterns were recorded on a powder x-ray diffractometer (CPS120 of Inel, France) machine equipped with a Co Kα = 0.178896nm and gas phase position sensitive detector. XAFS measurements on the FNS stacks were performed at the Ni K-edge at 8333 eV, in fluorescence mode at two different incidence geometries. The two incident angles are 2° and total reflection (TR) geometries. In total reflection geometry only few nanometers of top layer are monitored (10 to 20 nm). At 2 degree incident angle, the entire thickness of the film is under focus. The radiation source was the European Synchrotron Radiation Facility (ESRF, Grenoble, France) and XAS measurements were performed at the BM08 (GILDA) beamline, with an average storage ring current of 180 mA. Data are collected in fluorescence mode using a 13- element hyper pure Ge detector.

Figure 4 shows the X-ray diffraction pattern (XRD) of FS/ *a*-Si / Ni (FSN) stack deposited at different substrate temperatures. The FSN stack deposited at 200 and 300 °C, shows similar XRD pattern. These XRD patterns, indexed according to the PCPDF file no -701849 and refined using PowderX software, are identified as belonging to face centred cubic (FCC) structure of metallic Nickel. The refined FCC unit cell parameter is *a*=3.53Å. No change in the unit cell parameter is observed for the samples deposited at 200 and 300 °C. In the FSN stack deposited at 400 °C, it is observed that, a new set of XRD peaks are grown along with the metallic Ni XRD peaks. These new set of peaks belong to the face centre cubic structure of $NiSi_2$ indexed according to the PCPDF file no-652974. The FCC unit cell parameter of metallic Ni calculated using (111) peak is found to be 3.52 Å and that of $NiSi_2$ using (111) peak is 5.36 Å. There is no change in the unit cell parameter of metallic Ni for the samples deposited at 300 and 400 °C. The small change of 0.01Å is considered within the range of instrumental measurement error. The disagreement of our result with that of I. H. Hong et al. (Hong et al., 2006) work is expected to be due to the ultra high vacuum pressure deposition conditions, which may facilitate the reaction of the diffused metallic Ni atoms at lower temperature to form different silicide phases.

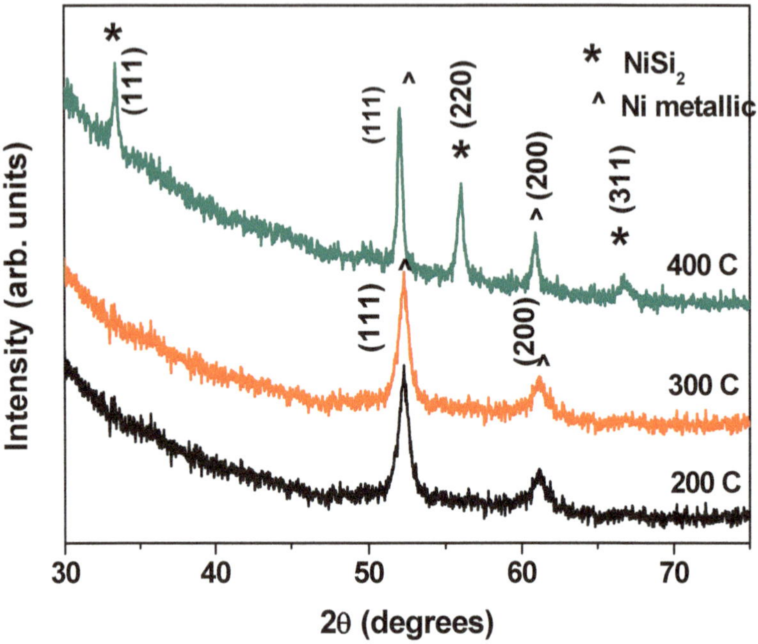

Fig. 4. XRD pattern of FS/Si/Ni stack deposited at different substrate temperature

Figure 5(a) compares the experimental XAFS spectrum of Ni K edge in the FNS stack, which were measured in two different incident angles (TR and 2 degree), along with the reference metal Ni foil spectra. The first derivative near the XANES region of the spectra is presented in the inset of the Fig. 5(a). The observation of the inset figure shows that the XANES spectra of FNS stack deposited at 200 °C and measured in TR and 2 degree has a resemblance with that of the metallic Ni reference spectra. The qualitative analysis of the XAFS spectra shows that the nature of Ni impurity measured in TR and 2 degree are similar and has close resemblance with that of the metallic Ni. Thus at 200 °C annealing temperature the metallic Ni, which is deposited at the bottom, is diffused through 400 nm a-Si layer. The similar observation is found in the EXAFS part of the spectra. Figure 5 (b) compares the Fourier Transform (FT) of the EXAFS signal at the K edge of Ni presented for the FNS stack deposited at 200 °C and measured in two different incident angles, along with the metallic Ni. It shows that the FNS stack deposited at 200 °C and measured in two different incident angles has the features of metallic Ni. The detailed quantitative analyses were carried by FEFF8 and FEFFIT theoretical code and are presented elsewhere (Mohiddon et al., 2011). The conclusions of the work are in good agreement with that of the qualitative analysis presented above. Thus the EXAFS work carried on the FNS spectra support the Ni diffusion assisted MIC mechanism.

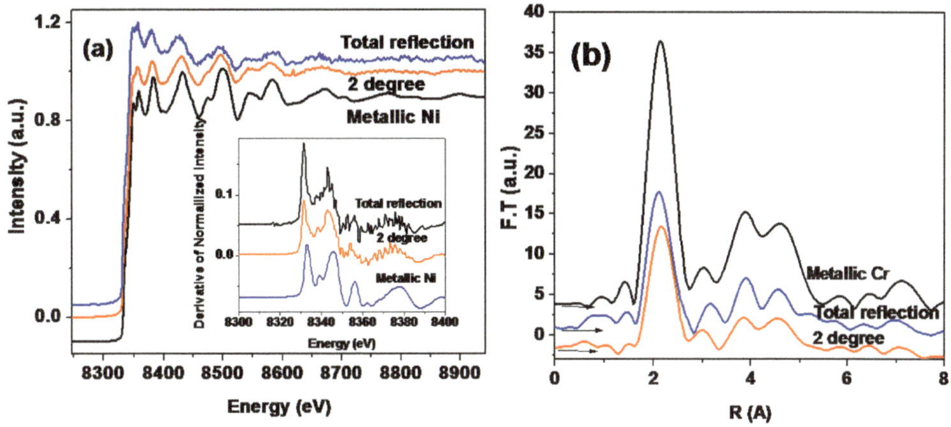

Fig. 5. (a) X-ray absorption spectra along with its derivative near XANES in the inset (b) F.T of EXAFS part of FS/Ni/a-Si stack deposited at 200 °C measured in total reflectance and 2 degree along with metallic Ni reference compound.

4.2 Ni induced Lateral Crystallization (NILC)

As discussed in Sec 2.3, in case of MILC first metal silicides are formed at the interface between metal and Si layers at the lower annealing temperatures. On further annealing, nucleation of c-Si on metal silicide precipitate starts, which is then followed by migration of the metal silicide precipitates throughout the a-Si thin film, resulted in crystallization of the entire a-Si film. We have carried a detailed transmission electron microscope study and found similar observation. The details of the experiment are as follows. The Ni and Si films are deposited using electron beam evaporation at $4x10^{-6}$ Torr and ambient temperature onto Si (311) substrates. The starting materials were granular pure silicon powder (99.999%) and nickel powder (99.99% pure). The (311) oriented Si wafer was well cleaned and subjected to heat treatment of 1000 °C for 2 hours in normal furnace atmosphere to form a thick SiO_2 layer, before the deposition is started. The SiO_2 layer is expected to act as a barrier to stop the diffusion of Ni atom into the Si substrate. X-ray diffraction patterns were recorded in grazing incidence of 0.5 degree on a powder x-ray diffractometer (CPS120 of Inel, France) machine. A 50 nm film of Ni was deposited on the *c*-Si/SiO_2 substrate, followed by the deposition of the 400 nm Si film without breaking vacuum in the deposition chamber. The thickness of the films was measured after deposition using a surface profilometer (model XP-1 Ambios Tech., USA). The films were subsequently annealed in a furnace (in air) at 600 °C for 1 hr. Transmission electron micrographs were obtained by a Tecnai 20 G2 STwin, FEI electron microscope, operated at 200 kV. Electron diffraction patterns (EDPs) were recorded with a Gatan CCD camera. A 10 nm gold film deposited on the grid was used for purpose of camera length calibration. The samples for TEM measurement were prepared by scratching the film and transferring it on to the grid.

Figure 6 is a bright-field transmission electron micrograph obtained from a region of the Ni/Si thin film crystallized by the MIC process. The typical microstructure shown in Fig. 6 consists of dark region with an irregular shape. The microstructure consists of dark dendrite like nanowire structures spread over the bright looking matrix. To find the nature of nanostructures at two different regions of the sample, selected area electron diffraction (SAED) patterns were recorded at both the dark and bright regions of the sample. A typical diffraction pattern from the dark part of the image is shown in the inset of the Fig. 6. The diffraction pattern consists mainly of bright diffraction spots radially positioned on rings. The diffraction spots arise from the crystalline part of the sample, whereas the ring originates from the polycrystalline parts of the sample. The inset of the Fig 6 shows that, the diffraction spots A, B and C lie on the polycrystalline ring of radius 0.314 nm, which is indexed to the (111) plane of the Silicon diamond cubic phase. The spot D has a d=0.185 nm which can be indexed as belonging to the $(2\,\bar{2}\,0)$ plane of the diamond cubic system [JCPDS-892955]. However, an alternative assignment is possible for the same diffraction spots (and the higher order diffraction spots), because they can be equally assigned to the face centred cubic *c*-$NiSi_2$, with d values of 0.312 nm, 0.191 nm for (111) and $(2\,\bar{2}\,0)$ planes, respectively [JCPDS-652974]. Hence, the diffraction taken in dark region of the sample shown in Fig 6 may belong either to diamond cubic Si or to face centre cubic $NiSi_2$.

The diffraction pattern from the bright area region is shown in other inset of Fig.6. Even at the higher magnifications, the microstructure of this bright region is seen to contain dark dendritic lines. It was thus inpossible to eliminate the presence of the darker regions. Hence, the diffraction pattern shown in second inset of Fig 6 has a combined effect, with a small contribution from dark nanostructures and large contribution from the bright area. The diffraction pattern is slightly different from that of dark region diffraction pattern. It consists of one diffuse diffraction rings with d=0.314 nm and a bright diffraction spots with d = 0.192 nm (indicated by E). There are, however, no diffraction spots disposed on the ring, thus suggesting that the dark region is better crystallized compared to bright region although both the regions show similar diffraction patterns.

To exactly identify the material(s) that are contributing to the diffraction from the bright and dark regions of the samples, Z (atomic number)-contrast scanning transmission electron microscopy (STEM) study was carried out. Fig. 7 is a Z-contrast STEM image of the same selected area that is shown in Fig. 6. The change in contrast in the image is due to change in Z value of Ni (Z= 28) and Si (Z= 14). This observation suggests that the central high contrast region is due to the contribution from Ni component, i.e. $NiSi_2$ and the low contrast region is from Si. Hence, it can be inferred that the $NiSi_2$, after its formation, starts spreading through the matrix of *a*-Si like a dendrite and leading to its crystallization. The results presented in this part of the study supports the mechanism of movement of $NiSi_2$ dendrites in the *a*-Si matrix as the cause for crystallization. Similar movement of Cu_3Si dendrites with time of annealing was reported using in-situ bright field transmission electron microscopy, but no diffraction study was reported (Russell et al.,1991). It is evident from the study of SAED and STEM, that when Ni makes contact with the Si layer at high temperature it forms $NiSi_2$. This silicide starts diffusing as dendrites into the *a*-Si matrix and crystallizes the silicon at the nanoscale. This strongly supports the mechanism of silicide movement as the basis for crystallization of *a*-Si by the MILC process using silicide forming metals.

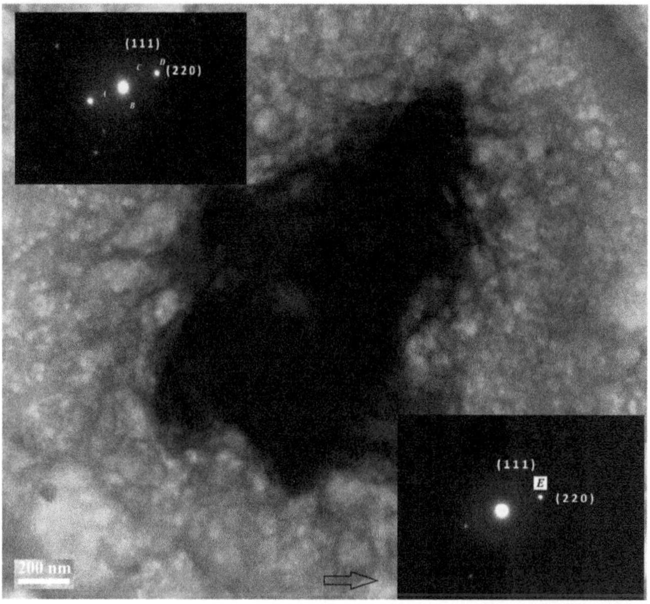

Fig. 6. Selected Bright field image of Ni/Si film annealed at 600 °C for 1 hr along with the diffraction pattern in the dark and bright regions (Mohiddon et al., 2012).

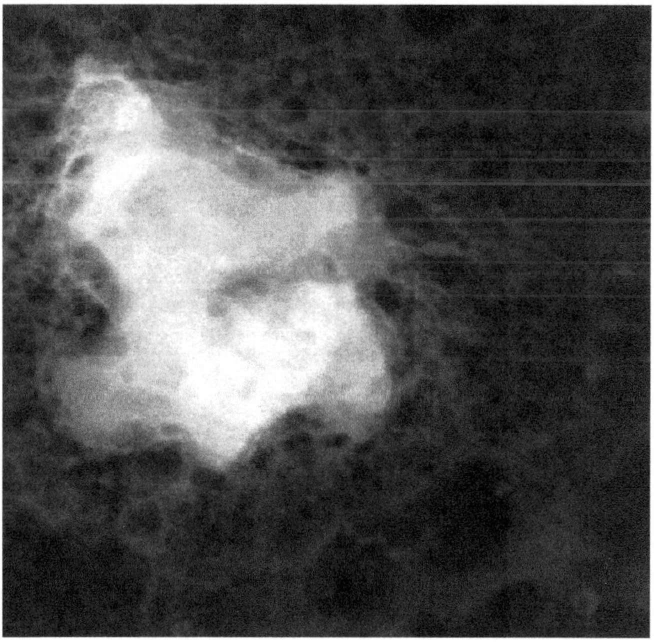

Fig. 7. STEM image of Ni/Si film annealed at 600 °C for 1 hr, selected the same region as that selected for bright field image in Fig. 6 (Mohiddon et al., 2012).

4.3 Stabilization of wurtzite Si

Silicon (Si) usually crystallizes in the cubic diamond structure with fourfold coordinated symmetry. Numerous high-pressure experiments have been performed, revealing no less than 12 different polymorphs of silicon between the well characterized diamond cubic phase and theoretically intractable amorphous phase. With the release of pressure several metastable phases are observed (Wu et al., 2000). For example, a nonmetal to metal transition of cubic-diamond phase, which occur by changing pressure from ambient atmosphere pressure to 10–13 GPa to form a β-tin structure, which further transforms to rhombohedral phase on slow pressure release. The latter transforms reversibly at a pressure of 2 GPa to a body-centered-cubic phase (bc8). Hexagonal-wurtzite silicon (*w*-Si) can be formed by heating the bc8 phase to above 200 °C, or directly from the cubic-diamond phase in the presence of shear stresses at twin intersections, or a nonhydrostatic stress of 8 GPa in indentation experiments (Kailer et al., 1997). *w*-Si material is rarely studied by spectroscopic measurements, because it cannot be obtained in a stable phase. Zhang et al. (Zhang et al., 1999) produced stable *w*-Si phase by laser ablation. Its identification by electron diffraction has been confirmed by micro Raman spectroscopy. Bandet et al. (Bandet et al., 2002) deposited *w*-Si during elaboration of SiO_2 thin films and reported that oxygen seems to play a crucial role in the stabilization of the uncommon metastable structure. Kim et al. (Kim & Lee, 1996) reported that the micrometer-sized diamond cubic silicon (*c*-Si) crystals contain minority part of hexagonal silicon, when *a*-Si films crystallized by a pulsed laser. In the present work we have crystallized the silicon film with a metal contact and found that silicon nanocrystals are in wurtzite phase. The experimental details are as follows.

Nickel and Silicon films were deposited by electron beam evaporation on to BoroSilicate Glass (BSG) substrates in high vacuum of the order of 10^{-6} Torr. First a 50 nm thin nickel blanket bottom layer was deposited on BSG substrates maintained at ambient temperature. This is followed by the deposition of a 700 nm thick Si film, without breaking vacuum, to form a BSG/Ni/Si stack. The thickness of the films is measured after deposition using a surface profilometer (model XP-1 Ambios Tech., USA). The films were annealed in a furnace atmosphere (in air) at 550 °C for 1 hr. Transmission electron micrographs were obtained by a Tecnai 20 G2 STwin, FEI electron microscope, operated at 200 kV. Electron diffraction patterns (EDPs) were recorded with a Gatan CCD camera. A 10 nm gold film deposited on the grid was used for camera length calibration purposes. The Raman spectra were recorded in air using an Nd-YAG 532nm laser in the back scattering geometry in a CRM spectrometer equipped with a confocal microscope and 100× objective (1 μm diameter focal spot size) with a CCD detector (model alpha 300 of WiTec Germany). The phase content with in the samples was investigated in a spectral region 200–1500 cm^{-1}.

Figure 8 shows the electron diffraction pattern along with a bright-field transmission electron micrograph obtained from a region in the silicon thin film crystallized by a MIC process in the inset. The typical microstructure shown in inset of Fig. 8 consists of spherical particles with approximate diameter of 40 nm. The diffraction spots can be identified as belonging to well defined orientations. The calculated interplanar distances of the lattice planes at different spots of the diffraction pattern assigned by symbols A, B and C in Fig. 8 are A=0.256nm, B=0.237nm and C=0.237nm respectively. From the calculated interplanar distances the diffraction spots were indexed as (1 0 0), ($1\ \bar{1}\ 1$), and ($O\ \bar{1}\ 1$), which belong to the $[1\bar{2}13]$ zone axis. These diffraction spots were diffracted from a grain which has a hexagonal structure and oriented in the $[1\bar{2}13]$ direction and the spots A, B and C were diffracted from the (1 0 0), (1 $\bar{1}$ 1), and (O $\bar{1}$ 1) planes, respectively. The corresponding lattice parameters were calculated as a=0.296nm, c=0.629nm with c/a ratio of 2.13. The angle calculated between these planes using c/a value is found to be exactly matching with the measured values suggesting that Si crystallizes in the hexagonal structure. For diamond cubic Si crystals the interplanar distances of the lattice planes should be 0.33 1 nm, 0.33 1 nm and 0.331 nm, respectively and the angles between transmitted beam and the spots will be 60°, 60° and 60°, with the diffraction pattern showing the expected six fold symmetry. But in our case the angles between the dots are ∠AOB=65°, ∠BOC=55° and ∠AOC=120° respectively, clearly not cubic Si. Thus, the TEM analysis shows that the BSG/Ni/Si stack after heat treatment consists of 40 nm particles of hexagonal Silicon with c/a ratio of 2.11.

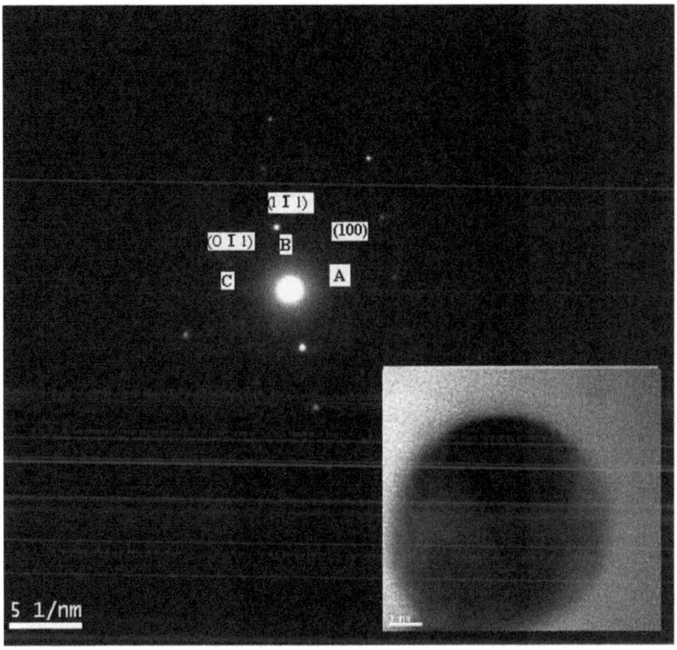

Fig. 8. Electron diffraction pattern along with indexing of selected w-Si part of BSG/Ni/Si stack, inset shows the bright field image of the above diffraction area (Mohiddon et al., 2011).

Additional evidence for the formation of *w*-Si is provided in the form of Raman spectra of the films. It has earlier been used to characterize porous Si (Sui et al., 1992) and Si nanostructures (Kozlowski et al., 1991). The Raman shift and the shape of the Raman peak yield information on the degree of crystallinity achieved in Si by MIC process. Fig. 9 shows Raman spectra from the Silicon film annealed at 550 °C for 1 hrs. A sharp peak at 504 cm^{-1} and weak broad peaks around 300 and 950 cm^{-1}, are associated with optical phonons, two transverse acoustic (2TA) phonons, and two transverse optical (2TO) phonons of *w*-Si (Mohiddon & Kirshna, 2011) respectively. The asymmetric nature of the peak at 504 cm^{-1} can be attributed to the presence of small amounts of *a*-Si, which has its broad peak at 480 cm^{-1}. In general, when Si crystallizes in cubic diamond structure the Raman peak occurs at 520 cm^{-1}. The broad peaks around 283 cm^{-1} in our observation is may belong to NiSi, $NiSi_2$ or to wurtzite TA mode. However, this observation implies that, at 550 °C nickel silicides are present in the samples. Tan et al. (Tan et al., 1981) reported a stress-induced metastable form of hexagonal silicon, which has the wurtzite structure with the c/a ratio close to 1.63, the interplanar distances and angular relationships obtained from our experiment were compared with those of silicon which has the wurtzite structure. However, for silicon with the wurtzite structure, no exact coincidence was obtained. Kim et al. reported a similar observation and suggested a hexagonal phase of Si with *a*=0.382nm, *c*=1.024nm and *c*/*a* ratio of 2.68 (Kim & Lee, 1996). Parson and Hoelke suggested similar hexagonal structure for Ge with *c*/*a* ratio of 2.17 (Parsons & Hoelke, 1983). Zhang *et al.* (Zhang et al., 1994) have reported wurzite silicon by laser ablation, and they observed Raman peaks at 516 and 518 cm^{-1} due to the hexagonal silicon. Kumar et al. observed *w*-Si in chromium induced Si crystallization evidenced by a peak at 495 cm^{-1} (Kumar & Krishna, 2008). Kailer et al. (Kailer et al., 1997) performed indentation Raman investigations and reported that, shear deformation leads to *w*-Si phase formation. They also stated that high pressure phases leads to the formation of *w*-Si at moderate temperatures and to the reversal to the original diamond structure at higher temperatures.

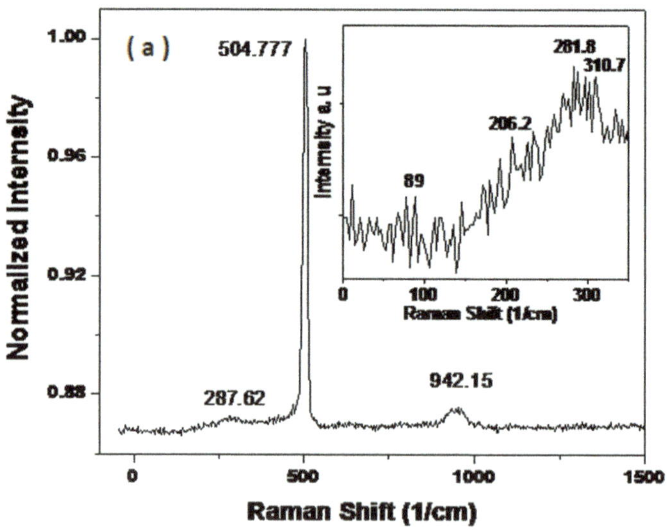

Fig. 9. Raman spectra of BSG/Ni/Si stack annealed at 550 °C for 1 hr (Mohiddon et al., 2011).

4.4 Chromium induced crystallization

The search for new metal in the MIC process to reach the best device needs is still continuing. Chromium is another member of the silicide forming group. Chromium induced MIC has attracted our attention due to its ability to crystallize the *a*-Si into wurtzite phase (Kumar et al., 2008). Our recent work on Cr induced MIC has revealed that Cr deposited at ambient substrate temperature, remains in metallic form up to an annealing temperature of 500 °C (Mohiddon et al.,2012). As an extension of this work we deposited Cr/Si stacks at higher substrate temperature. The details of the experiment are as follows. A Cr(200nm)/Si(400nm) stack was deposited over a fused silica substrate at 400 °C substrate temperature and a similar stack of Cr(50nm)/Si(400nm) was deposited at ambient substrate temperature by electron beam evaporation. We have carried out a detailed EXAFS analysis on these set of samples (Mohiddon et al., 2012) and concluded that, in case of film deposited at ambient substrate temperature, major part of Cr is in metallic form apart from being oxidized. In the case of the stack deposited at 400 °C, the entire Cr is turned into either Cr_2O_3 or $CrSi_2$. After the EXAFS measurements were completed, these samples were heat treated at different temperatures. Samples were annealed at 600, 650 and 700 °C for 1 hr. in excess Nitrogen flux. At each heat treatment, the samples were cooled to room temperature and then the Raman spectra were collected at three to four different position of the each sample. The details of the Raman measurements are discussed in the section 4.3 of the chapter.

Figure 10(a) and 10(b) compares the Raman spectra of the FS/Cr(200nm)/Si(400nm) stack deposited at 400 °C substrate temperature and FS/Cr(50nm)/Si(400nm) stack deposited ambient substrate temperature then annealed at different temperature respectively. Fig. 10(a) shows that the as deposited stack contains a broad diffuse peak around 480 cm^{-1} that can be assigned to the a-Si phase. After the heat treatment at 600 °C, a sharp peak, around 520 cm^{-1}, appears that can be assigned to c-Si. The intensity of the c-Si peak increases with annealing temperature up to 650 °C. No such change in the Raman spectra is observed in the Fig. 10 (b). By combining the EXAFS and Raman results we conclude that the FCS stack deposited at 400°C possesses $CrSi_2$ in the as deposited condition. When this stack is further heat treated the $CrSi_2$ acts as a seed for the Si crystallization. In the FCS stack deposited at minimum substrate temperature, no $CrSi_2$ is observed from the EXAFS analysis. Thus even after heat treatment up to 700 C, no Si crystallization is observed.

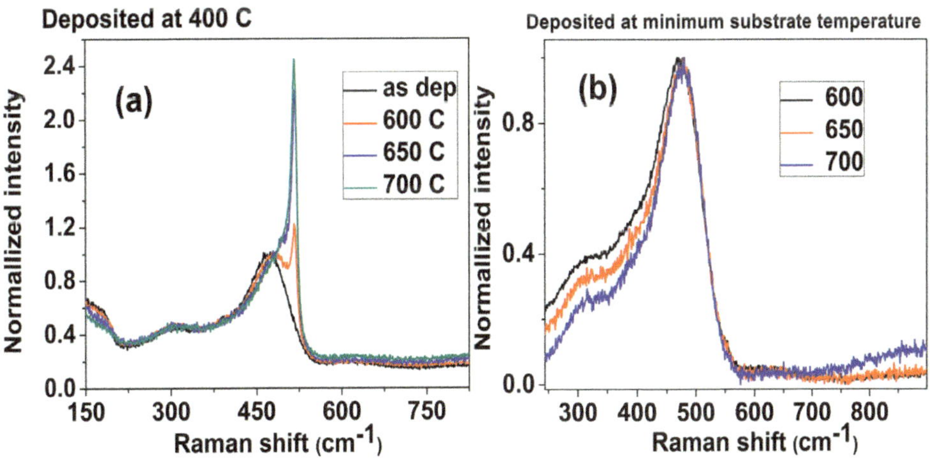

Fig. 10. Raman spectra of (a) FS/Cr(200nm)/Si(400nm) stack deposited at 400 C; (b) FS/Cr(50nm)/Si(400nm) stack deposited at minimum substrate temperature

5. Conclusions

An overview of the metal induced crystallization of amorphous Silicon films has been presented. The possible mechanisms of crystallization have been discussed. Two metals Sn and Cr, which have not been used frequently for MIC, have been shown to be as attractive as Ni for metal induced crystallization of a-Si. Our results clearly show that these are very useful materials for the microelectronics industry. The crystallization occurs at temperatures ranging from 200 to 600 °C, which is much lower than the normal crystallization temperature of Si (1100 °C). The stabilization of the wurtzite form of Si has also been demonstrated.

6. Acknowledgements

The authors acknowledge discussions with Prof. F. Rocca and Prof. G. Dalba of University of Trento and funding for this work through the DST-ITPAR program. Facilities provided by the CAS programme of the School of Physics and DST funded Centre for Nanotechnology at the University of Hyderabad are also acknowledged.

7. References

Choi, J.H., Cheon, J.H., Kim, S.K., & Jang, J. (2005). Giant-grain silicon (GGS) and its application to stable thin film transistor. *Displays* 26, 3, (June 2005), pp. (137-142), ISSN 0141-9382

Bandet, J., Despax, B., & Caumont, M. (2002). Vibrational and electronic properties of stabilized wurtzite like silicon. *Journal of Physics D: Applied Physics*, 35, 3, (January 2002), pp. (234-239), ISSN 1361-6463

Ferri, F.A., Zanatta, A.R., & Chambouleyron, I. (2006). Metal-induced nanocrystalline structures in Ni-containing amorphous silicon thin films. *J. Appl. Phys.*, 100, 9, (November 2006), pp. (094311-094317), ISSN 0021-8979

Grisenti, R., Dalba, G., Fornasini, P., Rocca, F., Kumar, K.U.M., & Krishna, M.G. (2008). XAFS study on Ni surroundings in metal induced crystallization of thin film amorphous silicon. Solid State Comm., 147, 9-10, (September 2008), pp. (401-404) ISSN 0038-1098

Hayzelden, C., & Batstone, J.L. (1993). Silicide formation and silicide-mediated crystallization of nickel-implanted amorphous silicon thin films. J. Appl. Phys., 73, 12, (February 1993), pp. (8279-8289), ISSN 0021-8979

Hiraki, A. (1980). A model on the mechanism of room temperature interfacial intermixing reaction in various metal-semiconductor couples: what triggers the reaction?. *J. Electrochem. Soc.* 127, 12, (December 1980), pp. (2662-2665), ISSN 1945-7111

Hong, I., Hsu, T.C., Yen, S.C., Lin, F.S., Huang, M.L., & Chen, C.H. (2006). Proc. 8th Int. Conf. X-ray Microscopy, IPAP Conf. Series 7 (2006) 270.

Hwang, J.D., & Lin, J.W. (2005). Nanostructural silicon films prepared by metal induced growth using RTCVD system. *J. Electrochem. Soc.*, 152, 1, (Novermber 2005), pp. (G35-G39), ISSN 1945-7111

Jeon, M., Jeong, C., & Kamisako, K. (2010). Tin induced crystallization of hydrogenated amoprphous silico, K., Kowk, H.S., & Wang, M. (1999), The effects of extended heat treatment on Ni induced lateral crystallization of amorphous silicon thin films. *IEEE Trans. Electron Devices*, 46, 1, (January 1999), pp. (78-82), ISSN 0018-9383

Jin, Z., Bhat, G.A., Yeung, M., Kwok, H.S., & Wong, M. (1998). Nickel induced crystallization of amorphous silicon thin film. *J. Appl. Phys.* 84, 1, (April 1998), pp. (194-200) ISSN 0021-8979.

Kailer, A., Gogotsi, Y.G., & Nickel, K.G. (1997). Phase transformation of silicon caused by contact loading. *J. Appl. Phys.*, 81, 7, (January 1997), pp. (3057-3063), ISSN 0021-8979

Kim, J.H., & Lee, J.Y. (1996). Hexagonal silicon formation by pusled laser beam annealing. *Materials Letters* 27, 6, (August 1996), pp. (275-279), ISSN 0167-577X

Kozlowski, F., Petrova-Koch, V., Kux, A., Stadler, W., Fleischmann, A., & Sigmund, H. (1991). Structural instability of Si-nanocrystals in a $(SiH_2)_n$. *J. Non-Cryst. Solids*, 137-138, 1, (January 1991), pp. (91-94), ISSN 0022-3093

Kumar, K.U.M., Brahma, R., Krishna, M.G., Bhatnagar, A.K., & Dalba, G. (2007). An optical study of Ni induced crystallization of a-Si thin films. *J. Phys. Condens. Matter* 19, 49, (December 2007), pp. (496208-1-11), ISSN 1361-648X

Kumar, K.U.M., & Krishna, M.G., (2008). Chromium-induced nanocrystallization of a-Si thin films into the wurtzite structure. *J. Nanomater.* 2008, (February 2008), pp. (736534-1-6)

Lee, C.J., Lee, J.B., Chung, Y.C., & Choi D.K. (2000). Lateral crystallization of amorphous silicon films. *Jpn. J. Appl. Phys.* 39, (2000), pp. (6191-6195), ISSN 1347-4065

Leonard, R.T., & Koch, C.C. (1992). Nanoscale composites of Si/Sn and Ge/Sn synthesized by mechanical attrition. *Nanostructured Materials*, 1, 6, (December 1992), pp. (471-478), ISSN: 0965-9773

Mahan, A.H., Ahrenkiel, S.P., Schropp, R.E.I., Li, H., & Ginley, D.S. (2008). A comparison of grain nucleation and grain growth during crystallization of HWCVD and PECVD a-Si:H films. *Thin Solid Films*, 516, 5, (January 2008), pp. (529-532), ISSN 0040-6090

McCaldin, J. O., & Sankur, H. (1971). Diffusivity and solubility of Si in the Al metallization of integrated circuits. *Appl. Phys. Lett.*, 19, 12, (July 1971), pp. (524-527), ISSN 0003-6951

Mohiddon, M.A., & Krishna, M.G. (2011). Nanocrystalline wurtzite Si-nickel silicide composite thin films with large band gap and high resistivity. *J Mater. Sci.*, 46, 8, (April 2011), pp. (2672-2677), ISSN 1573-4803

Mohiddon, M.A., & Krishna, M.G. (2012) Growth and optical properties of Sn-Si nanocomposite thin films *J. Material Science*, DOI: 10.1007/s10853-012-6647-0

Mohiddon, M.A., & Yadav, K. L. (2008). Effect of 90° domain on ferroelectric properties of alkali modified SBN. *J. Phys. D: Appl. Phys.*, 41, 22, (October 2008), pp. (225406-1-7), ISSN 1361-6463

Mohiddon, M.A., Naidu, K.L., Dalba, G., Rocca, F., & Krishna, M.G. (2012) unpublished

Mohiddon, M.A., Naidu, K.L., Dalba, G., Rocca, F., & Krishna, M.G. (2012) Cr induced nanocrystallization of *a*-Si thin films: its mechanism *Physica Status Solidi C*, 9 (April 2012), pp. (1493–1495)1493–1495

Mohiddon, M.A., Naidu, K.L., Dalba, G., Rocca, F., & Krishna, M.G. (2011) Growth, optical, and electrical properties of silicon films produced by the metal-induced crystallization process, *J Nanopart Res* 13 (June 2011), pp. (5999–6004).

Mohiddon, M.A., Krishna, M.G., Dalba, G. & Rocca, F. (2012) Transmission electron microscopy study of Ni-Si nanocomposite films *Mate. Sci. Eng. B*, DOI: 10.1016/j.mseb.2012.05.018

Murray, J. L., & McAlister, A.J. (1984). Al-Si. *Bull. Alloy Phase Diagram*, 5, 1 (January 1984), pp. (74-84) ISSN 0197-0216.

Nast, O., & Wenham, S.R. (2000). Elucidation of the layer exchange mechanism in the formation of polycrystalline silicon by aluminum induced crystallization. *J. Appl. Physics*, 88, (January 2000), pp. (124-131), ISSN 0021-8979

Oliver, N., & Hartmann, A. J. (2000). Influence of interface and Al structure on layer exchange during aluminum induced crystallization of amorphous silicon. *J. Appl. Phys.* 88 (April 2000), pp. (716-724) ISSN 0021-8979

Park, S.H., Jun, S.I., Song, K.S., Kim, C.K., & Choi, D.K. (1999). Field aided lateral crystallization of amorphous silicon thin film. *Jpn. J. Appl. Phys.*, 38, 2, (January 1999), pp. (L108-L109), ISSN 1347-4065

Park, K.C., Song, I.H., Jeon, J.H., & Han, M.K. (2001). Excimer laser annealing effect on Nickel-induced crystallized polycrystalline silicon film. *J. Electrochem. Soc.*, 148, 10, (August 2001), pp. (G563-G565), ISSN 1945-7111

Parr, A.A., Gill, K., Gardiner, D., Holyland, J., Sands, D., Brunson, K., & Carline, R. (2002). A comparison of laser- and furnace- annealed poly-silicon structure. *Semicond. Sci. Technol.* 17, 1, (January 2002), pp. (47-54), ISSN 1361-6641

Parsons, J.R., & Hoelke, C.W. (1983). Crystallography Of hexagonal germanium. *Nature*, 30, 5901, (1983), pp. (591-592), ISSN 1476-4687

Russell, S. W., Jian, Li., & Mayer, J. W. (1991). In situ observation of fractal growth during a-Si crystallization in a Cu_3Si matrix. *J. Appl. Phys.*, 70, 9, (August 1991), pp. (5153-5155), ISSN 0021-8979

Schropp, R.E.I., & Zeman, M. (1998). *Amorphous and Microcrystalline Silicon Solar Cells:Modeling, Materials and Device Technology*, Kluwer Academic Publishers, ISBN 978-0-7923-8317-8, Boston.

Spinella, C., Lombardo, S., & Priolo, F. (1998). Crystal grain nucleation in amorphous silicon. *J. Appl. Phys.*, 84, 10, (July 1998), pp. (5383-5415), ISSN 0021-8979

Sui, Z., Leong, P.P., Herman, I.P., Higashi, G.S., & Temkin, H. (1992). Raman analysis of light emitting porous silicon. *Appl. Phys. Lett.*, 60, 17, (February 1992), pp. (2086-2088), ISSN 0003-6951

Tan, T.Y., Foll, H., Hu, S.M. (1981). On the diamond-cubic to hexagonal phase transformation in silicon. *Philos. Mag. A*, 44, 1, (July 1981), pp. (127-140), ISSN 1478-6443

Wang, M., & Wong, M. (2001). Characterization of an individual grain boundary in metal-induced laterally crystallized polycrystalline silicon thin-film device. *IEEE Trans. Electron Devices*, 48, 8, (August 2001), pp. (1655-1660), ISSN 0018-9383

Wu, B.R. (2000). First- principles study on the high-pressure behavior of the zone-center modes of lonsdaleite silicon. *Phys. Rev. B*, 61, 1, (January 2000), pp. (5-8), ISSN 1550-235X

Yoon, S. Y., Park, S. J., Kim, K. H., & Jang, J. (2001). Metal induced crystallization of amorphous silicon. *Thin Solid Films*, 383, 1-2, (February 2001), pp. (34-38), ISSN 0040-6090

Zener, C. (1949). Theory of growth of spherical precipitates from solid solution. *J. Appl. Phys.*, 20, 10, (March 1949), pp. (950-953), ISSN 0021-8979

Zhang, Y., Iqbal, Z., Vijayalakshmi, S., & Grebel, H. (1999). Stable hexagonal wurtzite silicon phase by laser ablation. *Appl. Phys. Lett.*, 75, 18, (September 1999), pp. (2758-2760), ISSN 0003-6951

Zhang, S.L., Wang, X., Ho, K., Li, J.J., Diao, P., & Cai, S. (1994). Raman spectra in a broad frequency region of p- type porous silicon. *J. Appl. Phys.* 76, 5, (May 1994), pp. (3016-3019), ISSN 0021-8979

10

Crystallization of Ge:Sb:Te Thin Films for Phase Change Memory Application

J. J. Gervacio Arciniega, E. Prokhorov, F. J. Espinoza Beltran and G. Trapaga
CINVESTAV, Unidad Queretaro, Juriquilla, Querétaro, Mexico

1. Introduction

Chalcogenide glasses are a chemical compound consisting of at least one chalcogen element, sulphur, selenium, or tellurium, in combination with other elements. These glasses obtained great attention after discovery between 1962 and 1969 by Kolomiets, Eaton, Ovshinsky and Pearson of the S-shape current-voltage characteristic in chalcogenide glasses and the switching phenomenon from high to low resistivity states (Popescu, 2005). In 1968 S. R. Ovshinsky demonstrated very short (about of 10^{-10} seconds) reversible electrical switching phenomena in $Te_{81}Ge_{15}Sb_2S_2$ thin films due to amorphous-crystalline phase transition (Ovshinsky, 1968). This work opened a new area of phase change technology, which now is one of the most important technologies for memory devices and applications in computers, CD, DVD, phase-change random access memories, etc.

The first electrical phase-change memory devices used various binary, ternary and quaternary Te-based films with compositions made up of Ge:Te, Si:As:Te, Ge:As:Si:Te, etc., systems (Stand, 2005). The early phase change materials used in optical storage comprised simple alloys based primarily on compositions in the vicinity of the tellurium-germanium eutectic. Antimony was primarily used, although other elements including selenium, arsenic and bismuth were all shown to have beneficial effects. Based on the results obtained from films with $Ge_{15}Sb_4Te_{81}$ composition (Ovshinsky, 1971) the first application of Ge:Sb:Te alloys was reported for phase change rewritable optical disks.

In the early 1990s, a second generation of high speed phase change materials based on Ge:Sb:Te alloys was reported by several optical memory research groups (Ohta et al 1989, Yamada et al, 1991, Gonzalez-Hernandez et al, 1992). These alloys have stoichiometric compositions along the GeTe–Sb_2Te_3 pseudobinary line of phase diagram such as $Ge_2Sb_2Te_5$, $Ge_1Sb_2Te_4$ and $Ge_1Sb_4Te_7$.

Ge:Sb:Te stoichiometric alloys have three phases: one amorphous and two crystal structures. The first crystalline phase is the rock salt NaCl-like and the second more stable phase is hexagonal. When the amorphous films are heated, the transition from amorphous to rock salt-like structure occurs at around 120-170^{0}C; subsequent heating transforms this phase into a stable hexagonal structure at temperatures around 200-250^{0}C. The exact transition temperature depends on the composition of the film. The hexagonal structure remains stable over a wide temperature range from 200-250^{0}C to around the melting point (593-630 ^{0}C, depending on composition) (Gonzalez-Hernandez et al, 1992). The principle of phase-change memory operation is based on a reversible phase-transformation from the amorphous (high resistance/low reflectivity state) to crystalline NaCl-type (low resistance/high reflectivity state) under short laser or electrical pulses. It is necessary to note that for phase-change memory applications only the amorphous-NaCl-type transition has been used, probably because the heating time produced by the laser or electrical pulse is too short to form the stable hexagonal structure (Yamada, 1991). This transformation is always accompanied by abrupt changes in reflection (about 20-30 %) and resistivity (about 3 orders of magnitude) in chalcogenide films. The reversible transformation from crystalline to the amorphous structurally disordered state can be obtained by increasing the local temperature of the Ge:Sb:Te layer above its melting point by short intense laser, or electrical pulses and the subsequent quenching with a cooling rate about 10^{10} deg/s (Yamada et al, 1991).

The phase transition between the amorphous and NaCl-type crystalline phase in stoichiometric materials is fast because atoms in the amorphous state do not need to travel long distances to take their position in the crystal lattice. On the other hand, non-stoichiometric compounds require long-range diffusion when they crystallize from an amorphous state (Yamada et al, 1991, Yamada 1996), which in general slows down the crystallization process. For example, 30 at. % of the deviation from $Ge_2Sb_2Te_5$ stoichiometric composition increased the crystallization time from 220 to 500 ns (Kyrsta et al., 2001).

For several years, Ge:Sb:Te alloys along the pseudobinary $GeTe–Sb_2Te_3$ line have been used as the main material for optical phase-change data storage devices and are currently being investigated for nonvolatile electronic storage purposes (which utilize the difference in the electrical resistance of the two phases) due to the high reflectivity and resistivity contrast between the amorphous and crystalline phases. Many investigations have been carried out on this type of phase-change materials with the purpose of improving their properties and to increase its storage capacity, stability, speed and versatility. These investigations have shown that the crystallization of the phase-change (or amorphous-NaCl-type transition) can be considered as a rate-limiting process to obtain a fast data transfer. That is why there are a considerable number of experimental and theoretical studies investigating the amorphous-to-crystalline NaCl-type phase transformation. In spite of the large number of publications, the activation energy for crystallization of Ge:Sb:Te materials reported in the literature shows a large discrepancy; the activation energy of crystallization for $Ge_2Sb_2Te_5$ reported in the literature varied between 0.8 and 2.9 eV, for example. Such dispersion in the kinetic parameters might depend on the differences in the deposition methods, type of substrate, dielectric cover layer, film thickness and/or parameters of deposition processes, which lead to differences in the crystallization temperature, activation energy, and the like. In addition, experimental data show relatively long incubation times for crystallization of Ge:Sb:Te films and a relatively large amount of crystallized material during this period of time (Morales-Sanchez et al 2010). Incubation time manifests itself as the time necessary to reach critical values of nucleus for the crystallization to occur (Senkader et al, 2004). The existence of a non-negligible amount of crystallite during the incubation time could also be responsible for the dispersion in the kinetic parameters reported in the literature.

Based on such antecedents, we aimed in this chapter at investigating the crystallization kinetics of thin films on a glass substrate obtained by the same deposition process (DC sputtering), with the same thickness (around 200 nm), and composition along the GeTe–

Sb_2Te_3 pseudobinary line ($Ge_2Sb_2Te_5$ $Ge_1Sb_2Te_4$ and $Ge_1Sb_4Te_7$) including GeTe and Sb_2Te_3 (which lie at the end of GeTe-Sb_2Te_3 pseudobinary line), taking into account the processes during the incubation time. It is necessary to note that GeTe and Sb_2Te_3, as well as Ge:Sb:Te, can be used in phase-change data storage. Additionally, the crystallization kinetics of $Ge_4Sb_1Te_5$ films will be present in this chapter. The structure of this alloy is unclear. In contrast to other Ge:Sb:Te ternary compounds, it does not belong to the GeTe - Sb_2Te_3 homologous series, although it lies on (or very close to) the GeTe-Sb_2Te_3 pseudobinary line in the Ge:Sb:Te phase diagram between GeTe and $Ge_2Sb_2Te_5$. It is suggested (Coombs et al., 1995) that this alloy is a solid solution between GeTe and a compound with a composition close to $Ge_4Sb_1Te_5$. This material, in contrast to stoichiometric alloys, demonstrates only one amorphous-NaCl-type phase transition, but it has the largest optical (Kato et al., 1999) and electrical (Morales-Sanchez et al., 2005) contrast between the crystalline and amorphous states, compared with other Ge:Sb:Te ternary alloys.

Necessary to note that as-prepared and melt-quenched Ge:Sb:Te amorphous materials show different crystallization kinetics (Nobukuni et al., 1999, Park et al., 1999, Khulbe et al., 2000, Wei et al., 2003, Kalb et al., 2004, Raoux et al., 2008). It has been proposed in the literature that for melt-quenched amorphous materials may exist: embryos following the condensation and evaporation of embryos to form crystalline clusters (Khulbe et al., 2000), preexisting clusters (Park et al., 1999), crystal nuclei (Wei et al., 2003), sinks and voids after repeated overwriting (Nobukuni et al., 1999), locally ordered regions with structure similar to that of crystalline Sb (Naito et al., 2004) or a crystalline amorphous border (Raoux et al., 2008). In spite of different models, crystallization in the melt-quenched amorphous material can start from these nucleation centers, which decrease incubation time and increase crystallization speed. But in this chapter only crystallization of as-prepared materials will be discussed.

Studies of crystallization kinetics of phase change materials are mostly analyzed using the Johnson-Mehl-Avrami Kolmogorov (JMAK) model for isothermal annealing (see for example: (Weidenhof et al., 2001, Ruitenberg et al., 2002, Trappe et al., 2000, Morales-Sanchez et al., 2010)), which permitted to determine the activation energy for the crystallization process. This method will be used for investigation of the crystallization kinetics in the compositions mentioned above.

Figure 1 shows in-situ optical reflection (using a laser diode emitting at 650 nm) as a function of temperature with a heating rate of $5^0C/min$ for all investigated films with composition indicated on the graph. An abrupt change in reflection is associated with the onset of phase crystallization. Stoichiometric $Ge_2Sb_2Te_5$, $Ge_1Sb_2Te_4$ and $Ge_1Sb_4Te_7$ materials demonstrate two changes: the first corresponds to amorphous-NaCl-type transition and the second to NaCl-type-hexagonal transformation. In contrast, GeTe, $Ge_4Sb_1Te_5$ and Sb_2Te_3 show only one amorphous-crystalline transition. The highest crystallization temperature (T_c) has been observed in GeTe, the lowest in Sb_2Te_3. The crystallization kinetics of the GeTe-Sb_2Te_3 pseudobinary line will be analyzed according to this Figure from the highest to the lowest crystallization temperature.

2. Deposition of phase change films

An important aspect to develop thin films of phase change materials is the deposition method. It is well known from the literature that deposition technology strongly affects the microstructure and as a results the optical, electrical and crystallization (especially the crystallization temperature) properties of phase-change materials. For example, the crystallization temperature of the films with the same composition but obtained by different deposition methods can differ for more than 20ºC, and no suitable explanation has yet been proposed (Morales-Sanchez et al., 2005). That is why, to compare crystallization properties all studied films have to be obtained with the same deposition process (DC sputtering), with the same thickness (around 200 nm), and composition along the GeTe–Sb_2Te_3 pseudobinary line. This method has been chosen because sputtering can produce films with the desired stoichiometry. Compared, for example, with thermal evaporation, in sputtering the thickness and chemical composition of thin films are easier to control.

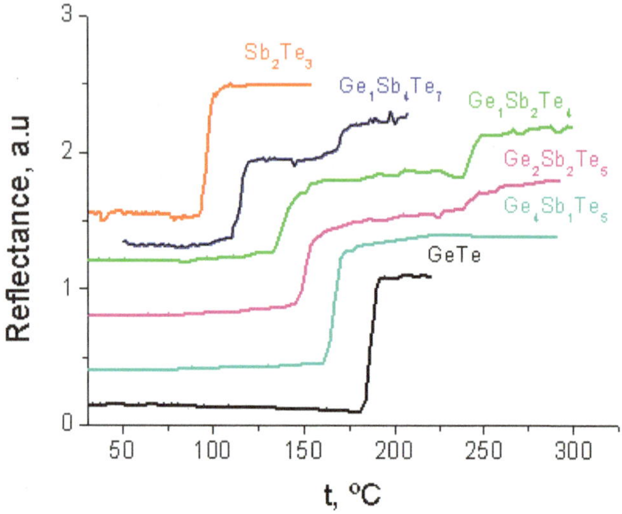

Fig. 1. Dependencies of optical reflection on temperature for films with compositions indicated on the graph.

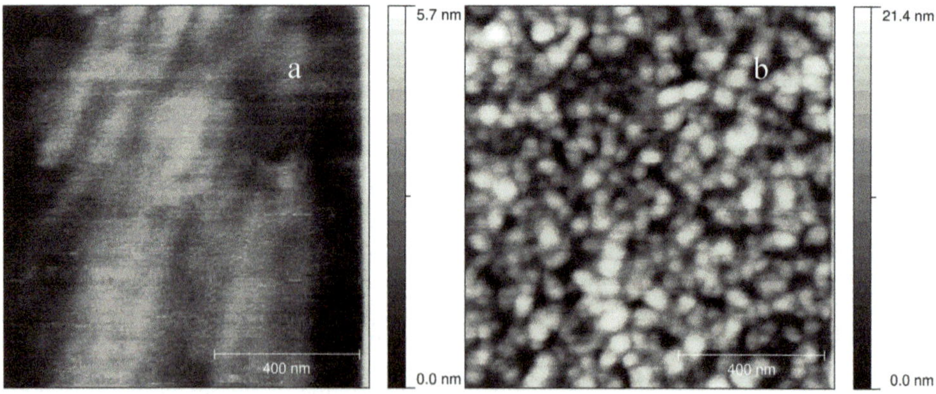

Fig. 2. Surface topography of as-prepare amorphous (a) and crystalline (b) $Ge_2Sb_2Te_5$ films.

The thin films have been prepared on glass and Si substrates by DC magnetron sputtering from one composite target. The deposition conditions were: $4.5x10^{-6}$ mbar base pressure, $2.0x10^{-5}$ mbar work pressure, 180 cm^3/min Ar gas input flow, 0.49 W/cm^2 DC power density, and 10 min deposition time. Before films sputtering, the targets were cleaned for 10 minutes by sputtering to remove the oxide from the surface. According to XRD measurements all films were obtained in amorphous state. Energy dispersive spectroscopy has shown that deviations of films compositions from the targets were approximately 2%.

Figure 2a shows typical surface topography obtained from AFM measurements of as-prepared amorphous thin films. All amorphous films demonstrated vary smooth surface with the surface roughness between 1 and 3 nm. In contrast, in the crystalline films (annealed to the temperature of 200^{0}C) the topography image shows grains with average size between 20-120 nm (Fig. 2b). The dimensions of grains are dependent on the films composition.

3. Crystallization of GeTe

GeTe is a promising candidate for the application of in phase-change technology due to its higher crystallization temperature (onset of crystallization temperature at about 184^{0}C) when compared to other alloys, since it offers a significant improvement in data-retention at high temperature (Perniola et al., 2010, Fantini et al., 2010). Amorphous films, in contrast to Ge:Sb:Te ternary alloys, crystallize in the rhombohedral phase. This phase can be viewed as a rock salt structure distorted along the [111]-direction (Caravati et al., 2010). Additionally, the central atom is displaced along the [111]-direction from the center of the rhombohedron. Upon crystallization, in a rhombohedral (distorted rock salt) structure appeared about 10% of vacancies occurring on Ge sites (Kolobov 2004). GeTe is a classical ferroelectric material that shows displacive type ferroelectric-paraelectric transition from a rhombohedral ferroelectric phase to a rock salt type structure with paraelectric properties. Two sublattices form this rock salt structure: Ge atoms compose one of the lattices and tellurium atoms the other (Rabe et al., 1987).

The NaCl type crystalline structure of GeTe is an unstable phase. This crystalline structure exists for temperatures above 300°C. The ferroelectric-paraelectric transition occurs in the interval of 327-427^{0}C depending on various factors, such as exact composition, carrier density, etc. For example, changes in Te composition from 50 to 50.7 at. % shift the transition temperature from 427^{0}C to 356^{0}C (Okura, 1992).

Activation energy of amorphous to rhombohedral phase crystallization reported in the literature demonstrates large dispersion: 1.7 eV (Libera et al., 1993), 1.77 eV (Lu et al., 1995), 1.96 eV (Fantini et al., 2010), 2.5 eV (Fan et al., 2004), 3.9 eV (Matsushita et al., 1989). Such dispersion can be dependent on the differences in the deposition methods, type of substrate, method of results interpretation, etc. According to the literature and to our measurements (Fig. 1), GeTe has the highest crystallization temperature for materials on the GeTe-Sb_2Te_3 pseudobinary line and must demonstrate the highest activation energy of crystallization.

Figure 3 shows X-ray patterns of GeTe film measured at the temperature indicated on the graph. At temperatures below 180^{0}C, the material shows only wide bands, which are characteristic of amorphous materials. At higher temperatures, a rhombohedral GeTe phase appears. At a temperature of 350^{0}C two phases have been observed: rhombohedral and

NaCl-type, and at higher temperatures the material transforms into a NaCl-type phase. This phase exists only at temperatures above 400^{0}C and in the process of cooling, again transforms into the stable rhombohedral phase.

Crystallization kinetics in GeTe, as in other materials, has been investigated using optical reflection during isothermal measurements. Reflection measurements were made with a 650 nm wavelength laser diode and a PIN diode as detector. In the reflectance measurements, the generally employed assumption has been used; the reflection signals are linearly related to the total transformed crystalline volume fractions x (Weidenhof et al., 2001):

$$x = [R(t) - Ra] / [Rc - Ra], \quad (1)$$

where R(t), the experimental measurement value of reflectance, Ra and Rc are reflectance of amorphous and fully crystalline phases, respectively.

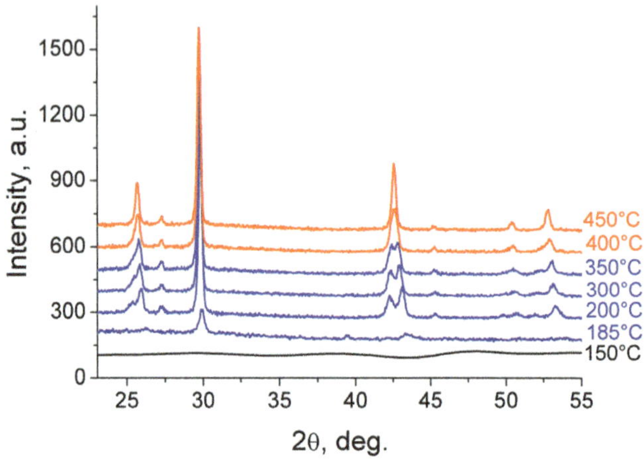

Fig. 3. XRD patterns of GeTe film measured at the temperature indicated on the graph. Blue patterns correspond to rhombohedral phase. Red patterns correspond to NaCl-type phase.

Figure 4 shows the evolution of the total crystalline volume fraction x, for GeTe samples, calculated from reflection measurements for films isothermally annealed at the temperatures indicated on the graph.

In this chapter, crystallization kinetics will be analyzed using the Johnson-Mehl-Avrami-Kolmogorov (JMAK) model for isothermal annealing. According to the classical JMAK model, the transformed volume fraction x can be determined by the following expression:

$$x(t) = 1 - \exp(-Kt^n), \quad (2)$$

where $K=\gamma exp(-E/kT)$, γ is the frequency factor, E is the effective activation energy, t is the annealing time, and n the Avrami exponent, which provides information about the mechanisms of crystallization. The value of n can be evaluated from the slope of the $ln[-ln(1- x)]$ versus $ln(t)$ plot, which in materials with random nucleation and isotropic growth should be linear. The JMAK model is based on several assumptions such as:

homogeneity of the system, random and uniform nucleation, and nucleation rate taking place at the very beginning of the transformation is time independent and applied for isotropic growth at a constant rate.

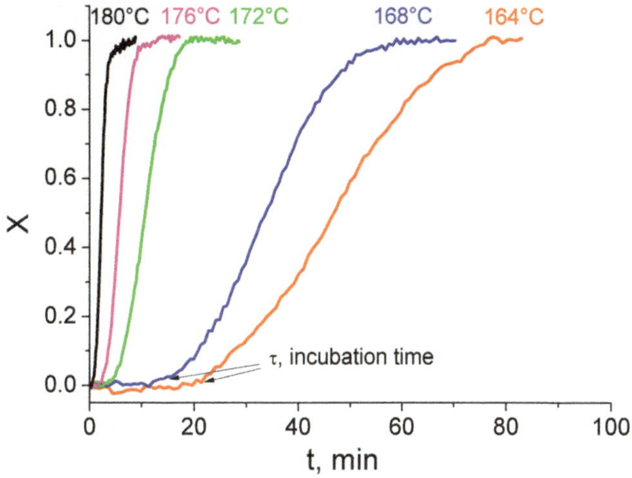

Fig. 4. Dependencies of the volume fraction x versus time obtained for GeTe samples from reflection measurements at the temperature indicated on the graph.

The value of activation energy for GeTe obtained using Equation (2) was very high: close to 12 eV. Such high value of activation energy has been obtained because GeTe, as other Ge:Sb:Te materials, shows long incubation times τ for crystallization, namely, the annealing time required to reach a critical nuclei size or to observe an abrupt increase in the crystalline volume fraction. In this case, the nucleation rate cannot be considered to be time independent for the entire crystallization process as is considered in classical JMAK models (Senkader et al., 2004).

In the case of GeTe, the volume fraction of crystalline phase approximately equals zero during all the incubation time (see Fig. 4). In such cases than the volume fraction of crystalline phase does not demonstrate a substantial increase during the incubation time it is possible to define the beginning of the transformation after the incubation time (Weidenhof et al., 2001). The JMAK equation can now be expressed as:

$$x(t) = 1 - \exp[-K(t-\tau)^n]\,, \tag{3}$$

where τ is the incubation time. According to Equation (3) the plot $ln[-ln(1- x)]$ versus $ln(t-\tau)$ must be a straight line with slope n. Figure 5 shows such modification of an Avrami plot for GeTe. The insert shows the rate constant as a function of the reciprocal temperature.

The calculated activation energy for GeTe as determined from the modification of the Avrami Equation (3) using the Arrhenius type relation for K was 3.98±0.12 eV. The value of the Avrami exponent was about 1.6, which corresponds to diffusion controlled growth from small dimension grains with decreasing nucleation rate (Christian, 1975).

In addition, X-ray measurements have shown that during isothermal annealing from amorphous phase only one rhombohedral phase appeared. As time increases, the volume fraction of amorphous phase decreases (Figure 6); in the same time, the volume fraction of crystalline phase increases. No shift in the position of diffraction peaks was observed. This observation will be important in the analysis of crystallization processes of $Ge_2Sb_2Te_5$ and $Ge_1Sb_2Te_4$ materials.

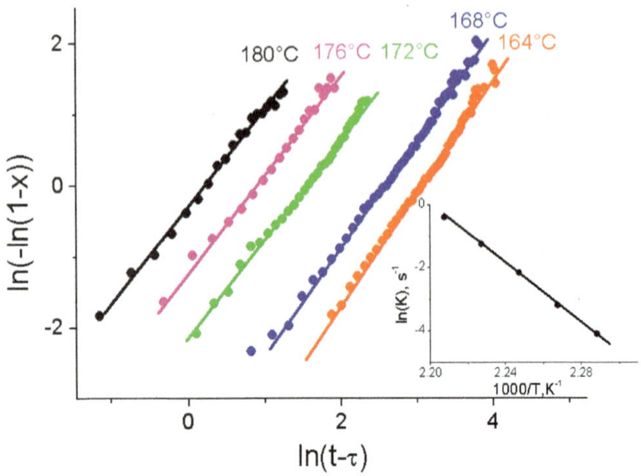

Fig. 5. Avrami plot of *ln[−ln(1−x)]* vs *ln(t−τ)* for GeTe. Points - experiment, lines - result of fitting using JMAK equation (3). The insert shows the rate constant *K* as function of the reciprocal temperature.

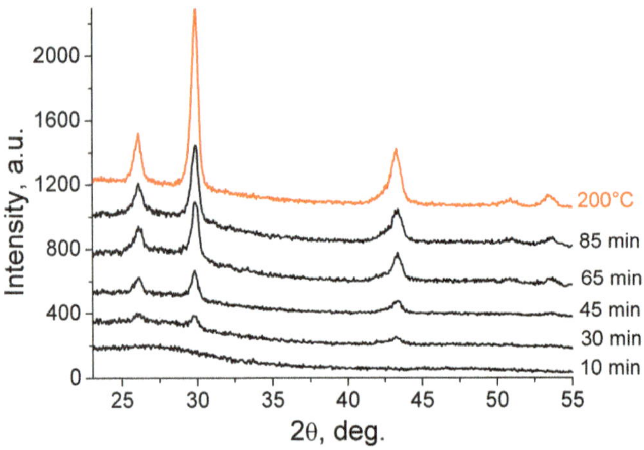

Fig. 6. X-ray diffraction spectra for a GeTe film obtained in the process of isothermal annealing at a temperature of 164^{0}C during the time indicated on the plot. The upper pattern corresponds to film annealing at 200^{0}C.

4. Crystallization of $Ge_4Sb_1Te_5$

As mentioned above, the structure of $Ge_4Sb_1Te_5$ is not known. It lies on or is close to the GeTe-Sb_2Te_3 pseudobinary line in the Ge:Sb:Te phase diagram between GeTe and $Ge_2Sb_2Te_5$. Because of such position, the onset of crystallization temperature is about 160⁰C, less than in GeTe but higher than in another Ge:Sb:Te ternary alloys. This alloy crystallizes in NaCl-type structure (Gonzalez-Hernandez et al., 1992, Wamwangi et al., 2002, Ruiz Santos et al., 2010). For $Ge_4Sb_1Te_5$ films the reported values of the activation energy varied in a wide range: 1.13 eV (Kato et al., 1999), 3.09 eV (Morales-Sanchez et al., 2005), 3.48 eV (Wamwangi et al., 2002).

In addition, similarly to GeTe, bulk material and thin $Ge_4Sb_1Te_5$ films demonstrate at a temperature of approximately 327⁰C ferroelectric-paraelectric transition (Bahgat et al., 2004, Ruiz Santos et al., 2010). But in contrast to GeTe, in which ferroelectric-paraelectric transition relates to a transformation from a rhombohedral phase to a rock salt type structure, in $Ge_4Sb_1Te_5$ the ferroelectric-paraelectric transition relates to a transformation between two different rock salt type structures.

Figure 7 shows DSC measurements on a crystalline $Ge_4Sb_1Te_5$ bulk alloy and on the amorphous material obtained by removing the as-deposited films from the glass substrates. In the bulk alloys, a nonsymmetrical endothermic peak at a temperature of 329⁰C has been observed. In contrast, DSC measurements on amorphous material showed an exothermic peak at 176⁰C, which is associated with the crystallization of the sample and a broad endothermic peak, with a maximum at approximately 327⁰C, and that corresponds to the same temperature as in the bulk alloy. The observed endothermic peak can be associated with the ferroelectric-paraelectric transition.

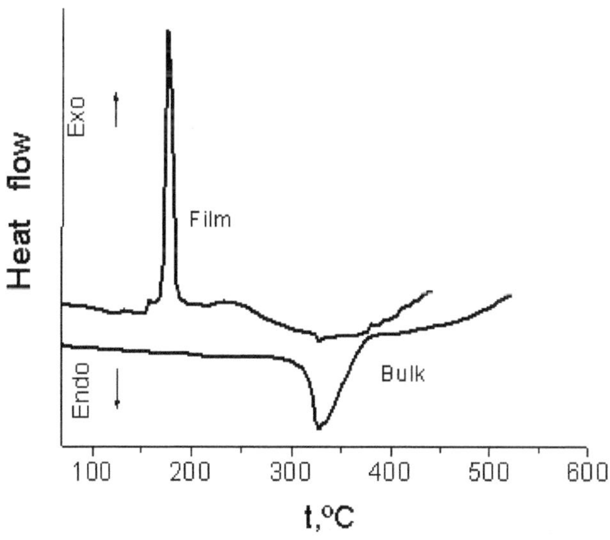

Fig. 7. DSC measurements on a bulk crystalline alloy and on the amorphous material, obtained by removing the as-deposited films from the glass substrates.

Additional confirmation of ferroelectric-paraelectric transition in $Ge_4Sb_1Te_5$ films has been obtained by temperature capacitance measurements (Ruiz Santos et al., 2010). In this work, it has been shown that the reciprocal capacitances (which are proportional to the reciprocal dielectric constant) as a function of temperature show a typical Curie-Weiss behavior for temperatures above 327°C.

Specific data of crystallization process and ferroelectric-paraelectric transition in $Ge_4Sb_1Te_5$ films can be obtained by in situ XRD measurements at different temperatures. Figure 8 shows that below the crystallization temperature the film is in an amorphous state (pattern at 140°C). As the temperature is increased, the material crystallizes in the NaCl type structure. As the temperature continue increasing, the position of all $Ge_4Sb_1Te_5$ peaks shift to lower 2θ values, reaching saturation at a temperature of approximately 327°C at positions close to the NaCl type of GeTe structure (ICSD cart #602124), which demonstrate paraelectric properties. The position of the NaCl type of GeTe is marked on the graph with the horizontal line. This shift can be clearly seen on the insert of Figure 8, which shows the temperature dependence of the (200) diffraction peak in $Ge_4Sb_1Te_5$. The horizontal line on the insert indicates the position of the (200) peak in crystalline GeTe.

Figure 9 shows the evolution of the total crystalline volume fraction x, for $Ge_4Sb_1Te_5$ samples, calculated from reflection measurements for films isothermally annealed at temperatures indicated on the graph. As in the case of GeTe, the volume fraction of crystalline phase of $Ge_4Sb_1Te_5$ is approximately equal to zero during all the incubation time. The same as for GeTe, kinetics parameters for $Ge_4Sb_1Te_5$ can be calculated using Eq. (3).

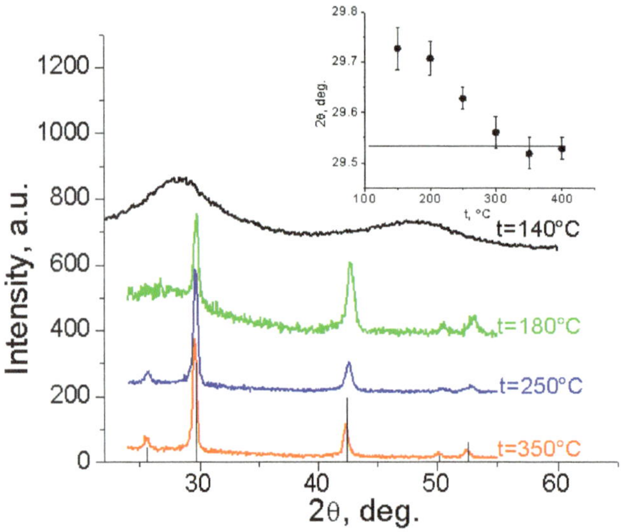

Fig. 8. XRD patterns of $Ge_4Sb_1Te_5$ film measured at the temperature indicated on the graph. The insert shows temperature dependence of (200) diffraction line in $Ge_4Sb_1Te_5$.

Figure 10 shows the modification of an Avrami plot for $Ge_4Sb_1Te_5$. The insert shows the rate constant as a function of the reciprocal temperature. The effective activation energy, determined from the Avrami plot for different temperatures using the Arrhenius type relation for *K*, was 3.46±0.22 eV (insert in Fig. 10). This value is in good agreement with other values reported in the literature (3.48 eV±0.12 eV) obtained using the Kissinger analysis (Wamwangi et al., 2002). The Avrami exponent for $Ge_4Sb_1Te_5$ was close to 1.8, which corresponds to a diffusion controlled growth from small dimension grains with decreasing nucleation rate, an exponent similar to that of GeTe (Christian 1975).

Additional X-ray measurements have shown that during isothermal annealing from amorphous phase only one NaCl-type phase appeared.

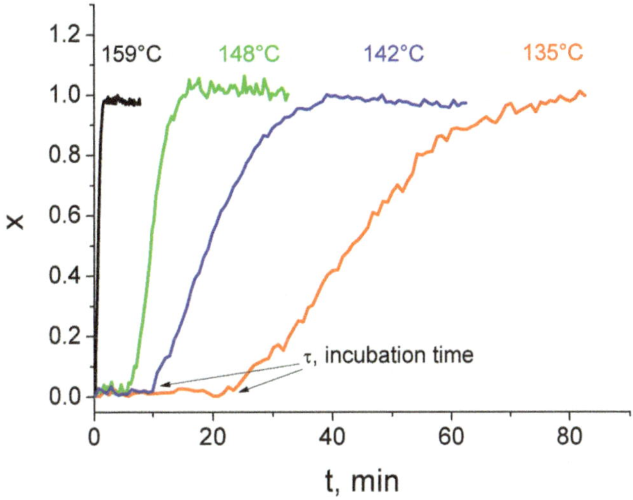

Fig. 9. Dependencies of the volume fraction *x* versus time obtained for $Ge_4Sb_1Te_5$ samples from reflection measurements at the temperature indicated on the graph.

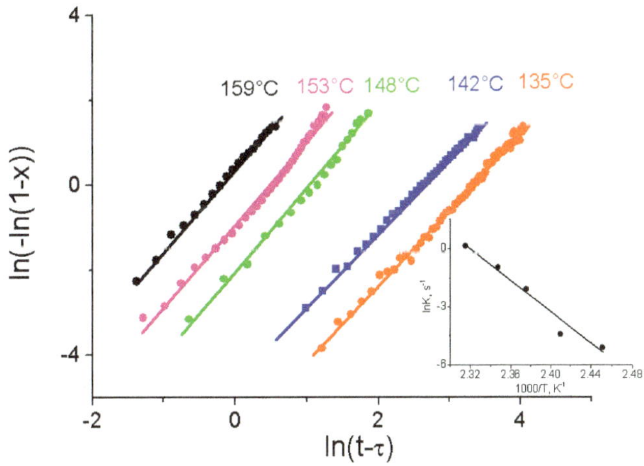

Fig. 10. Avrami plot of *ln[−ln(1−x)]* vs *ln(t−τ)* for $Ge_4Sb_1Te_5$ films. Points - experiment, lines - results of fitting using JMAK equation (3). The insert shows the rate constant *K* as function of the reciprocal temperature.

5. Crystallization of $Ge_2Sb_2Te_5$

$Ge_2Sb_2Te_5$ thin films are the most commonly employed materials for phase-change memory technology application due to its high crystallization speed and relatively high crystallization temperature (but less than in GeTe and $Ge_4Sb_1Te_5$), which lead to high thermal stability. Because of this, many extensive experimental and theoretical studies have been conducted to understand the structure and crystallization phenomena in this material (see for example (Yamada et al, 1991, Weidenhof et al., 2001, Matsunaga et al., 2004, Matsunaga et al., 2006, Paesler et al., 2007, Im et al., 2008, Claudio et al 2006)).

As mentioned above, $Ge_2Sb_2Te_5$ material demonstrates two phase changes: first at a temperature close to 145^{0}C there is an amorphous-NaCl-type transition and second (at about 240^{0}C) a NaCl-type-hexagonal transformation.

Under heating, the amorphous $Ge_2Sb_2Te_5$ films crystallize at around 130-170^{0}C, depending on the preparation method and heating rate, into a phase with a NaCl-type structure ($Fm\bar{3}m$). In this structure, the 4(a) site is fully occupied by Te atoms, whereas the 4(b) site is randomly occupied by Ge and Sb atoms and vacancies. The composition of Ge:Sb:Te ternary system, which lie on the GeTe-Sb_2Te_3 pseudobinary line, can be described as $(GeTe)_x$+ $(Sb_2Te_3)_{1-x}$ ($0<x<1$). In such case the site occupancy of the vacancy varies continuously according to $(1-x)/(3-2x)$ (Matsunaga et al., 2004, Matsunaga et al., 2006). But Ge and Sb atoms deviate from the ideal rock-salt positions in $Ge_2Sb_2Te_5$ not in a random way but in a strongly correlated manner with respect to the neighboring Te atoms (Kolobov et al., 2004, Kolobov et al., 2006). The off-center location of Ge atoms means that there is a net dipole moment and suggests that a NaCl-type phase of the $Ge_2Sb_2Te_5$ is a ferroelectric material (Tominaga et al., 2004). The ferroelectric properties in $Ge_2Sb_2Te_5$ NaCl-type phase have been observed using capacitance-temperature measurements. The temperature dependence of the capacitance shows an abrupt change with a maximum at the temperature that corresponds to the end from a NaCl-type to a hexagonal transition. In addition, the reciprocal capacitance for temperatures above this transition shows the Curie–Weiss dependence, which is typical of ferroelectric materials (Gervacio Arciniega et al., 2010).

In spite of the large number of publications about crystallization phenomena in $Ge_2Sb_2Te_5$, the activation energy of crystallization reported in the literature shows a large discrepancy in the range between 0.8 and 2.9 eV and between 1.2 and 4.4 for the Avrami exponent (Morales-Sanchez et al., 2010, Liu et al., 2009). Such parameter dispersion can be related not only to the difference in the preparation methods but to the specific crystallization process.

Experimental data in $Ge_2Sb_2Te_5$ films, as in all materials along the pseudobinary GeTe-Sb_2Te_3 line, show relatively long incubation times during isothermal crystallization (Weidenhof et al., 2001, Laine et al., 2004, Zhang et al., 2008, Morales-Sanchez et al., 2010). But in GeTe and $Ge_4Sb_1Te_5$ the amount of crystallized material is close to zero during the incubation. In contrast, in $Ge_2Sb_2Te_5$ films large amount of crystallized material has been observed during this period of time (ranging from 50% in Ref. (Zhang et al., 2008), 10% in Ref. (Sian et al., 2008), 9% in Ref. (Laine et al., 2004) and 8% in Ref. (Weidenhof et al., 2001)). It is necessary to note that the amount of crystalline phase material during incubation time depends on the annealing temperature. The existence of a non-negligible amount of crystallite during the incubation time, as will be shown below, is the most important factor responsible for the dispersion in the kinetic parameters reported in the literature.

Figure 11 shows the evolution of the total crystalline volume fraction x, for $Ge_2Sb_2Te_5$ samples, calculated from reflection measurements for films isothermally annealed at the temperatures indicated on the graph. In contrast to GeTe and $Ge_4Sb_1Te_5$, the volume fraction of crystalline phase of $Ge_2Sb_2Te_5$ does not equal zero during incubation time.

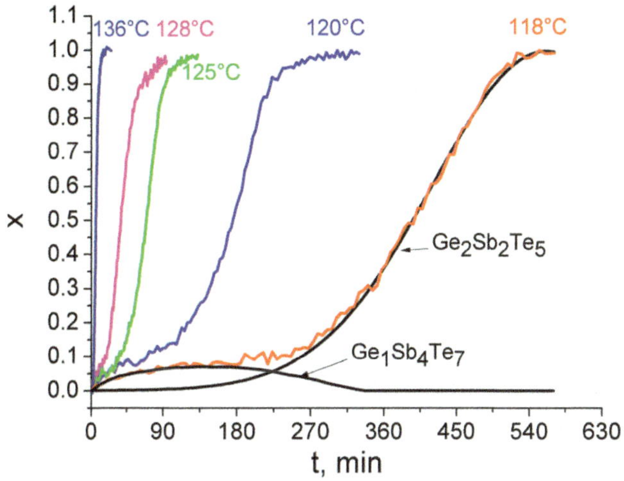

Fig. 11. Dependencies of the volume fraction x versus time obtained for $Ge_2Sb_2Te_5$ samples from reflection measurements at the temperature indicated on the graph. Black curves results from fittings using equation (7).

A typical JMAK plot for $Ge_2Sb_2Te_5$ films is shown in Figure 12. If we evaluate the slope only on a linear behavior, which is observed after incubation time τ, using a classical equation (2), the effective activation energy and the Avrami exponent will be equal to 5.61 eV and 3.3, respectively. The evaluation using a modification of a JMAK equation (3), which takes into account the incubation time, renders values for effective activation energy and an Avrami exponent equal to 2.07 eV and 1.5, respectively. This value of activation energy correlates well with what is reported in the literature (2.0 eV) and is obtained from the same interpretation of the reflection measurements (Weidenhof et al., 2001). Thus what comes into question is the pertinence of using these two approximations.

As has been mentioned above, the JMAK model is based on the following assumptions: the system is homogeneous, nucleation is random and uniform, the nucleation rate is time independent and takes place at the very beginning of the transformation, the isotropic growth maintains a constant rate, etc. (Henderson, 1979). However, nucleation in $Ge_2Sb_2Te_5$ films is not random nor uniform (Senkader et al., 2004) and due to partial crystallization during incubation time the nucleation rate cannot be considered to be time independent for the entire crystallization process as is considered in the classical model (equation 2). To overcome this limitation it has been proposed to avoid using a modification of the JMAK model (equation 3): after incubation time the rate constant can be considered as independent in time (Weidenhof et al., 2001). But in the case of $Ge_2Sb_2Te_5$ a large amount of crystallized material has been observed during incubation time. Moreover, experimental investigation has shown that nuclei, which appeared in $Ge_2Sb_2Te_5$ films during incubation time, have a composition corresponding to $Ge_1Sb_4Te_7$ which differs from the nominal value for the amorphous matrix (Laine et al., 2004, Claudio et al., 2006). Figure 13 shows X-ray diffraction patterns for $Ge_2Sb_2Te_5$ film isothermal annealing at the temperature 118^{0}C and times indicated on the graph. The patterns of samples annealed during 50, 100, and 200 minutes show weak peaks, with positions corresponding to $Ge_1Sb_4Te_7$ NaCl-type phase. After annealing for more than 200 min, the positions of the peaks begin to change to those

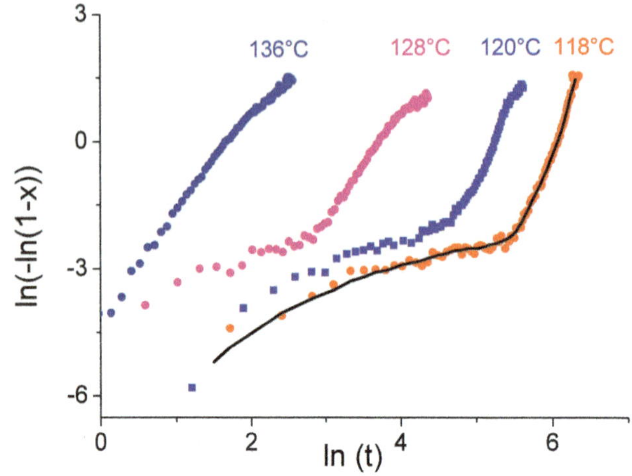

Fig. 12. Avrami plot of *ln[−ln(1−x)]* vs *ln(t)* for $Ge_2Sb_2Te_5$ films. Points - experiment, black line - results of fitting using equation (7).

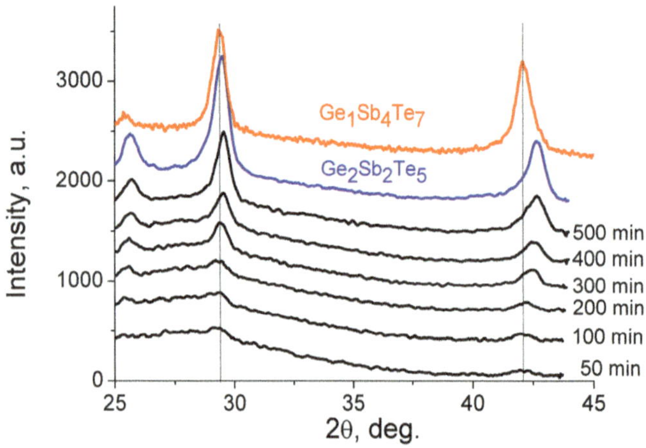

Fig. 13. X-ray diffraction spectra for a $Ge_2Sb_2Te_5$ film obtained in the process of isothermal annealing at a temperature of 118^{0}C during the time indicated in the plot. Upper patterns correspond to $Ge_1Sb_4Te_7$ (red curve) and $Ge_2Sb_2Te_5$ (blue curve) taken at 180^{0}C. The positions of the $Ge_1Sb_4Te_7$ peaks are indicated with the vertical lines.

corresponding to the NaCl-type phase of $Ge_2Sb_2Te_5$ composition. In this figure the blue pattern corresponds to material annealed at 180⁰C; the upper pattern corresponds to $Ge_1Sb_4Te_7$ film annealed at 180⁰C. Thus, in $Ge_2Sb_2Te_5$ films the formation of a stable rock salt type crystalline phase is preceded by the formation of a metastable $Ge_1Sb_4Te_7$ phase.

On the basis of these results, the crystallization of $Ge_2Sb_2Te_5$ material during isothermal annealing can be considered as a process which takes place in two stages: in the first stage nuclei of a metastable $Ge_1Sb_4Te_7$ phase appear; in the second stage, the nuclei transform into the equilibrium NaCl-type stoichiometric $Ge_2Sb_2Te_5$ structures.

In materials in which phase transformation starts with the formation of metastable phases the isothermal crystallization process cannot be simply described by JMAK theory and the plots of $ln(-ln(1 - x))$ versus $ln(t)$ are not linear, which implies that the Avrami exponent n does not remain constant during the crystallization process.

In ref. (Claudio et al., 2006) an analytical model that can describe the isothermal crystallization process of materials is proposed. In this process, the stable crystalline phase is preceded by the formation of a metastable phase. The model assumes that the volume fraction of metastable phase f_m grows up to a maximum value f_{max} and then stops growing when the stable phase that has nucleated into it overpasses the metastable grain boundaries. The kinetics of the metastable fraction can be represented by a JMAK-type equation subtracting the fraction of stable phase f_{sm} which grows inside the metastable phase:

$$f_m(t) = f_{\max}(1 - \exp(-K_m t^{n_m})) - f_{sm}(t), \tag{4}$$

where K_m is the crystallization rate constant and n_m the Avrami exponent.

The kinetic behavior of the stable phase into a metastable phase can be represented by the modification of classical JMAK formula, which takes into account the incubation time t_{sm} of the stable phase into the metastable phase:

$$\begin{gathered} f_{sm}(t) = f_{\max}(1 - \exp(-K_{sm}(t - t_{sm})^{n_{sm}})) \quad \text{for } t \geq t_{sm} \\ \text{and} \quad f_{sm}(t) = 0 \text{ for } t < t_{sm} \end{gathered} \tag{5}$$

where K_{sm} and n_{sm} are the crystallization rate constant and the Avrami exponent, respectively, which represent the kinetic parameters of the transformation from stable phase into metastable phase.

Furthermore, it is possible to propose that the stable phase can grow into the amorphous phase. The stable phase transformation kinetics in amorphous phase can be described by JMAK-type equation with its specific parameters:

$$\begin{gathered} f_{sa}(t) = (1 - f_{\max})(1 - \exp(-K_{sa}(t - t_{sa})^{n_{sa}})) \text{ for } t \geq t_{sa} \\ \text{and} \quad f_{sa}(t) = 0 \text{ for } t < t_{sa} \end{gathered} \tag{6}$$

where f_{sa} is the fraction of the stable phase in amorphous phase (from 0 to $1-f_{max}$), K_{sa} and n_{sa} are the JMAK parameters and t_{sa} is the incubation time of stable phase in amorphous phase.

The total fraction of transformed material is given by the equation:

$$f_{total}(t) = f_m(t) + f_{sm}(t) + f_{sa}(t) \tag{7}$$

This model allows fitting the experimental transformation data obtained in $Ge_2Sb_2Te_5$ films using a genetic algorithm. The continuous lines on Fig. 11 show the evolution on time of the volume fractions of metastable $Ge_1Sb_4Te_7$ and stable $Ge_2Sb_2Te_5$ phase obtained from fitting using Equation (7). Results of simulations have shown that the volume fraction of metastable $Ge_1Sb_4Te_7$ phase grows up to a maximum value and then decreases and disappears. After some incubation time, the $Ge_2Sb_2Te_5$ phase began to grow and all material is transformed into a stable crystalline phase.

In addition, the model is capable of predicting the three slopes clearly shown in the JMAK plot (Fig. 12) corresponding to three distinguishable stages in the crystallization process observed at dependencies obtained at low annealing temperature: the first one related to the metastable transformation, the second one (low value) with the step between metastable and stable transformation, and the last one with the stable phase nucleation and growth. It is quite evident from these simulated results that the first slope of the JMAK plot can be related to the kinetic behavior of the metastable phase.

The results of the investigation have shown that during the isothermal process of amorphous to crystalline phase transformation in $Ge_2Sb_2Te_5$ film a metastable phase with composition $Ge_1Sb_4Te_7$ and a stable NaCl-type $Ge_2Sb_2Te_5$ phase coexist within a certain time range. The appearance of nuclei of the $Ge_1Sb_4Te_7$ composition could be related to the local fluctuation in the film composition and to the fact that the crystallization temperature of $Ge_1Sb_4Te_7$ is lower than in $Ge_2Sb_2Te_5$ material.

6. Crystallization of $Ge_1Sb_2Te_4$

$Ge_1Sb_2Te_4$, as with $Ge_2Sb_2Te_5$ film, during annealing crystallize at a temperature of approximately 133^{0}C (lower than the crystallization temperature of $Ge_2Sb_2Te_5$) into a NaCl-type structure ($Fm\bar{3}m$) (Matsunaga et al., 2004) and at a higher temperature (about 235^{0}C) into a hexagonal phase. Because of the lower crystallization temperature, if compared with $Ge_2Sb_2Te_5$, this material may possess a low programming current in random access memory.

Crystallization kinetics, the same as in $Ge_2Sb_2Te_5$ material, is sufficiently complicated. Figure 14 shows the dependence of the total crystalline volume fraction x versus time for $Ge_1Sb_2Te_4$ samples calculated from reflection measurements for films isothermally annealed at the temperatures indicated on the graph and Figure 15 shows a typical JMAK plot. If we evaluate the slope only on a linear behavior, which is observed after incubation time τ, using classical equation (2), the effective activation energy will be equal to 5.45 eV. The evaluation using modification of JMAK equation (3), which takes into account incubation time, renders values of effective activation energy equal to 1.77 eV. But in $Ge_1Sb_2Te_4$, the same as in $Ge_2Sb_2Te_5$ films, a large amount of crystallized material has been observed during the incubation time. Moreover, XRD measurements (Figure 16) have shown that nuclei, which appeared in $Ge_1Sb_2Te_4$ films during the incubation period, have a composition corresponding to the $Ge_1Sb_4Te_7$ structure. The upper XRD pattern in this Figure corresponds to a fully crystallized film with the $Ge_1Sb_4Te_7$ composition; the peak positions of this phase are marked with the vertical lines. The pattern mark as $Ge_1Sb_2Te_4$ shows diffraction lines of $Ge_1Sb_2Te_4$ film annealing to 180^{0}C.

The results obtained have shown that in $Ge_1Sb_2Te_4$, the same as in $Ge_2Sb_2Te_5$ film, during the isothermal process of amorphous to crystalline phase transformation, a metastable phase appeared with composition of $Ge_1Sb_4Te_7$. Thus, for the interpretation of the crystallization process in $Ge_1Sb_2Te_4$ an analytical model can be used in which the formation during crystallization of metastable phase is taken into account (Equation 7).

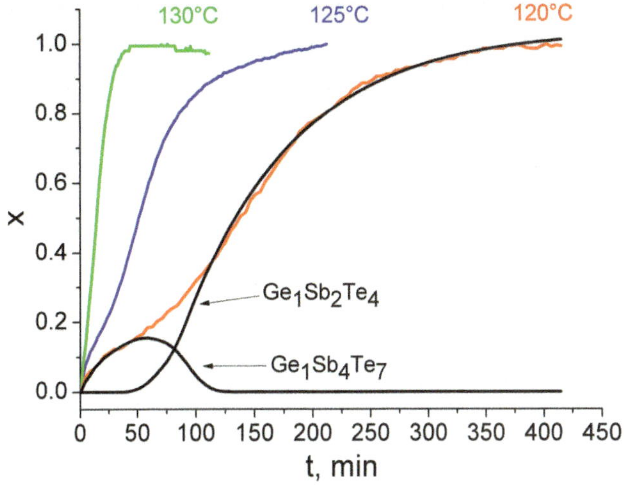

Fig. 14. Dependencies of the volume fraction x versus time obtained for $Ge_1Sb_2Te_4$ samples from reflection measurements at the temperature indicated on the graph. Black curves - results from fittings using equation (7).

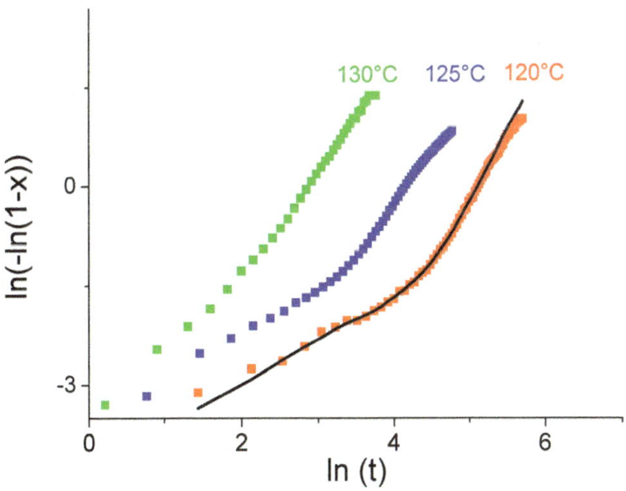

Fig. 15. Avrami plot of $ln[-ln(1-x)]$ vs $ln(t)$ for $Ge_1Sb_2Te_4$ films. Points - experiment, black curve - results from fitting using equation (7).

Continuous black curves on Fig. 14 show the calculated the volume fractions of metastable $Ge_1Sb_4Te_7$ and stable $Ge_1Sb_2Te_4$ phase obtained from fittings using Equation (7). Results of simulation have shown that the same as in $Ge_2Sb_2Te_5$, the volume fraction of metastable $Ge_1Sb_4Te_7$ phase grows up to a maximum value and then decreases and disappears. After some incubation time, $Ge_1Sb_2Te_4$ phase began to growth and all material was transformed into a stable crystalline phase.

Additionally, the model is capable to simulate the JMAK plot of Fig. 15, in which the first slope of the JMAK plot can be related to the kinetic behavior of the metastable $Ge_1Sb_4Te_7$ phase.

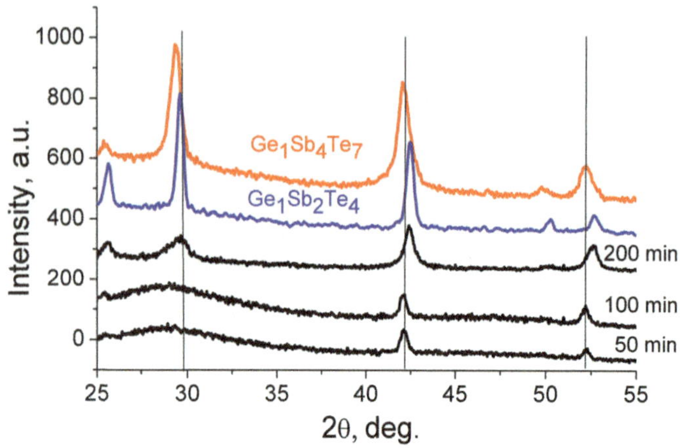

Fig. 16. X-ray diffraction spectra for a $Ge_1Sb_2Te_4$ film obtained in the process of isothermal annealing at a temperature of 120^{0}C during the time indicated on the plot. Upper patterns correspond to $Ge_1Sb_4Te_7$ (red curve) and $Ge_1Sb_2Te_4$ (blue curve) taken at 180^{0}C. The position of the $Ge_1Sb_4Te_7$ peaks is indicated with the vertical lines.

7. Crystallization of $Ge_1Sb_4Te_7$

Similarly, as with $Ge_2Sb_2Te_5$ and $Ge_1Sb_2Te_4$, $Ge_1Sb_4Te_7$ demonstrates a two phase transition: amorphous to the rock salt-like structure at a temperature of approximately 110^{0}C and a second transition to the hexagonal phase at the temperature close to 168^{0}C. This material demonstrates the fastest phase transformation among Ge:Sb:Te ternary alloys and phase-reversible transformations can be observed in the femtoseconds range (Huang et al., 2006). But because of the low crystallization temperature, $Ge_1Sb_4Te_7$ has lower thermal stability when compared with other ternary alloys (Miao et al., 2006).

Crystallization transformation from amorphous to NaCl-type structure in $Ge_1Sb_4Te_7$ differs from the transformations observed in $Ge_2Sb_2Te_5$ and $Ge_1Sb_2Te_4$ films. In contrast with $Ge_2Sb_2Te_5$ and $Ge_1Sb_2Te_4$, in-situ X-ray diffraction measurements show that amorphous films with the $Ge_1Sb_4Te_7$ composition crystallize in the NaCl-type phase with the same composition as at the beginning of the transformation (Figure 17). Moreover, isothermal reflection measurements (Figure 18) have shown practically zero value of crystalline phase during the much shorter incubation time compared with $Ge_2Sb_2Te_5$ films (Figure 11). In this case, as is for GeTe and $Ge_4Sb_1Te_5$, it is also possible to neglect the amount of crystallized material during τ, and describe the transformation using the modified JMAK equation (3).

Figure 19 shows a plot of *ln[-ln(1-x)]* versus *ln(t-τ)* for a $Ge_1Sb_4Te_7$ film, which demonstrates a linear dependence. This means that the crystallization process in $Ge_1Sb_4Te_7$ material displays random nucleation and isotropic growth with effective activation energy of 1.7±0.27 eV and an Avrami exponent n close to 1.94. According to (Christian, 1975), the values of n in the range $1.5 < n < 2.5$ correspond to a crystallization process dominated by all shapes growing from small dimensions with decreasing nucleation rate.

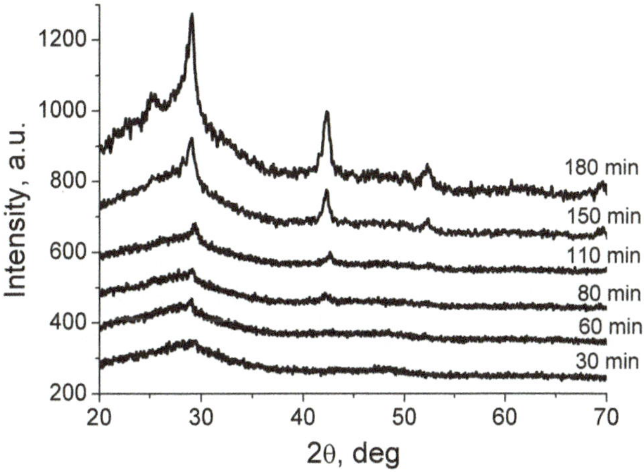

Fig. 17. X-ray diffraction spectra for a $Ge_1Sb_4Te_7$ film obtained in the process of isothermal annealing at a temperature of 90°C during the time indicated on the plot.

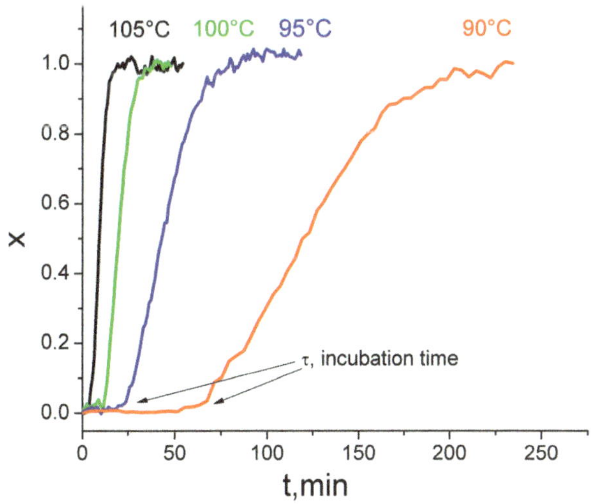

Fig. 18. Dependencies of the volume fraction x versus time obtained for $Ge_1Sb_4Te_7$ samples from reflection measurements at the temperature indicated on the graph.

8. Crystallization of Sb_2Te_3

Sb_2Te_3 lies at the end of the GeTe-Sb_2Te_3 pseudobinary line and has the lowest crystallization temperature but the highest crystallization speed. This material is an attractive candidate for phase change random access memory due to its rapid crystallization speed, but the crystallization temperature of Sb_2Te_3 (about 94°C in material investigated in this chapter and between 90-100°C reported in the literature (Yin et al., 2007, Kim et al., 2010)) is too low for it to be of practical use.

Upon annealing, amorphous Sb_2Te_3 films crystallize in rhombohedral Sb_2Te_3 (Kim et al., 2008, Lv et al., 2010) or fcc structure, which at a temperature above 200°C transforms into an hexagonal phase (Yin et al., 2007, Zhu et al., 2011). In the investigated Sb_2Te_3 films within the temperature range comprised between 95-200°C, only the rhombohedral phase (JCPDS data #15-0874) has been observed.

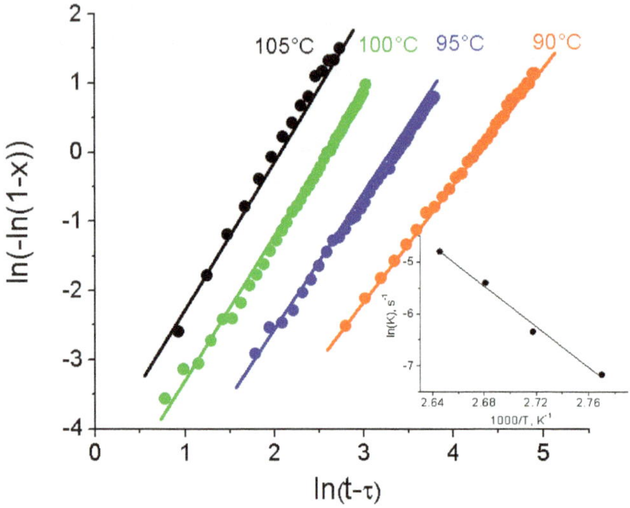

Fig. 19. Avrami plot of *ln[−ln(1−x)]* vs *ln(t−τ)* for $Ge_1Sb_4Te_7$ films. Points - experiment, lines - results from fittings using JMAK equation (3). Insert shows the rate constant *K* as a function of the reciprocal temperature.

Isothermal reflection measurements (Figure 20) have shown the same as for $Ge_1Sb_4Te_7$: practically zero value of crystalline phase during the incubation time. In this case, the same as for GeTe, $Ge_4Sb_1Te_5$ and $Ge_1Sb_4Te_7$ it is possible to neglect the amount of crystallized material during τ, and describe the transformation using the modified JMAK equation (3).

Figure 21 shows a plot of *ln[-ln(1-x)]* versus *ln(t-τ)* for a Sb_2Te_3 film, which demonstrates a linear dependence. This means that the crystallization process in Sb_2Te_3 films, the same as in $Ge_1Sb_4Te_7$ material, display random nucleation and isotropic growth with an effective activation energy of 1.54±0.15 eV and an Avrami exponent *n* close to 1.1, which according to ref. (Christian, 1975), corresponds to a crystallization growth of particles of appreciable initial volume. It is necessary to note that the values obtained for crystallization activation energy is in good agreement with values reported in the literature (1.51 eV) obtained using the Kissinger analysis in the films deposited on the glass substrates by the same DC magnetron sputtering (Zhai et al., 2009).

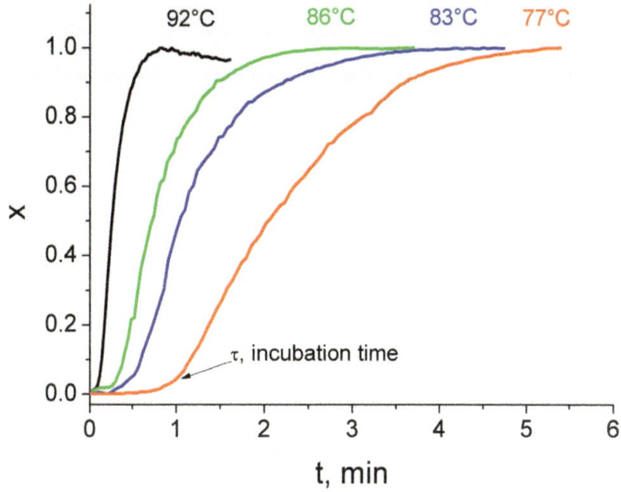

Fig. 20. Dependencies of the volume fraction x versus time obtained for Sb_2Te_3 samples from reflection measurements at the temperatures indicated on the graph.

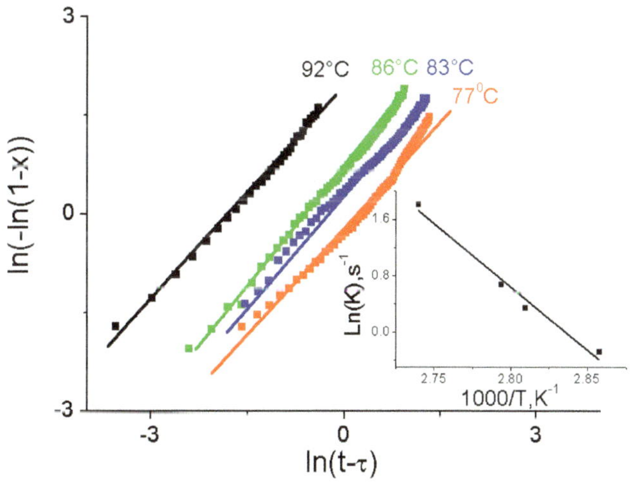

Fig. 21. Avrami plot of $ln[-ln(1-x)]$ vs $ln(t-\tau)$ for Sb_2Te_3 films. Points - experiment, lines - results from fittings using JMAK equation (3). Insert shows the rate constant K as function of the reciprocal temperature.

9. Conclusions

This chapter presents some of the results concerning the crystallization properties of the most commonly employed materials on the GeTe-Sb_2Te_3 pseudo-binary line for phase change memory application. Despite the fact that these alloys have found application in already commercialized optical disc and workable electrical phase change memory devices, their properties are not well known so far. This leads to many contradictory and controversial results in the literature.

As has been shown in this chapter, the crystallization properties of GeTe, $Ge_4Sb_1Te_5$, $Ge_1Sb_4Te_7$, and Sb_2Te_3 cannot be analyzed using the classical Johnson-Mehl-Avrami-Kolmogorov model for isothermal annealing due to the long incubation times. This limitation can be avoided by using a modification of the JMAK model, which takes into account that after incubation time the rate constant of crystallization can be considered as independent from time. Such model gave reasonable values of crystallization energy, which proved compatible with crystallization energy obtained by other methods. The difference in interpretation of isothermal measurements, in addition to the difference in preparation methods, can be responsible for the dispersion of crystallization parameters reported in the literature. But this modified model can be applied if during incubation time non-neglible amounts of crystalline phase material are observed. In the case of $Ge_2Sb_2Te_5$ and $Ge_1Sb_2Te_4$ films during incubation times, the nuclei fraction is of about 10% and has a composition corresponding to $Ge_1Sb_4Te_7$, which is different from the nominal value for the amorphous matrix. That is why the crystallization in $Ge_2Sb_2Te_5$ and $Ge_1Sb_2Te_4$ materials during isothermal annealing can be considered as a process that takes place in two stages: in the first stage, nuclei of a metastable $Ge_1Sb_4Te_7$ phase appear; in the second stage, the nuclei transforms into the equilibrium NaCl-type stoichiometric structures correspondent to the material composition. Such crystallization process can be described by a model in which the stable crystalline phase is preceded by the formation of a metastable phase. The most questionable issue is to explain these crystallization properties.

It is necessary to note that crystallization with phase separation has been observed in materials close to the GeTe–Sb_2Te_3 pseudo-binary line with excess of Ge ($Ge_{2+x}Sb_2Te_5$ x=0.5 (Privitera et al., 2003) and Sb ($Ge_2Sb_{2+x}Te_5$, $0<x<1$, (Yamada et al., 2000), and $Ge_2Sb_{2.3}Te_5$ (Yao et al., 2003)). In all of these materials, separation occurs during the crystallization phase, with segregation of small amounts of the excess elements, which remain in the amorphous state at the grain boundaries. By increasing the annealing temperature, the residual amorphous material can convert into another polycrystalline NaCl-type structure with a slightly lower lattice parameter. As a result, after the formation of the first phase, the crystallization rate is strongly reduced and a further conversion of the film into another crystalline structure that crystallizes at higher temperatures can occur (Privitera et al., 2003).

Furthermore, $Ge_2Sb_2Te_5$, $Ge_1Sb_2Te_4$ and $Ge_1Sb_4Te_7$ materials crystallize into the same NaCl-type ($Fm\bar{3}m$) structure (Matsunaga et al., 2004, Matsunaga et al., 2006) with lattice constants of 6.001, 6.044 and 6.0876 Å respectively (Morales-Sanchez et al., 2005). In materials with excess Sb, as the content of Sb increases, the lattice constant also increases (Yamada et al., 2000). Thus, during crystallization, atoms in the amorphous state must travel less distance to take their position in the $Ge_1Sb_4Te_7$ crystal lattice than in any other ternary alloys. This effect is responsible for the highest crystallization speed of $Ge_1Sb_4Te_7$ when compared with $Ge_2Sb_2Te_5$ and $Ge_1Sb_2Te_4$ materials. Also, $Ge_1Sb_4Te_7$ has the lowest crystallization temperature. All these factors and the possible existence of a local composition fluctuation can be responsible for the appearance of $Ge_1Sb_4Te_7$ nuclei in the process of crystallization in $Ge_2Sb_2Te_5$ and $Ge_1Sb_2Te_4$ materials.

The clarification of crystallization mechanisms in alloys that lie on the GeTe–Sb_2Te_3 pseudo-binary line will allow the production of phase-change materials with better recording properties.

10. References

Bahgat, A. A., Mahmoud, E. A., Abd, Rabo, A. S. & Mahdy, I. A. (2006). Study of the ferroelectric properties of Ge–Sb–Te alloys. *Physica B: Condensed Matter*, 382 No. 1-2, (April 2006) 271-278, ISSN 0921-4526.

Caravati, S., Bernasconi, M. & Parrinello, M. First principles study of the optical contrast in phase change materials. *Journal of Physics: Condensed Matter*, 22, No. 31, (July 2010) 315801, ISSN 0953-8984.

Claudio, D., Gonzalez-Hernandez, J., Licea, O., Laine, B., Prokhorov, E. & Trapaga, G. (2006) An analytical model to represent crystallization kinetics in materials with metastable phase formation. *Journal of Non-Crystalline Solids*, 352, No. 1, (January 2006) 51–55, ISSN 0022-3093.

Christian, J. W. *The Theory of Transformation in Metals and Alloys*, (1975), Pergamon Press, Oxford. ISBN 978-0-08-044019-4, United Kingdom.

Coombs, J. H., Jongenelis, A. P. J. M., Van Es-Spiekman, W. & Jacobs, B. A. (1995). Laser induced crystallization phenomena in GeTe based alloys. II. Composition dependence of nucleation and growth. *Journal of Applied Physics*, 78, No. 8, (October 1995) 4918-4956, ISSN 0021-8979.

Fan, Z., Wang, L. & Laughlin, D. E. (2004). Modeling of Crystallization Activation Energy for GeTe-Sb_2Te_3 Based Phase Change Materials. *Proceedings of SPIE*, ISBN 0-8194-5293-9, Monterey CA. USA, April 2004.

Fantini, A., Sousa, V. Perniola, L., Gourvest, E., Bastien, J. C., Maitrejean, S., Braga, S., Pashkov, N., Bastard, A., Hyot, B., Roule, A., Persico, A., Feldis, H., Jahan, C., Nodin, J. F., Blachier, D., Toffoli, A., Reimbold, G., Fillot, F., Pierre, F., Annunziata, R., Benshael, D., Mazoyer, P., Vallee, C., Billon, T., Hazart, J., De Salvo, B. & Boulanger, F. (2010). N-doped GeTe as Performance Booster for Embedded Phase-Change Memories. *Proceedings of International Electron Device Meeting (IEDM)*. ISBN 978-1-4424-7418-5, San Francisco, CA. USA, December 2010.

Gervacio, Arciniega, J. J., Prokhorov, E., Espinoza, Beltran, F. J. & Gonzalez-Hernandez, J. (2010). Ferroelectric properties of $Ge_2Sb_2Te_5$ phase-change films. *Applied Physics Letters*, 97, No. 6, (August 2010) 063504, ISSN 0003-6951.

Gonzalez-Hernandez, J., Chao, B. S., Strand, D., Ovshinsky, S. R., Pawlik, D. & Gasiorowski, P. (1992). Crystallization studies of GeSbTe optical memory materials. *Applied Physics Communications*, 11, No. 4, (April 1992) 557-581, ISSN 0277-9374.

Henderson, D.W. (1979). Thermal analysis of non-isotermal crystallization kinetics in glass forming liquits. *Journal of Non-Crystalline Solids*, 30, No. 3, (July 1979) 301-315, ISSN 0022-3093.

Huang, S. M., Huang, S. Y., Zhao, Z. J. & Sun, Z. (2006). Investigation of phase changes in $Ge_1Sb_4Te_7$ films by single ultra-fast laser pulses. *Applied Physics A*, 82, No. 3, (March 2006) 529–533, ISSN 0947-8396.

Im, J., Eom, J. H., Park, C., Park, K., Suh, D. S., Kim, K., Kang, Y. S., Kim, C., Lee, T. Y., Khang, Y., Yoon, Y. G. & Ihm, J. (2008). Hierarchical structure and phase transition of $(GeTe)_n(Sb_2Te_3)_m$ used for phase-change memory. *Physical Rewiew B*, 78, No. 20, (November 2008) 205205, ISSN 1098-0121.

Kalb, J., Spaepen, F. & Wuttig, M. (2004). Atomic force microscopy measurements of crystal nucleation and growth rates in thin films of amorphous Te alloys. *Applied Physics Letters*, 84, No. 25, (June 2004), 5240–5242, ISSN 0003-6951.

Kato, N., Takeda, Y., Fukano, T., Motohiro, T., Kawai, S. & Kuno, H. (1999). Compositional Dependence of Optical Constants and Microstructures of GeSbTe Thin Films for Compact-Disc-Rewritable (CD-RW) Readable with Conventional CD-ROM Drives *Japanese Journal of Applied Physics*, 38, No. 3B, (March 1999) 1707-1708, ISSN 0021-4922.

Khulbe P. K., Wright, E. M. & Mansuripur, M. (2000). Crystallization behavior of as-deposited, melt quenched, and primed amorphous states of $Ge_2Sb_{2.3}Te_5$ films. *Journal of Applied Physics*, 88, No. 7, (October 2000) 3926-3933, ISSN 0021-8979.

Kim M. S., Cho, S. H., Hong, S. K., Roh, J. S. & Choi. D. J. (2008). Crystallization characteristics of nitrogen-doped Sb_2Te_3 films for PRAM application. *Ceramics International*, 34, No. 4, (May 2008) 1043–1046. ISSN 0272-8842.

Kim, M.Y. & Oh, T. S. (2010). Crystallization behavior and thermoelectric characteristics of the electrodeposited Sb_2Te_3 thin films. *Thin Solid Films*, 518, No. 22, (September 2010) 6550-6553, ISSN 0040-6090.

Kolobov, A. V., Fons, P., Frenkel, A. I., Ankudinov, A. L., Tominaga, J. & Uruga, T. (2004). Understanding the phase-change mechanism of rewritable optical media. *Nature Materials*, 3, (September 2004) 703-708, ISSN 1476-1122.

Kolobov, A. V., Fons, P., Tominaga, J., Ankudinov, A. L., Yannopoulos, S. N. & Andrikopoulos, K. S. (2004). Crystallization-induced short-range order changes in amorphous GeTe. *Journal of Physics: Condensed Matter*, 16, No. 44, (October 2004) S5103–S5108, ISSN 0953-8984.

Kolobov, A. V., Fons, P., Tominaga, J. & Uruga, T. (2006). Why DVDs work the way they do: The nanometer-scale mechanism of phase change in Ge–Sb–Te alloys. *Journal of Non-Crystalline Solids*, 352, No. 9-20, (March 2006) 1612–1615, ISSN 0022-3093.

Kyrsta, S., Cremer, R., Neuschutz, D., Laurenzis, M., Bolivar, P. H. & Kurz, H. (2001). Deposition and characterization of Ge–Sb–Te layers for applications in optical data storage. *Applied Surface Science*, 179, No.1-4, (July 2001) 55-60, ISSN 0169-4332.

Laine, B., Trapaga, G., Prokhorov, E., Morales-Sanchez, E. & Gonzalez-Hernandez, J. (2003). Model for isothermal crystallization kinetics with metastable phase formation. *Applied Physics Letters*, 83, No. 24, (December 2003) 4969–4971, ISSN 0003-6951.

Libera M. & Chen, M. (1993). Time-resolved reflection and transmission studies of amorphous Ge-Te thin-film crystallization. *Journal of Applied Physics*, 73, No. 5, (March 1993) 2272-2282, ISSN 0021-8979.

Liu, D., Xu, L., Liao, Y. B., Dai, M., Zhao, L., ...Xu, J., Wu, L. C., Ma, Z. Y. & Chen, K. J. (2006). Comparison of Thermal Stability and Electrical Characterization between $Ge_2Sb_2Te_5$ and $Ge_1Sb_2Te_4$ Phase-Change Materials. *Japanese Journal of Applied Physics*, 48, No. 12, (December 2009) 121104, ISSN 0021-4922.

Lu Q. M. & Libera, M. (1995). Microstructural measurements of amorphous GeTe crystallization by hot-stage optical microscopy. *Journal of Applied Physics*, 77, No. 2, (January 1995) 517-521, ISSN 0021-8979.

Lv, B., Hu, S., Li, W., Di, X., Feng, L., Zhang, J., Wu, L., Cai, Y., Li, B. & Lei, Z. (2010). Preparation and Characterization of Sb_2Te_3 Thin Films by Coevaporation. *International Journal of Photoenergy*, Article ID 476589 (June 2010), ISSN 1110-662X.

Matsunaga, T., Kojima, R., Yamada, N., Kifune, K., Kubota, Y., Tabata, Y. & Takata, M. (2006). Single Structure Widely Distributed in a GeTe-Sb_2Te_3 Pseudobinary System: A Rock Salt Structure is Retained by Intrinsically Containing an Enormous Number of Vacancies within Its crystal, *Inorganic Chemistry*, 45, No. 5, (February 2006) 2235-2241, ISSN 0020-1669.

Matsunaga, T. & Yamada, N. (2004). Structural investigation of $GeSb_2Te_4$: A high-speed phase-change material. *Physical Rewiew B*, 69, No. 10, (March 2004) 104111, ISSN 1098-0121.

Matsushita, T., Nakau, T., Suzuki, A. & Okuda, M. (1989). Measurements of activation energy in the initialization process of amorphous GeTe films. *Journal of Non-Crystalline Solids*, 112, No. 1-3, (October 1989) 211-214, ISSN 0022-3093.

Miao, X. S., Shi, L. P., Lee, H. K., Li, J. M., Zhao, R., Tan, P. K., Lim, K. G., Yang, H. X. & Chong, T. C. (2006). Temperature Dependence of Phase-Change Random Access Memory Cell. *Japanese Journal of Applied Physics*, 45, No. 5A, (May 2006), 3955–3958, ISSN 0021-4922.

Morales-Sanchez, E., Lain, B., Prokhorov, E., Hernarndez-Landaverde, M. A., Trapaga, G. & Gonzalez-Hernandez, J. (2010). Crystallization process in $Ge_2Sb_2Te_5$ amorphous Films. *Vacuum*, 84, No. 7, (March 2010) 877–881, ISSN 0042-207X.

Morales-Sanchez, E., Prokhorov, E., Gonzalez-Hernandez, J. & Mendoza-Galvan, A. (2005). Structural, electric and kinetic parameters of ternary alloys of GeSbTe. *Thin Solid Films*, 471, No. 1-2, (January 2005) 243-247, ISSN 0040-6090.

Naito, M., Ishimaru, M. & Hirotsu, Y. (2004). Local structure analysis of Ge-Sb-Te phase change materials using high-resolution electron microscopy and nanobeam diffraction. *Journal of Applied Physics*, 95, No. 12, (June 2004) 8130-8135, ISSN 0021-8979.

Nobukuni N., Takashima, M., Ohno, T. & Horiea, M. (1995). Microstructural changes in GeSbTe film during repetitious overwriting in phase change optical recording. *Journal of Applied Physics*, 78, No. 12, (December 1995) 6980-6988, ISSN 0021-8979.

Ohta, T., Inoue, K., Uchida, M., Yoshida, K., Akiyama, T., Furakawa, S., Nagata, K. & Nakamura, S. Phase change disk media having rapid cooling structure. *Japanese Journal of Applied Physics*, 28, Suppl. 28, No. 3, (March 1989) 123-128, ISSN 0021-4922.

Okura, M., Naito, H. & Matsushita, T. (1992). Discussion on the Mechanism of Reversible Phase Change Optical Recording. *Japanese Journal of Applied Physics*, 31, No. 2B, (February 1992) 466-470, ISSN 0021-4922.

Ovshinsky S. R. (1968). Reversible electrical switching phenomena in disordered structures, *Physical Review Letters*, 21, No. 20, (November 1968) 1450–1453, ISSN 0031-9007.

Ovshinsky, S. & P. Klose, Reversible High-Speed High-Resolution Imaging in Amorphous Semiconductors, *Society for Information Display International Symposium Digest of Technical Papers*, First Edition, 58-61 (1971).

Paesler, M. A., Baker, D. A., Lucovsky, G., Edwards, A. E. & Taylor, P.C. (2007). EXAFS study of local order in the amorphous chalcogenide semiconductor $Ge_2Sb_2Te_5$. *Journal of Physics and Chemistry of Solids*, 68, No. 5-6, (March 2007) 873-877, ISSN 0022-3697.

Park, J., Kim, M. R., Choi, W. S., Seo, H. & Yeon, C. (1999). Characterization of Amorphous Phases of $Ge_2Sb_2Te_5$ Phase-Change Optical Recording Material on Their Crystallization Behavior. *Japanese Journal of Applied Physics*, 38, No. 8, (May 1999) 4775–4779, ISSN 0021-4922.

Perniola, L., Sousa, V., Fantini, A., Arbaoui, E., Bastard, A., Armand, M., Fargeix, A., Jahan, C., Nodin, J.-F., Persico, A., Blachier, D., Toffoli, A., Loubriat, S., Gourvest, E., Betti, Beneventi, G., Feldis, H., Maitrejean, S., Lhostis, S., Roule, A., Cueto, O., Reimbold, G., Poupinet, L., Billon, T., De Salvo, B., Bensahel, D., Mazoyer, P., Annunziata, R., Zuliani, P. & Boulanger, F. (2010). Electrical Behavior of Phase-Change Memory Cells Based on GeTe. *Electron Device Letters IEEE*, 31, No. 5, (April 2010) 488-490, ISSN 0741-3106.

Popescu, M. A. Ovonic Materials. (2005). *Journal of Ovonic Research*, 1, No. 6, (December 2005), 69-76, ISSN 1584–9953.

Privitera, S., Rimini, E., Bongiorno, C., Zonca, R., Pirovano, A. & Bez, R. (2003). Crystallization and phase separation. in $Ge_{2+x}Sb_2Te_5$ thin Films. *Journal of Applied Physics*, 94, No. 7, (October 2003) 4409-4413, ISSN 0021-8979.

Rabe, K. & Joannopoulos, J. (1987). Theory of the structural phase transition of GeTe. *Physical Rewiew B*, 36, No. 12, (October 1987) 6631-6639, ISSN 1098-0121.

Raoux, S., Shelby, R. M., Jordan-Sweet, J., Munoz, B., Salinga, M., Chen, Y. C., Shih, Y.H., Lai, E. K. & Lee, M.H. (2008). Phase change materials and their application to random access memory technology. *Microelectronic Engineering*, 85, No. 12, (August 2008) 2330–2333, ISSN 0167-9317.

Reddy, G. B., Dhar, A., Malhotra L. K. & Sharmila E. K. (1992). Comparative study of crystallization processes in Sb,Te, films using laser and thermal annealing techniques. *Thin Solid Films*, 220, No. 1-2 , (1992) 111-115, ISSN 0040-6090.

Ruitenberg, G., Petford-Long, A. K. & Doole, R. C. (2002). Determination of the isothermal nucleation and growth parameters for the crystallization of thin $Ge_2Sb_2Te_5$ films. *Journal of Applied Physics*, 92, No. 6, (September 2002) 3166-3123, ISSN 0021-8979.

Ruiz, Santos, R., Prokhorov, E., Espinoza, Beltran, F. J., Trapaga, Martınez, L. G. & Gonzalez-Hernandez, J. (2010). Crystallization and ferroelectric properties of $Ge_4Sb_1Te_5$ films. *Journal of Non-Crystalline Solids*, 356, No. 52-54, (July 2010) 3026-3031, ISSN 0022-3093.

Senkader S. & Wright, C. D. (2004). Models for phase-change of $Ge_2Sb_2Te_5$ in optical and electrical memory devices. *Journal of Applied Physics*, 95, No. 2, (January 2004) 504-511, ISSN 0021-8979.

Sian, T. S., Mittal, S., Goel, A., Chakraborty, S., Gautam, S., Budhwar, A., Kapil, R., Achar, P. C. & Nyati, G. (2008). Correlations between electrical parameters of rewritable digital versatile disc and optical and thermal properties of phase change alloy: effect of sputter power. *Japanese Journal of Applied Physics*, 47, No. 2, (February 2008) 936–941, ISSN 0021-4922.

Strand, D. (2005). Ovonics: from science to products. *Journal of Optoelectronics and Advanced Materials*, 7, No. 4, (August 2005) 1679–1690, ISSN 1454-4164.

Tominaga, J., Shima, T., Kuwahara, M., Fukaya, T., Kolobov, A. &Nakano, T. (2004). Ferroelectric catastrophe: beyond nanometre-scale optical resolution. *Nanotechnology, 15, No. 5, (January 2004) 411-415, ISSN 0957-4484.*

Trappe, C., Bechvet, B., Hyot, B., Winkler, O., Facso, S. & Kurz, H. (2000). Recrystallization Dynamics of Phase Change Optical Disks with a Nitrogen Interface Layer. *Japanese Journal of Applied Physics*, 39, No. 2B, (February 2000) 766-769, ISSN 0021-4922.

Wamwangi, D., Njoroge, W. K. & Wuttig, M. (2002). Crystallization kinetics of $Ge_4Sb_1Te_5$ films. *Thin Solid Films*, 408, No. 1-2, (March 2002) 310–315, ISSN 0040-6090.

Weidenhof, V., Friedrich, I., Ziegler, S. & Wuttig, M. (2001). Laser induced crystallization of amorphous $Ge_2Sb_2Te_5$ films. *Journal of Applied Physics*, 89, No. 6, (March 2001) 3168–3176, ISSN 0021-8979.

Wei J. & F. Gan. (2003). Theoretical explanation of different crystallization processes between as-deposited and melt-quenched amorphous $Ge_2Sb_2Te_5$ thin films. *Thin Solid Films*, 441, No. 1-2, (September 2003) 292-297, ISSN 0040-6090.

Yamada, N. (1996). Erasable phase-change optical material, *MRS Bulletin*, 21, No. 9, (October 1996) 48-50, ISSN 0883-7694.

Yamada, N. & Matsunaga, T. (2000) Structure of laser-crystallized $Ge_2Sb_{2+x}Te_5$ sputtered thin films for use in optical memory. *Journal of Applied Physics*, 88, No. 12, (December 2000) 7020-7028, ISSN 0021-8979.

Yamada, N., Ohno, E., Nishiuchi, K., Akahira, N. & Takao, M. (1991). Rapid-phase transitions of GeTe-Sb, Te, pseudobinary amorphous thin films for an optical disk memory, *Journal of Applied Physics*, 69, No. 5, (March 1991) 2849-2856, ISSN 0021-8979.

Yao, H. B., Shi, L. P., Chong, T. C., Tan, P. K. & Miao, X. S. (2003). Optical Transition of Chalcogenide Phase-Change Thin Films. *Japanese Journal of Applied Physics*, 42, No. 2B, (February 2003) 828-831, ISSN 0021-4922.

Yin, Y., Sone H. & S. Hosaka. (2007). Characterization of nitrogen-doped Sb_2Te_3 films and their application to phase-change memory. *Journal of Applied Physics*, 102, No. 6, (September 2007) 064503, ISSN 0021-8979.

Zhai, F., Wang, Y., Wu, Y. & Gan, F. (2009). Phase Transition Kinetics of Sb_2Te_3 Phase Change Thin Films. *Proceedings of SPIE*, ISBN 0-8194-5293-9 Wuhan, China, November 2008.

Zhang, T., Cheng, Y., Song. Z., Liu, B., Feng, S., Han, X., Zhang, Z. & Chen, B. (2008). Comparison of the crystallization of Ge–Sb–Te and Si–Sb–Te in a constant-temperature annealing process. *Scripta Materialia*, 58, No. 11, (February 2008) 977–980, ISSN 1359-6462.

Zhu, M., Wu, L., Rao, F., Song, Z., Peng, C., Li, X., Yao, D., Xi, W. & Feng. S. Phase Change Characteristics of SiO_2 Doped Sb_2Te_3 Materials for Phase Change Memory Application. *Electrochemical and Solid-State Letters*, 14, No. 10, (July 2011) H404-H407, ISSN 1099-0062.

11

Oriented Lateral Growth and Defects in Polycrystalline-Silicon Thin Films on Glass Substrates

Kuninori Kitahara[1] and Akito Hara[2]
[1]*Shimane University*
[2]*Tohoku Gakuin University*
Japan

1. Introduction

Silicon (Si) thin films on glass substrates have been extensively developed as a semiconductor material for electronic devices. This material is especially useful for large-area panel devices such as thin-film transistors (TFTs) on active-matrix flat panel displays. The most widely used Si films are hydrogenated amorphous Si (a-Si:H), which can be deposited at temperatures lower than the strain point of the substrate. However, improved electronic properties are required to achieve higher device performance. Using polycrystalline Si (poly-Si) films instead of a-Si:H films enhances carrier mobility by two or three orders of magnitude; thus, driver circuits can be incorporated into display panels, as shown in Fig. 1. The application of poly-Si will be extended to mobile displays with large pixel density, microprocessor–display combined panels, and thin-film solar cells.

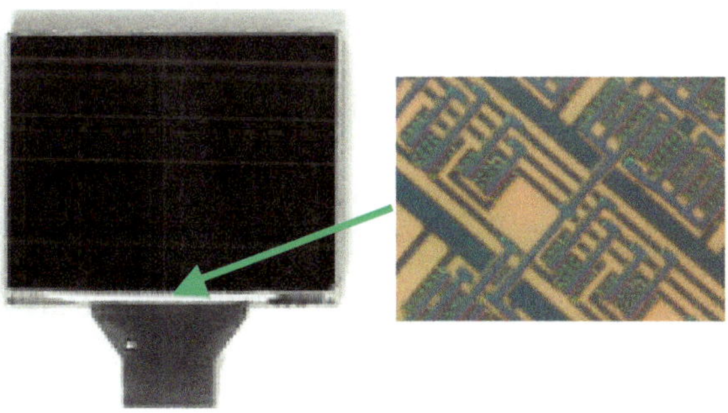

Fig. 1. Low-temperature poly-Si liquid crystal display (2 in. diagonal). Driver circuits are integrated at the periphery of the panel.

The poly-Si used for TFT must be ≤ 50 nm thick to ensure the desired device performance. Furthermore, the crystalline fraction should be almost 100%. Such thin films cannot be deposited directly on glass; they must be formed by recrystallization of a-Si precursor films. For this purpose, manufactures have employed solid-phase crystallization (SPC) and excimer-laser crystallization (ELC) techniques, as shown in Fig. 2 (Hayashi et al., 1984; Sameshima et al., 1986).

The grain boundaries in poly-Si thin films typically feature random configurations. However, random grain boundaries were reported to severely degrade device performance (Blake et al., 1997). Therefore, grain enlargement is desirable. To achieve this, lateral growth during crystallization is necessary. The lateral growth descrived in this manuscript corresponds to oriented overgrowth of the crystalline film on a non-crystalline layer (i.e., the glass substrate), which is regarded as a type of epitaxy (Givargizov, 1991). Epitaxy proceeds so as to minimize the free energy and is expected to effectively reduce the defect density.

In this work, we investigate the relationship between lateral growth and defects using various characterization techniques. First, we give an overview the crystallization process on non-crystalline substrates and the characterization of defects in poly-Si thin films. Next, we describe previous studies and our work on SPC, ELC, and flow-shaped growth of poly-Si on glass from the perspective of lateral growth. Finally, the growth of a quasi-single-crystal SiGe thin film on a glass substrate, that is super-lateral growth, is demonstrated.

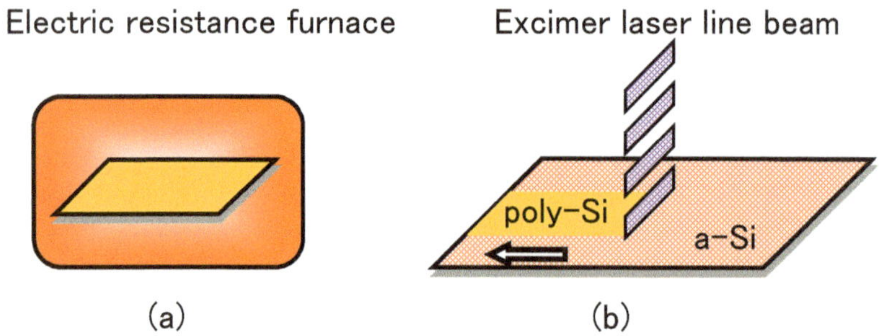

Fig. 2. Techniques for recrystallization of a-Si films on glass substrates using (a) solid-phase crystallization and (b) excimer-laser crystallization.

2. Overview of lateral crystallization on non-crystalline layer

The basic technology for poly-Si formation on a glass substrate is Si on insulator (SOI), which was developed for applications such as low power consumption large-scale integrated circuits (LSIs), three-dimensional LSIs, radiation-hardened electronics and solar cells. SOI technologies have been reviewed in detail elsewhere (Givargizov, 1991; Colinge, 2004). The substrate used for SOI is typically single-crystalline Si (c-Si) coated with a dielectric film such as SiO_2 or Si_3N_4. Although SOI technology includes a wide range of methods, we focus on the recrystallization of a-Si thin films. The principle SOI techniques used for this purpose are zone-melting recrystallization and lateral SPC.

Zone-melting recrystallization is performed by applying a heat source to an a-Si film, which induces melting followed by solidification of the film. Common heat sources are line-shaped electric resistors, electron beams, and continuous-wave (CW) Ar ion lasers. Lateral SPC is also performed by zone heating; its advantages include low T_s (500–700 °C) and consistent surface smoothness.

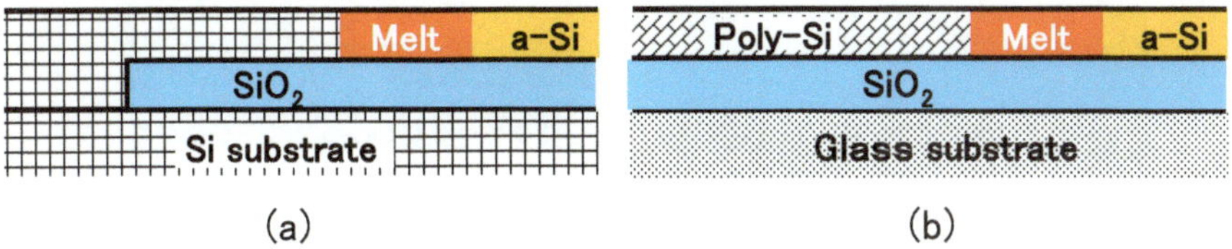

Fig. 3. Structures of Si layers on non-crystalline layers: (a) Si on insulator, where Si substrate is coated by SiO_2 film with windows for seed areas: (b) excimer-laser crystallized Si, where glass substrate is fully coated by SiO_2 film.

Control of nucleation during the initial stage of lateral growth and crystalline orientation is essential for the growth of single or quasi-single crystals. To achieve this control, the non-crystalline underlayer is partially opened to reveal the crystalline Si substrate, which acts as the seed area, as shown in Fig. 3(a). Alternatively, the fundamental layers are trenched in order to form a grating-like shape to guide the crystalline orientation; this is known as artificial epitaxy or graphoepitaxy. However, the structures thus created require additional photolithographic processing. In contrast, the laser crystallization process enables control of the crystalline orientation without seeding or photolithography; however, the obtained crystalline quality is not always adequate for the submicron-scale processes required for LSI.

For applications in electronic displays, the substrate should be glass or plastic rather than c-Si. On these substrates, Si thin films become essentially polycrystalline because of the seedless growth, as shown in Fig. 3(b). Poly-Si on glass has been applied in devices called giant microelectronics or large-area electronics (Kuriyama et al., 1992; Sameshima, 2009), which have areas several orders of magnitude larger than those of SOI devices. The use of glass or plastic substrates strongly restricts the upper limit of T_s. Technologies are classified as high-temperature poly-Si or low-temperature poly-Si depending on T_s (Blake, 1997). The border between them is determined by the strain point of a low alkali glass substrate at ~590 °C (Corning's technical glass catalog).

SPC requires a T_s that is higher than 600 °C. Thus, fused quartz glass is used as the substrate because its strain point is as high as 990 °C. SPC poly-Si is used mainly in 1-in. diagonal or smaller liquid crystal displays (LCDs) for multimedia projectors.

Laser crystallization technology decreases T_s to below the strain point of low alkali glass substrates and in many cases to room temperature (RT) owing to selective heating of the Si layer. Progress in laser annealing has been reviewed elsewhere (Sameshima, 2009). Poly-Si films for medium-sized LCD panels suitable for mobile information terminal applications have been formed.

There are two opposing requirements for laser crystallization: ensuring uniformity in order to integrate TFTs, and eliminating grain boundaries in order to improve the performance of individual TFTs. The former and latter requirements are met by the growth of grains that are notably smaller and larger than the TFT channels, as shown in Fig. 4(b) and 4(c), respectively. Those opposing needs are expected to be reconciled by the growth of flow-shaped grains, as shown in Fig. 4(d).

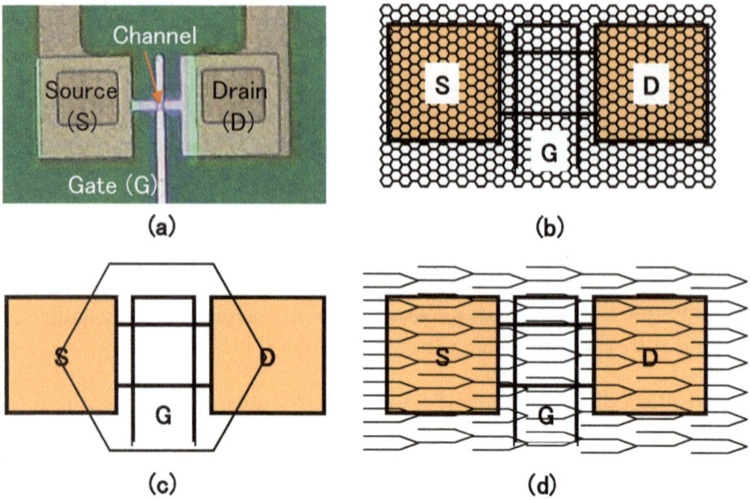

Fig. 4. Structures of thin-film transistors (TFTs) on glass substrates: (a) test device for characterizing electrical performance; geometry of grains and TFT channel for (b) grains notably smaller than the channel region, (c) a large grain containing the channel region, and (d) flow-shaped grains running parallel to the drain current.

3. Characterization of poly-Si films on glass

The principal terms for the characterization of poly-Si thin films on glass substrates are summarized in Table 1. The dangling bonds in a-Si and poly-Si films were qualitatively analyzed via the spin density of unpaired electrons using electron-spin resonance. The spin density was found to be reduced by hydrogenation (Nickel et al., 1997; Spinella et al., 1998), which causes termination of the dangling bonds with H atoms. The geometry and crystal orientation of the grains were investigated using images and diffraction patterns obtained by scanning electron microscopy (SEM), transmission electron microscopy (TEM), and electron backscattering diffraction (EBSD). When growth accompanies melting, the surface morphology can vary with the displacement of the melt. The surface morphology is observed using atomic force microscopy (AFM) or cross-sectional TEM. The electronic characteristics were estimated by fabricating test n-channel and p-channel TFT devices, as shown in Fig. 4(a). The devices exhibit characteristics such as the field-effect electron and hole mobilities (μ_{FEn} and μ_{FEp}, respectively), threshold voltage, subthreshold voltage, and off-current.

The characterization techniques described above require the specimen be processed. Moreover, they are not simple enough to be used to characterize large numbers of films. In contrast, optical techniques are suitable for simple, non-destructive characterization. We analyzed the defects and stress in a poly-Si/glass system by Raman scattering spectroscopy (Kitahara et al., 2002, 2003, 2011a). Micro-Raman spectroscopy was conducted by using a Renishaw System 1000 in the backscattering geometry under excitation with a 514.5 nm Ar ion laser. Details of the setup and band component analysis were described elsewhere (Frost & Shurvell, 1997). The full width at half maximum (FWHM) values of spectra obtained for poly-Si (described below) were calibrated according to the spectral resolution of the setup. The peak frequency of the spectra was exactly determined by fitting to a Lorentzian curve.

Typical first-order Raman scattering for c-Si, Si on sapphire (SOS), and poly-Si on glass are shown in Fig. 5. In the crystalline phase, the Raman spectrum exhibits an isolated band corresponding to the degeneration of a single transverse-optical (TO) phonon mode and two longitudinal-optical (LO) phonon modes; the c-Si band exhibits a linewidth as narrow as 2.5–3.5 cm^{-1} at RT (Temple & Hathaway, 1973; Menendez & Cardona, 1984) owing to the q = 0 (q: wave vector) selection rule. In the amorphous phase, four continuous phonon bands are observed; they are broad because the $q = 0$ selection rule is loosened as a result of a decrease in the phonon correlation length.

The features of the optical phonon mode (OPM) of the crystalline phase are represented by the peak frequency (ω), FWHM, intensity, and in some cases asymmetry. When the grains are smaller than a few tens of nanometers, the spatial correlation of the phonon decreases, which causes the peaks to shift to a lower frequency ($\Delta\omega$) and increases the FWHM (Richter et al., 1981). Compressive and tensile stress in the films cause the peak to shift to higher and lower frequencies, respectively. Moreover, $\Delta\omega$ is proportional to the magnitude of the stress (Englet, 1980). Therefore, the stress can be quantitatively estimated only when the impact of the other factors is negligible. Note, however, that the intensity of the stress depends on the crystallization technique and conditions used. Therefore, an analysis of individual crystallization techniques should be developed by comparing the Raman spectroscopy results with those obtained by SEM and TEM.

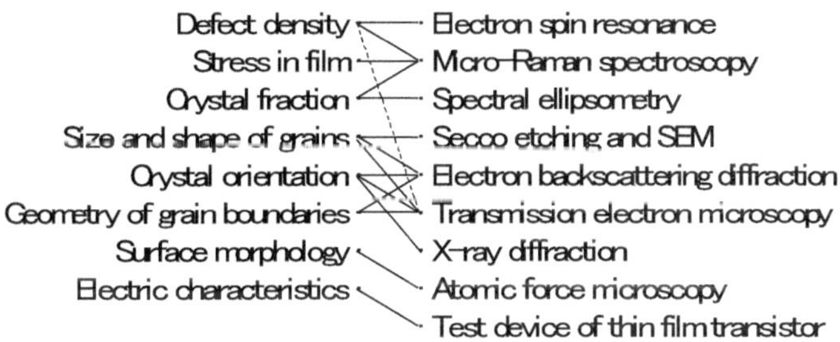

Table 1. Principal terms for characterization of poly-Si thin films on glass substrates.

Defects in poly-Si are frequently observed in SEM images. Grain boundaries and some defect clusters are clearly revealed by Secco etching (Secco d'Aragona, 1972). These defects are evidently electrochemically active, which is confirmed by the fact that they disappear upon hydrogenation before etching. On the other hand, inactive defects such as twin boundaries are observed only as shallow contrasts and are independent of hydrogenation (Kitahara et al., 2009a).

To summarize, various electron microscopies provide direct knowledge of defects and the geometric configuration of grains in poly-Si/glass systems. However, defects cannot always be detected by those methods. Raman microscopy is useful for easy macroscopic (submicron-scale) defect analysis and quantitative analysis of stress in films. However, semi-empirical analysis aided by other techniques is required to simultaneously characterize the defect density and stress in films by Raman spectroscopy.

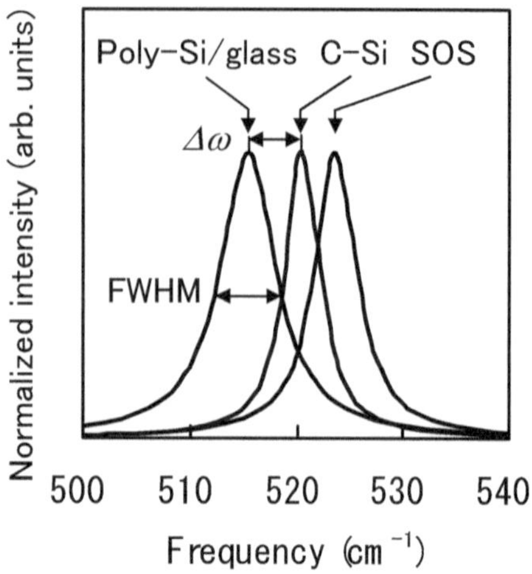

Fig. 5. Typical optical-phonon mode observed in the Raman scattering spectra for Si on sapphire (SOS), single crystalline Si (c-Si) bulk containing no stress, and poly-Si thin film on glass. Intensity is normalized to the individual peak values. Peaks are shifted to higher and lower frequencies by compressive and tensile stresses, respectively.

4. Solid-phase crystallization

The heat source for SPC typically consists of an electric resistance furnace and an infrared lamp annealing furnace. The force that drives crystallization of a-Si in the metastable state is the lowering of the Gibbs free energy through the crystalline phase change. SPC poly-Si technologies are reviewed in detail elsewhere (Hatalis & Greve, 1988; Spinella et al., 1998).

The typical grain size of solid-phase crystallized poly-Si for TFTs is a few hundred nanometers, which is at least several times larger than the film thickness. This suggests that SPC on non-crystalline substrates is dominated by lateral growth rather than columnar growth which is frequently seen in chemical vapor deposition.

SPC is initiated by homogeneous nucleation after incubation. In pure Si, the activation energy of nucleation is larger than that of growth by 0.25 eV (Spinella et al., 1998). Therefore, growth proceeds immediately after nucleation. The lower limit of the nucleation temperature depends on the quality of the precursor film and is ~560 °C for highly pure a-Si. The grain size decreases with increasing crystallization temperature (T_c), which is associated with increasing nucleation frequency. Lateral grain growth is first interrupted by collisions among grains. Then, the grain size increases due to secondary grain growth; the secondary growth rate increases with the grain boundary energy, surface energy anisotropy, grain boundary mobility, and T_c (Thompson, 1985). Furthermore, twin formation at the early stage of crystallization preferentially accelerates the lateral growth rate in the <112> direction, producing grains with an ellipsoidal outline (Nakamura et al., 1989). Growth along the minor axis of the ellipsoid, i.e., in the <111> direction, proceeds with simultaneous formation of many micro-twins (Drosed & Washburn, 1980). Although solid-phase crystallized poly-Si films have a considerable density of defects, hydrogenation of TFTs effectively inactivates those defects, producing a μ_{FEn} of 24 cm^2/Vs (Little et al., 2001).

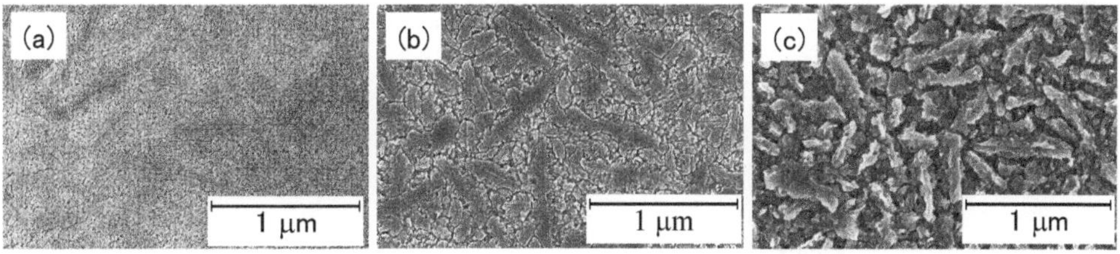

Fig. 6. Scanning electron microscopy images for solid-phase crystallized poly-Si heated at 900 °C for 1.5 h. Images were taken (a) before and after Secco etching for (b) 20 s and (c) 60 s.

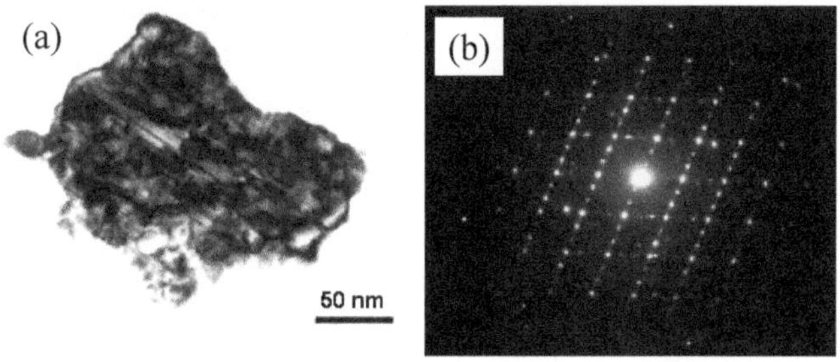

Fig. 7. Transmission electron microscopy image and diffraction pattern for solid-phase crystallized film heated at 900 °C for 3 h.

We investigated the microscopic geometry of the grains and defects by SEM and TEM. We then studied the phase variation at the initial stage macroscopically using Raman spectroscopy and spectroscopic ellipsometry. SPC was performed in a N_2 flow at atmospheric pressure. Figure 6 shows SEM images before and after Secco etching after SPC at T_c = 900 °C for 1.5 h. The surface of the as-crystallized film is relatively smooth. Grain boundaries were not clearly revealed by the short etching time (20 s) owing to the fine structures at the grain boundaries and the overlapping of grains (Spinella et al., 1998). The longer etching time (60 s) revealed feather-like grains lying parallel to the substrate. In addition, major twin boundaries appeared in the individual grains as shallow lines. The plan-view TEM image shown in Fig. 7 indicates that the twin boundaries extend along the major axis. Furthermore, a large density of micro-twins is distributed in the grains. The subspots and streaks evident in the diffraction pattern also indicate the presence of twins and {111} microfacets in the grains.

Figure 8 shows the OPMs in the Raman spectra during the initial stage of SPC at T_c = 580 °C. Only the amorphous phase is detected at 1 h. The crystal component begins to appear in the spectrum after a latent time of ~3 h. Figure 9 shows the crystal fraction in the 580 °C SPC film as a function of the heating time. In Raman spectroscopy, the volume fraction of the crystal was estimated by the area-intensity ratio of the amorphous (I_a) and crystal (I_c) components, i.e., $I_c/(I_c + I_a)$, which yields a relative value. In ellipsometry, the volume fraction of the crystal was analyzed using the Tauc–Lorentz model (Jellison, 1998), which yields an absolute value. Although no crystal component was found by Raman spectroscopy until 3 h had passed, the volume fraction estimated by ellipsometry began to increase at 1 h. The spectra of the imaginary dielectric function observed by ellipsometry show that the E1 and E2 bands were somewhat visible even in the latent time. This suggests that atomic reconstruction toward ordering before the appearance of the crystal was detected by ellipsometry.

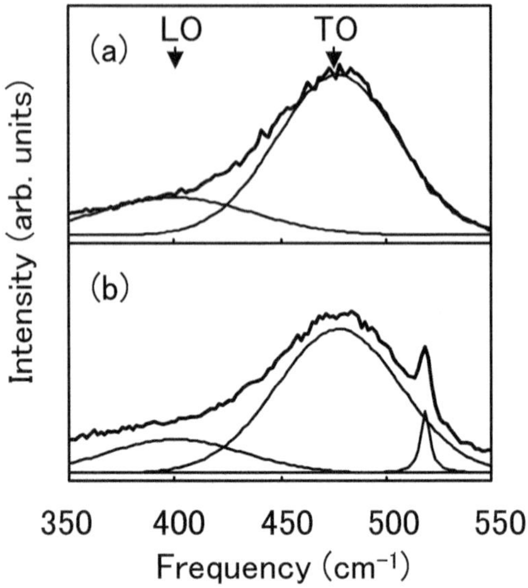

Fig. 8. Optical phonon modes in the Raman spectra for the initial stage of solid-phase crystallized film heated at 580 °C for (a) 1 h and (b) 3 h. Thick lines and thin lines are experimental values and fitting curves, respectively.

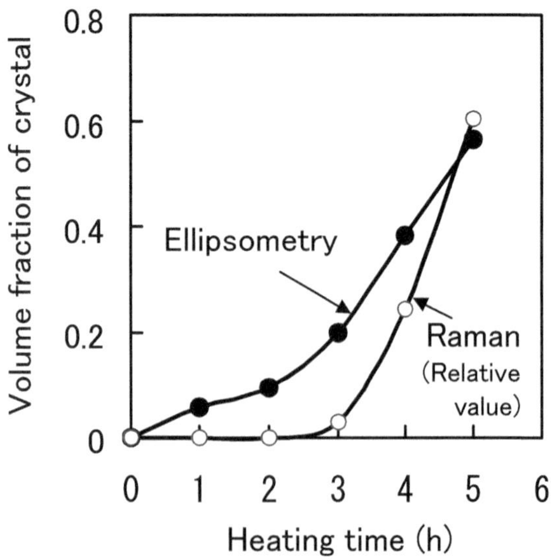

Fig. 9. Relationship between crystal fraction and heating time for solid-phase crystallized film heated at 580 °C. Crystal fraction was estimated using spectral ellipsometry and Raman spectroscopy.

Figure 10 shows the relationship between the FWHM of the OPM and T_c in the range of 625–900 °C with annealing times of 3–8 h; the times were varied in order to achieve sufficient crystallization. The FWHM evidently decreases with increasing T_c. Thus, the defect density decreases with increasing T_c, which is a universal tendency for crystal growth under essentially stable thermal equilibrium. The plots for metal-induced lateral crystallization (MILC) will be described in the next section.

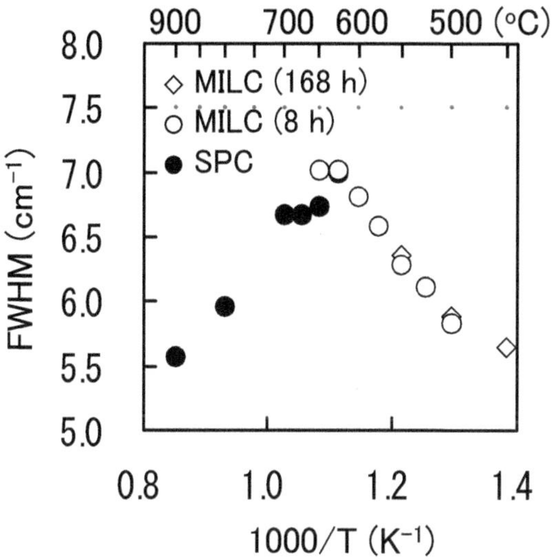

Fig. 10. Relationship between the full width at half maximum (FWHM) of the optical phonon mode and the crystallization temperature (T_c). Plots for poly-Si films fabricated by metal-induced crystallization (MILC) and solid-phase crystallization (SPC) are shown. Note that the dependence on T_c of MILC is opposite to that of SPC.

To summarize, Si atoms cause reconstruction toward ordering in the amorphous phase in the initial stage of annealing. SPC begins at homogeneous nucleations after the incubation time. Growth begins immediately after nucleation. The grain size increases with acceleration of lateral growth by the major twin boundaries formed at the early stage and the coalescence of grains by secondary grain growth. Growth proceeds with simultaneous formation of many micro-twins, resulting in many defects. Increasing T_c decreases the defect density in grains as the grain size is reduced.

5. Metal-induced lateral crystallization

The nucleation temperature of a-Si is decreased by the addition of a catalytic metal; the process is referred to as MILC and catalyst-assisted SPC (Kawatsu et al., 1990; Cammarata & Thompson, 1990; Takayama et al., 2000). Although metals such as Al, Mo, Ni, Pd, and Ti have been applied to cause MILC, Ni is the most frequently used catalyst. Ni can be supplied to the seed area of an a-Si film by evaporation, sputtering, or application of an acetate solution. Annealing of the film causes lateral diffusion of Ni atoms into the amorphous area before the silicide reaction occurs. $NiSi_2$ begins to segregate at $T_c > 420$ °C. Although $NiSi_2$ has the fluorite crystal structure, which is not the same as that of Si, the mismatch of their lattice constants can be as small as 0.4%. Thus, $NiSi_2$ operates as an excellent nucleus for Si growth. $NiSi_2$ seeds then migrate laterally, leaving needle-like Si crystals in the a-Si. The activation energies reported for the growth rate of Ni-MILC, which range from 1.3–2.3 eV (Makihara et al., 2003; Kitahara et al., 2009b), are smaller than those for SPC (3.4 eV) and solid-phase epitaxy (2.7 eV) (Roth et al., 1990; Spinella et al., 1998). Lower values of T_c lead to straighter crystallization. Annealing at 450 °C reportedly results in crystals 10 μm in length and 160 nm wide (Makihara et al., 2003). Needle-like crystals coalesce and form larger grain when the remaining a-Si in the interstitial area is crystallized via SPC.

The application of MILC to devices is not easy because of the long processing time required and the existence of silicide in the crystallized region. However, TFTs have been fabricated by combining MILC with ELC. In this method, the silicide was gettered to ion-implanted regions corresponding to the source and drain electrodes and achieved a μ_{FEn} value of 320 cm^2/Vs (Mizuki et al., 2004).

The density and geometry of the residual defects in MILC films are expected to differ from those in SPC films because MILC is more dominated by lateral growth than SPC is. We investigated the difference in defects between SPC and MILC (Kitahara et al., 2009b). MILC was performed by evaporating Ni onto an a-Si/SiO_2/fused quartz glass substrate through a metal mask with rectangular windows. The thicknesses of Ni and a-Si were 13 and 110 nm, respectively. The plan-view TEM image after MILC at 450 °C for 168 h is shown in Fig. 11(a). The image exhibits needle-like crystals with a {110} plane extending toward <111>. Note that the grains exhibited little contrast in the strain corresponding to defects. The coexistence of sharp diffraction spots and a halo in Fig. 11(a′) indicate that the grain is a single crystal with a surrounding amorphous component. The TEM image after MILC at 500 °C for 168 h [Fig. 11(b)] shows accumulated needle-shaped grains with strain contrast due to defects. Small subspots in the diffraction pattern shown in Fig. 11(b′) also indicate the presence of defects in the grains. However, the streaks corresponding to microfacets that were seen for the SPC films are not observed here. These results suggest that the number of defects in the grains decreases with decreasing T_c, which is opposite to the trend for SPC.

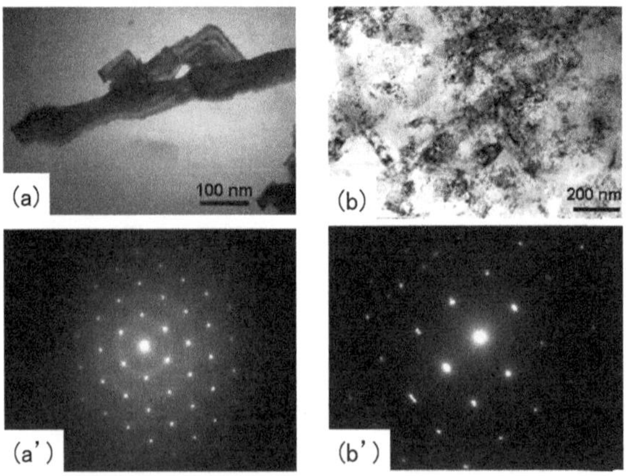

Fig. 11. Transmission electron microscopy images for (a) metal-induced crystallization (MILC) at 450 °C for 168 h and (b) MILC at 500 °C for 168 h. Individual diffraction patterns are also shown in (a′) and (b′).

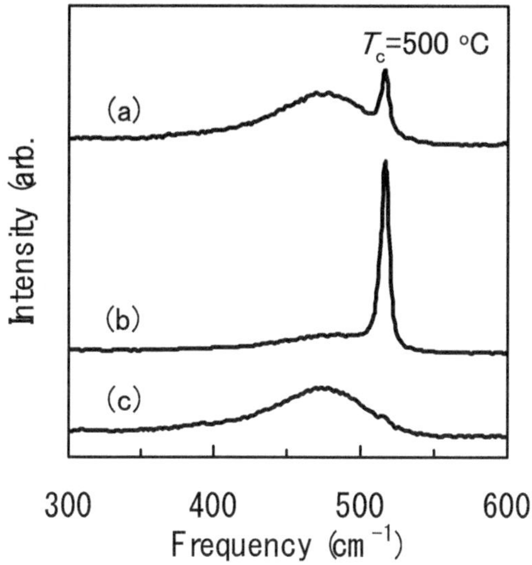

Fig. 12. Typical Raman spectra for (a) Ni-evaporated source region, (b) lateral growth region, and (c) uncrystallized region.

Typical Raman spectra for a Ni-evaporated source region, lateral growth region, and uncrystallized region are shown in Figs. 12(a)-(c). Here, MILC was conducted at 500 °C for 8 h. The spectrum for the Ni-evaporated region indicates the presence of a large amorphous fraction. In this area, the Ni density is likely too high to effectively induce lateral crystallization. In contrast, in the majority of the lateral growth region, the OPM is accompanied only by a weak amorphous-like mode. The uncrystallized region naturally remains amorphous. These results suggest that the amorphous-to-polycrystalline transition during MILC requires both a Ni supply and the lateral growth process.

The relationship between T_c and the FWHM of the OPM for the lateral growth region is shown in Fig. 10. The FWHM decreases with decreasing T_c and reaches 5.7 cm^{-1} at 450 °C, which is nearly equal to the FWHM of SPC at 900 °C, 5.6 cm^{-1}. A factor other than thermal energy is responsible for the T_c dependence of MILC because the dependence on T_c is opposite to that of SPC. In MILC, the Ni atoms exhibit a lateral gradation of density in a-Si due to diffusion from the source region. In particular, at $T_c < 500$ °C, the observed lateral growth length is smaller than the calculated Ni diffusion length in a-Si.

We concluded that the lateral growth at relatively low T_c is directed by the gradation in Ni density. Furthermore, the growth at T_c as low as 450 °C is dominated by needle-like lateral epitaxy and results in a low defect density. Increasing T_c restricts the length of straight growth and also enhances random growth due to SPC; these two effects increase the defect density despite the increase in T_c.

6. Excimer-laser crystallization

Crystallization is performed primarily by KrF or XeCl excimer lasers, which supply intense pulsed light with durations of ~30 ns and wavelengths of 249 and 308 nm, respectively. The luminous flux is typically shaped as a linear beam. Overlapping irradiation by the scanning pulsed light enables uniform crystallization over a wide area. The laser energy density is adjusted to be slightly lower than that at which complete melting of the Si film occurs. In practical use, T_s is maintained at around RT during crystallization; this forms grains a few hundred nanometers in size with high reproducibility.

Because of the short pulse duration, T_s does not almost increase. However, rapid cooling after laser irradiation is an obstacle to lateral growth. Some techniques, including double-beam irradiation, have been proposed to decrease the cooling rate. Applying multiple laser irradiations at $T_s < 400$ °C reportedly increases the grain size to 4.5 μm and strongly aligns the surface orientation to the <111> direction (Kurimiya et al., 1993).

One of the features of ELC is a rough surface morphology containing a large density of hillocks. The hillocks are generated by a positive feedback effect between the beam-induced periodic surface roughness pattern and the interference in subsequent pulses (McCulloch & Brotherton, 1995). Another feature of ELC is that μ_{FEn} is considerably higher than that of SPC. The μ_{FEn} value of an excimer-laser crystallized poly-Si TFT increases with increasing grain size and reaches 320 cm^2/Vs at an average grain size of 700 nm, at which the dominant factor determining μ_{FEn} varies from grain boundary scattering to lattice scattering (Hara et al., 2002a). Furthermore, a μ_{FEn} value of 914 cm^2/Vs was reportedly obtained by position-controlled large grain growth (Mitani et al., 2008).

The large μ_{FEn} value of excimer-laser crystallized poly-Si has been attributed to a low defect density in grains owing to a regrowth procedure through the liquid phase. In practice, TEM images typically exhibit few defects in the grains. The average dislocation density was reportedly 8×10^6 cm^{-2} for large (2–4 μm) grains and 10^8–10^9 cm^{-2} even for defective grains (Christiansen et al., 2001). However, the cooling velocity of ELC can be as high as ~10^{10} K/s (Sameshima & Usui, 1993). Therefore, recrystallization is expected to proceed under significant deviations from thermal equilibrium; consequently, large numbers of defects are frozen in the grains. In defects in bulk c-Si, impurity complexes generated at high temperature were shown to be frozen by quenching at a rate of ~10^3 K/s (Takahashi et al., 1995). The cooling rate of ELC is far larger than that in the quenching experiment. Hence, in ELC, defects undetectable by TEM are presumed to reside in the grains.

We studied defects in excimer-laser crystallized poly-Si (Kitahara et al., 2007, 2009a). We irradiated a-Si films (50 nm thick) on SiO_2-coated fused quartz substrates with a XeCl excimer laser with 95% overlap. Figure 13 shows the surface of the excimer-laser crystallized poly-Si observed by SEM after Secco etching. Many hillocks appear as bright spots, as shown in Fig. 13(a); they are located at the junctions of grain boundaries. The hillock interval under the optimum conditions for ELC is nearly equal to the wavelength of the laser. AFM images indicated that ridges also lay along grain boundaries.

The grain boundaries were more evident after a longer etching time, as shown in Fig. 13(b). In contrast, grain boundaries did not appear in the hydrogenated film even after a long etching time, as shown in Fig. 13(c). This result implies that the grain boundaries were electrochemically inactivated by hydrogenation. On the other hand, structures other than grain boundaries were detected as shallow lines independently of hydrogenation, which could be due to the lack of electrochemically active dangling bonds.

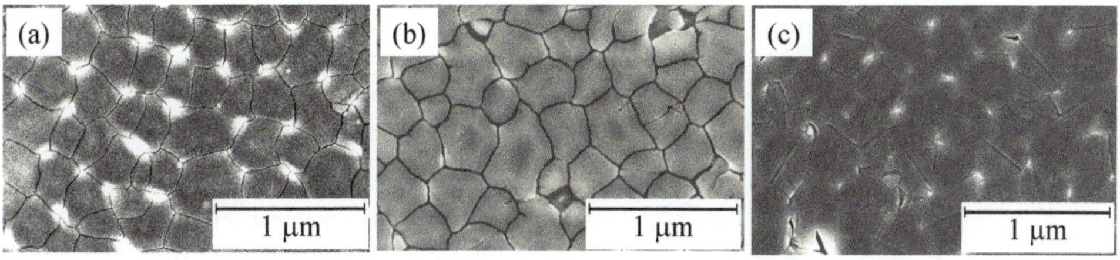

Fig. 13. Scanning electron microscopy images for (a, b) non-hydrogenated and (c) hydrogenated excimer-laser annealed poly-Si films. Secco etching was performed for (a) 15 s and (b, c) 40 s. Hydrogenation interfered with the appearance of grain boundaries after etching.

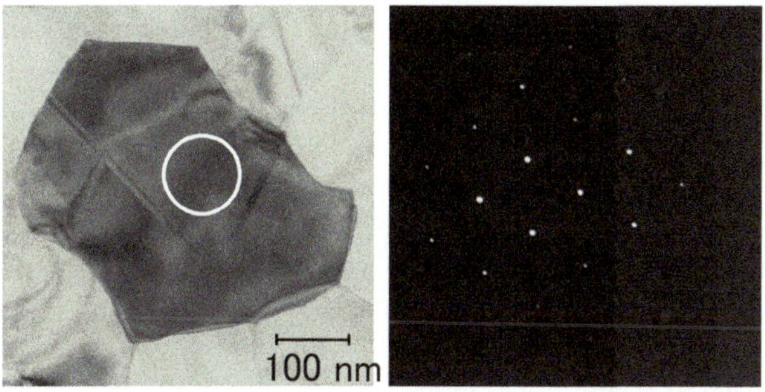

Fig. 14. Typical transmission electron microscopy image and diffraction pattern for a grain of excimer-laser crystallized poly-Si; diffraction pattern is that in the circled region..

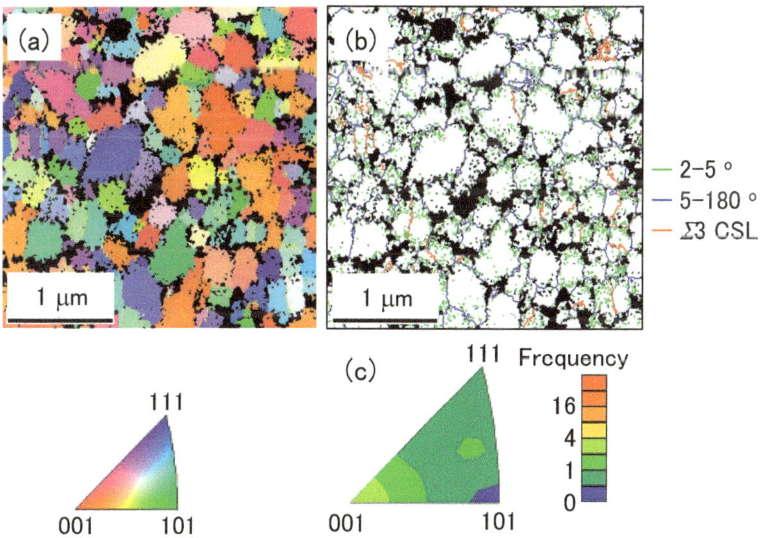

Fig. 15. (a) Crystalline orientation map in terms of the normal direction, (b) grain boundary map, and (c) inverse pole diagrams of the excimer-laser crystallized poly-Si, as observed by electron backscattering diffraction. Surface exhibits weak orientation to the {001} plane. Twin boundaries appear for some grains.

A typical plan-view TEM image and diffraction pattern of a grain of excimer-laser crystallized poly-Si are shown in Fig. 14. The major twin boundary lies across the grain. However, unlike the results for SPC, no micro-twin was found in the grain. Dark contrast due to dislocation, defect clusters, and stacking faults tends to appear at the periphery of the grain.

The EBSD pattern of the surface normal direction (ND) is shown in Fig. 15. Although the surface orientation is scattered over a wide range, the {001} orientation has the highest frequency. A low index plane tends to exist owing to the small growth rate. The grain boundary map indicates that most grain boundaries are random and a quarter of the grains contain twin boundaries corresponding to the Σ3 coincident-site lattice (CSL). The twin boundaries in other grains might be eliminated by grain coalescence due to secondary grain growth.

Raman spectroscopy was applied in order to obtain macroscopic and microscopic characterization of excimer-laser crystallized poly-Si. Figure 16 shows the variation in the FWHM and $\Delta\omega$ of the OPM with the energy density of the laser used for crystallization. The space correlation model (Richter et al., 1981) is useful for analysis of Raman spectra in order to deduce the decrease in the regularity of the crystalline structure, i.e., the presence of nanocrystals and large density defects. The relationship between the FWHM and $\Delta\omega$ of the OPM is plotted in Fig. 17, which also shows a line calculated using the space correlation model. The experimental plots exhibit the same tendency as the calculated values but are shifted to a lower frequency by ~3 cm^{-1}. The magnitude of that deviation is close to the frequency shift caused by tensile stress in the film, which is induced by shrinkage of the film during solidification. The large values of both the FWHM and $\Delta\omega$ at low laser energies are due to the formation of nanosized crystals. At higher laser energies, the space correlation is determined by the defect density because the grains are sufficiently large. This suggests that the defect density is greater than 10^{12} cm^{-2}, which is estimated on the basis of the space correlation length deduced from the FWHM.

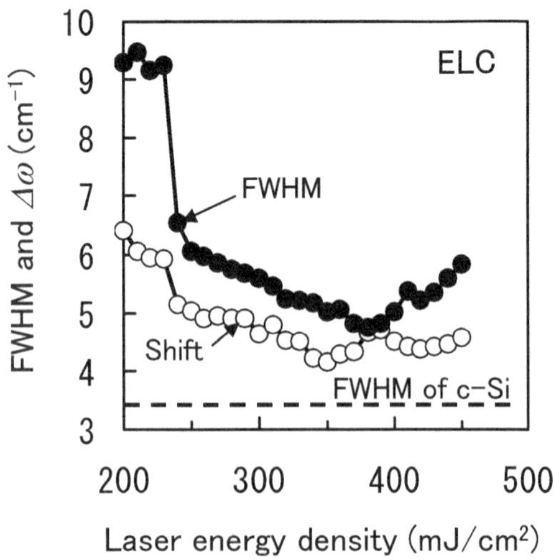

Fig. 16. Full width at half maximum (FWHM) and peak shift ($\Delta\omega$) of the optical-phonon mode as a function of the laser energy density used for crystallization.

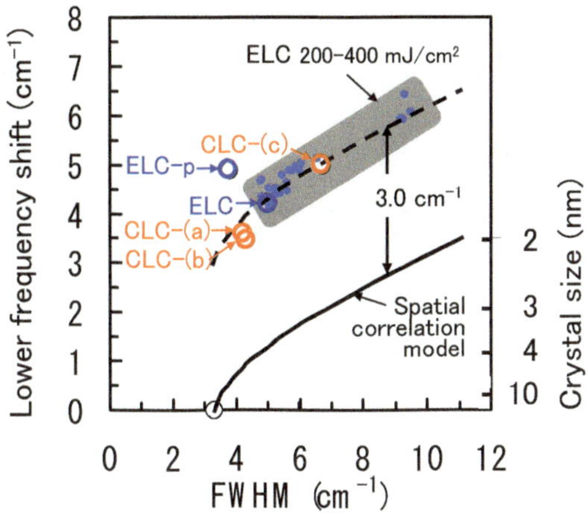

Fig. 17. Relationship between full width at half maximum (FWHM) and lower frequency shift ($\Delta\omega$) for excimer-laser crystallized poly-Si films irradiated at various energy densities. Plots for the film crystallized under the optimum conditions (excimer-laser crystallization, ELC) and that polished to remove hillocks and ridges on the surface (ELC-p) are also shown. Solid line is calculated from the space correlation model; broken lines are values shifted to larger $\Delta\omega$ by 3 cm^{-1}. Plots for continuous-wave laser lateral crystallization (CLC) will be described in the next section.

The Raman spectra for excimer-laser crystallized poly-Si exhibit remarkable features. For example, the intensity of Raman scattering is more than ten times that for SPC films with the same thickness and even that of bulk c-Si. This enhanced Raman scattering has been reported for a roughened semiconductor surface, Si nanocones, and Si nanotubes (Sridharan et al., 2003; Jayavel et al., 2006; Cao et al., 2006). Figure 18 shows a two-dimensional map of the OPM intensity for excimer-laser crystallized poly-Si. The observed area consists of a few micrometer-size grains enlarged by super-lateral growth (SLG) and nanocrystals adjacent to the SLG region (Im & Kim, 1993). The OPM intensity is clearly enhanced at grain boundaries. Thus, the large intensity for the excimer-laser crystallized films is attributed to the enhancement of Raman scattering by hillocks and ridges around the grain boundaries. Therefore, the Raman spectra of excimer-laser crystallized poly-Si reflect mainly the situation around grain boundaries. Accordingly, polishing the sample to remove the hillocks and ridges decreases the intensity to nearly one-tenth. Then, the Raman spectra begin to reflect the region inside the grain. The FWHM changes from 5.0 to 3.8 cm^{-1} after polishing. This supports the conclusion that Raman spectra of unpolished specimens reflect mainly the situation around grain boundaries, where the high defect density increases the FWHM.

The other feature of ELC is the hydrogenation effect. Excimer-laser crystallized poly-Si films with and without hydrogenation were examined in terms of the OPM. Figure 19 shows the variation in $\Delta\omega$ with Secco etching time; the results of SPC are also shown for comparison. The non-hydrogenated film exhibits relaxation of the tensile stress after a short etching period; the relaxation is due to penetration of the etching solution through clusters of defects in the grains to the poly-Si/SiO_2 interface. Defects inside the grain tend to accumulate rather than remain as point defects. In contrast, the hydrogenated excimer-laser crystallized film exhibits almost no stress relaxation. Therefore, a large number of electrochemically active defects are assumed to be present in the grains even though they were not detected by TEM. In the solid-phase crystallized films and excimer-laser crystallized films annealed at 1000 °C, the reduction in stress with etching was not as abrupt, which implies that the thermal treatment reduced the number of dangling bonds in the grains.

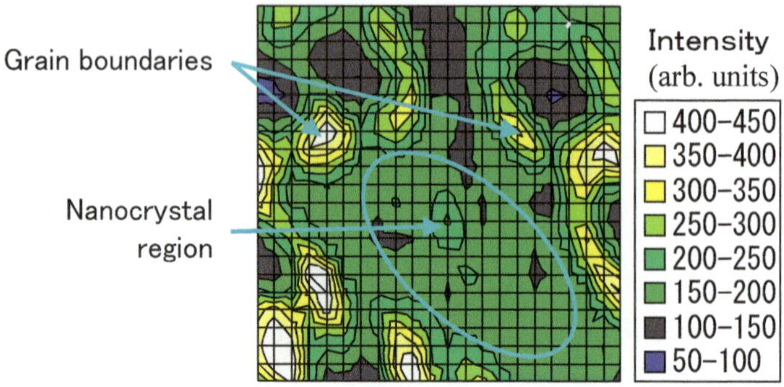

Fig. 18. Two-dimensional map of optical-phonon mode intensity for excimer-laser crystallized poly-Si consisting of a few micrometer-sized grains enlarged by super-lateral growth. View area is 4.2 × 4.2 μm². The intensity is apparently large at grain boundaries.

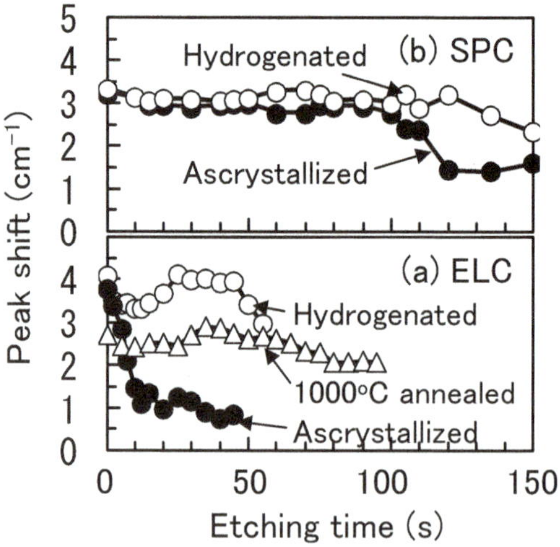

Fig. 19. Variation in peak shift with Secco etching time observed for (a) excimer-laser crystallization (ELC) and (b) solid-phase crystallization (SPC). Only the as-crystallized excimer-laser crystallized film exhibits a rapid decrease in the peak shift, which implies stress release in the film.

On the basis of the results shown above, ELC is considered to progress as follows. Excimer-laser irradiation under optimum conditions melts the film, leaving an adequate seed density. Because solidification occurs extremely quickly, a large number of defect clusters are frozen in the grains. Multiple irradiations affect the merging of relatively high-energy grains with relatively low-energy grains. At a practical grain size of a few hundred nanometers, the lateral growth of individual grains at random orientation progresses, forming high-energy grain boundaries. A large number of the dangling bonds at grain boundaries and the accumulated defects in the grains are electrochemically and probably electronically active.

7. Flow-shaped growth

Free carriers in the channels of poly-Si TFTs are frequently scattered by grain boundaries. The presence of flow-shaped crystals aligned parallel to the channel current is expected to improve the electrical performance. The advantage of this configuration is that it does not require photolithography or complicated optical laser setup.

Flow-shaped growth of Si films has been achieved via techniques such as sequential-lateral solidification (SLS), selectively enlarging laser crystallization (SELAX), and CW laser lateral crystallization (CLC) (Im et al., 1997; Crowder et al., 2000; Hatano et al., 2002; Hara et al., 2002b; Fujii et al., 2007). SLS and SELAX employ pulsed excimer lasers and pulse-modulated diode-pumped solid-state (DPSS) CW lasers (a Nd:YVO_4 laser with a wavelength of 532 nm), respectively, as heat sources. In these methods, the region crystallized by the previous laser pulse acts as the seed for the next laser irradiation. Grain boundaries and subgrain boundaries run approximately parallel to the motion of the solidification interface. The general pattern consists of a branching river-like or wishbone structure (Crowder et al., 2000). The CLC procedure is based on the laser annealing method used in SOI technology. A compact CW DPSS laser or blue-ray semiconductor laser diode (Noguchi et al., 2010) is used as the heat source. Flow-shaped lateral growth is achieved by adjusting the laser scanning velocity and output power. TFTs fabricated by SELAX and CLC exhibit μ_{Fn} values of 440 and 566 cm^2/Vs, respectively; these values are evidently larger than those of TFTs fabricated by conventional SPC and ELC.

We investigated the grain geometry and defects in grains produced by CLC (Kitahara et al., 2009a, 2011b). CLC was performed on 150 nm thick a-Si. Figure 20 shows SEM images of CW laser lateral crystallized poly-Si after Secco etching. The growth geometry varies from the center to the periphery of the laser beam and includes flow-shaped crystals, granular crystals, and leaf-shaped crystals extending to the periphery. The grain boundaries are distinctly visible as sharp etched lines. In the flow-shaped region, most of the etched lines were generated in the crystal, and the number of lines was diminished by coalescence with other lines. Lateral growth processes repeated the generation and reduction of boundaries, as in SLS. Note that the outlines of grains in the leaf-shaped region resemble those of the solid-phase crystallized poly-Si except for the length, which extends toward the outside of the beam. It is likely that the temperature gradient enhances lateral growth in SPC. Figure 20(a′)–(c′) show the surface of a hydrogenated specimen after Secco etching. Most of the grain boundaries in all regions became undetectable, which implies that they were electrochemically active.

Figure 21 shows the crystalline orientation maps and inverse pole diagrams of the Si film in terms of the ND and reference direction (RD) determined by EBSD; the RD is parallel to the laser scanning direction. The ND is distributed and tends to exhibit a high index, whereas the RD tends to be oriented toward <101> or <100>. This suggests that the growth direction is stabilized by the liquid-solid interface rather than the surface or film–substrate interface. Many of the grains began to exhibit ND orientation to <100> with increasing film thickness

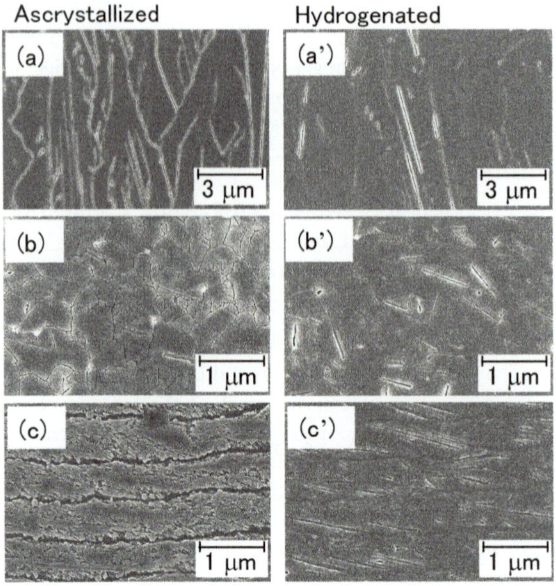

Fig. 20. Scanning electron microscopy images of continuous-wave laser lateral crystallized poly-Si. Secco etching was performed for 25 s. Observations were made in areas from the center to the periphery of the laser beam, as follows: (a) flow-shaped along the laser scanning direction, (b) granular, and (c) leaf-shaped extending to the periphery. (a′)–(c′) Images of the etched film after hydrogenation.

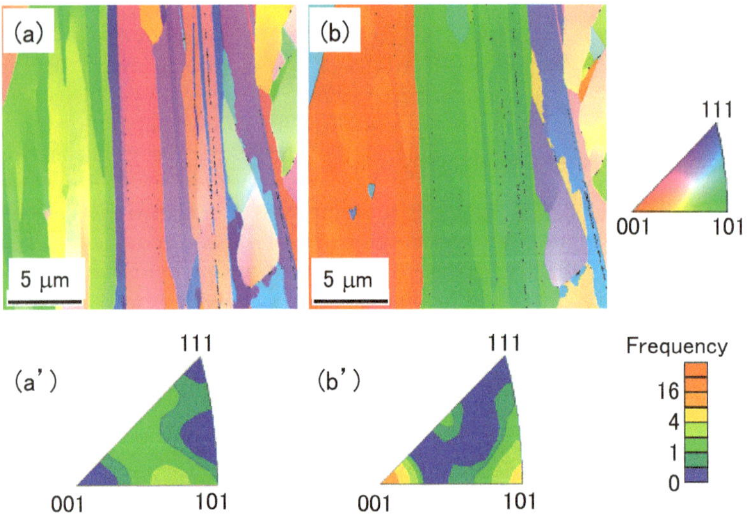

Fig. 21. Crystalline orientation maps for (a) normal direction and (b) reference direction and (a′, b′) individual inverse pole diagrams. Poly-Si film was formed by continuous-wave laser lateral crystallization.

(Hara et al., 2002b). The grain boundary map in Fig. 22(a) indicates that the Σ3 CSL corresponding to the first-order twin boundary lies between the high-angle grain boundaries and extends in the laser scanning direction. Figure 22(b) shows the number fractions of the boundaries in terms of their misorientation angles. The Σ3 CSL with a tilt angle of 60° occupies a fraction of ~0.4. First-order twin boundaries reportedly are not electrically active, whereas second-order twin boundaries act as strong recombination centers (Cunningham et al., 1982). Therefore, the present Σ3 CSL is regarded as electrically inactive. The remaining boundaries, which occupy a fraction of ~0.6, are distributed over a wide range of angles.

Figure 17 shows the relationship between the FWHM and $\Delta\omega$ for three regions in the CW laser lateral crystallized film: (a) flow-shaped growth, (b) granular growth, and (c) SPC-like growth. Regions (a) and (b) exhibit very similar FWHM and $\Delta\omega$ values, which are somewhat smaller than those obtained by ELC. The difference between CLC and ELC can be attributed to the cooling rate. That of CLC is three to four orders of magnitude smaller than that of ELC; the cooling rate of ELC is determined by the scan speed and energy gradation at the edge of the laser beam.

As described above, flow-shaped crystals were successfully grown by CLC. The crystalline boundaries in the resulting films, except for the Σ3 CSL, are electrochemically active, like those of excimer-laser crystallized films. The defect cluster density in the grains of the CW laser lateral crystallized films is smaller than that of the excimer-laser crystallized films; this is explained by the fact that the cooling rate of CLC is several orders of magnitude smaller than that of ELC.

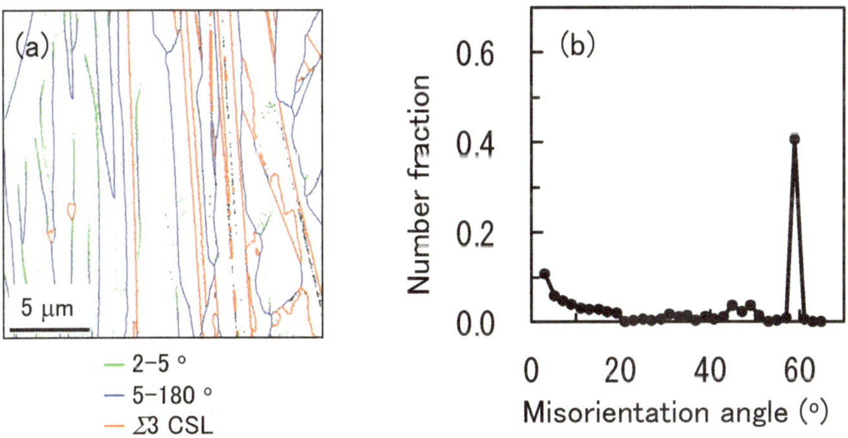

Fig. 22. (a) grain boundary maps and (b) number fractions of the boundaries in terms of the misorientation angle for the poly-Si films shown in Fig. 21. Σ3 coincident site lattice dominates the boundaries.

8. Growth of quasi-single crystalline SiGe thin films

Compounding Si with germanium (Ge) not only extends the application of band engineering but also reduces the melting point, which is advantageous for growth on glass substrates. Rapid melt growth on sapphire substrates and ELC on glass substrates have been reported as techniques for the growth of SiGe thin films (Sameshima et al., 2005; Weizman et al., 2005; Koh et al., 2010; Tanaka et al., 2010) Strong lateral segregation reportedly caused the local Ge content to differ by as much as 40% from the average value (Weizman et al., 2005). Therefore, it is necessary to examine the segregation of Ge and the appropriate growth mode for alloys in order to form crystalline SiGe on glass substrates.

We investigated the crystal configuration, Ge segregation, and growth modes specific to the alloy (Kitahara et al., 2011b). SiGe thin films were crystallized on glass substrates using CLC with a DPSS CW laser. The precursor film was amorphous $Si_{0.7}Ge_{0.3}$ with a thickness of 100 nm.

A typical Raman spectrum of the $Si_{0.7}Ge_{0.3}$ film is shown in Fig. 23. The crystallization of the SiGe films was confirmed by the appearance of sharp optical phonon modes at ~300 cm^{-1} for the Ge–Ge bond, ~400 cm^{-1} for the Si–Ge bond, and ~500 cm-1 for the Si–Si bond.

Figure 24 shows a reflection electron image. In the region corresponding to the central position of the laser beam, multiple areas of bright contrast that extend for more than 100 µm run parallel to the laser scanning direction at intervals of ~1 µm. The coalescence and generation of grain boundaries observed in pure Si films were not evident here, which indicates a type of super-lateral growth. Because the reflection electron intensity depends on the atomic number, the contrast suggests strong segregation of Ge along the bright lines.

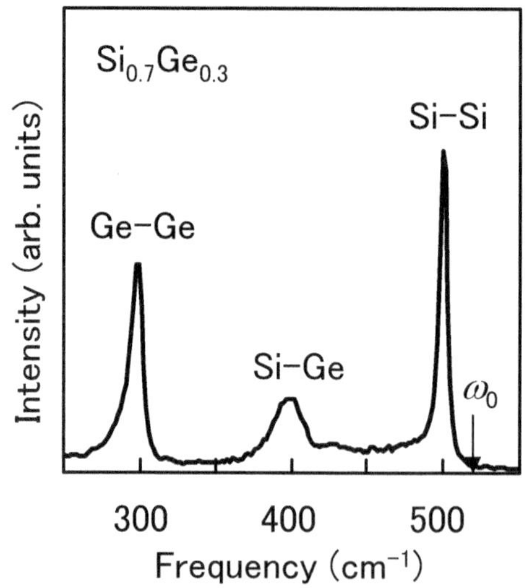

Fig. 23. Typical Raman spectrum for the optical phonon modes of Ge–Ge, Si–Ge, and Si–Si bonds in a crystallized $Si_{0.7}Ge_{0.3}$ film. Peak position for unstrained Si is indicated by ω_0. The lower shift of the Si–Si band from ω_0 depends primarily on the Ge content.

Figure 25 shows crystalline orientation maps and inverse pole diagrams of the film determined by EBSD. "ND" indicates that the crystals are strongly aligned with the <111> plane. The RD, which corresponds to the laser scanning direction, tends to align with the <101> plane. Figure 26 shows the grain boundary map and number fractions of boundaries in terms of their misorientation angles. The map shows that there were very few high-angle grain boundaries. Twin islands were aligned parallel to the laser scanning direction and were outlined by the Σ3 CSL. Although low-angle grain boundaries with a misorientation of <5° appeared, their population was relatively small. Thus, these findings indicate that a quasi-single crystal was formed in the film.

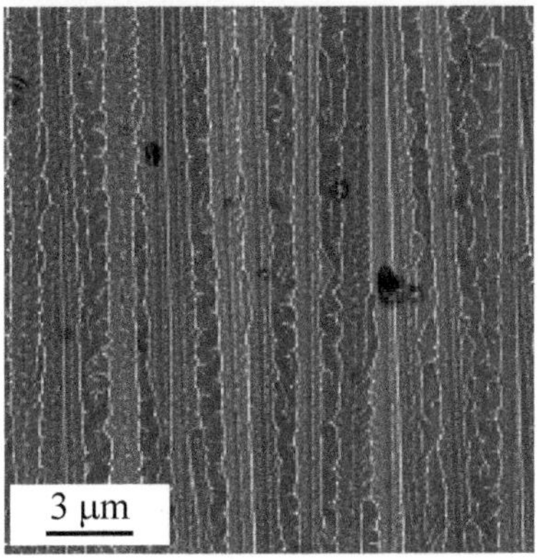

Fig. 24. Reflection electron image of the surface of the $Si_{0.7}Ge_{0.3}$ film. Bright areas corresponds to regions of Ge accumulation.

Figure 27 shows a TEM image and an energy-dispersive X-ray (EDX) profile of Si and Ge. Multiple areas of dark contrast run nearly parallel to the laser-scanning direction in the TEM image. The EDX profile indicates that Ge is strongly segregated in the dark regions of the TEM image. This finding is consistent with the Ge segregation estimated from the reflection electron image in Fig. 24. Thus, Ge segregation was found to occur at many Σ3 CSLs and at both low- and high-angle grain boundaries.

The Σ9 CSL for pure Si reportedly can construct a dangling-bond-free stable structure through the formation of 5- and 7-membered rings (Kohyama & Yamamoto, 1994). In SiGe alloys, the energy at the boundaries can be further reduced by locally varying the Ge content and thus adjusting the lattice constant.

The alloy's characteristics should be considered when discussing the proper growth of SiGe. In pure materials, the solid-liquid interface under a positive temperature gradient is essentially flat during solidification. In contrast, the solid phase in alloys extends into the melt owing to the instability of the solid-liquid interface, resulting in oriented cellular growth; this corresponds to the constitutional undercooling model (Chalmers, 1964). Si and Ge form a complete solid solution. From the phase diagram and the calculation of the equilibrium partition coefficient, the undercooled area was estimated to be on the order of submicrometers in front of the growth region (Kitahara et al., 2011b).

As described above, the growth mode of CLC was changed from flow-shaped growth to super-lateral growth by compounding Ge with Si. This result was attributed to the constitutional undercooling that is peculiar to alloys. At crystalline boundaries running along the laser scanning direction, the energy was lowered by Ge segregation, resulting in the suppression of high-angle grain boundaries. As a result, quasi-single-crystalline films were formed on glass substrates.

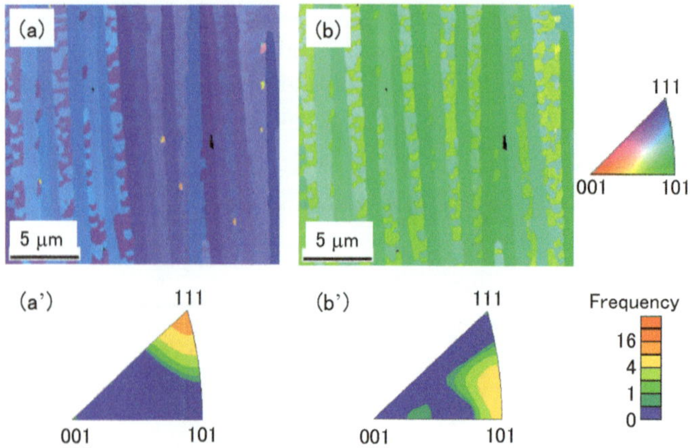

Fig. 25. Crystal orientation maps and inverse pole diagrams for (a, a') normal direction and (b, b') reference direction of $Si_{0.7}Ge_{0.3}$ film. Surface is strongly oriented to <111>.

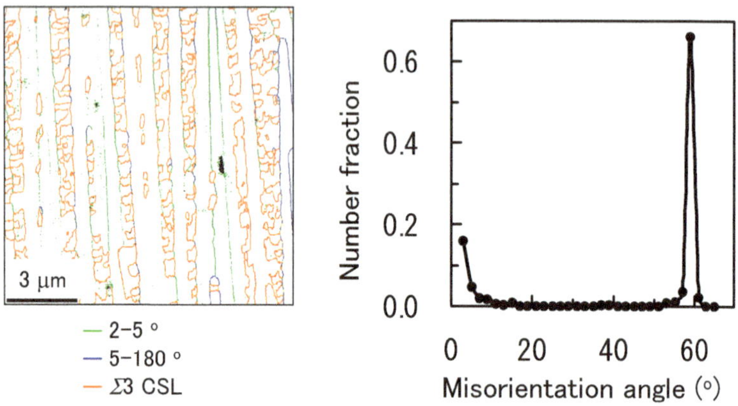

Fig. 26. Grain boundary map and number fractions of the boundaries in terms of the misorientation angle for $Si_{0.7}Ge_{0.3}$ film. Boundaries consist mainly of Σ3 coincident site lattice.

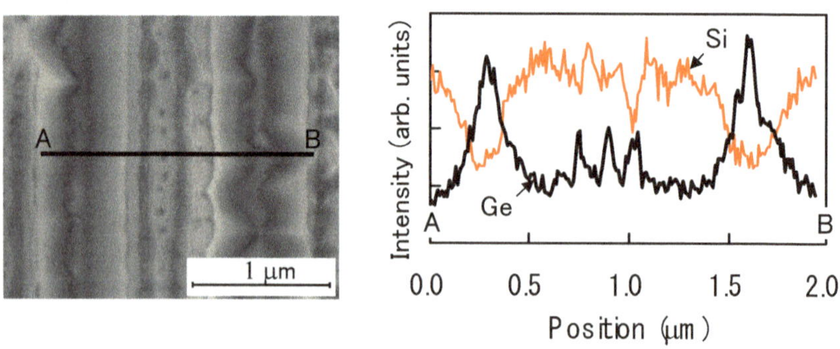

Fig. 27. Transmission electron microscopy image and electron dispersion X-ray profile of Si and Ge atoms in crystallized $Si_{0.7}Ge_{0.3}$ film. Profiles were taken between A and B in the image.

9. Conclusions

Lateral crystallization is indispensable for the growth of poly-Si films on glass substrates for application in large-area devices such as electronic displays and solar cells. In this paper, an outline of the various crystallization techniques was introduced, and our study of lateral growth and defects was described.

When SPC was used, lateral growth proceeded along the major twin boundary. However, a large density of micro-twins appeared in the grains. Oriented lateral growth is expected to decrease the number of defects. In practice, MILC exhibited a decreased defect density with lowered growth temperature, which is attributed to enhanced directivity of needle-like growth.

In ELC, growth is accompanied by melting and recrystallization of the film. TEM revealed only low-density defects in grains. Those defects consisted primarily of dislocations. However, Raman spectroscopy and chemical etching indicated the presence of a considerable density of defects in the grains and at the grain boundaries. The existence of defects in the grains was attributed to the high cooling velocity during recrystallization.

CLC is advantageous for its low cooling velocity and ability to produce oriented lateral growth. Relatively small defect densities were observed in pure Si films. Lateral growth proceeds by generation and disappearance of sub-grain boundaries along the laser scanning direction. Lateral growth was significantly enhanced by compounding Ge with Si, which results in a type of super-lateral growth. Such growth was attributed to the constitutional undercooling effect that is characteristics of alloys. The undercooling effect and the segregation of Ge decrease the energy at the boundaries running parallel to the laser scanning direction, resulting in the growth of quasi-single-crystalline films on glass.

Not only Si and Si-related films but also metal oxides and organic films are beginning to be extensively developed as materials for electronic devices on glass or plastic films. Si will share this role in the development of high-performance devices. One of the remarkable achievements of Si technology is the growth of single-crystalline films on large-area substrates, although such films suffer from cracking, which must be avoided in heterostructures. Both the advance of laser technologies and the control of the physical properties of materials, such as alloying and stacking of different layers, are more promising for the growth of single-crystal films on non-crystalline substrates.

10. Acknowledgements

The authors would like to thank Prof. Tsuda H. in Osaka Prefecture Univ. for TEM observation of SPC and MILC poly-Si. This study was funded in part by the Japan Society for the Promotion of Science [Grants-in-Aid for Scientific Research, (B) 19360165 and (C) 21560329].

11. References

Blake, J. G.; King, M. C.; Stevens III, J. D. & Young R. (1997). Low-Temperature Polysilicon Reshapes FPD Procudtuin, *Solid State Technology*, Vol, 40, No. 151, pp. 151-161

Cammarata, R. C.; Thompson, C. V.; Hayzelden, C & Tu, K. N. (1990). Silicide Precipitation and Silicon Crystallization in Nickel Implanted Amorphous Silicon Thin Films, *Journal of Materials Research*, Vo. 5, No. 10, (October 1990), pp. 2133-2138

Cao, L.; Nabet, B. & Spanier, J. E. (2006), Enhanced Raman Scattering from Individual Semiconductor Nanocones and Nanowires, *Physical Review Letters*, Vol. 96 (April 2006), pp. 157402-1-4.

Chalmers, B. (1964). *Principles of Solidification*, John Wiley and Sons Inc., New York

Christiansen, S.; Lengsfeld, P.; Krinke, J.; Nerding, M.; Nickel, N. H. & Strunk, H. P. (2001). Nature of Grain Boundaries in Laser Crystallized Polycrystalline Silicon Thin Films, *Journal of Applied Physics*, Vol. 89, No. 10 (May 2001), pp. 5348-5354

Crowder, M. A.; Limanov, A. B. & Im J. S. (2000). Sub-Grain Boundary Spacing in Directionally Crystallized Si Films Obtained via Sequential Lateral Solidification, *Materials Research Society Symposium Proceedings*, Boston, 2000, Vol. 621, pp. Q9.6.1-Q9.6.6

Colinge, J.-P. (2004). *Silicon-on-Insulator Technology: Materials to VLSI*, Springer, New York, ISBN: 1-4020-7773-4

Cunningham, B.; Strunk, H. & Ast, D. G. (1982). First and Second Order Twin Boundaries in Edge Defined Film Growth Silicon Ribbon, *Applied Physics Letters*, Vol. 40, No. 3 (February 1982), pp. 237-239

Drosed, R. & Washburn, J. (1980). A New Technique for Observing the Amorphous to Crystalline Transformation in Thin Surface Layers on Silicon Wafers, *Journal of Applied Physics*, Vol. 51, No. 8 (August 1980), pp. 4108-4110, ISSN 0021-8979

Englert, Th.; Abstrerrer, G. A.; & Pontcharra, J. (1980). Determination of Existing Stress in Silicon Films on Sapphire Substrate Using Raman Spectroscopy, *Solid-State Electronics*, Vol. 23, pp. 31-33

Frost R. L. & Shurvell H. F. (1997). Raman Microprobe Spectroscopy of Halloysite, *Clays and Clay Minerals*, Vol. 45, No. 1, pp. 58-72

Fujii S., Kuroki S., Kotani K. & Ito T. (2007). Enlargement of Crystal Grains in Thin Silicon Films by Continuous-Wave Laser Irradiation, *Japanese Journal of Applied Physics*, Vol. 46, No. 4B (April 2007), pp. 2501-2504

Givargizov, E. I. (1991). *Oriented Crystallization on Amorphous Substrate*, Premium Press, New York, ISBN: 0-306-43122-X

Hara, A.; Takeuchi, F. & Sasaki, N. (2002a). Mobility Enhancement Limit of Excimer-Laser-Crystallized Polycrystalline Silicon Thin Film Transistors, *Journal of Applied Physics*, Vol. 91, No. 2, (January 2002) pp. 708-714, ISSN 0021-8979

Hara, A.; Takeuchi, F.; Takei, M.; Suga, K.; Yoshino, K.; Chida, M.; Sano, Y. & Sasaki, N. (2002b). High Performance Polycrystalline Silicon Thin Film Transistors on Non-Alkali Glass Produced Using Continuous Wave Laser Lateral Crystallization, *Japanese Journal of Applied Physics*, Vol. 41, Part 2, No. 3B, (March 2002), pp. L311-L313

Hatalis, M. K. & Greve, D. W. (1988). Large Grain Policrystalline Silicon by Low-Temperature Annealing of Low-Pressure Chemical Vapor Deposited Amorphous Silicon Films, *Journal of Applied Physics*, Vol. 63, No. 7, (April 1988), pp. 2260-2266, ISSN 0021-8979

Hatano, M.; Shiba, T. & Ohkura, M. (2002). Selectively Enlarging Laser Crystallization Technology for High and Uniform Performance Poly-Si TFTs, *SID Symposium Digest of Technical Papers*, Boston, May 2002, Vol. 33, Issue 1, pp. 158-161

Hayashi, H.; Noguchi, T. & Oshima, T. (1984). Polysilicon Super-Thin Film Transistor (SFT), *Japanese Journal of Applied Physics*, Vol. 23, No. 11, (November 1984), L819-L820

Im, J. S. & Kim, H. J. (1993). Phase Transformation Mechanisms Involved in Excimer Laser Crystallization of Amorphous Silicon Films, *Applied Physics Letters*, Vol. 63, No. 14, (October 1993), pp. 1969-1971, ISSN 0003-6951

Im, J. S.; Sposili, R. S. & Crowder, M. A. (1997). Single-Crystal Si Films for Thin-Film Transistor Devices, *Applied Physics Letters*, Vol. 70, No. 25, (June 1997), pp. 3434-3436, ISSN 0003-6951

Jayavel, P. ; Nakamura, S.; Kesavamoorsly, R.; Srivastava, G. P.; Tomoda, W.; Koyama, T. & Hayakawa, Y. (2006). Surface Morphology Effects on the Optical Phonon Modes in $InAs_xSb_{1-x}$ Epilayers on GaAs(001), *Physica Status Solidi* (b), Vol. 243, No. 4, (February 2006), pp. R19-R21

Jellison, Jr. G. E. (1998). Spectroscopic Ellipsometry Data Analysis: Measure Versus Calculated Quantities, *Thin Solid Films*, Vol. 313-314, (February 1998), pp. 33-39, ISSN 0040-6090

Kawazu, T.; Kudo, H.; Onari, S. & Arai, T. (1990). Low-Temperature Crystallization of Hydrogenated Amorphous Silicon Induced by Nickel Silicide Formation, *Japanese Journal of Applied Physics*, Vol. 29, No. 12, (December 1990), pp. 2698-2704

Kitahara, K.; Yamazaki, R.; Kurosawa, T.; Nakajima, K. & Moritani, A. (2002). Analysis of Stress in Laser-Crystallized Polysilicon Thin Films by Raman Scattering Spectroscopy, *Japanese Journal of Applied Physics*, Vo. 41, Part 1, No. 8, (August 2002), pp. 5055-5059

Kitahara, K.; Ohnishi, K.; Katoh, Y.; Yamazaki, R. & Kurosawa, T. (2003). Analysis of Defects in Polycrystalline Silicon Thin Films Using Raman Scattering Spectroscopy, *Japanese Journal of Applied Physics*, Vol. 11, Part 1, No. 11, (November 2003), pp. 6742-6747

Kitahara, K.; Ohashi, Y. & Matsumoto, S. (2007). Hydrogenation Effects on Chemical Activity of Defects in Excimer-Laser-Annealed Polycrystalline Silicon Thin Films, *Japanese Journal of Applied Physics*, Vol. 46, No. 19, (May 2007), pp. L448-L450

Kitahara, K.; Ohashi, Y.; Yamamoto, K. & Sasaki, N. (2009a). Electrochemical and Raman-Scattering Characterizations of Defects in Polycrystalline Silicon Thin Films Formed by Excimer-Laser Annealing, Solid-Phase Crystallization, and Continuous-Wave Lateral Crystallization, *Japanese Journal of Applied Physics*, Vol. 48, (February 2009), pp. 021205 (6 pages)

Kitahara, K.; Kambara, J.; Kobata, M. & Tsuda, H. (2009b). Suppression of Defects during Metal-Induced Lateral Crystallization of Polycrystalline-Silicon Thin Films by Directed Lateral Growth, *Japanese Journal of Applied Physics*, Vol. 48, pp. 091203 (5 pages)

Kitahara, K.; Ishii, T.; Suzuki, J.; Bessyo, T.; Watanabe, N. (2011a). Characterization of Defects and Stress in Polycrystalline Silicon Thin Films on Glass Substrates by Raman Microscopy, *International Journal of Spectroscopy*, Vol. 2011, (August 2011), pp. 632139 (14 pages), doi:10.1155/2011/632139

Kitahara, K.; Hirose, K.; Suzuki, J.; Kondo, K. & Hara, A. (2011b) Growth of Quasi-Single-Crystal Silicon-Germanium Thin Films on Glass Substrates by Continuous Wave

Laser Lateral Crystallization , *Japanese Journal of Applied Physics*, Vol. 50, (October 2011), pp, 115501 (6 pages)

Koh, H.-Y. S.; Chen, S.-L.; Griffin, P. B. & Plummer, J. D. (2010). High Quality Single-Crystal Laterally Graded SiGe on Insulator by Rapid Melt Growth, *Electrochemical and Solid State Letters*, Vol. 13, Issue 8, (May 2010), pp. H281-H283

Kohyama, M. & Yamamoto, R. (1994). Theoretical Study of Grain Boundaries in Si: Effects of Structural Disorder on the Local Electronic Structure and the Origin of Band Tails, *Physical Review* B, Vol. 50, No. 12, (September 1994), pp. 8502-8522, ISSN 0163-1829

Kuriyama, H.; Kuwahara, T.; Ishida, S.; Nohda, I.; Sano, K. & Iwata, H. (1992). Improving the Uniformity of Poly-Si Films Using a New Excimer Laser Annealing Method for Giant- Microelectronics, *Japanese Journal of Applied Physics*, Vol. 31, Part 1, No. 12B, (December 1992), pp. 4550-4554

Kuriyama, H.; Nohda, T.; Ishida, S.; Kuwahara, T.; Noguchi, S.; Kiyama, S.; Tsuda, S. & Nakano, S. (1993). Lateral Grain Growth of Poly-Si Films with a Specific Orientation by an Excimer Laser Annealing Method, *Japanese Journal of Applied Physics*, Vol. 32, Part 1, No. 12B, (December 1993), pp. 6190-6195

Little, T. W.; Takahashi, K.; Koide, H.; Nakazawa, T.; Yudasaka, I. & Ohshima, H. (2001). Low Temperature Poly-Si TFTs Using Solid Phase Crystallization of Very Thin Films and an Electron Cyclotron Resonance Chemical Vapor Deposition Gate Insulator, *Japanese Journal of Applied Physics*, Vol. 30. No. 12B, (December 1991), pp. 3724-3728

Makihara, K.; Nozaki, H.; Asano, T. & Miyasaka, M. (2003). Metal-Induced Lateral Crystallization of Amorphous Silicon under Reduced Nickel Supply, *Solid State Phenomena*, Vol. 93, pp. 207-212

McCulloch, D. J. & Brotherton, S. D. (1995). Surface Roughness Effects in Laser Crystallized Polycrystalline Silicon, *Applied Physics Letters*, Vol. 66, No. 16, (April 1995), pp. 2060-2062, ISSN 0003-6951

Menendez, J. & Cardona, M. (1984). Temperature Dependence of the First-Order Raman Scattering by Phonons in Si, Ge, and a-Sn: Anharmonic Effects, *Physical Review* B, Vol. 29, No. 4, (February 1984), pp. 2051-2059

Mitani, M.; Endo, T.; Taniguchi, Y.; Katou, T.; Shimoto, S.; Ohno, T.; Tsuboi, S.; Okada, T.; Azuma, K.; Kawachi, G. & Matsumura, M. (2008). Ultrahigh-Performance Polycrystalline Silicon Thin-Film Transistors on Excimer-Laser-Processed Pseodo-Single Crystal Films, *Japanese Journal of Applied Physics*, Vol. 47, No. 12, (December 2008), pp. 8707-8713

Mizuki, T.; Matsuda, J. S.; Nakamura, Y.; Takagi, J. & Yoshida, T. (2004). Large Domains of Continuous Grain Silicon on Glass Substrate for High-Performance TFTs, *IEEE Transactions on Electron Devices*, Vol. 51, No. 2, (February 2004), pp. 204-211, ISSN 0018-9383

Nakamura,A.; Emoto, F.; Fujii, E.; Yamamoto, A.; Uemoto, Y.; Senda, K. & Kano, G. (1989). Analysis of Solid Phase Crystallization in Amorphized Polycrystalline Si Films on Quartz Substrates, *Journal of Applied Physics*, Vo. 66, No. 9, (November 1989), pp. 4248-4251, ISSN 0021-8979

Nickel, N. H.; Anderson, G. B. & Johnson R. I. (1997). Grain-Boundary Defects in Laser-Crystallized Polycrystalline Silicon, *Physical Review* B, Vol. 56, No. 19, (November 1997), pp. 12062-12068, ISSN 0163-1829

Noguchi, T.; Chen, Y.; Miyahira, T.; Mugiraneza, J. D.; Ogino, Y.; Iida, Y.; Sahota, E. & Terao, M. (2010). Advanced Micro-Polycrystalline Silicon Films Formed by Blue-Multi-Laser-Diode Annealing, *Japanese Journal of Applied Physics*, Vol. 49, (March 2010), pp. 03CA10 (3 pages)

Richter, H.; Wang, Z. P. & Ley, L. (1981).The One Phonon Raman Spectrum in Microcrystalline Silicon, *Solid State Communications*, Vol. 39, pp. 625-629, ISSN 0038-1098

Roth, J. A.; Olsen, G. L.; Jacobson, D. C. & Poate, J. M. (1990). Kinetics of Solid Phase Epitaxy in Thick Amorphous Si Layers Formed by MeV Ion Implantation, *Applied Physics Letters*, Vol. 57, issue 13, pp. 1340-1342

Samesima, T.; Usui, S. & Sekiya, M. (1986). XeCl Excimer Laser Annealing Used in the Fabrication of Poly-Si TFT's, *IEEE Electron Device Letters*, Vol. EDL-7, No. 5, (May 1986), pp. 276-278, ISSN 0741-3106

Sameshima, S. & Usui, S. (1993). Pulsed Laser-Induced Melting Followed by Quenching of Silicon Films, *Journal of Applied Physics*, Vol. 74, No. 11, (December 1993, pp. 6592-6598, ISSN 0021-8979

Sameshima, T.; Watakabe, H.; Kanno, H.; Sadoh, T. & Miyao, M. (2005). Pulsed Laser Crystallization of Silicon-Germanium Films, *Thin Solid Films*, Vol. 487, pp. 67-71, ISSN 0040-6090

Sameshima, T. (2009). Laser Crystallization for Large-Area Electronics, *Applied Physics* A, Vol. 96, (April 2009), pp. 137-144

Secco d' Aragona, F. (1972). Dislocation Etch for (100) Planes in Silicon, *Journal of Electrochemical Societies: Solid State Science and Technology*, Vol. 119, No. 7, (July 1972), pp. 948-951

Sridharan, M.; Narayandass, Sa. K.; Mangalaraj, D. & Lee, H. C. (2003). Effect of Boron Ion Implantation on the Structural and Optical Properties of Polycrystalline $Cd_{0.96}Zn_{0.04}Te$ Thin Films, *Nuclear Instruments and Methods in Physics Research* B, Vol. 201, pp. 465-474

Spinella, C.; Lombard, S. & Prioro, F. (1998). Crystal Grain Nucleation in Amorphous Silicon, *Journal of Applied Physics*, Vol. 84, No. 10, (November 1998), pp. 5383-5413, ISSN 0021-8979

Takahashi, H.; Suezawa, M. & Sumino, K. (1995). Generation and Dissociation of Complexes of Iron and Phosphorus Atoms in Silicon, *Journal of Applied Physics*, Vol. 78, No. 5, (September 1995), pp. 3078-3082, ISSN 0021-8979

Takayama, T.; Ohtani, H.; Miyanaga, A.; Mitsuki, T.; Ohnuma, H.; Nakajima, S. & Yamazaki, S. (2000). Continuous Grain Silicon Technology and Its Application for Active Matrix Display, *Digest of Technical Papers: AM-LCD 2000*, Tokyo, July 2000, pp. 25-28

Tanaka, T.; Toko, K.; Sadoh, T. & Miyao, M. (2010). High Quality Single-Crystalline Ge-Rich SiGe on Insulator Structures by Si-Doping Controlled Rapid Melting Growth, *Applied Physics Express*, Vol. 3, (February 2010), pp. 031301 (3 pages)

Temple, P. A. & Hathaway, C. E. (1973). Multiphonon Raman Spectrum of Silicon, *Physical Review* B, Vol, 7, No. 8, (April 1973), pp. 3685-3697

Thompson, C. V. (1985). Secondary Grain Growth in Thin Films of Semiconductors: Theoretical Aspects, *Journal of Applied Physics*, Vol. 58, No. 2, (July 1985), pp. 763-772, ISSN 0021-8979

Weizman, M.; Nickel, N.H.; Sieber, I.; Bohne, W.; Rohrich, J.; Strub, E. & Yan, B. (2005). Phase Segregation in Laser Crystallized Polycrystalline SiGe Thin Films, *Thin Solid Films*, Vol. 487, (March 2005), pp. 72-76, ISSN 0040-6090

12

ArF Excimer Laser Annealing of Polycrystalline Silicon Thin Film

Noriah Bidin and Siti Noraiza Ab Razak
Laser Research Group, Advanced Photonic Science Institute, Faculty of Science, Universiti Teknologi Malaysia Johor Malaysia, Malaysia

1. Introduction

Crystallization of amorphous silicon (a-Si) using excimer laser annealing (ELA) has been reported since 1994 by Watanabe group. It is known as the best method to fabricate a good poly-silicon because it can heat the film up to the melting point and, at the same time no thermal damage occur into the glass substrate (Carluccio et al., 1997; Matsumura and Oh, 1999). ELA technique is widely used to increase the grain size and changes the microstructure of polysilicon thin film which is the most important characteristics of excellent built-in polysilicon devices (Palani et al., (2008). The principle advantage of excimer lasers is the strong absorption of ultraviolet (UV) light in silicon beside having larger beam size and high energy density than other laser light sources (Watanabe, et al. 1994). Another major advantage of the excimer lasers is it low-temperature polysilicon annealing. Excimer laser with 308 nm wavelength for example has been reported can transform 50-nm-thin layers of amorphous silicon into high-quality polycrystalline silicon with greatly enhanced electron mobility, for use in flat-panel displays for mobile phones and flat-screen televisions (Delmdahl, 2010). In the low-temperature annealing of polysilicon, excimer lasers with UV output energies of over 1 J per pulse and output powers of 600 W are used to manufacture liquid-crystal and organic LED backplanes at a rate of 100 cm^2 s^{-1}. A new VYPER/LB750 line beam annealing system enables volume production of low-temperature polysilicon (LTPS) on large generation 6 glass panels. LTPS is the key material for high-resolution liquid crystal displays (LCDs) and organic light-emitting diode (OLED) displays for smartphones, tablet PCs and TVs.

Today, the most prominent applications of excimer lasers are in semiconductor chip manufacturing. This relies on the deep-UV emission wavelength of 193 nm, with semiconductor chip manufacturing capitalizing on the optical resolution of excimer lasers. For example, high repetition rate line-narrowed excimer laser models with 10 mJ per pulse, pulse frequencies of up to 6 kHz and narrowed bandwidths of 0.35 pm are used in advanced photolithography to produce computer chips with feature sizes of 45 nm.

Ongoing miniaturization in microelectronics and the trend towards thin-film technologies demands increasing lateral resolution and selective machining. Functional structures and active layers are often only tens of nanometres thick and has to be annealed, patterned and removed in a selective manner without damaging underlying layers or substrates. Because excimer lasers provide the shortest wavelength of all laser technologies, they will continue to have a crucial role in many industries over the next decade. In years to come we will see high-power applications of excimer lasers, with output energies reaching 2 J and output power levels of 1,000 W and above (Delmdahl, 2010).

The aim of this chapter is to demonstrate the crystallization of polysilicon thin film by excimer laser annealing. Various aspects of excimer laser annealing process will be covered in this chapter including principle and experimental results. The definition and the history of excimer laser annealing are described in introduction section. Since excimer laser is the key component in this annealing process, the detail regarding excimer laser system is discussed. Mechanism of laser annealing is explained under third section of laser ablation which includes the ablation phenomena and photochemical mechanism. The detail sample preparation using solid phase crystallization technique is described under section 4. Section 5 will give information regarding polycrystalline silicon. The basic annealing process relies on transformation temperature in which three important parameters, comprising critical point, recalescence effect and super lateral growth energy are desired to be identified. The crystallization of silicon thin film is characterized via atomic force microscope analysis. Finally the excimer laser annealing on silicon thin film is summarized in section 8.

2. Excimer laser

2.1 History and application

An excimer laser is a form of ultraviolet laser which is commonly used in the production of microelectronic devices (semiconductor integrated circuits or "chips"), eye surgery, and micromachining. The excimer laser was invented in 1970 by Nikolai Basov, V. A. Danilychev and Yu. M. Popov, at the Lebedev Physical Institute in Moscow, using a xenon dimer (Xe_2) excited by an electron beam to give stimulated emission at 172 nm wavelength. A later improvement, developed by many groups in 1975 was the use of noble gas halides (originally XeBr).

The excimer laser typically uses a combination of a noble gas (argon, krypton, or xenon) and a reactive gas (fluorine or chlorine). Under the appropriate conditions of electrical stimulation and high pressure, a pseudo-molecule called an excimer (or in the case of noble gas halides, exciplex) is created, which can only exist in an energized state and can give rise to laser light in the ultraviolet range.

Laser action in an excimer molecule occurs because it has a bound (associative) excited state, but a repulsive (dissociative) ground state. This is because noble gases such as xenon and krypton are highly inert and do not usually form chemical compounds. However, when in an excited state (induced by an electrical discharge or high-energy electron beams, which produce high energy pulses), they can form temporarily-bound molecules with themselves (dimers) or with halogens (complexes) such as fluorine and chlorine. The excited compound can give up its excess energy by undergoing spontaneous or stimulated emission, resulting in a strongly repulsive ground state molecule which very quickly (on the order of a picosecond) dissociates back into two unbound atoms. This forms a population inversion. Excimer lasers are usually operated with a pulse repetition rate of around 100 Hz and a pulse duration of ~10 ns, although some operate at pulse repetition rates as high as 8 kHz and some have pulsewidth as large as 30 ns.

Excimer lasers are widely used in high-resolution photolithography machines, one of the critical technologies required for microelectronic chip manufacturing. Current state-of-the-art lithography tools use deep ultraviolet (DUV) light from the KrF and ArF excimer lasers with wavelengths of 248 and 193 nanometers (the dominant lithography technology today is thus also called "excimer laser lithography), which has enabled transistor feature sizes to shrink below 45 nanometers. Excimer laser lithography has thus played a critical role in the continued advance of the so-called Moore's law for the last 20 years.

The most widespread industrial application of excimer lasers has been in deep-ultraviolet photolithography, a critical technology used in the manufacturing of microelectronic devices (i.e., semiconductor integrated circuits or "chips"). Historically, from the early 1960s through the mid-1980s, mercury-xenon lamps had been used in lithography for their spectral lines at 436, 405 and 365 nm wavelengths. However, with the semiconductor industry's need for both higher resolution (to produce denser and faster chips) and higher throughput (for lower costs), the lamp-based lithography tools were no longer able to meet the industry's requirements. This challenge was overcome when in a pioneering development in 1982, deep-UV excimer laser lithography was proposed and demonstrated at I.B.M. by Kanti Jain. With phenomenal advances made in equipment technology in the last two decades, and today microelectronic devices fabricated using excimer laser lithography totaling $400 billion in annual production, it is the semiconductor industry view that excimer laser lithography has been a crucial factor in the continued advance of Moore's law, enabling minimum features sizes in chip manufacturing to shrink from 0.5 micrometer in 1990 to 32 nanometers in 2010. This trend is expected to continue into this decade for even denser chips, with minimum features approaching 10 nanometers. From an even broader scientific and technological perspective, since the invention of the laser in 1960, the development of excimer laser lithography has been highlighted as one of the major milestones in the 50-year history of the laser.

For micro- and nano-technologies, laser micromachining is currently used in a large number of R&D and industrial applications. The range of applications to which laser methods are applied is continuously expanding, supported also by the development of novel processing techniques. Over the last decades the excimer laser has obtained the key position among lasers in various sectors of micromachining. Excimer lasers have developed into powerful manufacturing tools mainly because of two reasons: (i) The short wavelengths of the excimer laser offers excellent quality of machining and a great versatility in features which can be produced. (ii) The progress in basic excimer laser technology has made the excimer lasers to reliable machines suitable for the industrial environment.

There are certainly not many types of lasers which have found such broad markets as the excimer laser. Over the last years the main growth results from increasing industrial use followed by medical applications while new sales into R&D applications stay nearly constant. Today the largest known industrial applications of excimer lasers are (i) based on micromaching of different materials as polymers, ceramics and glasses, applied for example in the production of ink jet cartridges by drilling the nozzles, (ii) excimer laser radiation is being used for changing the structure and properties of materials as oxides, silicon or glass in bulk or thin films, as applied for the production of active matrix LCD monitors, fiber Bragg gratings in telecommunication, and high temperature superconducting films, (iii) employing the excimer laser as "short wavelength light bulb" in optical microlithography for the production of computer chips with critical dimensions below 0.25 μm (the largest homogeneous market for excimer lasers).

All the widespread applications of excimer lasers in micromachining and medicine are based on the early use of excimer lasers leading to the discovery of the ablation of materials under intense illumination with ultraviolet laser pulses by R. Srinivasan.

2.2 Crystallization source

The source of energy used for crystallization in this work is an argon fluoride excimer laser. The excimer laser generates deep ultraviolet light measuring 193 nm when high voltage energy is discharged into a cavity containing a mixture of a rare gas argon and a halogen (i.e., fluoride) Almost, 90% or more of the mixture contain buffer rare gas (typically helium) that does not take part in the reaction. This gas mixture is pre-ionized by a set of electrodes before a high-voltage current (about 30,000 eV is applied) resulting in formation of highly unstable rare-gas halide molecules, which rapidly dissociate, emitting UV light whose wavelength is determined by the particular gas mixture chosen. Excimer lasers use in this particular project is comprised of a mixture of argon and fluoride gas. The term excimer is derived from the two words "excited" and "dimer" which are used to describe the reaction in which the laser transfer energy through an ultraviolet beam of light (Schneider, 1998). All emitted powerful pulses lasting in 10 nanoseconds at wavelength of ultraviolet region of 193 nm.

2.3 Excimer laser structure

The gain of excimer lasers is extremely high, so the output is superradiant. A single rear mirror is employed and an output coupler of 4 to 8% which transmit in the region of interest (which is in UV region) is used (Figure 1). Divergence of the beam is reduced when a full optical cavity is used, and alignment is easy since the laser operates even when cavity mirrors are completely misaligned. The beam is rectangular profile. Quartz cannot be used as an output coupler for argon fluoride (ArF) Excimer laser since fluorine attacks the material, hence magnesium fluoride are used to replace quartz as it absorbs less UV at most wavelength compare to quartz (Csele, 2004). In discharge excitation, electric current flows through the laser medium, typically ranging from a kilovolt (kV) to over tens of kilovolts deliver energy to the laser gas (Hecht, 1992).

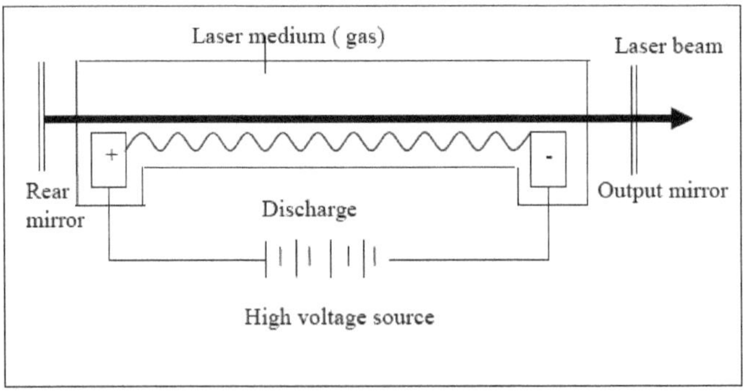

Fig. 1. An electrical discharge exciting a gas laser (Hecht, 1992)

Excimer laser is excited by passing a short, intense electrical pulse through a mixture of gases containing the desired rare gas and halogen. However, the molecule were found more stable when it is on excited state and become less stable on ground state and this properties can be found on argon fluoride (ArF), krypton fluoride (KrF_2) and xenon chloride ($XeCl_2$). This phenomenon could be explained in detail by referring on Figure 2.

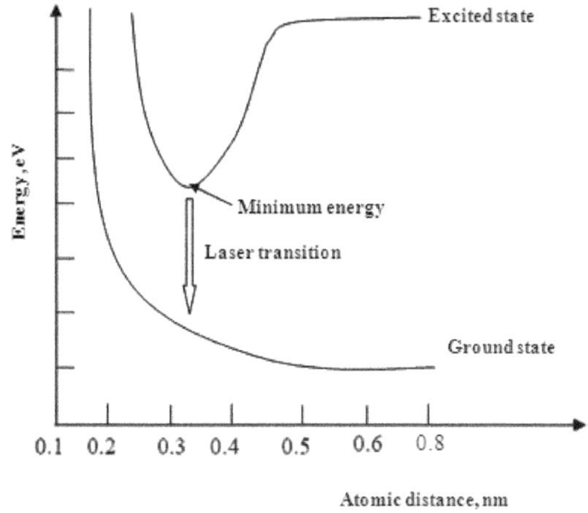

Fig. 2. Excimer laser potential energy curve.

Figure 2 shows a graph of potential energy curve as a function of atomic distance between the molecules. Generally, the potential energy is minimized at equilibrium molecule distance. However, for ArF, KrF_2 and $XeCl_2$ material, these properties did not occur. They are found not stable even when they are in their ground state. Thus, excitation process cannot directly be done on ground state. Hence, indirect excitation in discharge system is employed. For example, in ArF excitation process, electron reaction process is as followed,

$$e + F^2 \rightarrow F^- + F \qquad (1)$$

The negative ion produced will combine with positive ion to produce excited molecule

$$Ar^+ + F^- \rightarrow (ArF)^* \qquad (2)$$

This reaction can produce an efficient excited dimer molecule. The unstable molecules at ground state leads to a small number of population. Hence, it causes the population inversion to easily happen and laser transition will trigger. The nature of dimer material that is not stable on their ground state causes the atom to easily breaks-up (Shah, 2009). Because of the ground state essentially does not exist, there is a population inversion as long as there are molecules in the excited state. This process is performed again and again; this is how the pulse excimer laser is trigger out.

2.4 External triggering

An argon fluoride ArF excimer laser, manufactured by GAMLaser model EX5/200-110, is employed as a source of ablation. The wavelength of the laser is 193 nm with pulse duration of 10 ns and repetition rate which can be varied from 20 Hz to 200 Hz. The beam size of the laser output when it hits the sample is 4 mm x 6 mm. The size of the beam also depends on the output energy of the laser beam (GAMLaser Inc., 2003).

The excimer laser is internally operated via the aids of computer using 32 bit Windows software (GAMLaser Inc., 2003). The laser parameters including the frequency, high voltage and number of pulse can be commendable according to the requirement of the experiment. The output energy of the laser can be varied by manipulating the capacitor voltage from 10 kV to 15 kV. The dose number of laser exposure is not limited; however, the minimum number of pulses can be set at 100 pulses. However, to trigger the number of pulse as minimum as 1 pulse, a Sony Tektronix arbitrary function generator model AFG 310 is employed. Figure 3 shows photograph of the excimer laser, which coupled to an arbitrary function generator in the real field. In this particular experiment, only a few pulses of laser exposure are required to obtain the desired optical surface treatment.

Fig. 3. An excimer laser connected to a function generator for external triggering.

The function generator can be used to control the number of pulses lower than 100 pulses. As an external trigger unit, the function generator is set up at '5 Volts', operated at 'pulse' output function, and with 'burst' mode (Sony Tektronix Inc., 2003). The frequency of the function generator is synchronized with the excimer laser system.

2.5 Excimer laser annealing

A GAM LASER Argon Fluoride ArF excimer laser is used to anneal the hydrogenerated silicon thin film. The ArF excimer laser generates ultraviolets laser light at 193 nm with 10 ns pulse duration. The lasers parameters such as number of pulses, repetition rates and pumping voltages can be controlled via a computer program. In this experiment, the laser capacitor voltage is kept constant at 12 kV with repeatition rate of 20 Hz, operating at room temperature. The gas pressure is set at 3200 Torr.

The excimer laser annealing experimental setup is shown in Figure 4. The laser energy (in mJ) for each laser pulses is measured using 13PEM001 Broadband power and energy Meter.

The laser is controlled externally using a AFG310 function generator for obtaining laser pulse below than 100 pulses. In this present works, the number of pulse is varied from one to ten pulses. The operational voltage of the function generator is set at 5 Volts 'pulse' output function, and with 'burst' mode. The laser pulse is generated after 2.5 μs upon application of the trigger pulse rising edge with DC charging is operating in external trigger mode. The frequency of the function generator and the excimer laser system is fixed at 20.0 Hz. The silicon copper thin film placed on *x* and *y* translational stage in and vertical position with respect to the excimer laser beam axis. The number of pulses of excimer laser beam exposed on the thin film target is, controlled by the external trigger unit.

Maximum laser performance can be achieved by aligning the laser mirrors. The mirrors are alligned via the guidance of a visible of helium-neon He-Ne laser. The He-Ne laser is aligned to coaxis with excimer laser beam. The excimer laser is optimized by adjusting the output coupler via the aids of power/energymeter. A bright and rectangular dimension of beam spot is recorded permanently using beam profiler. The images are arranged in order of 3D, 2D and the bright beam spot in the real field such as shown in Figure 5 a, b and c respectively. The dimension of excimer laser beam spot is 4 × 6 mm^2. The a-Si films is recrystalized at different energy densities. Finally, nanostructure of the crystallized silicon is examined via atomic force microscope AFM.

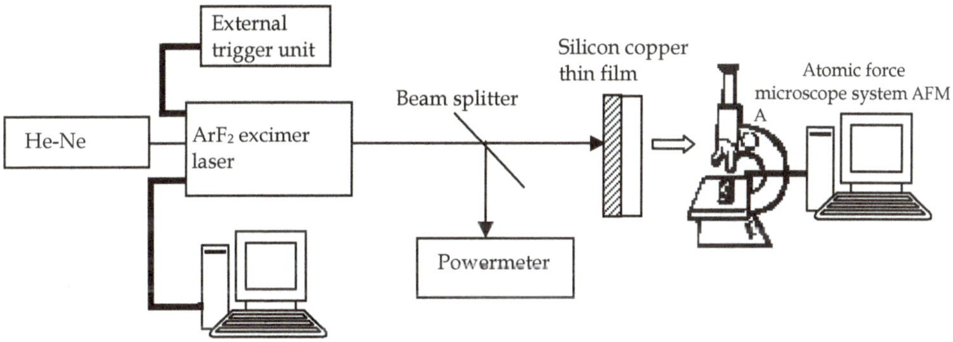

Fig. 4. The schematic diagram of excimer laser annealing process.

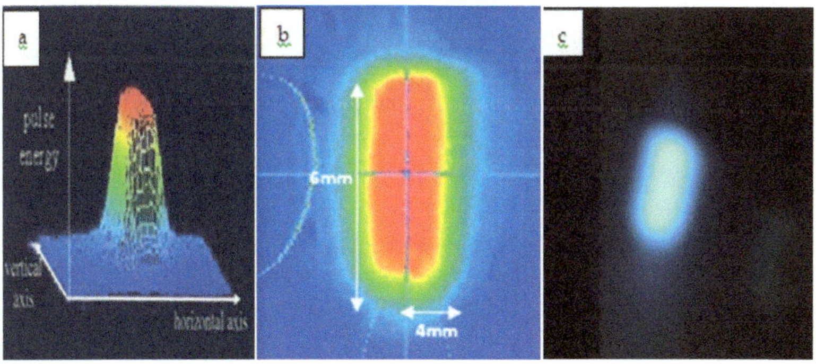

Fig. 5. Beam profile and beam spot of excimer laser. a. 3D beam profile, b. 2D beam profile, and c. The beam spot in the real field.

3. Laser ablation

In laser annealing process the main key for its successful work is the ablation process between optical material and UV light from argon fluoride (ArF) excimer laser. In order to understand the laser-matter interaction, it is better to understand the fundamental mechanism of laser ablation and optical material behavior with respect to the ultraviolet (UV) light illumination. In this section, the photochemical process occurred during ablation process will be discussed.

Ablation is usually described in terms of physical mechanism such as vaporization and shock effects. It is often performed in vacuum or air. Examples of ablation application include polymer ablation, removal of thin metal films from an insulating substrate, and deposition of high-temperature superconducting thin films by ablation of bulk targets (Brannon, 1997).

The laser ablation process requires an intense UV light (λ = 193 nm- 351 nm) produced by an excimer laser. Because of the high absorption of UV light and relatively poor conductivity of many materials, particularly silicon, the energy is deposited in a very thin layer. Whenever the energy density or fluence exceeds the ablation threshold value for the material, chemical bonds are broken, fracturing the material into energetic fragments. The fragments are atoms, groups of atoms, ions and electrons. Because the fragments leave the reaction zone as an energetic gas and solid debris, the ablation process resembles explosive evaporation of the material (Speidell, 1997).

Generally, there are two classes of laser ablation mechanisms: thermal and electronic (non-thermal). Thermal processes rely on an intense laser pulse to very rapidly heat a surface at rates of the order of 10^{10} K/sec. These processes can produce expansion and vaporization from solid and melted regions.

Laser ablation of given material can involve both thermal and non-thermal mechanisms. For example, during the interaction of the leading edge of the laser pulse with the solid, non-thermal processes may predominate because significant heating has not yet occurred. As the rest of the pulse is absorbed, the temperature rise rapidly and thermally activated mechanism may commence. It should be noted that laser excitation of a strongly absorbing solid will always result in substantial heating regardless of the ablation mechanism. Although non-thermal process may occur, there is no way to "turnoff "the thermal processes. The dominant process is ultimately determined by the detailed chemistry and physics of the solid.

3.1 Photochemical

In the photochemical processes the photon absorption event break a chemical bonding within a molecule forming molecules with smaller number of atoms. The resulting photoproducts occupy a large volume and create pressure inside the irradiated volume that can then convert to the translational energy of ablation (Sato et al., 2001). During this process, thermal and mechanical damage to the surrounding material is minimized, therefore achieving more precise control over the ablated region.

There has been much un-certainty and debate over the fundamental ablation mechanisms in material. Several photochemicals and photothermals models have been developed to explain the ablation mechanisms. In photochemical mechanisms, the photon break the chemical bonds of the material directly where as in photothermal mechanism the material is ablated by heating, melting and vaporizing the material (Chen et al., 1995). It is found that the material removal by laser ablation approximately obeys the Beer lamberts Law at lower laser fluences (Srinivasan, 1982). For photochemical ablation to occur, energy of the photons at that wavelength should overcome the intermolecular bond energies of the target material. The relation between the photon energy of light and laser wavelength is given by (Sauerbrey, 1989),

$$E = 1.245 / \lambda \tag{3}$$

Where λ is wavelength of light (193 nm), E is energy of photon (eV). The photon energy of the lights depends on the wavelength of the light and as the wavelength increases, the energy decreases. An UV laser with a wavelength of 193 nm has photon energy of 6.45 eV. It can be seen that for photochemical ablation to occur in thin film the photon energy of the light should be greater than the bond energy of the material. While excimer laser does not produce as high average powers as longer wavelength infrared lasers such as Nd:YAG and CO_2, their shorter wavelength carried high photon energy. As for example, argon fluoride (ArF) excimer laser with wavelength of 193 nm have photon energy of 6.45 eV, while Nd:YAG with wavelength of 1064 nm have photon energy of 0.18 eV. Hence, only laser with short wavelength can initiate photochemical reaction while the laser with longer wavelength may only lead to heating the material only.

Apparently, the advantage of photochemical reaction, is that the mechanism can change the molecule structure as well as the refractive index of the material, but not involve in external damage unless the energy deliver is too high. As a result the photochemical mechanism is important effect in the nucleation and the growing of grain size or in the crystallization process.

4. Sample preparation

Target used to be annealed is an amorphous silicon. In this works silicon is chosen to be the main material and deposited on the glass substrate as the first layer of thin film. Silicon is widely used in semiconductors industry. It can stand at higher temperatures compared to germanium. Furthermore its native oxide is easily grown in a furnace. Plus it can form a better dielectric interface rather than any other materials.

An ordinary glass substrate is conducted to make it possible for low annealing temperature. The glass substrate used in this research has maximum operation temperature in the range of 350°C to 400°C. The thickness of the glass substrate is in between 1.0 mm to 1.2 mm. The glass substrates are cut into small pieces using diamond cutter. Each of them having dimension of 10 × 10 mm^2. All glass substrates are cleaned using Ultrasonic Branson 3210 cleaning machine to ensure the glass surface is free from any dust, oil, or contaminations.

4.1 Thin film fabrication

Solid phase crystalization (SPC) is prepared by using a vacuum chamber technology. An Edwards 360 thermal evaporation is a vacuum chamber used to fabricate the silicon thin film. Silicon is deposited on a glass substrate. Initially, the source material to be deposited or evaporant is loaded into a molybdenum boat. The glass substrate is placed 10 cm above the evaporant. The chamber is set at a vacuum condition until the pressure achieved down to 10^{-6} mbar. The current is increased slowly until the evaporant on the molybdenum boat begin to melt. The thickness of film is detected by quartz crystal thickness detector inside the vacuum chamber. The thickness is measured via an Edwards FTM 7 which connected to the vacuum system. To deposit the silicon for about 100 nm thick, 0.3 g of silicon powder is needed. A schematic diagram of the thin film is shown in Figure 6.

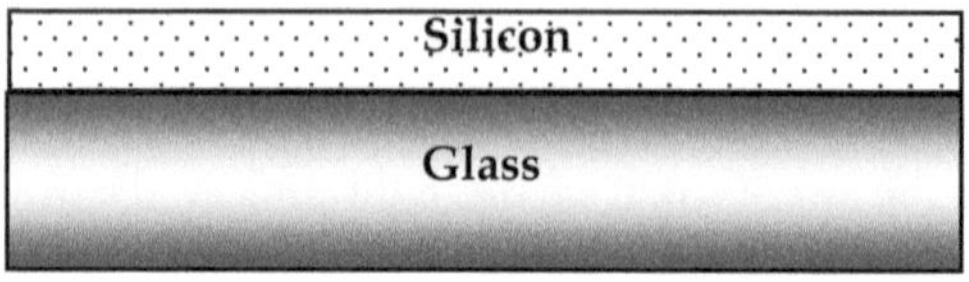

Fig. 6. Silicon thin film on glass substrate

4.2 Annealing process

The annealing of silicon thin film is a combination from heat treatment and excimer exposure. Immediately after fabricated the thin film, it is heated via conventional Tube Furnace type F21100. This is important procedure in order to dry out the water inside the thin film so that the hydrogen percentage can be lowered. The hydrogenated thin film is then annealed by using an argon fluoride (ArF) excimer laser.

4.2.1 Heat treatment

The prepared solid phase crystallization of silicon thin film is initially under-went heat treatment. As mention earlier the aim is to extract the water contained in the film during deposition. An encapsulated annealing technique is utilized for this heat treatment in order to avoid vaporization of thin film material. The silicon thin film is placed in a carbon block (both the container and its cover are made from the same element). The capsulated thin film is then placed in the tube furnace. The annealing process is carried out under atmospheric pressure for four hours in air. The annealing is set at a fixed temperature of 350 °C. This furnace has a single set point control which enables the end user to control the furnace up to a preset temperature and remained constant at the set temperature.

4.2.2 Laser annealing

Nearly all optical materials show moderate to intense absorption in the ultraviolet region. This absorption is usually ascribed to electronic transition from a ground singlet to the first excited state. The unique feature in the UV laser ablation of optical material is encountered only in those wavelength region in which such electronic absorption exist (Srinivasan et al., 1988). Material ablation depends on two key conditions, as shown in Figure 7. First, the material must absorb light strongly at the laser's wavelength. Clean, precise ablation usually requires linear absorption coefficients of at least 10^4 cm^{-1}. Secondly, ablation occurs only after the material has absorbed a minimum energy per unit volume, i.e., the laser intensity must exceed a threshold value (Brannon et al., 1985).

The laser is absorbed in amorphous silicon thin film surface without heating the substrate. A homogenized excimer laser beam spot is exposed on the thin film surface. Within the laser pulse duration the silicon layer is rapidly heated and melted as shown in Figure 7(a, b). As it cools down the crystallization into poly-silicon occurs (Figure 7c, d). In the process of excimer laser annealing, the amorphous silicon thin film is exposed by multiple pulses of laser. The exposed area is partially melted. The period for heating as well as for freezing is the same as the pulse duration of the excimer laser. The heat release in high speed freezing is more compared to the slow rate once. As a result during cooling the heat released by convection force-liquid flow which break the planar symmetry so that the crystal grow develops along the columns or finger such as shown in Figure 7c and d. In Figure 7c, there is partially melted and partially crystallized growth. The totally crystal growth is illustrated in Figure 7d. The crystal grows in the finger formation or poly-crystallization.

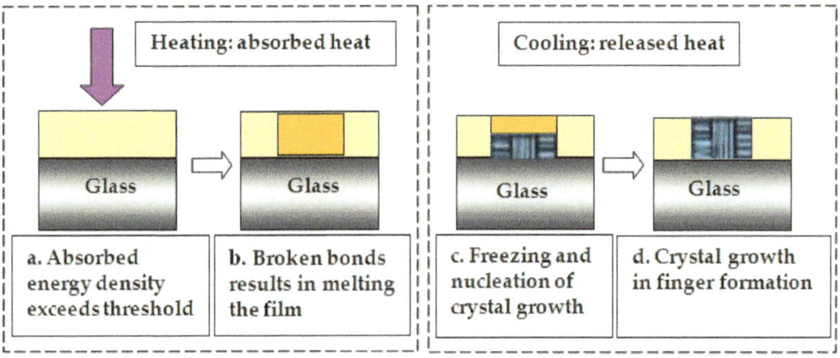

Fig. 7. Schematic diagram of excimer laser annealing on amorphous silicon copper thin film.

5. Polycrystalline silicon

Silicon films are fabricated with a microstructure tailored to the application depending on economic and performance requirements. Silicon film is typically divided into three categories: monocrystalline, amorphous, and polycrystalline.

Monocrystalline silicon is a form in which the crystal structure is homogenous throughout the material; the orientation, lattice parameter, and electronic properties are constant throughout the material (Green, 2004). Dopant atoms such as phosphorus and boron are often incorporated into the film to make the silicon n-type or p-type respectively. Monocrystalline silicon is fabricated in the form of silicon wafers, usually by the Czochralski Growth method, and can be quite expensive depending on the radial size of the desired single crystal wafer (Green, 2004). This monocrystalline material is one of the chief expenses where approximately 40% of the final price of the product is attributable to the cost of the starting silicon wafer used in cell fabrication (Campbell, 2001).

Amorphous silicon has no long-range periodic order. The application of amorphous silicon as a standalone material is somewhat limited by its inferior electronic properties (Streetman and Banerjee, 2000). When paired with microcrystalline silicon in tandem and triple-junction solar cells, however, higher efficiency can be attained than with single-junction solar cells (Shah et al., 2003). This tandem assembly of solar cells allows one to obtain a thin-film material with a bandgap of around 1.12 eV (the same as single-crystal silicon) compared to the bandgap of amorphous silicon of 1.7-1.8 eV bandgap (Shah et al., 2003). Tandem solar cells are then attractive since they can be fabricated with a bandgap similar to single-crystal silicon but with the ease of amorphous silicon.

Polycrystalline silicon is composed of many smaller silicon grains of varied crystallographic orientation (Figure 8). This material can be synthesized easily by allowing liquid silicon to cool using a seed crystal of the desired crystal structure. Additionally, other methods for crystallizing amorphous silicon to form polycrystalline exist such as high temperature chemical vapor deposition (CVD). Each grain is crystalline over the width of the grain. The grain boundary separates the grains where the adjoining grain is at a different orientation than its neighbor. The grain boundary separates regions of different crystal structure thus serving as a center for recombination. '*d*' here is a characteristic grain size, which should be maximized for maximum film efficiency. Typical values of *d* are about 1 micrometer.

Presently, polysilicon is commonly used for the conducting gate materials in semiconductor devices such as MOSFETs; however, it has potential for large-scale photovoltaic devices (Streetman and Banerjee, 2000; Ghosh et al., 1980). The abundance, stability, and low toxicity of silicon, combined with the low cost of polysilicon relative to single crystals makes this variety of material attractive for photovoltaic production (Ghosh et al. 1980). Grain size has been shown to have an effect on the efficiency of polycrystalline solar cells. Solar cell efficiency increases with grain size. This effect is due to reduce recombination in the solar cell. Recombination, which is a limiting factor for current in a solar cell, occurs more prevalently at grain boundaries as shown in Figure 8 (Ghosh et al., 1980).

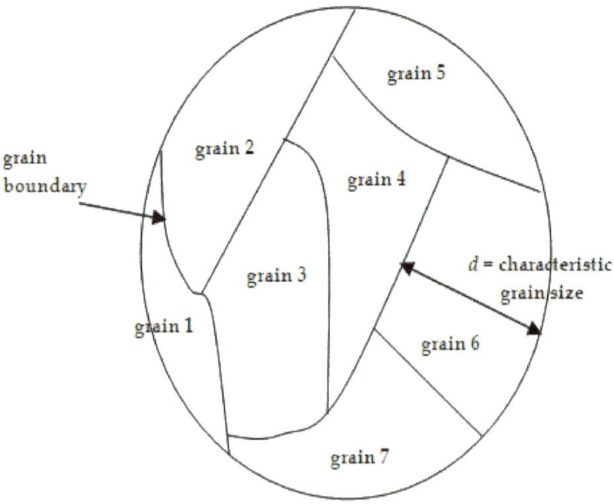

Fig. 8. Grain boundaries of polysilicon

The resistivity, mobility, and free-carrier concentration in monocrystalline silicon vary with doping concentration of the single crystal silicon. Whereas the doping of polycrystalline silicon does have an effect on the resistivity, mobility, and free-carrier concentration, these properties strongly depend on the polycrystalline grain size, which is a physical parameter that the material scientist can manipulate (Ghosh et al., 1980). Through the methods of crystallization to form polycrystalline silicon, a scientist can control the size of the polycrystalline grains which will vary the physical properties of the material.

5.1 Novel ideas for polycrystalline silicon

The use of polycrystalline silicon in the production of solar cells requires less material and therefore provides for higher profits and increased manufacturing throughput. Polycrystalline silicon does not need to be deposited on a silicon wafer to form a solar cell, rather it can be deposited on other-cheaper materials, thus reducing the cost. Not requiring a silicon wafer alleviates the silicon shortages occasionally faced by the microelectronics industry. An example of not using a silicon wafer is crystalline silicon on glass (CSG) materials (Basore, 2006).

A primary concern in the photovoltaics industry is cell efficiency. However, sufficient cost savings from cell manufacturing can be suitable to offset reduced efficiency in the field, such as the use of larger solar cell arrays compared with more compact or higher efficiency designs. Designs such as CSG are attractive because of a low cost of production even with reduced efficiency (Basore, 2006). Higher efficiency devices yield modules that occupy less space and are more compacts however the 5-10 % efficiency of typical CSG devices still makes them attractive for installation in large central-service stations, such as a power station (Basore, 2006). The issue of efficiency versus cost is a value decision of whether one requires an "energy dense" solar cell or sufficient area is available for the installation of less expensive alternatives. For instance, a solar cell used for power generation in a remote location might require a more highly efficient solar cell than one used for low-power applications, such as solar accent lighting or pocket calculators, or near established power grids.

Thin film silicon photovoltaics are typically produced by chemical vapor deposition processes yielding an amorphous, polycrystalline, or nanocrystalline film. Conventionally, amorphous silicon thin films are most common. Silicon is usually deposited on glass, plastic, or metallic substrates coated with a transparent conducting oxide material. While chalcogenide-based Cadmium-Tellurium (CdTe) and Copper-Indium-Selenium (CIS) polycrystalline thin films cells have been developed in the lab with great success, there is still industry interest in silicon-based thin film cells. Silicon-based devices exhibit fewer problems than their CdTe and CIS counterparts such as toxicity and humidity issues with CdTe cells and low manufacturing yields of CIS due to material complexity. Additionally, due to political resistance to the use non-"green" materials in solar energy production, there is no stigma in the use of standard silicon. Three major silicon-based module designs dominate: amorphous silicon cells, amorphous or microcrystalline tandem cells, and thin-film polycrystalline silicon on glass (Green, 2003).

Amorphous or microcrystalline silicon consists of a mixed phase of small crystalline regions surrounded by amorphous material. This material typically behaves more like crystalline silicon than the amorphous variety. A 3-month field study has shown that hybrid amorphous or microcrystalline cells degrade roughly to the same degree as triple-junction amorphous cells while maintaining higher conversion efficiencies (7.0% versus 5.0% as measured at the conclusion of the study). This result suggests hybrid designs of this type may supplant traditional amorphous-based modules (Green, 2003).

A new attempt to fuse the advantages of bulk silicon with those of thin-film devices is thin film polycrystalline silicon on glass. These modules are produced by depositing an antireflection coating and doped silicon onto textured glass substrates using plasma-enhanced chemical vapor deposition (PECVD). The texture in the glass enhances the efficiency of the cell by approximately 3% by reducing the amount of incident light reflecting from the solar cell and trapping light inside the solar cell. The silicon film is crystallized by an annealing step, temperatures of 400 - 600 °C, resulting in polycrystalline silicon.

These new devices show energy conversion efficiencies of 8% and high manufacturing yields of >90%. Crystalline silicon on glass (CSG), where the polycrystalline silicon is 1-2 micrometers, is noted for its stability and durability; the use of thin film techniques also contributes to a cost savings over bulk photovoltaics. These modules do not require the presence of a transparent conducting oxide layer. This simplifies the production process twofold; not only can this step be skipped, but the absence of this layer makes the process of constructing a contact scheme much simpler. Both of these simplifications further reduce the cost of production. Despite the numerous advantages over alternative design, production cost estimations on a per unit area basis show that these devices are comparable in cost to single-junction amorphous thin film cells (Green, 2003).

5.2 Low temperature induced crystallization of amorphous silicon

Amorphous silicon can be transformed to crystalline silicon using well-understood and widely implemented high-temperature annealing processes. This typical method is the typical method used in industry but requires high-temperature compatible materials, such as special high temperature glass that is expensive to produce. However, there are many applications for which this is an inherently unattractive production method. Flexible solar cells have been a topic of interest for less conspicuous-integrated power generation than solar power farms. These modules may be placed in areas where traditional cells would not be feasible, such as wrapped around a telephone pole or cell phone tower. In this application a photovoltaic material may be applied to a flexible substrate, often a polymer. Such substrates cannot survive the high temperatures experienced during traditional annealing. Instead, novel methods of crystallizing the silicon without disturbing the underlying substrate have been studied extensively. Aluminum-induced crystallization (AIC) and local laser crystallization are common in the literature, however not extensively used in industry. In both of these methods, amorphous silicon (a-Si or a-Si:H) is grown using traditional techniques such as plasma-enhanced chemical vapor deposition (PECVD). The crystallization methods diverge during post-deposition processing.

Another method of achieving the same result is the use of a laser to heat the silicon locally without heating the underlying substrate beyond some upper temperature limit. An excimer laser or, alternatively, green lasers such as a frequency-doubled Nd:YAG laser is used to heat the amorphous silicon, supplying energy necessary to nucleate grain growth. The laser fluence must be carefully controlled in order to induce crystallization without causing widespread melting. Crystallization of the film occurs as a very small portion of the silicon film is melted and allowed to cool. Ideally, the laser should melt the silicon film through its entire thickness, but not damage the substrate. Toward this end, a layer of silicon dioxide is sometimes added to act as a thermal barrier (Yuan et al., 2009). This allows the use of substrates that cannot be exposed to the high temperatures of standard annealing, polymers for instance. Polymer-backed solar cells are of interest for seamlessly integrated power production schemes that involve placing photovoltaics on everyday surfaces.

A third method for crystallizing amorphous silicon is the use of thermal plasma jet. This strategy is an attempt to alleviate some of the problems associated with laser processing - namely the small region of crystallization and the high cost of the process on a production scale. The plasma torch is a simple piece of equipment that is used to thermally anneal the amorphous silicon. Compared to the laser method, this technique is simpler and more cost effective (Lee et al., 2009).

Plasma torch annealing is attractive because the process parameters and equipment dimension can be changed easily to yield varying levels of performance. A high level of crystallization (~90%) can be obtained with this method. Disadvantages include difficulty achieving uniformity in the crystallization of the film. While this method is applied frequently to silicon on a glass substrate, processing temperatures may be too high for polymers.

6. Transformation temperature

Laser annealing on the thin film is a high-speed process. It depends on the pulse duration of the excimer laser normally in nanoseconds region and the threshold energy to nucleate and optimum energy to stop crystallization. Such short duration of annealing process is referred as transient crystallization of amorphous silicon film (Viatella and Singh, 1997). Within transient duration, the annealing process still needs to pass through several critical points. In the next section, such phase transformation point will be described in detail.

6.1 Critical points

In order to figure-out the physical change involve in laser annealing it is better to understand the basic concept of phase transformation during heating and cooling of hardening process. There are several important stages in heating and cooling of material. The crucial parameter is temperature at several critical points (Figure 9). The "critical points" are the temperatures at which certain changes in the chemical composition of the material take place, during both heating and cooling. Steel for example at normal temperatures has its carbon (which is its chief hardening element) in a certain form called *pearlite* carbon, and if the steel is heated to a certain temperature, a change occurs and the pearlite becomes *martensite* or hardening carbon. If the steel is allowed to cool slowly, the hardening carbon changes back to pearlite. The points at which these changes occur are the decalescence and recalescence or critical points, and the effect of these molecular changes is as follows: When a piece of steel is heated to a certain point, it continues to absorb heat without appreciably rising in temperature, although its immediate surroundings may be hotter than the steel. This is the *decalescence* point. Similarly, steel cooling slowly from a high heat will, at a certain temperature, actually increase in temperature, although its surroundings may be colder. This takes place at the *recalescence* point. The recalescence point is lower than the decalescence point, and the lower of these points does not manifest itself unless the higher one has first been fully passed. These critical points have a direct relation to the hardening of steel. Unless a temperature sufficient to reach the decalescence point is obtained, so that the pearlite carbon is changed into a hardening carbon, no hardening action can take place; and unless the steel is cooled suddenly before it reaches the recalescence point, thus preventing the changing back again from hardening to pearlite carbon, no hardening can take place. The critical points vary for different kinds of material, and must be determined by tests in each case.

Similar phase transformation also experience by other materials during heating and cooling process. The different may be arisen in the aspect of temperature and the time taken to complete the same process. If the steel is replaced with other material like any amorphous silicon thin film and take into account the duration involves in nanosecond time domain then the whole event is considered as a high-speed phenomenon. This includes the rate of absorption and release of heat which occurs in transient time and enough energy to cause the phase change. In case excimer laser ablation which posses in nanoseconds pulse duration and fluence more than 500 J/cm^2, having high potential to cause the target thin film experience the same process as hardening steel. This implies excimer laser annealing on thin film should passing through the decalescence and recalescence points to ensure the formation of crystallization.

In general when amorphous silicon thin film is heated and cooling, normal phase change is occurred including the melting and re-solidification transformation. The phenomenon of this phase transformation is similar as illustrated in Figure 9. The red line is indicated the heating process, while the blue line represents the cooling process. In heating process, there are two important points that are decalescence and critical points. Decalescence is a phenomenon that occurs when a material is being heated and there is a sudden slowing in the rate of temperature increase. The slowing rate is due to the change in the internal crystal structure of the material. Meanwhile during cooling down or re-solidification, the phase change is passing through a recalescence point. A sudden spontaneous increase in the temperature of cooling resulting from an exothermic change in crystal structure occurs. This is the critical state in cooling process whereby the formation of crystallization is taken place.

Furthermore, a single pulse of the excimer laser, carried out temperature of 6.42 eV which is higher than threshold to break the bonding of silicon molecules. Once the excimer laser pulse is exposed on the film, it must passing through the decalescence and recalescence points in order to form crystallization. If a lower energy than threshold material or too high energy achieve up to the peak temperature in the transformation process, no crystallization will be formed on silicon film.

In both temperatures, there is latent heat of fusion is liberated from the solid during heating and from liquid during cooling. The latent heat is liberated during both processes to delay either the process of heating or cooling. The coexistent both solid and liquid associated in heating and cooling, resistant the rate of transformation. The advantage of this drawback is given an opportunity to nucleate and grow the crystallization in the silicon.

If the heating is over than optimum energy by exposing with higher number of pulses, the thin film is totally melted. No solid form is left to be as a seed for crystallization. In this case the treatment is achieved up to the peak temperature, passing over the decalescence point. Subsequently re-solidification occurs faster without passing through the recalescence point. Hence no formation of crystallization and end up with homogenous or no change in crystal structure in the excimer laser annealing process.

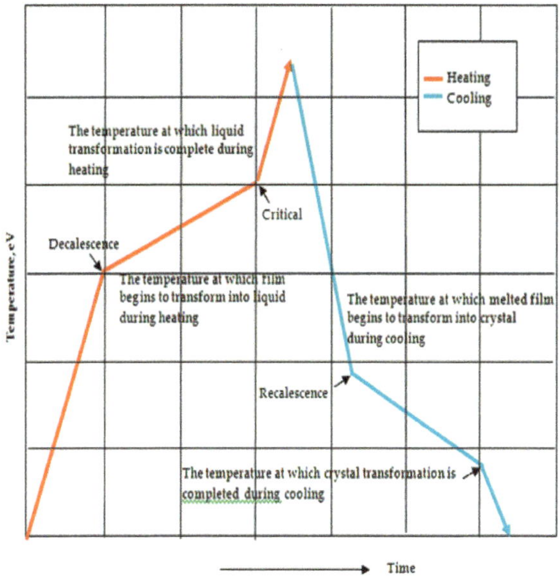

Fig. 9. Transformation temperature

6.2 Recalescence effect

There are several researches on the recalescence phenomenon have been reported. Among them are Zhang and Atren, (1992), Armengol et al., (2003) and Mariucci et al., (2003). Zhang and Atren, (1992) have studied the recalescence characteristics in rapid solidification of copper by using a thermokinetic model. They have investigated the effects of the heat transfer coefficient, the melt thickness and the nucleation temperature. Their results had shown that the lower nucleation temperature and thinner melt lead to a longer recalescence effect while larger heat transfer coefficient results in a weaker recalescence effect. A dimensionless number was derived to measure the extent of recalescence.

Mariuci group (2003) had claimed that crystallization less than complete melting point will experience the recalescence effect. Structural properties of thin polycrystalline silicon films, crystallized by single shot excimer laser annealing at different laser energy densities, have been investigated by various researchers (Armengol, et al., 2003,; Mariucci et al., 2003; Razak and Bidin, 2010; Bidin and Razak, 2011a, b). Formation of disk structures has been observed in a wide range of energy densities, from complete melting down to 180 mJ/cm^2. These structures have been correlated to the lateral growth of grains starting from the small grains present in the central regions of the disks. They had proposed a new crystallization scenario for energy densities below the complete melting. In this framework, the recalescence effect plays an important role while the super lateral growth-regime is no longer a particular crystallization condition but simply represents the upper energy density limit of partial melt crystallization regime.

Similar recalescence effect after bulk solidification in germanium films melted by nanosecond laser pulses had also reported by Armengol, et al., (2003). Rapid solidification dynamics in amorphous germanium films melted by nanosecond laser pulses has been analyzed by means of single-shot subnanosecond time resolved reflectivity measurements using a streak camera based setup. The results showed that once a minimum melt depth is induced, a bulk solidification process followed by the release of the solidification enthalpy dominates the solidification process. Moreover, the laser-melted material solidifies completely before being remelted as a consequence of the solidification enthalpy release, something only observed, up to date, upon irradiation with picosecond laser pulses (Armengol et al., 2003).

Bossuyt and Greer, (2004) have been investigated the effects of positive feedback on crystallization kinetics and recalescence. They described that in a series of alloys in the Cu-Ni-Ti-Zr system, with compositions close to the bulk glass forming alloy Cu47Ni8Ti34Zr11 (Vit101), the nucleation density has been shown to be spatially inhomogeneous; in an amorphous matrix there are spherical clusters with a high density of nanocrystals. The implied positive feedback in the nucleation rate was analyzed in terms of recalescence instability, where the latent heat released upon crystallization causes the nucleation rate to increase locally, if the nucleation rate is an increasing function of temperature and thermal diffusivity is low enough to avoid distributing the heat evenly over the sample. In deeply under cooled liquids, the first of these requirements is satisfied, but not for the second requirement.

Bossuyt and Greer have used numerical technique to trace the cooling rate with and without the feedback heat in crystallization formation. They identified a critical cooling rate for recalescence to occur. At slightly faster cooling rates, the heat released by crystallization causes the actual cooling rate to drop to nearly zero, but the heat output is never sufficient to actually raise the temperature. When that does occur, there is an abrupt transition from approximately 40% crystallization at cooling rates just above the critical cooling rate, to full crystallization below the critical cooling rate. Furthermore they claimed that disabling the heat of crystallization eliminates the recalescence effect from the temperature traces, and the transition to full crystallization is now much more gradual, without a clearly defined critical cooling rate.

Other feedback mechanisms have effects similar to those of recalescence: there is always a threshold condition where the acceleration due to feedback offsets the quenching effect of decreasing temperature. At cooling rates above this threshold, the crystallization kinetics did not deviate much from those in the absence of feedback. The effect of feedback becomes noticeable over a relatively narrow range of cooling rates just above the threshold, and then dominant below the threshold, abruptly increasing the degree of crystallization (of at least one of the phases). The decalescence and recalescence effect are shown in Figure 9. Further explanation about the critical points will be discussed on section 6.5.

However not many researchers have been realized about the decalescence effect in the re-crystallization process. Without determination the decalescence effect during heating process the recalescence point cannot be achieved. Therefore in order to crystallization of polycrystalline silicon film the emphasized is focusing on these both phenomena (decalescence and recalescence effects). In this project, silicon thin film is crystallization by controlling the number of laser pulses.

In most cases, silicon is deposited onto a corning glass substrate and annealed with high energy excimer laser which requires high fabrication cost for industrial purposes. In this attempt, low energy ArF (wavelength of 193 nm) excimer laser was employed together with conventional annealing method to anneal the silicon thin film. As reported by previous researchers (Carluccio et. al., 1997, Han et. al., 2003) combination annealing techniques between conventional and laser annealing would produce a better poly-si thin film. Figure 10 shows a different in terms of grain size structure between polysilicon thin film PSTF annealed by (a) conventional annealing and (b) combination of conventional and excimer laser annealing. Polycrystalline silicon thin film (PSTF) has been an interest currently due to high potential as electrical interconnections in microelectronics and microsensors.

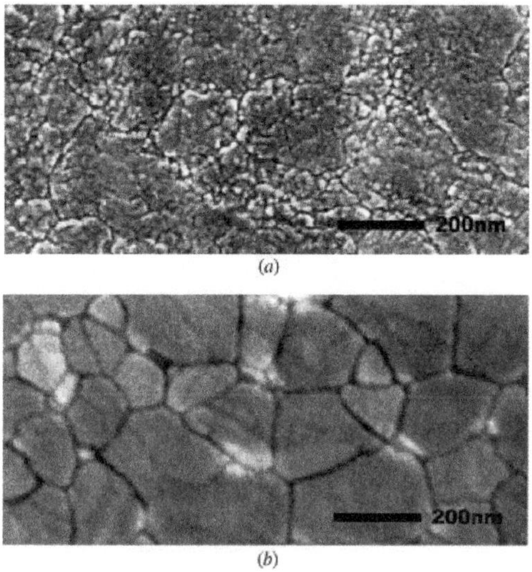

Fig. 10. FESEM images of poly-Si (a) annealed at 600 °C for 24 h (*a*) and the combination annealing with ELA at 470 mJ cm^{-2} (Han et. al., 2003)

6.3 Super lateral growth energy

In order to optimize the annealing process, a super lateral growth SLG energy density needs to be identified. The SLG point is known as the critical energy density or laser fluence desires to generate the maximum crystallization. Greater than this energy normally the crystallization will be reduced and soon eliminated. In other word the SLG energy density is equivalent with the critical point in the temperature transformation curve (Figure 9) during heating process. Beyond this energy, the silicon film will totally melt and cause elimination of spontaneous crystallized formation. This implies that calibration needs to be done in determining this crucial energy.

There are two parameters commonly used to quantify the degree of crystallization that are, grain size and surface roughness. The later parameter is due to the melting effect. Both parameters are utilized for measurement of crystallization. If the boundaries of the grain size is expanding or the surface roughness is increasing after being exposed by a series of excimer laser pulses that means the annealing process is within the decalescence and recalescence region. Once the optimum energy density that is super lateral growth energy is achieved that is at the critical point in the temperature transformation curve, no more number of pulses can be added or no more laser energy density can be increased.

Otherwise, the excess energy exposed that would cause the film totally melted and no solid form leaves at the interface between substrate and film to seed the crystallization.

Hence in excimer laser annealing the most crucial process is to identify the SLG energy. Under this particular SLG energy, the heating and the cooling are associated with the decalescence and recalescence effects. This implies that, the phase change is accompanied with the coexistence of liquid and solid forms. This is illustrated in Figure 11. At the interface between film and substrate, there are some particles remain in solid form whereas the rest in liquid phase. The left over the particles are used to seeding the crystallization. This region is similar as the condition achieved at recalescence point in the transformation temperature curve during quenching or cooling period.

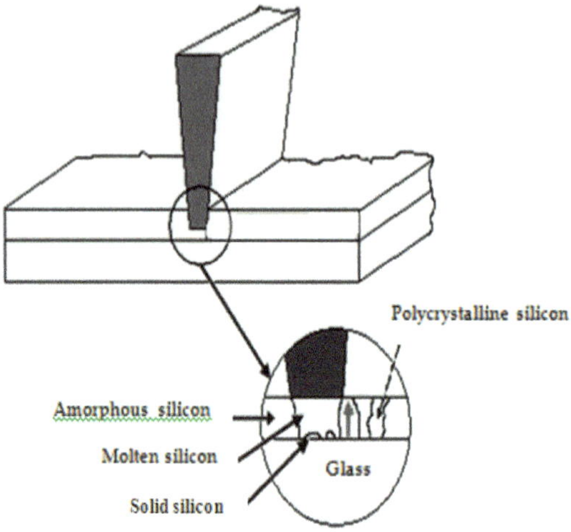

Fig. 11. Nearly complete melt of amorphous silicon layer. The liquid/solid interface seeds the crystallization process

It is well known that laser ablation involves photothermal and photochemical process, depending on the nature of the material used and the experimental conditions, for example laser fluence, wavelength, and pulse duration (Sato et al., 2001; Serafetinides et al., 2001; Hahn et al., 1999). A 193 nm wavelength radiation can cause a neat and clean etch region because at this wavelength the molar extinction coefficient of the material is high (~ 10^4 L mol^{-1} cm^{-1}) so that 95% of the light is absorbed in the first 300 nm of the substrate. On the other hand, a laser with longer wavelength (~ 500 nm) light melts and damage nearby regions of the sample as a longer wavelength light vibrationally excites the irradiation sites. This process can eventually cause material to desorb from the heated region (Barbara et al., 1984).

7. Atomic force microscope analysis

The crystallization result is quantified based on metallurgical technique. The annealed silicon thin film is examined under the atomic force microscope. The scan area for each sample is fixed to 5000 nm X 5000 nm and zoomed until 1000 nm X 1000 nm to get clearer image of the thin film surface. The typical of the nucleation and the growth of silicon thin film is shown in Figure 12. The frames of crystallization are arranged in the increasing order of the energy density. Prior annealing, the grain size of the amorphous silicon thin film is measured. The average nanocrystal size, G_{av} of the amorphous silicon film before doping and heating as shown in Figure 12(a) is approximately 17 nm. After experienced four hours heat treament via conventional annealing, a new crystal structure is realized. The crystallization pattern almost uniform with an increment up to 56 nm as depicted in Figure 12(b). Immediately after crystalization with a single shot of excimer laser corresponding to energy density of 65.50 mJcm^{-2}, another new pattern nanostructure is realized. The crystallization of the silicon film is accelerated further to a grain size of 75 nm as illustrated in Figure 12(c). The crystallization of thin film is continuously increasing after received the sequential number of pulses from argon fluoride ArF excimer laser. The maximum crystallization of silicon film is notified after received five number of pulses as depicted in Figure 12(g). The enlargement achieved approximately to 143 nm corresponding to the super lateral growth energy of 345 mJcm^{-2}. Beyond the critical energy as shown in Figure 12(h) the crystallization reduced drastically.

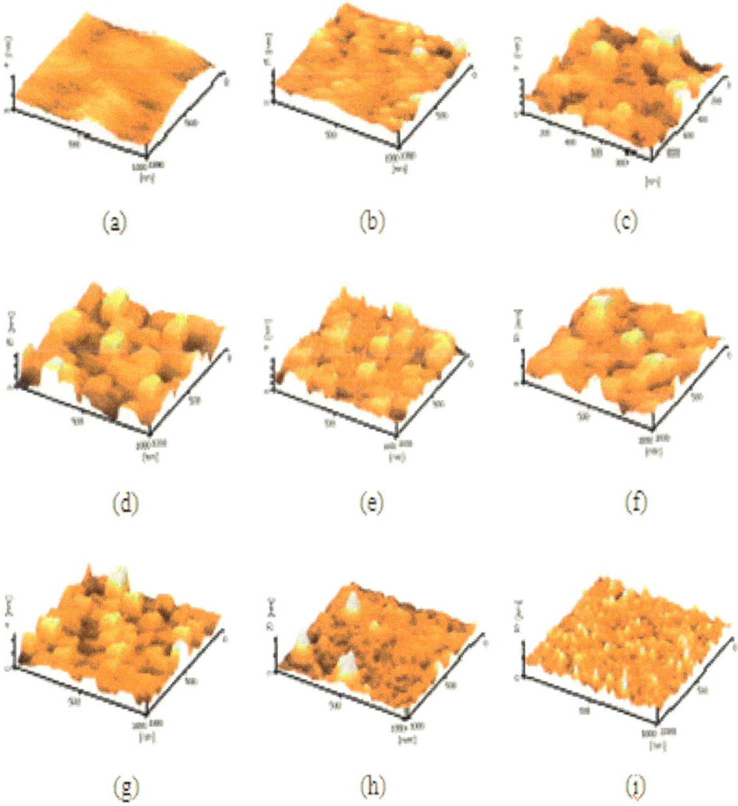

Fig. 12. AFM images of silicon thin film with magnification area of 1000 nm x 1000 nm at different energy density.

The stress developed during solidification due to the melting effect of silicon thin film is quantified based on surface roughness measurement. A typical example of the surface roughness of the silicon copper film is shown in Figure 13. Figure 13(a) indicates the roughness before annealing process. The measurement of surface roughness for the amorphous silicon thin film is 0.3 nm. After annealing the intergration of crystalization formation comprised of the agglomeration and coalesence of grain. Simialr results have also reported by He et al., (2007). The surface roughness of the annealed silicon thin film then increased up to 3.0 nm. The profile of the grain sized and surface roughness measurement are presented in Figure 14 and 15 respectively. However the maximum roughness is not overlapping with maximum grain size. The possible reason for such different may be due to the measured surface roughness is not matched with measured area of the grain size.

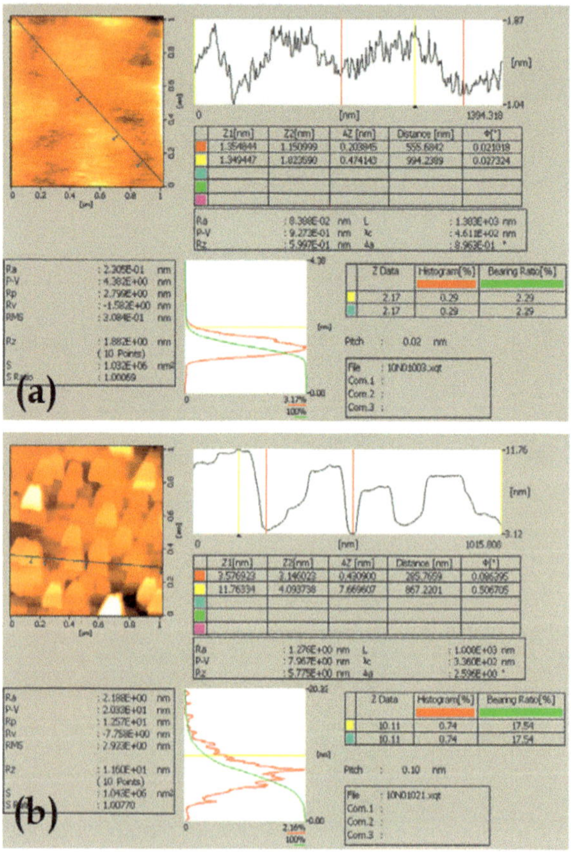

Fig. 13. Surface roughness of silicon at different number of pulses: a. o pulses, b. 3 pulses.

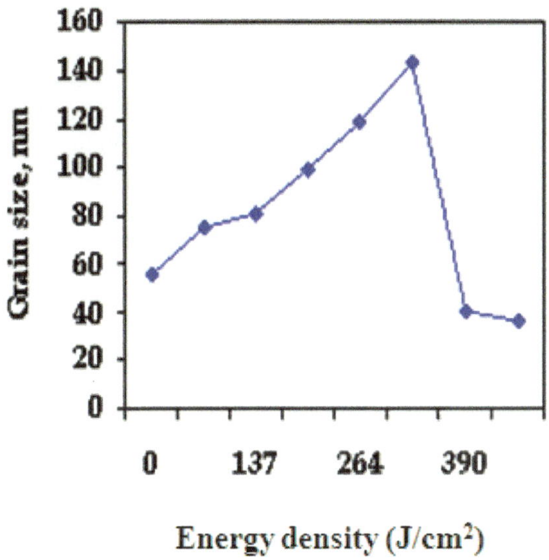

Fig. 14. Grain size as function of energy density

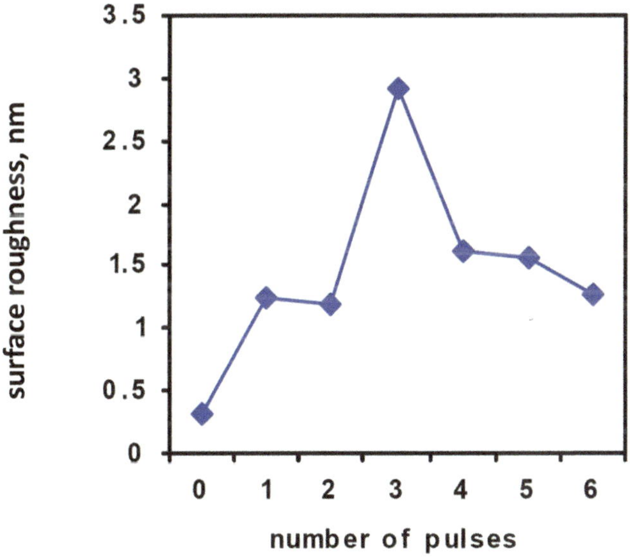

Fig. 15. Surface roughness of silicon thin film with respect to number of pulses

8. Conclusion

In general, the excimer laser annealing is capable to crystallize the silicon thin film based on the increment of the grain size as well as the surface roughness. In the process of annealing, the film experience transient heating and freezing. The UV light of excimer laser carried out energy of 6.42 eV which is equivalent with 7.5 × 10^4 K that is more than enough to melt the thin film. Considering the temporal length of pulse duration each of laser pulse is only 10 ns, this means the silicon copper thin film is experience an ultra-speed heating and freezing. The energetic excimer is enough to control the phase transformation with the decalesence point in which the melted is associated with some solid form to nucleate the crystal growth. The decalescence effect is controlled via the number of pulses. In contrast during rapid re-solidification the liquid form is associated with recalescence effect which allows the formation of crystallization. The latent heat is released to delay the transformation from liquid to solid. Inherently the recalescence effect responsible to spontaneously crystallize the silicon film. When the energy density greater than the end limit of decalescence point which equivalent with the super lateral growth energy, the heating achieved the peak temperature where the film totally melted. Consequently no recalescence effect occur which result homogenous solidification without crystallization formation during cooling. Thus, in annealing it is crucial to determine the super lateral energy in order to ensure the process is within the decalescence and recalescence points to allow the formation of crystallization.

In addition, laser annealing process of silicon thin film promising well-arranged and large poly-Si grains. As far as we can see, laser annealing has become the most technique used to fabricate a good quality of poly-Si film for ultra-large scale integrated circuit devices, thin film transistors, active matrix liquid crystal displays, solar cells and other optoelectronic devices for industrial purposes. The employment of excimer laser for low temperature poly-Si (LTPS) thin film fabrication makes it possible to replace glass substrate with plastic substrate. This would certainly reduce the cost for large scale productions. Furthermore the high speed crystallization process will shorten the time to anneal as compare to the conventional method. The combination of time for annealing and the low energy consumption will give advantage for the production of high quality silicon thin film.

9. Acknowledgement

This work has been supported from the government of Malaysia via FRGS grant vote 4F001 and GUP vote 00H10. We would like to acknowledge the support given by all technicians during experimental performances.

10. References

Armengol, J.; Vega, F.; Chaoui, N.; Solis, J.; Afonso, C. N. (2003). Recalescence after bulk solidification in germanium film melt by ns laser pulses. J. of Appl. Phys. 93(3):1505

Barbara, J. and Srinivasan, R., (1984). Microscopic Model for the Ablative Photodecomposition of Polymers by Far-Ultraviolet Light Radiation, Appl. Phys. Lett. 44(9):849

Basore, P. A. (2006). "CSG-2: Expanding the production of a new polycrystalline silicon PV technology", Proc. of the 21st European Photovoltaic Solar Energy Conference, http://www.csgsolar.com/downloads/CSG-2%20Basore.pdf.

Bidin N. and Razak S.N. (2011a). Crystallization of Poly-Silicon Film by Different Annealing Techniques. Proc. of Saudi Int. Electronic, communication and Photonic conference SIECPC-2011 Riyadh Saudi Arabia, 24-26 April, 2011

Bidin N. and Razak S.N. (2011b). Nucleation and Growth of Cu Thin Films on Silicon Base Induced By Excimer Laser Annealing (Invited paper). Proc. of International symposium on Functional materials, Bali Indonesia, April 27-28, 2011

Bossuyt, A. S. and Greer L. (2004). Effects of Positive Feedback on Crystallization Kinetics and Recalescence. Mat. Res.Soc. Symp. Proc. Vol. 806, Materials Research Society

Brendel K.; Nickel N. H.; Lengsfeld P.; Schöpke A.; Sieber I.; Nerding M.; Strunk H. P. and Fuhs W. (2003). Excimer laser crystallization of amorphous silicon on metal coated glass substrates. Thin Solid Films. 427(1-2): 86

Brannon, J.; Lankard, J.R.; Baise, A.; Burns I. and Kaufman F. (1985). Excimer Laser Etching of Polymide. J.Appl. Phys. 58:2036

Brannon, J. (1997). Excimer Laser Ablation and Etching: Making a Mark in Many Applications. *Circuit & Devices*, 97: 11

Campbell, S. A. (2001).The Science and Engineering of Microelectronic Fabrication (2nd ed.), New York: Oxford University Press, ISBN 0195136055

Chen,Y.H.; Zheng, H.Y.; Wong, K.S. (1995). Excimer Laser Drilling of Polymers, Proc. SPIE , 3184:202

Csele, M. (2004). Fundamental of Light Sources and Lasers. New Jersey: Wiley-Intersciece

Carluccio R.; Cina S.; Fortunato G.; Friligkos S.; and Stoemenos J. (1997). Structure of Poly-Si Films Obtained By Excimer Annealing. Thin Solid Films. 296: 57

Delmdahl, R. (2010). The excimer laser; precision engineering. Nature photonic. 4: 286

GAMLaser Inc. (2003). EX5 Excimer Laser Manual. Florida: User Manual

Ghosh, A. K.; Fishman, C. and Feng, T. (1980). Theory of the electrical and photovoltaic properties of polycrystalline Silicon. Journal of Applied Physics. 51 (1): 446. doi: 10.1063/1.327342

Green, M. A. (2003). Crystalline and thin-film silicon solar cells: state of the art and future potential. Solar Energy 74 (3): 181. doi: 10.1016/S0038-092X (03)00187-7

Green, M. A. (2004). Recent Developments in Photovoltaics. Solar Energy 76 (1–3): 3. doi:10.1016/S0038-092X(03)00065-3

Hahn, C. H.; Lippert, T. and Wokaun, A. (1999). Comparison of the Ablation Behaviour of Polymer Films in the IR and UV with Nanosecond and Picosecond pulse. J. Phys. Chem. B. 103: 1287

Han, W. H.;Kang, C.J. and Yong, S.K (2003). A Novel Structured Polycrystalline Silicon Thin Film Transistors tha increases the On/Off Current ratio. Semicond. Sci. Technol. 18 (845-849).

He M.; Ishihara R.; Metselaar W. and Beenakker K. (2007). Agglomeration of amorphous silicon film with high energy density excimer laser irradiation. Thin Solid Films. 515(5): 2872
Hecht, J. (1992). Understanding Laser: An Entry –Level Guide. New York: IEEE Press

Lee, H. S.; Choi, S.; Kim, S. W.; Hong, S. H. (2009). Crystallization of Amorphous Silicon Thin Film by Using a Thermal Plasma Jet. Thin Solid Films 517 (14): 4070. doi: 10.1016/j.tsf.2009.01.138

Mariucci L.; Pecora A.; Fortunato G.; Spinella C. and Bongiorno C. (2003). Crystallization mechanisms in laser irradiated thin amorphous silicon films. Thin Solid Films. 427(1-2): 91

Matsumura M. and Oh C. H. (1999). Advanced Excimer-Laser Annealing Process for Quasi Single-Crystal Silicon Thin-Film Devices. Thin Solid Films. 337: 123

Palani I. A., Vasa N. J., and Singaperumal M. (2008). Crystallization and Ablation in Annealing of Amorphous-Si Thin Film On Glass and Crystalline-Si Substrates Irradiated By Third Harmonics of Nd^{3+}:YAG Laser, Materials Science in Semiconductor Processing. 11: 107

Razak S.N and Bidin N. (2010). Effect of Multipulses Exposure On Excimer Laser Annealed Polycrystalline Silicon Thin Film. Proc. of 2010 International Conference on Enabling Science and Nanotechnology. Escienano2010, 1-3 December, 2010, KLCC, Malaysia

Sato H. and Nishio S. (2001). Polymer laser photochemistry, ablation, reconstruction, and polymerization, J. Photochem. Photobiol. 2:139

Sauerbrey, R. and Petit, G. H. (1989). Theory of Etching Organics Materials by Ultraviolet Laser Pulses. Applied Physics Letter, 55(5): 421

Schneider, T. L. (1998). Refractive Surgery and the Excimer Laser. Jacksonville Medicine

Serafetinides, A. A.; Makropoulou M. I.; Skordoulis, C. D; and Kar, A. K. (2001). Ultrashort Pulsed Laser Ablation of Polymers. Appl. Surf. Sci. 180: 42

Shah, A.; Meier, J.; Vallat-Sauvain, E.; Wyrsch N.; Kroll U.; Droz C.; Graf U. (2003). Material and solar cell research in microcrystalline silicon", Solar Energy Materials and Solar Cells. 78 (1–4): 469. doi: 10.1016/S0927-0248(02)00448-8

Shah, Z. A. (2009). Surface Treatment of Polymethylmethacrylate by Excimer Laser Ablation. MSc Thesis. Universiti Teknologi Malaysia 2009

Speidell, J. L.; Pulaski D.P.; Patel R.S. (1997). Mask for laser ablation technology: New requirement and challenges, IBM Journal and research development. 8618-8646.

Srinivasan, R. and Mayne-Banton, V. (1982). Self Developing Photoetching of Poly(ethyleneterephtalate) Film by Far Ultraviolet Excimer Laser Radiaion.Appl. Phys. Lett. 41(6): 576

Streetman, B. G. & Banerjee, S. (2000). Solid State Electronic Devices (5th ed.), New Jersey: Prentice Hall. ISBN 0130255386

Takaoka G. H.; Fujita K.; Ishikawa J., and Takagi T. (1989). Grain Boundary Diffusion Control of Cu Films Prepared By The ICB Technique. Nuclear Instruments and Methods in Physics Research B37/38: 882

Viatella J. and Singh R. K. (1997). Transient recrystallization of amorphous silicon films. Materials Science and Engineering B. 47(1): 78

Watanabe, H.; Miki H.; Sugai, S.; Kawasaki K.; Kioka, T. (1994). Crystallization process of polycrstalline silicon by KrF_2 Excimer laser annealing. Jpn. J. Appl. Phys. 33(8): 4491

Yuan, Z.; Lou, Q.; Zhou, J.; Dong, J.; Wei, Y.; Wang, Z.; Zhao, H.g; Wu, G. (2009). Numerical and experimental analysis on green laser crystallization of amorphous silicon thin films. Optics & Laser Technology 41 (4): 380. doi: 10.1016 /j.optlastec.2008.09.003

Zhang X and Atrens A. (1992), Rapid solidification characteristics in melt spinning. Materials Science and Engineering: A. 159(2): 243. doi:10.1016/0921-5093(92)90295.c

13

Crystallization, Fractionation and Solidification of Co-Magmatic Alkaline Series Sequentially Emplaced in the Carbonatite Complex of Tiruppattur, Tamil Nadu, India

R. Ramasamy
Ocean Engineering, IITM, Chennai, India

1. Introduction

A mineral is naturally occurring substance with a characteristic crystal structure and chemical composition. It occurs in rocks. Frequently, it has homogeneous crystalline structure in which one or more types of atoms or molecules may be partially substituted for the original atoms and molecules without changing the structure. Thus, most rock-forming minerals are formed in solid solutions. Aggregates of minerals produce characteristic rock-textures which shed light on their trends of magmatic evolution. They crystallize from molten rock, called magma. Magmas are a natural material of hot silicate / carbonate /oxide / phosphate/ sulphide and sulphur melts from which igneous rocks form. They develop by the partial melting of deep-seated rocks at depth. Primary melts are composed of suspended crystals and dissolved gases. The crystallization of a primary melt and the isolation of the rest liquid freed from suspended solids are called the parent magma, which then produces a series of residual magmas of a secondary nature with varying compositions through differentiation and fractionation. The molten material is necessarily forced to the surface, either by hydrostatic head in the mantle where the encasing rock is denser than the melt or else via gas by propulsion. Pressure enhances the solubility of H_2O and other volatiles, which lower the equilibrium temperature of solidification. These volatile substances may largely escape during the course of the ascension of the magma as well as during the course of crystallization. Rising toward the Earth's surface, the magma enters zones of lower temperatures and pressures. Decreasing temperatures tend to bring about crystallization, which produces solid crystals suspended in the liquid. Other solid fragments are incorporated from the walls and roof of the conduit through which the magma is rising and the magma changes its composition by assimilation. A closed magmatic chamber is formed by the filling up of magma in cavities at intermittent stopping places. Over the course of the magmatic crystallization of the parent magma in a closed magmatic chamber, volatiles may be concentrated and interacted with minerals during slow cooling. Magma is generally composed of eight major oxides SiO_2, Al_2O_3, Fe_2O_3, FeO, MgO, CaO, Na_2O and K_2O, as well as from lesser proportions of TiO_2, P_2O_5, H_2O, CO_2, S and other volatiles. In addition to these there are other substances constituting various gases and trace elements. As crystallization progresses, the volatiles and the more soluble silicate components are concentrated in the remaining liquid. They increase the fluidity of the magma. Moreover, they do not enter appreciably into the earlier minerals forming from the magma, but are instead finally concentrated in an aqueous solution as hydrothermal fluid. Minerals initially crystallized from the magma are denser than the magma and buoyancy forces lead to gravity settling the crystals. In some rare and denser iron-rich or carbonatite magmas, feldspar crystals may float. The viscosity of silicate magma generally increases with a decrease in the temperature - with decreases in dissolved H_2O - with an increase in crystal fraction and an increase in the degree of SiO_2 content.

2. Polymerization

The polymerization of silicate melt takes place by the reaction of a silicate melt to form three dimensional networks of SiO_4 chains. This structure of a silicate melt provides information on its structure and physical, chemical and thermal properties. From this information, it is possible to understand the conditions of crystallization and the evolution of the silicate minerals from the melt /magma. In magmatic processes, phase equilibria in melt-crystal-vapour systems, diffusion in melts and the thermodynamic, electrical and rheological properties of magma systems are of particular importance. Silicate minerals essentially contain silicon and oxygen in the form of a network of tetrahedra in which a Si ion is surrounded by 4 oxygen ions. The tetrahedra occur as isolated $(SiO_4)^{4-}$ or else joined together in various ways, such as $(Si_2O_7)^{6-}$ and rings $(Si_6O_{18})^{12-}$. Ortho-silicates have isolated $(SiO_4)^{4-}$ tetrahedra that are connected only by interstitial cations. Sorosilicates have isolated double tetrahedral groups with $(Si_2O_7)^{8-}$. Cyclosilicates have linked tetrahedra with $(Si_xO_{3x})^{2x-}$. Inosilicates have interlocking chains of silicate tetrahedra with either SiO_3 for single chains or Si_4O_{11} for double chains. Sheet-silicates form parallel sheets of silicate tetrahedra with Si_2O_5. Tectosilicates have a three-dimensional framework of silicate tetrahedra with SiO_2. As a silicate melt cools, minerals crystallize that are in equilibrium with the melt. Ortho-silicates like olivine generally crystallize from a silicate melt at a very high temperature in a disordered state; on the other hand, quartz and alkali feldspar crystallize from the melt at a relatively low temperature state. At the ends of the scale, the structural states of silicate melts vary widely. Similarly, within the crystallizing minerals, the olivine crystallizing at the early stage of magmatic evolution and the quartz crystallizing at very high temperature states at the late magmatic stage are deficient in Si ions in their tetrahedral co-ordinations. Thus, the structural states of co-existing minerals and the silicate melt from which they crystallize vary widely. In studying such variations, it is possible to see the trend of magmatic evolution. Magmatic evolution may reveal the cooling history of minerals and their accompanying rocks. Volcanic rock which has been suddenly quenched also reveals genetic relationships between phenocrysts and their groundmass matrix, which represents a quenched silicatemelt. There exists a genetic relationship between the structure of a silicate melt and the structure of the silicate mineral crystallizing stemming from it. For example, under an increase of alkalis and alkaline earth elements in the silicate melt, tetrahedrally co-ordinated Al^{3+} displays a strong preference for three-dimensional network units in the form of feldspars in the silicate melts with an increase of the Al/(Al+Si) of the melt and a decrease of SiO_4^{4-} units in the melt (Mysen et al., 1985). The structure of a silicate melt depends upon the viscosity of the melt and the subsolidus structure of the crystallizing silicate mineral at a high temperature state. Under these favourable conditions, the crystal-melt equilibrium of the partition of one element between them sets. Magmatic evolution can be traced by studying the changes of the structural state of crystallizing minerals and the changes in structure of residual melts in spatial and temporal conditions. Slow cooling permits extensive large crystal growth under plutonic magmatic crystallization. As melting and crystallization are reversible processes, different types of phase diagrams are needed to understand how melts crystallize. Magmatic liquids have a sub-solidus structure which is created largely by varying the degrees of bonding among SiO_4 tetrahedra (polymerization). The difference between a silicate liquid and a silicate mineral is that the mineral has a definite long-range order structure that is the same throughout while a silicate melt shows different types of the short-range order of polymerization throughout the melt. Additionally, the degree of the polymerization of the melt controls the viscosity of the melt. Mafic magmas that have relatively low Si contents have depolymerized melts that produce depolymerized crystals - like olivine or pyroxene - and these melts tend to have a high temperature and a low viscosity. In contrast, felsic magmas have abundant Si and Al, whereby the melts are highly polymerized owing to their high viscosity such that they crystallize as sheet and framework silicates. Because of these consequences, Mg-Fe-rich basaltic magmas erupt easily and the lavas flow several kilometres from their vents, while the Si-rich rhyolite lava erupts explosively without escaping out of its inherent gas bubbles. Volatiles are elements that dissolve in magmas but transform to gas when magma crystallizes or because of a sudden decrease in pressure. The eolution of magmatic rocks may be traced out based on the concept of "Bowen's Reaction Series". The order of crystallization of common plagioclase feldspars from cooling magma is evolved from Ca-rich to more Na-rich plagioclase feldspars as a continuous series during decreasing temperature. The high temperature olivine would react with residual magma and change to the next mineral pyroxene in the series. Pyroxene continues to cooling, it would convert to amphibole and then biotite by adjusting the mineral crystalline lattice to achieve stability of different temperatures as discrete minerals. At lower temperatures both continuous and discrete series merge together and orthoclase, muscovite and quartz tend to crystallize. The difference in crystallization temperatures for different kinds of minerals plays a major role in the differentiation of rock composition as magma cools. Thus, various types of magmatic rocks from the early formed ultamafic peridotites to granitic rocks at the late magmatic stages are evolved (Bowen, 1956). Bowen's reaction series show that one homogeneous body of magma can form more than one kind of igneous rock. It reveals the relationship between cooling magma and the formation of minerals that make up igneous rocks.

Some generalizations:

1. The compositional paths followed by both the melts and the growing crystal depend on the initial bulk composition of the melt.
2. The composition of both the melt and the crystal at any point in the crystallization process is dependent on the extent to which the crystallization process has been completed.
3. For minerals that exhibit a solid solution, crystallization proceeds in a continuous manner with the composition of the mineral changing along with that of the magmatic liquid.
4. For minerals that do not exhibit a solid solution, melting or crystallization will proceed in discrete discontinuous steps. As a melt cools, such discrete minerals will appear suddenly when the melt reaches the appropriate temperature.
5. Minerals that crystallize at high temperatures are rich in Mg and Fe and relatively poor in Si.

6. Minerals that crystallize at the lowest temperatures are rich in Si and Al.
7. Intermediate minerals that crystallize at intermediate temperatures are intermediate in relation to Si.
8. Another important observation in that those minerals which crystallize at high temperatures show less polymerization of the Si-tetrahedra than those which crystallize at lower temperatures.
9. The discrete mineral sequences starts with orthosilicates (olivine), single chain (pyroxene), double chain (amphibole) and sheet silicates (biotite), and exhibit a high degree of silica polymerization with incorporations of H_2O, F and other volatiles. The crystallization of plagioclase in a continuous series also indicates an increase of silica constituents, even though they crystallize in framework silicates.
10. Highly polymerized silica-rich tectosilicates are more stable than silica-poor tectosilicates. The degree of polymerization increases from orthosilicates to tectosilicates.
11. The initial crystallization of olivine, the residual magma, is enriched in silica.
12. The initial crystallization of pyroxene under dry conditions, the residual liquid, is deficient silica.
13. The initial crystallization of alkali pyroxene, alkali amphibole, biotite or magnetite, from a silica undersaturated volatile-rich basic alkali magma, residual magma slowly gets enriched with silica and finally at the late magmatic stages free quartz crystallizes in significant amount as an end product.
14. Rock textures are controlled by conditions of crystallization and cooling. Studying rock textures and the mode of the occurrence of rock types in the field, it is possible to correlate rocks and minerals.

The most common magma which is rich in silica is highly polymerized and has a low activity of oxygen ions and a basic magma rich in MgO and CaO has a high activity of oxygen ions. During the course of crystallization, the most abundant, least soluble mineral grows first followed by the less abundant, more soluble substances from the magma. During the course of crystallization, silicate magma displays progressive enrichments of both soda relative to lime and ferrous iron relative to magnesia in crystals and residual melts. Slowly ascending very hot magmas may cool before reaching the surface.

The crystallization of minerals from magma is a complex process because of changes in the temperature, pressure and composition (TPX). The TPX can have dramatic effects on the order of crystallization and the order-disorder structural state of various minerals in a sequence. The addition or loss of water, CO_2, H_2 and O_2 changes the course of magmatic differentiation. The initial crystallization and separation of early-formed minerals from the residual liquid plays a critical role in the trend of magmatic evolution. The general order of crystallization from the magma is as follows: a) low silica containing basic minerals, such as olivine and calcic plagioclase, together with non-silicate minerals, b) medium-silica-bearing minerals, such as clinopyroxene or labradorite, c) high silica-bearing minerals, such as orthoclase and quartz. Depending on the initial chemical composition of the common basaltic magma, either plagioclase or pyroxene may start to crystallize first, changing the composition of the residual magma until simultaneous crystallization occurs. Slight differences in the composition of the parent basaltic magma may result in the production of an end-magma which is either over-saturated or under-saturated in silica (Barth, 1962). The early crystallization and fractionation of olivine / plagioclase / diopside, the respective residual liquid, is progressively enriched in silica over-saturated rocks (Bowen, 1956) or else alumina poor or silica under-saturated rocks (Barth, 1962) or silica under-saturated alkali-aluminium silicates (Schairer and Yoder, 1960). If the crystal's settling and fractionation are irregular, the composition of the residual magma continuously changes over the course of time. The early crystallized and fractionated silicate or non-silicate mineral from the magma controls the chemical composition of the residual melt and its subsequent trend of crystallization. The composition of the parent magma and its volatile constituents decide which of the minerals start the initial crystallization. Only by studying the field relationship, petrographic features and compositions of the minerals in the rocks and their mode of occurrence and geological setting, along with experimental results, is it possible to trace the magmatic evolution of co-magmatic series of rocks.

3. Magmatic melt structure

The kinetic properties of magma depend upon the atomic mobility within variably polymerized silicate melts. Most polymerized silicic crystal fractions affect the properties of melts by resisting shear deformations. They provide bases for polymerized links to attach in melts' rich fluids and by providing porous networks for the melts and the chemical species within them to flow or diffuse from one location to another. Crystals with large interfacial angles (>60 degrees) represent a near collapse of the grain-supported structure, which can act to either press out or lock melt fractions from one another within a crystal - a melt mush at lower temperatures. Therefore, it may also be impossible to initiate the types of grain boundary movements continuously needed for both minerals in new solutions. This may have an impact on grain compositions, resulting in the zoning of the crystals that form silicic melts becoming more viscous. The magmatic melt structure is viewed as linkages of Si^{4+} and Al^{3+} ions in tetrahedral coordination with O^{2-}-ions. They are polymerized in the more Si-Al rich melts. Dissolved water breaks the Si-O chains and reduces the degree of polymerization and makes it less viscous. The presence of CO_2 rich fluid inclusions in calcites, feldspars, pyroxenes and olivines (Ramasamy and Shapenko, 1980) could be considered to be vapour trapped phases during the growth of these minerals. In addition to H_2O, the vapour phases of CO_2, SO_3, and P_2O_5 play a critical role in the viscosity of the melt.

The formation of crystal nucleus and crystal growth or the accretion of atoms onto the nucleus may take place in two or more stages, depending upon the kinetic conditions of the magma. Microlites are relatively associated with higher energies of surface tension and surface energies than their associated phenocrysts. The rates of growth increase with the increasing of under-cooling up to a maximum value and then diminish. High nuclei populations yield fine-grained rocks. The close associations of both fine-grained aplitic syenites and coarse-grained pegmatitic syenites with subvolcanic miarolitic crystallizations indicate wide variations in the textural pattern of the alkaline rocks in the area under study. The drastic cooling of highly viscous melts produces amorphous glass. Vitrophyric textures are produced by rapid cooling while emplacement occurs in some dolerites in this area. Complex minerals with solid solution relationships often grow when their growth rates are faster than their diffusion rates. The diffusion of chemical constituents is due to variations in the thermal energy in the magmatic chamber. Zoned plagioclases, clinopyroxenes, zoned alkaline rocks and zoned carbonatites are evolved in this manner. An increase of volatile proportions in late magmatic melts produces boiling, exsolution and / or vesiculation in the residual rocks. The sizes of comagmatic bodies and their volumes during fractionation in sequences and subsequent emplacements also play vital roles in evolution of melt structures at various levels of emplacements as well as in residual magmas. In studying the order / disorder relationships and partitioning elements, it is possible to trace the cooling history of the magma.

4. Carbonatite complex

The carbonatite complex of Tiruppattur (12° 00′00″-12°30′00″N and 78°25′00″-78°35′00″E) Tamil Nadu India is an ideal area to trace magmatic evolution and its magmatic zonal variations (Fig. 1). It is unique in its geological settings, in the fractional crystallization of various minerals crystallized during the course of a prolonged period of magmatic differentiation, and in the emplacement of co-magmatic zoned alkaline rocks and the immiscible separation of carbonatites and alkali syenites from common parent magma. The geological field settings present in this area indicate the occurrences of co-magmatic sequences of both silica undersaturated series of rocks and silica oversaturated series of rocks together. Both these series of rocks have co-magmatic relationships and continuous compositional variations among mineral assemblages. Progressive enrichment of alkali, silica and volatile constituents plays a critical role during the course of magmatic evolution from a highly silica undersaturated shonkinite magma to silica oversaturated alkali syenite or granite magma under favourable tectonic environment..

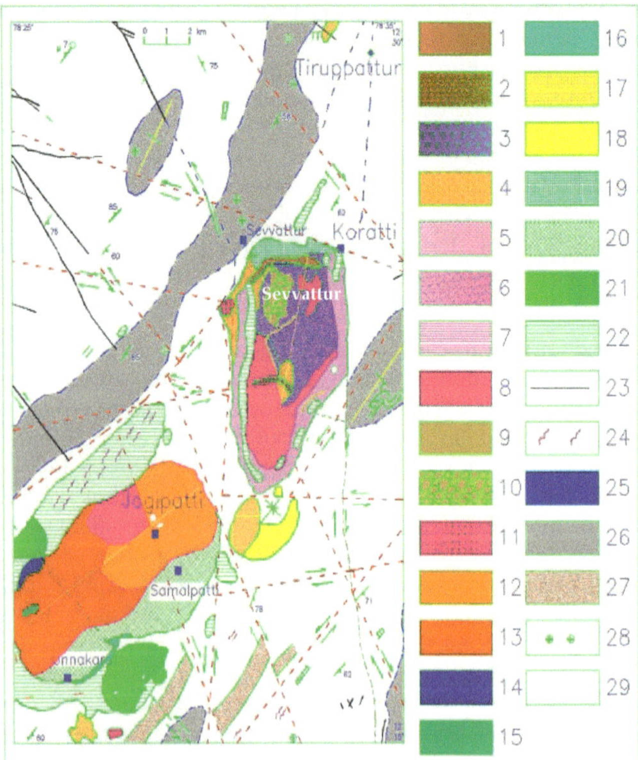

1. Carbonatite,
2. Shonkinite,
3. Vogesite,
4. Oligoclasite,
5. Hornblende syenite,
6. Hornblende porphyritic syenite,
7. Augite syenite,
8. Porphyritic augite syenite,
9. Foliated syenite,
10. Foliated porphyritic syenite
11. Hornblede biotite syenite,
12. Riebeckite syenite,
13. Garnetiferous syenite,
14. Microcline garnet syenite,
15. Wollastonite syenite,
16. Skarn,
17. Granite
18. Granodiorite,
19. Biotite oligoclase pyroxenite
20. Biotite pyroxenite,
21. Ultrabasics
22. Ultramafics,
23. Dolerite,
24. Pegmatite-aplite
25. Albitite
26. Charnockite,
27. Pyroxene granulite,
28. Norite,
29. Granite gneiss

Fig. 1. a. Geological map of the carbonatite complex of Tiruppattur, India

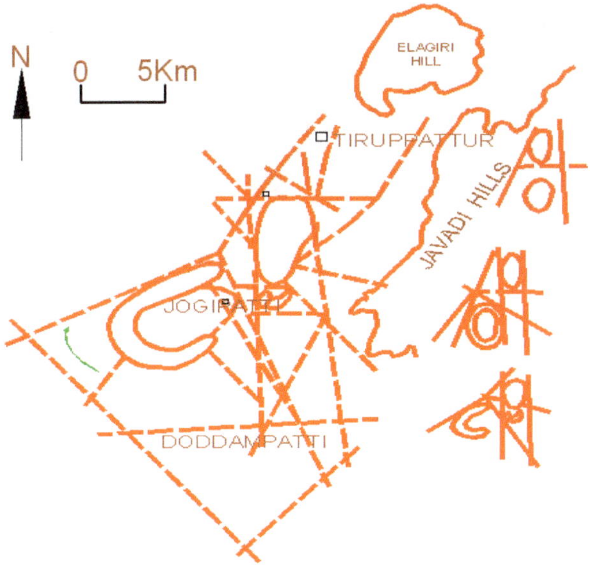

Fig. 1. b. Structural sketch map

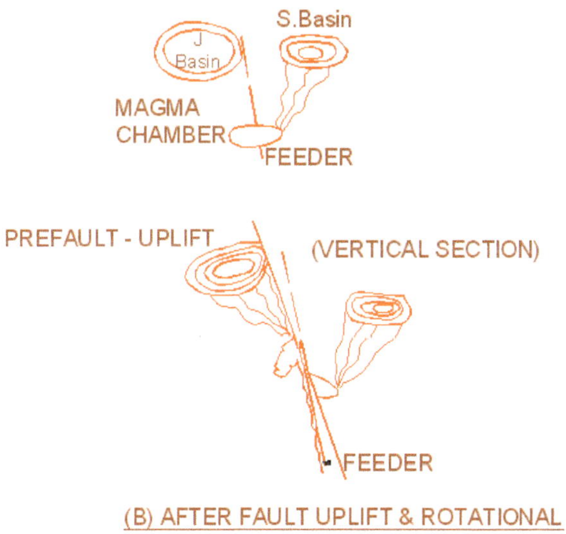

Fig. 1. c. Parent magma ascending from a deep-seated source filled in a magmatic chamber

5. Methodology

Several fieldtrips and observations were made during the course of geological mapping. Thin sections were prepared for various rocks collected in the field and studied under a universal stage (Naidu, 1958) attached polarizing microscope for the identification and determination of the volume proportions of the minerals.

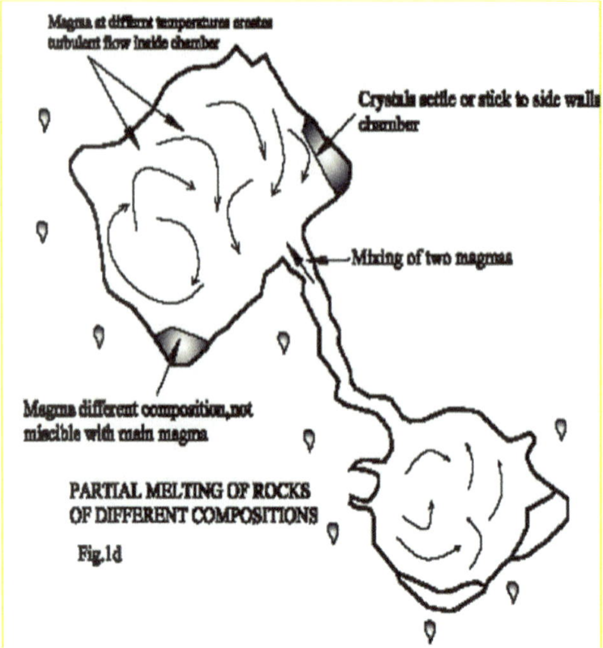

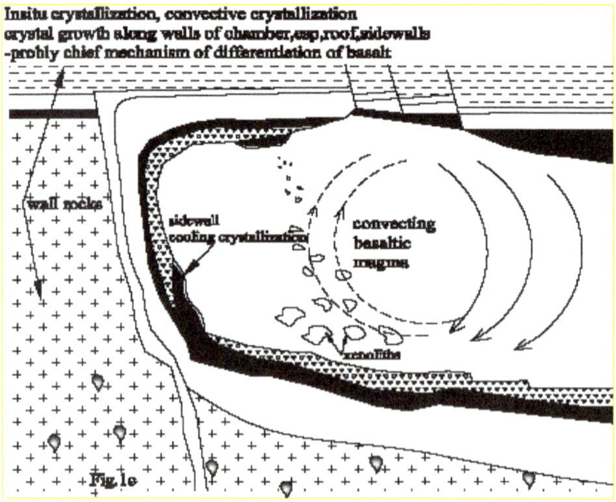

Fig. 1. d. The partial melting of rocks of different compositions concentrates in some large cavities of magmatic chambers. Hot magma rising from the feeder chamber mixes with the already existing magma at a relatively lower temperature and turbulently circulates inside the magmatic chamber. Some of early-formed crystals stick to the walls. The pre-existing magma may dissolve with the incoming magma and mixes with it. The undissolved immiscible magma of different compositions may settle at certain portions. Fig. 1e Magma ascending through fractures breaks the wall rock into pieces of rock fragments which are assimilated by the magma while the magma changes its composition.

i. Differentiation and fractionation of the parent magma for a quiescent prolonged period
ii. Emplacements of a series of soda-rich comagmatic syenites and carbonatites in the Sevvattur Basin
iii. Shifting of the magmatic emplacements of potash rich co-magmatic alkali series and carbonatites into the continuously uplifted and rotated Jogipatti Basin towards the SW
iv. The two basins are enclosed with ultrabasic rocks

By measuring the relationships between the crystallographic and optic axes of the chemical compositions of the minerals and their order-disorder, the structural features are estimated. The textural and structural features of the mineral assemblages, the order of crystallization of the crystallization of the minerals and their magmatic evolutions are traced. The composition of feldspar grains, Fe-Mg ratios in olivine, pyroxene, amphiboles and biotite were calculated by measuring the refractive indices and by finding the optical properties and using the appropriate nomograms given in the texts (Winchell, 1945; Deer et al., 1965). X-ray powder diffraction analyses were made to study the order-disorder relationship by measuring the triclinicity of the feldspars and the M2, M1 and T sites in the clinopyroxenes and the M_4 sites in the amphiboles. Staining techniques were adopted to distinguish calcite and ferro carbonates from dolomite, both in the laboratory as well as in the field (Dickinson, 1965). The chemical compositions of rocks were given by the conventional gravimetric analyses (author in the Geology Dept. Presidency College Chennai and Petrochemical Laboratory, Geology Faculty, Moscow State University) XRF and EDAX under a high resolution scanning electron microscope (IITM, Chennai) . The trace elements contents were determined by Atomic Absorption Spectrometry, Quartz -Grating Spectrograph and HR EDAX. These instrumental facilities were availed by the Department of Geology, Presidency College, Chennai, the Petrography Department, Faculty of Geology, Moscow State University, Russia, the Petrology laboratory, Department of Geology, Government of Tamil Nadu, Chennai and the Material Science Laboratory, Indian Institute of Technology Madras, Chennai.

6. Structure

The carbonatite complex is located in Tamil Nadu, India in a NE-SW trending rift valley formed between the Javadi Hills and the Elagiri Hills (Fig. 1 b), which are located within 5 km of one another. According to Grady (1971), the rift valley, bounded by fault planes, belongs to the category of the deep main faults of Peninsular India, extending more than 200 km in distance. Before rifting, the region was folded, faulted and uplifted by major tectonic deformations, trending with fold-axial plunges in the NW, N, NE, ENE, E and ESE directions. Owing to the number of re-worked fault systems occurring, since the Archaeans, the area has been dissected into several segments of block faults with rotational movements. Hence, it is very hard to analyse the age relationships of different fold styles. However, the regional pattern of fault systems present is seen around the carbonatite complex of Tiruppattur. The culmination of a NE-SW fold system caused the coeval and collinear formation of the Sevvattur and Jogipatti basins by the refolding of ultramafic, basic granulite, charnockite and granite gneiss (Fig. 1b & 1c). The regional NE-SW and N-S fault systems caused formation of a graben structure (Grady, 1971). Due to block faulting, the Jogipatti block was moved towards the west and then towards the north. The melting of the mantle rocks occurred at depth due to E-W crustal shear, which later provides an access route for the rising magma from a mantle source with intermittent stopping in a cavity, forming a closed magmatic chamber at depth. After a prolonged period of magmatic crystallization, differentiation and fractionation with the immiscible separation of carbonatite magmas, the sequences of magmatic rocks were emplaced, first, in the Sevvattur basin and the block was then down warped. A forceful southern movement of the Sevvattur block abutting over the Jogipatti block caused an uplift and a rotational effect over which the rotational axis remained normal to the fault plane trending E-W and resting on the NE-SW fault plane. By the geometrical analysis of this fault system (Ramasamy, 1982), it was estimated that the Jogipatti basin was uplifted by more than 1000 m and, apparently, rotated by about 16° towards the east, and that it caused the emplacements of continuous sequences of late co-magmatic rocks. The magmatic activities were shifted to the latter basin from the original feeding source at depth (Fig. 1 b and 1 c). The generalized features of the magmatic activities are described in Fig. 1 d and 1 e.

7. Field relationship, mineralogical and textural variations in the alkaline complex

The carbonatite complex of Tiruppattur (Borodin et al., 1971; Saravanan and Ramasamy, 1971; Ramasamy, 1982) belongs to the Proterozoic Period, being around 800 Ma old (Schieicher et al., 1998). It is emplaced in two adjacent structural basins amidst Pre-Cambrian granite gneisses and charnockites (Fig. 1). Both of the basins are bounded by ultramafic rocks. The outer shell of the northern Sevvattur body is composed of a fine-grained speckled hornblende syenite with chilled margins and sharp contact with ultramafics and granite gneisses. Towards the south, the specked hornblende syenite imperceptibly grades into mottled augite syenite. The syenite complex exceeds 30 km^2 in extent. These syenites imperceptibly grade inwards towards the centre into their porphyritic syenite variants. The mottled porphyritic syenite is composed of large plates of feldspars containing smaller grains of feldspars in a matrix with accessories of aegirine augite, ferro-hastingsite, hornblende and magnetite. Large plates of plagioclase exhibit normal zoning from andesine to sodic oligoclase. The anorthite contents of the plagioclases decrease from the portions from the north to the south and from the peripheral portions towards the centre. The normal zoning and twinning of the plagioclase lamellae can be clearly seen in large platelets, even in hand-specimens. These rocks are essentially composed of alkali feldspars measuring up to 10 cm x 7 cm x 1 cm. The accessory mafic minerals are diopside, hornblende, augite and ferro-hastingsite. Feldspathic phenocrysts are present in the primary foliation of the rock. However, those mafic minerals occurring in the south are slightly coarse-grained, subhedral to euhedral in form, and are characteristically enriched in alkali constituents and potash feldspars. In places, cumulate feldspar phenocrysts with primary flow orientations are seen in some syenite bodies. The plagioclase and potash feldspars are composed of inclusions of plagioclases with complex twinning lamellae belonging to two or three generations and oriented in different directions. Microcline micro-perthite is commonly present in mottled porphyritic syenite. Albite lamellae, as blebs, orient along crystallographic directions. The plagioclase laths are increasingly plate-like and inform towards the centre of the basin. Similarly, the sizes of the mafic minerals increase and their compositions change with increases in alkali and ferric iron contents. Some places among the syenites - owing to the concentration of mafic minerals in felsic rocks or else the mush of feldspar phenocrysts in hypo-melanocratic syenites - form cumulate textures. The flow textures of feldspar phenocrysts and amphiboles are common in most of the syenites and they orient in the same directions. The rocks of the inner portions are successively younger than the rocks of the peripheral portions. The core portions are occupied by highly differentiated magmatic layers derived from deeper levels. Small veins of fine-grained acmite-syenite are composed by a significant amount of magnetite and quartz and large plates of acmite are surrounded by the released products of magnetites. Abrupt variations from a porphyritic syenite to a non-porphyritic syenite are also seen. The impregnation of cumulates of feldspar phenocrysts into a syenite body is also seen. Many melanocratic streaks, thin slivers and patches of pyroxenites, are strewn as disseminated or as scattered material in the syenite body as mafic cumulates. The dissemination of mafic minerals is often crowded around mafic xenoliths. It seems to be the case that melanocratic rocks rich in mafic minerals are developed in this manner. Foliated fine-grained syenite occurs as a linear band in the porphyritic syenites in the Sevvattur basin. It exhibits crude foliation by the presence of abundant mica. In thin sections, it displays a xenomorphic granular texture. It is composed of oligoclase, anorthoclase, microcline, biotite, margarite and calcite. The grain size of the feldspar ranges from 0.1 mm to 0.3 mm. Discrete grains of microcline are formed as an interstitial to biotite. The replacement of oligoclase by potash feldspar is found along the cleavage planes. The development of patchy perthite is obvious: it imperceptibly grades into a pink coloured porphyritic syenite with a trachitoid texture. It occurs as a pale pink coloured rock with large plates of potash feldspar and oligoclase set in a fine-grained feldspathic matrix. Most of the feldspar platelets are square shaped with up to 5 cm sides. The breadth of the plates varies between 2 mm to 4 mm. In thin sections, feldspar displays an inequigranular poikilitic texture. Hornblende needles are present as inclusions within the plates of the feldspars and the needles orient in parallel towards the foliation planes of the rock. It is essentially composed of microcline, oligoclase and orthoclase, with accessories of chlorite, biotite, hornblende and epidote. The microcline and oligoclase are formed in two generations. Often, the feldspars are unoriented. They carry inclusions of fine needles of mafic minerals (with a helicitic texture) and show characteristic foliation. The core portion of the Sevvattur syenite body is occupied by oligoclasiteand albitite. The amphibole in the biotite-oligoclasite is altered into epidote granules and the amphibole is impregnated with felsic minerals producing a sieve texture. In some oligoclasites oligoclase and biotite, flakes exhibit a sheath texture with radiating platelets orienting towards a common centre. Along the peripheral portion of the carbonatite, albitite is exposed. In the albitite, the feldspar plates have inclusions of magnetite. The modal composition of potash feldspars varies widely and their content increases towards the core of the basin. Again, the volume percentage of the potash feldspars increases towards the south. Albitite, oligoclasite and biotite-bearing oligoclasites occur adjacent to the carbonatite body. A fine-grained monomineralic rock composed of more than 90% of albite in volume with accessories of magnetite and others exhibits the hypidiomorphic granular texture. Post magmatic growth of lueshite (sodium pyrochlore) in this rock produces expansion cracks in the surrounding albite. .

Syenite pegmatites and aplites are common in the carbonatite complex. They also occur as sheets along the fracture planes present in the syenites. Numerous anastomosing veins of pegmatites and aplites are seen in the syenites. Small bodies of pegmatites and aplites carry coarse-grained crystals along their borders. The mafic minerals grow towards it as shooting grains from the wall rocks and the core is only composed of felsic minerals. Massive bodies of pegmatites and aplites occupy over 50 km^2 in the Jogipatti basin. The northern portion of the syenite complex of this basin is composed of riebeckite-anorthoclase syenite while its southern portion is composed of orthoclase-bearing garnetiferous pegmatitic syenite. Garnetiferous syenite exhibits a hypidiomorphic granular texture. Euhedral to subhedral grains of garnets are embedded in the potash feldspar. The grossularite-andradite garnet gradually transforms into melanite towards the southwest of this region. Melanite-microcline syenite, wollastonite syenite, wollastonite-scapolite syenite and wollastonite-melanite carbonatite are some of the aplitic veins found in this complex. Fine-grained microcline melanite syenite displays a xenomorphic granular texture. Micro-grains of microcline are surrounded by accessories of anhedral albite and melanite. These microclines and anorthoclases are highly disordered feldspars, having significant trace amounts of ferric iron. The biotite-syenite has a xenomorphic granular texture with euhedral to subhedral biotite. The biotite syenite porphyry is composed of large phenocrysts of anorthoclase, displaying zonal variation from the core to the periphery and set in a micro-granular feldspathic matrix. Such syenite porphyries are also composed of potash feldspar phenocrysts with a cryptoperthitic texture at the core and with homogeneous peripheral portions. The feldspars belong to two or more generations and they occur as large plates of phenocrysts as well as very small anhedral grains in the matrix. Younger aplites trend in N85° crosscuts while older aplites trend in the N45° direction. Small pockets of zircon- and magnetite-bearing syenites are seen in this basin. The miarolitic texture is a commonly seen in riebeckite pegmatites and aplites. This texture is produced by a crystal growth mechanism that was initially governed under the influence of surface-controlled kinetics. Larger crystals tend to grow by absorbing the required amount of water from the pore fluids present in the surrounding fine-grained matrix. Therefore, the fine-grained grains in the matrix are unable to grow larger due to a lack of water. Moreover, owing to their rapid cooling, larger phenocrysts formed at high temperatures exhibit lognormal rates of growth and in sizes relative to the small grains present in the matrix, forming syenite porphyries or porphyritic rocks under sub-volcanic conditions. The sudden increase of the viscosity of the residual magma also plays a critical role in the crystallization process of felsic and mafic minerals of two or more generations. Large crystals enclosing inclusions of euhedral feldspars oriented in two or three different directions are seen in mottled augite porphyritic syenite and this feature indicates slow cooling of the syenite magma followed by rapid ascending and consolidation of the magma. On the other hand, the anorthoclase-bearing miarolitic syenitic aplites and pegmatites carry abundant miarolitic cavities in which radiating needles of riebeckite are grown towards the centre of the cavities. Cross-cutting anorthoclase-bearing aplites are seen in mottled porphyritic syenites. Two generations of magnetite, apatite and zircon crystallize, both as early-formed minerals as well as very late-formed minerals. The late-formed minerals tend to have comparatively high ratios of surface area to volume. The large feldspar phenocryst platelets increasingly thin in speckled and mottled porphyritic syenites. Moreover, feldspars which crystallize very late in the mottled porphyritic syenite appear to be formed as early crystallized mineral in some of the agpaitic syenite pegmatites and aplites in the Jogipatti basin. The concentration of garnet, aegirine-augite or magnetite along a particular zone produces a banded structure. Radiating bundles and prisms of dark green riebeckite are seen in the aplitic riebeckite syenite and in the carbonatites of Jogipatti. All these varieties of syenites are more or less interrelated to each other in their mineralogy and chemical composition. The syenites of the Sevvattur basin exhibit magmatic continuity with the syenites of the Jogipatti basin. They exhibit a co-magmatic relationship and gradational, zonal, spatial and temporal variations and continuities.

An arc-like vogesite outcrop with a lamprophyric pandiomorphic texture with black needles of katophorite set in a feldspathic matrix of two generations occurs in the Sevvattur basin with close proximity to carbonatite exposures. The vogesite imperceptibly grades to ferro-hastingsite syenite, speckled hornblende syenite and hornblende-biotite oligoclasite. An acmite-bearing syenite grades with an enrichment of cumulate needles of katophorite into lamprophyric vogesite, displaying a pandiomorphic texture. The shonkinite occurring in Jogipatti basin is a coarse-grained inequigranular rock composed of equal proportions of sanidine / anorthoclase and augite, with accessories of olivine, hastingsite, phlogopite, apatite and magnetite and exhibiting a lamprophyric pandiomorphic texture. In the field, it occurs as xenoliths or nodules in ultramafic rocks. The ultramafic rock which is the host rock of the shonkinite appears to be possessed of kimberlitic affinity (Ramasamy et al., 2010). Mineralogical gradation exists between the shonkinite and garnetiferous syenite. An abrupt mineralogical and compositional gradation exists between the vogesite and shonkinite. Similarly, there exists such gradation between vogesite and hornblende-biotite oligoclasites. Again, the shonkinite exhibits such a type of gradation between the garnetiferous-orthoclase syenite and the mottled augite syenite. On the whole, one syenitic member has genetic relationship with any other member in this zoned complex, either in mineralogical or compositional gradations. In a detailed petrographic and mineralogical study, it is shown that the shonkinite is considered as a parent magma for this carbonatite alkali-syenite complex (Saravanan and Ramasamy, 1995; Ramasamy et al., 2010). Dykes, ring dykes, sills, veins, cavity fillings, cone sheets and plugs of syenites and carbonatites show sharp contacts with chilled margins with their host rocks. Curvilinear tensional cracks are seen around the exposures of the syenites and carbonatites. Late magmatic aplites and pegmatites filled these cracks and show co-magmatic relationships with the adjoining plutons.

All of the carbonatites occurring in the three places in Sevvattur, Jogipatti and Onnakarai grade with zonal variation from ferro-carbonatite, beforsite and sovite. Calcite, dolomite and ankerite constitute the essential minerals in these rocks and biotite, phlogopite, magnetite, riebeckite, aegirine augite, acmite, wollastonite, garnet and the rare-earth minerals of pyrochlore, zircon and niobian rutile constitute the notable accessory minerals in these rocks. Two generations of magnetites, both in the form of opaque dusts and anhedral, euhedral and twinned crystals, are present in some of the carbonatites and in the shonkinite. The Fe- and Mg-rich carbonatites appear to be emplaced from relatively deeper horizons. By their characteristic minerals, the zoned cone-sheets of the carbonatites in Sevvattur exhibit various sheets and layers as hybrid sovite, biotite sovite, phlogopite sovite, aegirine sovite, pyrochlore-bearing para-ankeritic sovite, pyrochlore-rich para-ankeritic beforsite, magnetite beforsite and apatite soviet. The elongated rings and veins of the riebeckite-carbonatites are seen in the riebeckite syenites of the Jogipatti basin. The hybrid carbonatites of the Jogipatti basin are enriched with richterite, magnesio-arfvedsonite, acmite, wollastonite and scapolite. The carbonatites of Onnakarai appear to be younger than the Jogipatti carbonatite, which is intermediate and younger itself than the Sevvattur carbonatite (Ramasamy, 1973). Again, the zoned alkali syenites of Sevvattur are older than those in the Jogipatti basin. A carbonatite body crops out in arcuate cone-sheets with a sharp contact along the meeting between the ultramafic rocks and the syenites in the northern periphery of the Sevvattur body. Mineralogical variations occur as zones which are almost conformable with one another, starting from the convex side of the arc and proceeding to the concave side in the following order: a) micaceous zone, b) ankerite zone, c) beforsite zone and d) sovite zone. About 400 m west of the carbonatite exposure is evidence of the forceful injection of carbonatite, as manifested in a syenitic outcrop which has been thoroughly brecciated with feldspar grains which were twisted and ragged with the emplacement of fine-grained carbonatitic cementing media. Along the contacts of the carbonatite, the host rocks have been carbonatized by the intrusion of a thin vein in which lots of carbonates and replacements of silicate minerals into calcites are observed. Iron-enriched phlogopite is found at the contact of carbonatite with the ultramafics. Schiller-type inclusions are common and abundant along the cleavages and partings of the augite crystals. Hornblende wraps round the augite and is pleochroic, from green to yellow. Small prisms of biotite are found along the margins of the pyroxene and hornblende. An economical deposit of vermiculite has been formed at the contact between the ultramafics and the carbonatites. The carbonatite exhibits a well-developed concentric flow structure with linear crystals of apatite, alkali-amphibole and magnetite or else a schlieren of hornblende and biotite and streaky patches of ferromagnesian minerals. The carbonatite imperceptibly grades into sovite, beforsite and ankerite, depending upon an increase of dolomite or ferro-carbonates. In the ferro-carbonatites, which are relatively emplaced from deep-seated sources, pyrochlore, ilmenite and ilmenorutile concentrated in the Sevvattur, Jogipatti and Onnakarai villages, respectively. The broad spectrum of the evolution of the carbonatite complex is manifested in: a) introduction of the calcite into the host rocks, b) the formation of calcium-bearing minerals like grossularite-andradite, melanite, wollastonite, richterite paragonite, margarite, sphene and epidote, c) the alkalinization of clinopyroxene and amphiboles, d) phlogopitization and e) feldspathization. The carbonatite exhibits a xenomorphic granular texture by the presence of anhedral calcite or dolomite grains. The oligoclase and potash feldspars in the plates of the calcites of carbonatites are often corroded with an embayed outline. The micro-fractures in the host rocks are filled with calcites and some mafic minerals are transformed into margarite. The potash feldspar is found as inclusion along the contact of carbonatites and syenites and the host rock is enriched with turbid granular potash feldspar. Lumps of granular clinopyroxenes are developed in the host rock of ultramafics along the point of contact with the carbonatites. The grain size of the granular clinopyroxene has been transformed into well-developed clinopyroxene with larger dimensions. Some of them grow to over 10 cm in their dimensions as single crystals. Apatite is the more common accessory mineral in this contact zone. Clinopyroxene and amphiboles are converted into biotite. Wollastonite, garnet (melanite) or scapolite are formed as small prisms interstitial to acmite and appear to be formed by magmatic crystallization (Eckermann, 1966).

Both the basins are surrounded by ultramafic rocks. They include dunite, peridotites, kimberlites, pyroxenite, biotite pyroxenite, massive amphibolites and biotite. It is not possible to classify these rocks in the field owing to their scarce exposure on the surface and so they are together classified as ultramafic rocks. Magnesite and kankar are found on the weathered surfaces of the ultramafics. Dunite and peridotites are altered to talc-steatite and calcite-bearing rocks. Massive dunite exhibits a saccharoidal texture. However, the grain sizes of these rocks vary widely. Porphyritic pyroxenite with kankar veins is also seen in some well sections. Ultramafics rich in biotite and amphibole are very common. Veins and selvages of calcite-bearing riebeckite showing a schistose structure fill the cracks developed within the ultramafics. The ultramafics are essentially comprised of pyroxenites and peridotites. Thin veins of oligoclasites, albitites and carbonatites intrude into the ultramafics, which are transformed into biotite-pyroxenite and hornblende-oligoclase-calcite-bearing pyroxenite. Many varieties of hybrid carbonatites are developed at the point of contact of the ultramafic rocks with the carbonatites. Alkali pyroxenes like aegirine-augite and augite have developed in the ultramafic rocks and in the syenites. Alkali amphiboles like magnesio-riebeckite, magnesio-arfvedsonite and richterite have developed extensively in the aplitic syenites. Grossularite, melanite, scapolite, calcite and wollastonite were developed at the point of contact with the alkali syenites-carbonatite-ultramafics. Feldspathic selvages in ultramafics often display the development of biotite. At the contact between the garnetiferous syenites and the ultramafics, a skarn rock has been developed at around 2000 m length and with a width ranging from 10 to 200 m. It is comprised of calcite, wollastonite, grossularite and epidote as major minerals. In these rocks, no accessory minerals characteristic of carbonatites, such as apatite, magnetite, zircon, alkali pyroxenes, amphiboles and REE minerals, are seen. The individual minerals in the skarn rock do not have any mineral lineation steeply plunging towards the centre of the basin. However, evidence such as the partial transformation of diopside into calcite, garnet or wollastonite is present for the carbonate metasomatism of one mineral into another. The rock has been extensively folded, sheared and deformed, carrying ultrabasic nodules of varying dimensions from less than 1 cm to over 40 cm. These nodules are partially carbonatized and they exist discontinuously along the folded axial planes. The intricate folding pattern with layers and veins of calcite is even seen in hand specimens. Granulitic and gneissic layers occur within it as contorted and elongated lenses. These knotty inclusions stand out boldly on the weathered surfaces. The skarn rock exhibits a characteristic ribbed structure due to weathering. The ribs are parallel to the foliation direction and are developed along the trend of the elongation of the direction of the relics of mobilized ultramafic nodules towards the N 45° direction. Specks of native copper, bornite, covellite, chalcopyrite, pyrite, galena pyrrhotite are disseminated into the nodules. The modes of the occurrence of these nodules indicate that part of the ultramafic rock occurring along the contact of the garnetiferous syenite has been transformed into a skarn rock during the emplacement of the syenite body along its contacts. Moreover, along these contacts, riebeckite-sovites, acmite and riebeckite-bearing ferro-carbonatites, veins of ferro-carbonatite breccias are seen. The ferro-carbonatite breccia carrying angular fragments of ferro-carbonatite and riebeckite syenite, quartz carrying inclusions of riebeckite needles and euhedral magnetite crystals sets in very fine-grained carbonate matrix. Besides breccias, there are exposures of monazite bearing riebeckite syenite, benstonite carbonatite, barite veins, veins of galena and sulphide hosted aplites and pegmatites are emplaced just inside and on the western outside of the skarn rock exposures. The ultramafic body exposed to the west of the skarn rock is seen with the extensive development of biotite and with specks of sulphide minerals aligned characteristically along the foliation or schistose planes of the rock. It is subjected to vermiculitization with the emplacement of carbonatitic veins. In places, veins of ilmenorutile, barite, apatite-ilmenorutile rock, ankeritic carbonatite and riebeckite are seen in it. The biotite is extensively developed in the ultramafic rock along its contact with garnetiferous syenite. Small pockets of massive biotite-hornblende granites with a sharp point of contact with the country rocks of granite-gneisses occur in the north and south of the carbonatite complex. An arcuate fault-bounded aegirine granite exposure imperceptibly grades towards the east as the hornblende granodiorite in the east is seen in between the two syenite basins. The other granites occurring in this complex also have a gradational variation in their mineralogy and in their chemical compositions, and they also have a genetic relationship with that of the alkali syenites.

8. Zoned carbonatite-alkali complex

The mineral assemblages in the alkaline rocks occurring in the two adjacent basins exhibit continuous chemical variations from one end to the other in various members of co-magmatic alkaline rocks. Larger syenite bodies show imperceptible gradation to the adjacent syenite bodies. The syenites and carbonatites are composed of both early-formed and late-formed minerals (Saravanan and Ramasamy, 1995). All of these rocks are classified based on their wide variations in their textural features and mineralogical compositions. The mapping was carried out based on these features in the field. The anorthite content in plagioclase feldspars in the syenites varies between An28% and An35%, and often oligoclase is the predominant in the speckled syenite and it decreases to An4% in the places from the mottled porphyritic syenite which imperceptibly grade from that speckled porphyritic syenite with an oligoclase of An18%. Plagioclase growth twins are evident in the syenites. Carlsbad (twin axis {001}, penetration twin), acline(composition plane {001}) Baveno (contact twin, twin plane {021}) and Manebach (contact twin, twin plane {001} twins as well as periclines (twin axis {010}) and albite (twin plane {010}) twins are common in the feldspars of the syenites. Normal zoning (progressively sodic towards the rim) is common and reverse zoning (progressively more calcic towards the rim) is also found, occasionally, in the syenites. In places, the oscillatory normal zonings ranging between An34% and An24% are seen in the speckled syenites. Patchy zonings are found in the biotite syenite porphyries. The anorthite contents of the phenocrysts in the porphyritic syenites, vogesites and the syenite porphyries are slightly higher than those of the plagioclases in the matrix. The potash feldspars have the following mean variations of orthoclase and albite components: speckled hornblende syenite (Or_{77} Ab_{23}), speckled hornblende porphyritic syenite ($Or_{87}Ab_{13}$), mottled augite syenite (Or_{50} Ab_{50}), mottled augite porphyritic syenite (Or_{87} Ab_{13}), shonkinite (Or_{61} Ab_{39}), garnetiferous syenite (Or_{62} Ab_{38}) and riebeckite syenite (Or_{43} Ab_{57}). Microcline cryptoperthite is commonly found in mottled porphyritic syenite and in syenite porphyries. The plot of the percentages of or-ab-an in a trilinear diagram shows that most of alkali feldspars fall within the field of sanidine, anorthoclase, albite and oligoclase. The Al/Si ratios (0.11 in albitite, -0.49microcline-melanite-syenite) in the alkali feldspar appear to be related to the history of the unmixing of the alkali feldspar (Deer et al., 1965). These components are plotted in a diagram constructed for the experimental data under 5000 bars H_2O pressure; they indicate that the individual components fall within the range of 800°C to 950°C (Yoder et al., 1957). Microcline, microperthite, anorthoclase, sanidine and orthoclase are present in various syenite members and show a high degree of triclinicity (ranging between 0.6 and 1.0), indicating its disorder states and their high temperature formations and the emplacements (Ramasamy, 1986) of the syenite bodies. The microcline-microperthite, potassic host ranges from 48% to 67%, while the sodic guest ranges from 33% to 52%, indicating their high disorder relationships and high temperature formations. The feldspars display clouding owing to the presence of very fine dusty inclusions of iron oxides in some varieties of shonkinite and syenites. Such types of clouding are present in calcite, dolomite and in ankerite, with varieties of textural patterns in the carbonatites of Sevvattur. It is interpreted that dissociation of ankerite into calcite and magnetite during the course of the ascension of the magma during successive emplacements of cone sheets of carbonatites from deep-seated sources through the centre of the magmatic body at the late magmatic stages would have produced such textural variations (Ramasamy et al., 2001).

The diopside in pyroxenite and feldspar syenites adjacent to the carbonatite exposures are transformed into calcite. The ultramafic body in the Jogipatti basin is extensively deformed and carbonatized. the proportions of plagioclase content decrease towards the centre of the mottled porphyritic syenite. The colour index (Rittmann, 1973) for the syenites varies widely in the alkaline rocks, depending upon the degree of fractionation and the separation of the clinopyroxenes or mafic minerals during the course of magmatic differentiation. According to the volume proportion and mineral structure of the mafic minerals present in the various alkaline members, they are subjected to transformations of clinopyroxenes into amphibole, biotite or magnetite. There exists continuity in the compositional variations of the mafic minerals from one syenite member to another (Ramasamy, 1986a). The clinopyroxenes from the youngest alkali syenites fall into the fields of ferrosilite and the calcium-rich hedenbergite end. The clinopyroxene from the melanite-orthoclase syenite contains a high content of CaO owing to the partial transformation of the clinopyroxene into grossularite-andradite garnet (Saravanan and Ramasamy, 1995). During late magmatic stages, due to increase of agpaitic coefficient {(Na+K)/Al} with progressive depletion of Al, orthoclase transforms into highly disordered microcline incorporating ferric iron in its lattice. Release of Ti^{iv} and Fe^{iii} in the appropriate sites of orthoclase and grossularite-andradite garnet as per the exchange of $Ti^{iv}Fe^{iii} \leftrightarrow Ti^{iv}Al^{iii}$ from subsolidus garnetiferous orthoclasite transforms to melanite-microcline syentite (Ramasamy, 1986). Owing to increasing crystallization and the transformation of mafic minerals into alkaline mafic minerals, it is also dependent on the mineral proportion of mafic minerals present in the individual alkaline rocks. Replacing Ca ions from the amphiboles of $Ca_{76}Na_{15}K_9$ by the substitution of (Na+K) in the M4 sites of the amphiboles to $Ca_{27}Na_{65}K_8$ in the younger generation of syenites caused the crystallization of the riebeckite. Furthermore, the ratio of 100 Mg/(Mg+Fe+Mn) of the calciferous amphiboles decreases from 85 to 30% because of the changing composition from pargasite to ferrohastingsite in the younger syenites. The katophorite and richterite are restricted to vogesite and hybrid carbonatites respectively. The eckermannite, magnesio-arfvedsonite, arfvedsonite, magnesioriebeckite and riebeckite are widely distributed in potash rich syenites and carbonatites. Both the clinopyroxenes and amphiboles are deficient in Si in their tetrahedral sites and are compensated by Al and Ti ions. However, under increasing P_{H2O} in the late magmatic syenites, Si is saturated and Na replaces Ca with increasing silica activity. The "m" value of biotite Mg/(Mg+Fe) decreases from 0.91 to 0.29 in the late magmatic syenites. The Cs (4-10 ppm), Rb (30-345 ppm) and Li (1.5-7.5 ppm) contents increase as K increases in this complex.

The emplacement of younger alkali syenite bodies at the contacts of older syenite units plays a critical role in the transformation and metasomatism of mafic minerals into alkali-rich minerals. Again, feldsic minerals are also subjected to metasomatism, such as the transformation of oligoclase into potash feldspars. The oligoclase surrounded by the rims of reverse zoning in the mottled porphyritic syenite indicates that it is ascending to higher levels. The presence of peristerites formed by the sub-microscopic intergrowth of sodium-rich and calcium-rich phases is observed in the mica syenite porphyries. The compositional zoning indicates that with a decrease of the temperature, the sequence solidified from the margin inwards. However, such metasomatism appears to take place at sub-solidus phases of the phenocrysts through the escape of volatile constituents carrying the required ions in the vapour or liquid phases. Sheets and plugs emplaced at the inner portions appear to be derived from a deep-seated magmatic column. The solidification was interrupted repeatedly by surges of fluid core magma. The emplacements of the various plutons are structurally controlled. The isoclinal folding of the flow bands of the carbonatitic or syenitic layers were developed, indicating the stretching and spreading of the carbonatitic or syenitic magma along the narrow zones of the emplacements. The residual magmas formed through the differentiation and fractionation of a shonkinitic parent magma resulted in decreasing silica activity and steadily increasing oxygen fugacity and agpaicity due to the progressive enrichment of alkalis, ferric iron and volatile constituents in a closed magmatic chamber (Saravanan and Ramasamy, 1995). The relative rate of the development of agpaicity in the residual liquid and the crystallization of femic and felsic minerals also played a critical role in the alkaline magma's evolution. The silica deficiency developed in the residual magma due to the early crystallization and fractionation of the clinopyroxene and alkali feldspar, which may have been compensated for through the formation of amphibole, biotite, carbonates, phosphates and magnetites (Yagi, 1953). Therefore, all of these magmatic bodies exhibit compositional zonings. The compositional zoning indicates that, with a limited decrease in the temperature, the sequence solidified from the margin inwards. Sheets and plugs emplaced at the inner portions appear to be derived from a deep-seated magmatic column. The emplacements of the various plutons are structurally controlled.

The early-formed clinopyroxene in some of the syenites reacted with the residual magma and transformed into calcic-rich diopside and then into augite and aegirine-augite to acmite. Similarly, the acmite transformed into garnet which, again, transformed into melanite with an enrichment of titanium. the calcite reacted with silica-forming wollastonite in the garnetiferous syenite and it reacted with plagioclase forming scapolite, taking the required amount of SO_3 from the volatile phase enriched with this constituent (Tables 2 and 3), as revealed from the chemical analyses by the enrichment of SO_3 in bulk rock compositions. The quartz content decreases from the rocks outcropping at the outer shell to those rocks occurring towards the centre of the Sevvattur basin. On the other hand, such variation is remarkable given the formation of riebeckite, biotite, magnetite, wollastonite, garnet, melanite, scapolite and calcite by the heteromorphic transformation of mineral assemblages to compensate for the silica deficiency of the residual magma impoverished in the silica owing to the crystallization of sodaclase anorthoclase, sanidine and high temperature microclines in the syenites of the Jogipatti basin. The enrichment of volatile constituents like H_2O, CO_2, SO_3, P_2O_5 and F in the late magmatic residual magmas developed different varieties of oxide minerals, like magnetite, rutile, ilmenorutile, perovskite and zircon, carbonate minerals like calcite, dolomite, para-ankerite, benstonite, sulphate minerals like barite, gypsum and scapolite, phosphate minerals like apatite and monazite, and sulphide minerals like galena, chalcopyrite, pyrite and pyrrhotite. The very presence of these minerals indicates that the residual magma was enriched in these volatile constituents, which compensates for any silica under-saturation owing to the extensive development of feldspathic constituents. Most of the rocks in this complex have insufficient alumina to form adequate alkali feldspars. The emplacements of carbonatites and its co-magmatic alkali syenites are in a sequential order due to a series of pulses occurring during intermittent tectonic disturbances.

Olivine is present in significant proportions in ultramafic rocks, shonkinite and carbonatites. The compositions of olivine in the ultrabasic rock represent an intermediate position in the course of the magmatic evolution of the olivine occurring in the shonkinite. The olivine in the shonkinite exhibits peripheral zonal variation augite, aegirine-augite and jadeite, indicating a prolonged period of crystallization under the liquid stage. The high concentration of Ca-Na-K-Al in the olivine from the ultramafics indicates its kimberlitic affinity. These features indicate that shonkinite magma is the parent magma for the ultramafics. The early fractionated olivine from the parental shonkinite magma reacts with the residual magma and the peripheral portion of the olivine is transformed into diopside on slow cooling with release of volatile constituents into the residual magma, inducing magmatic pressure by incorporating CaO. The crystallization of the clinopyroxene in the place of the olivine, the residual magma, is depleted in the silica. Under increasing magmatic pressure and in order to meet the silica deficiency, more Na_2O and Al_2O_3 are incorporated with the development of aegirine augite. With the impoverishment of SiO_2 and Fe_2O_3, jadeite crystallizes from the residual magma along the peripheral portion of the olivine or the clinopyroxene. The jadeite develops along the peripheral portions of the olivine within the temperature range of 600 to 800°C with a pressure ranging from 10 to 20 kbar under progressive enrichments of Al_2O_3 and Na_2O and with decreasing silica activity in the residual magma (Ramasamy et al., 2010). The crystallization of jadeite (jd) in the place of albite (Ab) or nepheline (Ne) compensates the silica deficiency in the residual magma.

$$2jd \leftrightarrow Ne + Ab$$

The further crystallization of jadeite - the residual magma - is enriched with K_2O, Al_2O_3 and SiO_2 (Yagi, 1953).

Thus, the development of jadeite produces as end products silica under-saturated nepheline-free shonkinite with the crystallization of potassium feldspars from the residual magma. With an increase of the vapour pressure during the late magmatic processes, particularly under high P_{CO2}, the mafic cumulates become unstable and release more CaO into the melt. The Ca-enriched residual magma immiscibly separates into silicate and carbonate magmas under high P_{CO2}. The further differentiation of these magmas produces a series of alkali syenites and carbonatites, successively emplaced in a sequential order, first in the Sevvattur basin and then in the Jogipatti basin, and resulting in the formation of the zoned alkali carbonatite complex of Tiruppattur, Tamil Nadu.

Normally, a carbonatite complex is associated with alkaline rocks accompanied with significant amounts of alkali-rich ferro-magnesium minerals. Often, a fenitized aureole is present around the carbonatite complex owing to the alteration of country rocks by its reaction with alkali fluids escaping from the carbonatite body. The country rock may be changed into silica-impoverished nepheline-bearing fenites. In this complex, such a development of fenites is absent. As magmatic differentiation takes place within a closed magmatic chamber for a prolonged period of crystallization and differentiation, the alkaline fluid evolves during the course of magmatic evolution, reacting with the minerals and crystallizing from the magma and increasing the alkali constituents up to the development of ultrapotassic syenites. The transformation of diopside into acmite and jadeite during the course of magmatic evolution is adjusted with the development of acmite (Fig. 2) and the ensuing silica deficiency is compensated for by the reaction of albite with silica under-saturated residual magma forming a jadeite component. Under high oxygen fugacity and a volatile concentration, biotite and magnetite form in the place of jadeite. The distribution of volume proportions of alkali feldspar against quartz (Fig. 3a) and alkali feldspar against plagioclase is calculated on the basis of the Rittmann norm using the chemical analyses listed in Tables 1 and 2 for the plot in the Q-A-P-F (Rittmann, 1973) diagram, indicating a linear differentiation trend for co-magmatic alkali series of rocks (Fig. 4 a & b). In Fig. 5a-j, it is indicated that the distribution of major elements in binary variations indicate overlapping smooth linear trends for syenites, ultramafics, carbonatites and dolerites. The dolerites in this area also show a trend of alkali-enrichment during its magmatic evolution.

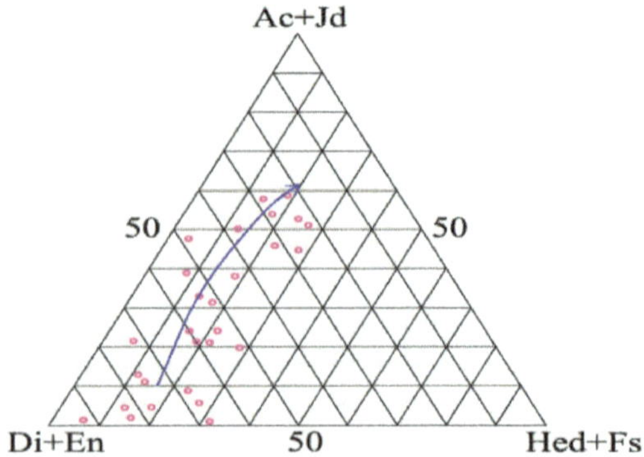

Fig. 2. The chemical compositions of alkaline rocks calculated on the basis of Acmite+Jadeite, Diopside+Enstatite and Hedenbergite+Ferrosalitse compositions, indicating a magmatic differentiation trend moving towards the Ac+Jd end members. The diagram indicates the enrichment of alkalis and iron during the course of differentiation

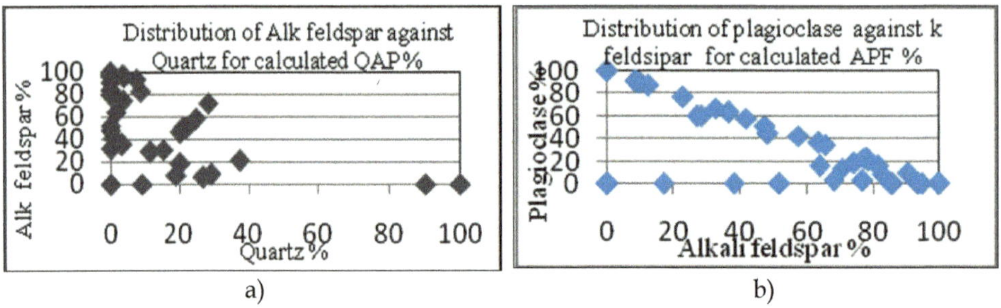

Fig. 3. a. & b. Streckeisen (1967) QAPF double triangle plots are presented in Qz-alk-felds and Alk-felds - Pl variation diagrams showing linear variations in co-magmatic series (The C.I. for syenites varies from 5-30; for hybrid rocks 30-70; for ultramafics 70-100).

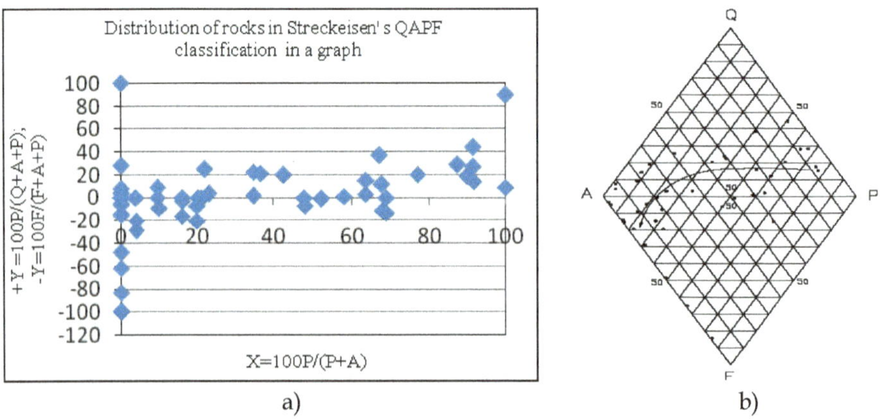

Fig. 4. a., b. Streckeisen graph and double triangle showing the linear trends of the magmatic evolution of alkaline rocks in the Tiruppattur carbonatite complex

Rock	No	SiO2	Al2O3	Fe2O3	FeO	MnO	MgO	CaO	Na2O	K2O	TiO2	P2O5	CO2	SO3	H2O
Bio oligoclasite	723	48.35	12.97	3.78	3.26		2.68	12.91	3.54	5.00	0.75	1.06	2.76		3.15
Bio fol aplite	1122	50.28	23.25	5.54	2.98	0.36	3.82	5.60	2.08	1.80	0.90	0.60	0.60		
Bihbpor syenite	428	51.53	18.22	7.53	2.79		4.00	7.50	3.85	2.70	0.65	0.22	0.39		0.44
Vogesite	733	53.07	13.03	4.28	5.09		3.75	11.75	4.00	3.93	0.69	0.48			0.36
Aughb syemite	464	53.17	14.06	5.32	5.42	0.35	3.62	9.06	3.80	2.25	1.30	1.50			0.68
Ac albitite	468	54.37	5.30	19.84	1.95	0.52	0.98	2.50	11.75	0.80	1.50	0.36			0.46
Bihbpor syenite	370	54.44	17.80	4.63	3.82	0.34	2.71	5.90	3.80	3.60	0.65	0.85			0.95
Augpor syenite	459	57.72	18.18	1.75	2.24		2.38	7.90	4.15	4.10	0.28	0.46			0.34
Augpor syenite	476	58.16	17.98	2.88	2.70		1.87	5.98	3.69	4.70	0.65	0.82			0.39
kato syenite	18	58.18	17.19	2.64	3.19	0.17	1.36	6.36	5.40	4.08	0.69	0.29			0.20
Augpor syenite	485	58.58	11.14	3.47	3.01	0.30	2.97	8.37	4.84	3.50	0.70	1.48	0.21		1.20
S.concave syenite	34	59.59	20.67	0.74	0.28		0.01	0.01	4.86	8.66	0.01	0.01	0.29		0.07
Albitite	461	60.87	14.09	0.32	2.62	0.40	2.51	4.14	8.00	2.57	0.20	1.81	0.51		1.58
Sev gar peg syenite	43	61.25	20.76	1.78	0.28		0.01	1.98	10.62	0.65	0.01	0.01	1.08		0.07
Kakangarai syenite	7	61.26	20.91	0.89	0.53		0.01	1.10	4.67	8.84	0.01	0.01	0.25		0.10
Sevvattur syenite	10	62.22	20.98	1.14	0.28		0.01	1.44	4.54	9.07	0.01	0.01	0.48		0.08
SevNE syeite	36	62.68	19.41	1.17	0.13		0.01	0.93	9.07	3.15	0.01	0.01	0.62		0.03
Kunnattur syenite	3	63.05	22.39	0.43	0.28		0.01	0.67	8.55	4.00	0.01	0.01	0.44		0.05
Koratti syenite	44	63.53	20.60	0.59	0.27		0.01	0.40	7.77	6.96	0.01	0.01	0.40		0.02
Karapattu syenite	6	63.54	22.27	0.74	0.27		0.01	0.75	8.55	2.86	0.01	0.01	0.30		0.09
PinkBihbporsy	463	64.24	14.74	2.84	2.02	0.83	1.13	2.42	4.93	3.34	0.62	0.52	0.44		1.43
Bi-ab-peg syenite	635	67.01	14.06	3.99	1.98		0.52	2.38	6.77	1.40	0.15	0.81	0.34		0.20
Bi hb syenite	369	70.30	13.35	1.94	2.63	0.20	1.17	4.05	3.36	1.58	0.45	0.62			
Richteite ultramafic	321	21.02	1.64	5.48	2.87	0.63	6.39	31.68	1.96	1.44	0.45		23.65	0.69	0.65
pl px ultramafic	319	41.26	3.65	10.18	8.26	0.29	9.67	20.18	0.41	1.20	1.48	0.73	0.41	0.34	1.15
bi-fels px ultramafic	320	44.12	9.10	3.64	1.26	0.06	14.20	21.78	1.15	1.24	0.88	0.10	1.12		0.60
gar pegmatite	572	44.84	11.35	12.14	0.50	0.40	0.70	19.01	3.95	4.98	0.40	1.20			0.40
px hb bi ultramafic	203	45.82	15.53	1.50	5.59	0.19	3.95	9.62	5.27	5.16	0.66		4.66	0.14	1.80
hb oligo ultramafic	204	49.16	21.03	2.57	5.16	0.08	5.43	12.37	2.43		0.83		0.14	0.16	0.30
Shonkinite	561	49.20	7.61	6.40	5.58	0.36	10.43	10.03	3.44	4.29	0.76	0.82			0.80
Olaipatti syenite	78	51.94	22.27	3.53	0.41	0.14	0.01	4.50	0.92	13.54	0.01	0.01	0.93		
gar aplite	39	52.28	19.59	2.82	1.73	0.13	0.25	5.15	0.14	17.16	0.56	0.13			0.44
bi px ultramafic	318	52.92	12.98	5.31	1.80	0.11	3.21	6.18	2.03	11.76	1.09	0.46	2.24	0.16	0.44
Olaipatti syenite	75	53.43	19.31	6.94	0.28		0.01	7.58	2.35	7.72	0.01	0.01	0.93		
Olaipatti syenite	77	54.25	22.99	2.83	0.56		0.01	2.12	2.44	12.05	0.01	0.01	1.04		
Olaipatti syenite	79	54.60	23.21	4.05	0.55	0.16	0.01	5.70	1.93	8.84	0.01	0.01	1.04		
Bi-mus oligoglacite	219	55.02	18.75	2.17	0.36	0.03	1.24	4.47	3.31	10.32	0.28		2.54	1.30	1.30
hb syeite	361	56.84	15.69	2.37	3.02	0.17	1.73	7.73	3.24	6.96	0.50	0.47	0.44	0.76	0.82
aplite	201	58.50	18.65	3.11	0.85	0.07	0.49	2.75	6.48	6.24	0.44		0.30	0.13	1.00
Jogipatti syenite	51	58.92	19.14	1.89	0.56		0.01	1.20	1.09	14.68	0.01	0.01	1.00		
anorthoclasite	349	58.98	16.20	2.51	2.72	0.12	2.22	5.15	6.75	3.48	0.56	0.17	0.22	0.74	0.50
Rieb gar pegmatite	360a	59.78	18.27	1.32	1.14	0.07	1.21	5.61	5.40	6.78	0.20	0.17			0.10
Bi pegmatite	340	61.82	16.90	0.61	0.34	0.02	0.49	2.06	0.27	16.80	0.09		0.42	0.25	0.32
Garigaipalli syenite	85	62.69	20.15	0.21	0.14		0.01	0.01	1.43	15.06	0.01	0.01	0.19		
bihb oligoclasite	206	63.18	15.02	2.73	4.39	0.14	2.72	5.50	3.98	1.32	0.66		0.30		0.38
rieb pegmatite	301	63.24	17.74	0.88	0.72	0.03	0.49	2.41	4.73	10.12	0.28			0.15	
rieb pegmatite	85	63.75	13.01	1.98	0.65		2.59	0.89	5.21	9.98	0.10	0.40			1.10
rieb aplite	541	64.01	15.44	2.44	0.25	0.40	1.62	0.21	4.04	9.15	0.14	0.22			1.58
Rieb diorite	80	64.97	16.75	2.61	1.31	0.40	0.53	3.75	5.04	2.70	0.30	0.48			0.82
Rieb sy pegmatite	36a	65.01	17.66	1.25	0.57	0.03	0.49	2.41	7.02	4.92			0.16	0.05	0.20
Pink granite	12	69.72	14.92	1.40	0.71	0.07	0.71	3.15	4.79	4.32	0.31				0.48
Pink granite	1	70.20	14.42	1.94	0.57	0.04	0.47	3.09	4.19	4.74	0.40	0.16			0.62
Rieb granite	78	71.40	13.60	2.53	0.63	0.50	0.72	2.67	3.80	2.93	0.15	0.32			0.52
ap-mt rock	498	0.48	0.73	54.17	18.07	0.49	0.73	6.66	0.14	0.42	13.20			0.77	
barite rock	452	0.50		0.10			0.05	0.50	0.05				0.53	33.40	0.80
apatite rock	450	2.06	6.91	2.89	2.06	0.07	1.46	41.36	0.47	0.10	10.01	31.77			0.26
Crocidolite rock	60	6.50	1.09	0.49	0.26	0.09	0.85	50.84	0.01	0.01	1.68	0.02	37.86		0.00
Ultramafic	460	18.84	0.91	28.94	17.60	0.19	5.42	15.42	0.06	0.14	3.15	3.00	4.50	0.30	1.30
Feore rock	13	20.36	0.07	27.60	37.85	0.46	0.01	11.16	0.71	0.34	1.57	0.04			
Ultramafic	492	27.72	0.80	16.25	7.15	0.78	6.37	19.09	4.04	0.28	0.60	0.21	15.26	0.23	0.36

Sevvattur syenites No723-369; Jogipatti syenites No 321-36a; Granites No12-78; Ultramafics No 498-47; Carbonatites No 310-495; Dolerites No 1176-1224

Table 1. Chemical composition of the alkaline rocks in the carbonatite complex of Tiruppattur

Rock	No	SiO2	Al2O3	Fe2O3	FeO	MnO	MgO	CaO	Na2O	K2O	TiO2	P2O5	CO2	SO3	H2O
Peridotite	31	30.09	2.00	19.09	6.95	0.79	9.91	15.68	3.67	0.01	0.63	1.29	8.93		
Biofels pyroxinite	9	36.88	1.47	15.49	10.72	0.10	8.00	17.76	1.53	1.88	1.74	1.97	1.10		
Kimberlite	29	37.71	0.12	8.46	3.29	0.15	8.71	21.10	0.32	2.93	0.75	0.09	16.60		
Samalpatti ultramafics	S3	38.09	5.17	1.05	14.10	0.18	14.18	15.52	0.23	3.44	1.66	3.69			1.90
Feore rock	21	39.30	0.07	49.84	10.36	0.02	0.01	0.01	0.16	0.17	0.33	0.14			
Kimberlite	7a	40.85	0.74	10.33	0.57	0.13	19.08	7.00	0.40	0.01	0.01	0.01	20.65		
Riebcc ultramafic	491	42.36	1.27	7.54	1.98	0.19	12.97	15.65	4.59	0.90	0.44	0.10	9.75		2.15
Olaugoligo ultramafic	40	42.58	4.62	6.32	5.25	0.15	10.62	25.08	0.34	0.06	0.93	3.10	0.60	0.33	0.60
apatite rock	451	42.82	1.27	21.86	2.51	0.04		14.76	1.95	0.06	1.56	7.90	2.46	1.65	0.24
Samalpatti ultramafic	S4	45.21	1.28	0.90	4.32	0.09	14.82	20.00	2.34	0.28	0.23	0.05	9.68		1.10
Ultramafic nodule	497	45.92	6.12	5.52	5.33	0.18	9.60	22.46	1.08	0.18	1.26	0.19	1.02		0.50
Rieb ultramafic	402	46.40	10.41	9.31	2.87	0.22	8.32	12.42	3.04	4.38	0.84	0.44	0.17	0.55	
Sam ultramafics	S5	46.41	2.09	1.64	2.90	0.10	14.46	18.90	2.55	0.74	0.17	0.74	8.73		1.08
Sev Ultramafics	S1	46.77	2.17	3.43	12.58	0.13	14.02	19.76	0.19	0.35	0.66	0.33			
wollastonite rock	491	49.66	1.22	2.01	0.63	0.67	1.52	41.86	0.34	0.30	0.07	1.15	0.30	0.30	0.20
Biopx ultramafic	736	52.40	12.26	2.99	6.75	0.02	6.01	8.48	3.11	4.26	0.99	1.53	0.49		0.00
Sam ultramafics	S2	54.65	1.01	0.85	4.12	0.15	17.17	21.37	0.44	0.12	0.22		0.02		
Fe ore ultramafics	47	62.14	0.17	33.58	1.22	0.01	0.01	1.07	0.01	0.34	1.30	0.01	0.39		
Sev carbonatite	310	0.20	0.18	3.60	2.97	0.55	17.40	29.54	0.10	0.60	0.10	0.05	43.56		0.50
Icelandspar	499	0.28	0.08		0.14		0.25	55.03	0.21	0.12	0.14		43.04		0.50
Sev.carbonatite	11	0.30		0.12		0.06	3.84	52.12	0.10	0.05			43.10		0.20
Sev. Carbonatite	59	0.50	0.29	1.25	1.38	0.48	5.00	45.59	0.46	0.07	0.40	1.30	41.45		1.10
Sev carbonatite	389	0.60	0.79	1.23	3.29		15.28	32.28	0.50	0.10	0.10	2.33	42.35		0.40
Sev carbonatite	250	1.00	0.70	4.08	0.60	0.06	0.45	50.46	0.34	0.22	0.20	0.82	39.28		0.50
Sev.carbonatite	410	1.26	0.42	5.00	1.33		4.52	45.46	0.68	0.43	0.15	1.80	39.10		0.28
Jogi. carbonatite	313	1.88	0.65	3.55	2.36	0.62	11.25	36.04	0.20	0.09	0.15	0.02	40.02	1.66	0.40
Jogi.carboatite	315	2.06	0.60	3.90	2.93	0.60	13.23	32.51	0.37	0.33	0.10	0.04	40.24	1.35	0.45
Jogi.carbonatiate	311	2.64	0.50	1.10	0.42	0.50	2.45	48.28	0.20	0.13	0.20	0.04	40.46	0.57	0.32
Jogi. carbonatite	990	3.24	0.94	2.30	0.41		1.06	48.43	2.30	0.50	0.20	1.48	38.52		0.52
Sev.carbonatite	369	4.90	2.95	4.62	0.95		12.29	32.24	0.73	0.08	0.20	1.66	38.14		0.44
Sev. carbonatite	20	9.97	5.06	0.88	2.35		3.12	46.90	1.10	0.57	0.30	2.01	26.32		1.00
Onna Bens carbonatite	500	12.51	5.73	2.05	0.13	0.16		24.39	1.05	0.98			27.17		0.59
Jogi carbonatite	321	21.02	1.64	5.48	2.87	0.63	6.39	31.68	1.96	1.44	0.45		23.65	0.69	0.65
Jogi. carbonatite	1010	24.70	2.01	2.54	2.00	0.30	7.06	33.51	2.92	0.90	0.20	3.68	19.45		1.20
Onna carbonatite	496	27.26	1.92	16.11	7.09	2.46	4.91	17.16	3.58	0.24	1.23	1.26	16.34		0.95
Onna.carbonatite	492	27.72	0.80	16.25	7.15	0.78	6.37	19.09	4.04	0.28	0.60	0.21	15.56	0.23	0.36
Onna.carbonatitae	495	41.36	1.45	11.16	5.30	0.32	8.32	13.61	5.20	0.24	0.98	0.52	10.95		0.20
dolorite	1176	47.22	18.44	1.30	9.76	0.38	7.19	11.09	1.68	0.68	0.75	0.40			0.80
dolorite	1155	48.08	19.06	2.30	9.41	0.15	5.42	10.01	2.52	0.90	1.25	0.50			0.38
dolorite	1159	48.58	15.49	1.60	12.80	0.40	4.68	7.80	2.92	0.90	2.60	0.93			0.84
dolorite	33	49.50	15.25	1.67	9.01		7.23	11.85	2.60	0.20	0.53	0.68	0.33		0.62
dolorite	1186	54.18	12.76	1.80	8.80	0.15	8.96	8.20	2.08	0.90	0.85	0.23			0.63
dolorite	1187	55.13	8.46	3.32	9.98	0.30	7.21	9.04	3.36	1.33	1.65	0.12			0.20
dolorite	1224	56.04	8.26	3.21	9.62	0.20	4.61	8.41	6.02	1.58	1.10	0.31			0.55

Table 1. Chemical composition of the alkaline rocks in the carbonatite complex of Tiruppattur (continued)

9. Geological setting and tectonomagmatic evolution

The location of the Precambrian carbonatite complex of Tiruppattur in the Indian peninsula along the Eastern Ghats Paleo-rift System (Ramasamy, 1982, 1987), extending over a stretch of 3000 km x 200 km from Cape Comorin / Palghat Gap to the Brahmaputra valley (Eastern Syntaxes of the Himalayas), is favourable site for a low degree of the partial melting of alkali-enriched upper mantle rocks from a low velocity zone (Schleicher et al., 1998). The rift system is a superimposed structure over which a series of older block-faulted horsts and graben structures are comprised of a number of magmatic emplacements of charnockites, anorthosites, alkaline rocks, carbonatites and volcanic effusives. It is also favourable for the genesis of under-saturated alkaline magma charged with relatively more anhydrous volatiles (such as CO_2, SO_3, P_2O_5, F, Cl and CH_4 etc.) with respect to H_2O vapour, which is the main constituent in the volatile phase derived from a deep-seated source. The Mg/(Mg+Fe^t) ratios of the magmatic melts are frequently used as an indicator of whether a melt could be a partial melting product of a mantle material. The high mg values between 0.76 and 0.86 in this area depend on the TPX and the volatile constituents by which they were formed; moreover, these ratios of the starting material of the partial melts from the mantle may fall between the ratios of 0.9-0.3 (Mysen, 1975). Accordingly, all of the rocks in this complex can be derived from the partial melts from the mantle horizons. The propagation and penetration of the Eastern Ghats Paleo-rift system by reactivations during the subsequent tectonic episodes from the Early Proterozoic Period may extend to different depth levels, causing certain low degrees of partial melting which have been attained only after some prolonged period of tectonic deformation. According to Ramasamy (1982), a span of charnockitic activities extends from 3100 Ma to 2600 Ma, with anorthositic events extending from 2000 Ma to 1100 Ma and alkali syenitic- carbonatitic activities commencing from 1200 Ma onwards (Ramasamy, 1981). It seems that successively younger magmatic episodes were formed under more anhydrous- and alkali-rich environments and at deeper levels than the older ones, owing to the deeper penetration of the rifted continental plates into the mantle. The emplacement of most of the carbonatites from the Proterozoic to recent periods occurring in various parts of the world is restricted along deep crustal fractures that were controlled by regional structures and tectonics. According to Macintyre (1975), the ages of carbonatite complexes younger than 200 Ma indicate that many of them are intimately related with major changes in plate motion which were globally synchronous. The continental plate formed by the propagation and penetration of the Eastern Ghats Paleo-rift system led to the later separation of Gondwana Land from Peninsular India. The wide-spread Deccan Trap volcanic activity covered more than 500000 km^2 and the thickness of the lava flows exceeds 2000 m in some places (Krishnan, 1962), with younger eruptions of olivine-tephrite, soda-trachyte and carbonatite eruptions at Kudangulam near Cape Comorin indicating that Peninsular India is prone to repeated volcanic and magmatic activities (Ramasamy, 1987) and are also bearing evidence of Indian plate movement towards the north after the break of Antarctica from Peninsular India. The ascension of magma and its rate of cooling, pressure and the volume of volatile constituents played a critical role in the magmatic evolution of the residual magma under a specific geological setting and tectonic movement. The influence of local variations in the TPX conditions during the course of the crystallization of minerals creates complexities in tracing the trends of magmatic evolution in a spatial order. There exists a compositional relationship between these new discrete minerals in younger plutons with the minerals in older plutons in this area. The distribution of incompatible HFSE of K, Ti, P, Zr, Nb, Ba, Sr (Table 4) in these rocks and high ratios of HFSE such as Ti/P, K/(K+Na), Nb/Ta, Zr/Hf, Sr/Ba, LREE/HREE and the presence of minerals like allanite, zircon, apatite, monazite, pyrochlore, niobian rutile, magnetite, galena and feldspars, indicate that the low degree of partial melts from the mantle horizon and the parent magma separated from the melt were derived from HFSE-enriched portions of the mantle during the folding and up-arching of the mantle rock at a depth there by enriching the above characteristic elements just prior to the initial stage of the rifting of the diverging continental plates of the Eastern Ghats Mobile Belt.

Mineral	SiO2	Al2O3	Fe2O3	FeO	MnO	MgO	CaO	Na2O	K2O	TiO2	BaO	SrO	P2O5	CO2	SO3	F	H2O	H20	Total
Wollastonite	49.66	1.22	2.01	0.63	0.67	1.52	41.86	0.34	0.30	0.07	0.00	0.00	1.15	0.20	0.00	0.00	0.00	0.00	99.63
Epidote	37.72	17.61	4.67	0.18	0.06	2.29	34.34	0.41	0.84	0.96	0.00	0.00	0.07	0.58	0.00	0.00	0.00	0.00	99.73
Apatite1	0.05	0.18	0.00	0.00	0.00	0.52	54.25	0.30	0.00	0.00	0.00	0.80	39.80	0.00	0.00	0.00	0.00	0.00	95.90
Apatite 2	0.22	0.06	0.00	0.00	0.00	0.17	54.15	0.29	0.00	0.00	0.00	1.51	40.03	0.00	0.00	0.00	0.00	0.00	96.43
apatite 1a	0.00	0.00	0.00	0.00	0.00	1.51	55.73	0.00	0.00	0.00	0.00	0.00	40.95	0.00	0.00	1.56	0.00	0.00	99.75
apatite2a	0.00	0.00	0.00	0.00	0.00	1.51	53.28	0.00	0.00	0.00	0.00	0.00	37.96	0.00	0.00	0.80	0.00	0.00	93.55
apatite 450	2.06	6.91	2.89	2.06	0.07	1.46	41.36	0.47	0.00	0.00	0.00	0.00	41.50	0.00	0.00	1.70	0.26	0.00	100.74
Magnetite 498I	0.10	0.18	67.64	22.57	0.39	0.25	1.05	0.00	0.00	8.10	0.00	0.00	0.00	0.00	0.00	0.00	0.00	0.00	100.28
magnetite 498ii	0.72	0.73	54.17	18.07	0.49	0.73	6.66	0.14	0.42	0.00	0.00	0.00	0.00	0.44	0.77	0.00	0.22	0.00	83.56
Iceladspar	0.36	0.00	0.14	0.00	0.00	0.25	55.03	0.21	0.12	0.14	0.00	0.00	0.00	43.04	0.34	0.00	0.00	0.00	99.63
calcite cc250	2.00	0.00	5.40	0.50	0.06	0.25	51.36	0.24	0.12	0.00	0.00	0.00	0.00	40.28	0.00	0.00	0.00	0.00	100.21
dolomite cc310	0.38	0.00	3.80	2.87	0.55	18.40	30.54	0.10	0.60	0.00	0.00	0.00	0.00	46.06	0.00	0.00	0.00	0.00	103.30
calcite cc311	3.74	0.00	1.40	0.32	0.50	2.25	48.52	0.10	0.03	0.00	0.00	0.00	0.00	40.46	0.57	0.00	0.00	0.00	97.89
Carbonatite cc31	36.40	0.00	2.00	0.25	0.37	1.50	31.93	0.17	0.18	0.00	0.00	0.00	0.00	26.62	0.41	0.00	0.00	0.00	99.83
Mg calcite 313	2.78	0.00	4.05	2.16	0.62	11.65	36.44	0.10	0.09	0.00	0.00	0.00	0.00	41.02	1.66	0.00	0.00	0.00	100.57
carbonatite 314	35.30	0.00	1.70	0.36	0.49	1.12	32.72	0.07	0.11	0.00	0.00	0.00	0.00	26.34	0.67	0.00	0.00	0.00	98.88
ferrocarbonate c	3.26	0.00	4.30	2.73	0.60	13.63	33.31	0.27	0.23	0.00	0.00	0.00	0.00	41.84	1.35	0.00	0.00	0.00	101.52
ferrocarbonate 4	53.02	0.00	6.25	1.15	0.20	3.37	19.09	0.04	0.08	0.00	0.00	0.00	0.00	15.56	0.29	0.00	0.00	0.00	99.05
ferrocarbonatecc	49.74	0.00	2.25	0.00	0.42	1.00	23.25	0.04	0.11	0.00	0.00	0.00	0.00	19.92	0.60	0.00	0.00	0.00	97.33
ferrocarboate494	49.82	0.00	2.50	0.00	0.23	1.25	23.95	0.04	0.11	0.00	0.00	0.00	0.00	20.78	0.39	0.00	0.00	0.00	99.07
Benstonite	12.51	5.73	2.03	0.13	0.16	0.00	24.39	1.08	0.98	0.00	20.69	4.33	0.00	27.17	0.00	0.00	0.59	0.00	99.79
Biotite	32.96	15.64	15.09	0.00	0.05	16.10	3.51	0.41	6.42	3.42	0.00	0.00	0.32	0.00	0.00	1.68	5.96	0.00	101.56
vermiculite1	30.92	8.17	23.56	0.00	0.00	10.15	6.60	4.95	4.14	2.88	0.00	0.00	0.00	0.00	0.00	0.00	7.54	1.09	100.00
vermiculite2	29.88	7.33	23.15	0.00	0.00	11.18	8.50	4.93	3.25	3.17	0.00	0.00	0.00	0.00	0.00	0.00	8.04	0.57	100.00
vermiculite3	32.45	24.47	4.09	0.00	0.00	9.96	11.20	1.27	4.08	2.19	0.00	0.00	0.00	0.00	0.00	0.00	7.97	1.61	99.29
vermiculite4	31.58	14.24	24.75	0.00	0.00	6.61	3.15	3.19	4.11	2.41	0.00	0.00	0.00	0.00	0.00	0.00	7.94	2.02	100.00
vermiculite5	31.64	10.47	21.36	0.00	0.00	12.26	3.15	2.39	5.05	3.17	0.00	0.00	0.00	0.00	0.00	0.00	8.18	2.33	100.00
vermiculite6	30.14	11.64	13.00	0.00	0.00	12.92	8.69	0.20	1.33	2.27	0.00	0.00	0.00	0.00	0.00	0.00	11.50	7.05	98.74
vermiculite7	22.44	13.99	3.42	0.00	0.00	9.02	23.97	2.39	0.88	1.22	0.00	0.00	0.00	0.00	0.00	0.00	19.93	2.82	100.08
vermiculite8	26.25	20.76	3.63	0.00	0.00	12.53	15.38	2.54	0.88	1.33	0.00	0.00	0.00	0.00	0.00	0.00	14.54	2.04	99.88
vermiculite9	26.41	20.48	3.76	0.00	0.00	8.24	20.46	2.27	0.82	1.61	0.00	0.00	0.00	0.00	0.00	0.00	13.72	1.88	99.65
vermiculite10	34.10	25.49	4.51	0.00	0.00	12.27	4.97	3.27	1.15	1.97	0.00	0.00	0.00	0.00	0.00	0.00	9.30	2.93	99.96
vermiculite11	21.82	13.70	3.68	0.00	0.00	7.39	24.93	1.41	0.60	0.99	0.00	0.00	0.00	0.00	0.00	0.00	22.01	3.83	100.36
vermiculite12	30.92	18.20	17.16	0.00	0.00	13.65	4.88	0.00	0.00	1.04	0.00	0.00	0.00	0.00	0.00	0.00	14.47	0.00	100.32
vermiculite13	32.15	21.58	12.51	0.00	0.23	4.36	4.89	0.18	1.52	3.01	0.00	0.00	0.00	0.00	0.00	0.00	18.12	0.00	98.55
Barite452	0.50	0.00	0.10	0.00	0.00	0.05	0.60	0.05	0.00	0.00	62.54	1.30	0.00	0.56	33.40	0.00	0.70	0.00	99.80
BariteAlan	4.56	0.17	0.06	0.00	0.00	0.00	1.00	0.00	0.00	0.00	58.74	4.54	0.00	0.05	30.66	0.00	0.08	0.00	99.86

Wo wollastonite from wollastonite-calcite-garnet syenite; Epi Epidote from skarn rock from Garigaipalli; Aps1, Aps2 and ap1 apatite from Sevvattur carbonatites; ap2,ap450 apatite from Onnakarai carbonatites; mt498 magnetitae from Sevvattur carbonatite; Ice Icelandspar from skarn rock of Garigaipalli; Cc250 Calcite from carbonaitte of Sevvattur; Cc310, cc311. cc312, cc313, cc314, Carbonates from carbonaites of Jogipatti; Cc492, cc493, cc494 carbonates from Carbonatites of Onnakarai; cc315; Ben Benstonite from Benstonite carbonatite from Onnakarai; Bi Biotite from Biotite pyroxenite; verm 1-13 Vermiculite from Sevvattur carbonatites; Ba452 Barite from Onnakarai; BaAlan Barite from Alangayam

The chemical compositions of the rocks are listed in Tables 1 and 2 on the basis of rock types and their places of occurrence. The chemical compositions of these rocks vary widely in their silica, alkalis and carbonate contents. CO_2-rich rock carbonatites, Alkali syenites vary in SiO_2 between 35% and 65%, Ultramafic rocks vary in SiO_2 between 40% and 52%. Hybrid rocks have both SiO_2 and CO_2 in an intermediate position.

Table 2. Compositions of various minerals occurring in the carbonatite complex of Tiruppattur

Rock	No	Ba	Sr	Nb	Zr	La	Ce	Nd	Y	Th	U	V	Cu	Zn	Cr	Ni	Co	Pb	Sc
Sev sovite	Sub21		5225		5180	2690	2930	1900	45	140		450							
Sev beforsite	Sub22	3570	9700	220		350	1500	450	90	10	110								
Sev ferrocarbonaite	Sub23		4250	635	40	180	1860	570	130	30	40								
Jogi sovite	Sub24	35	300		10	25	40	10		10	10	10							
Gari skarn	Sub25	220	350		20	15	30			10	10	10							
Onna ferrocarbontite	Sub26	1159	7795	60	50	290	760	240	105	10	25	40							
Sev syenite	Sub11		270	740	120	200	650	430	200	50	10		20	20	15			120	
Sev syenite	Sub12	2780	1120	15	120	60		30	10				25	104	15			100	
Sev syenite	Sub13		955	35	260	70	120	45	30	15	10		20	50	25			170	
Sev ultramafic	Sub1		375		12	10	10	10	10			510							90
Jogi ultramafic	Sub2		275		15	10	25	10	10	10		40							55
Jogi ultramafic	Sub3		900		30	80	180	50	35										
Gari ultramafic	Sub4		217		20	10	10					90							40
Sev granite gneiss	Sub32	2321	1424	60	155	55			25	10	10		45	70	10			20	
Jogi syenite	Sub33	1726	900	60	170	55			30	30	10		20	50	20			35	
Jogi syenite	Sub34	1945	1120	35	240	60			20	15	10		20	60	15			180	
Jogi albitite	Sub35	220	30	180	350	20			10	250	100		10	15	35			10	
Jogi syenite	Sub36		470		80	10	25	50	50		10		20	20	10			880	
Jogi shonkinite	8	1000	2000		600				15			2000	1000	1000	1000	300	200	10	300
Jogi Riebeckite	R	600		60	30				10			200	9	100	2000	30	10	10	20
Jogi hy carbonatite	319	200	200		10				8			100	100	100	8	20	30		10
Jogi hy carbonatite	320	200			20							60	60	200	8	8	8		8
Jogi hy carbonatite	321	2000	2000		8				8			60	30			8	8	8	20
Jogi Rieb carbonatite	401	200	300	60	8				8			30	8	200	600	30	20	20	20
Jogi Rieb carbonatite	402	600	100		20							6010	8		100	30	30		10
Jogi Rieb carbonatite	490	100			10				8			10	8		30	10	10		
Jogi Wollastonite rock	491	300	300		8							60	300	100			8		
Jogi ultramafic nodule	497	2000			10							300	100	300	100	60	30		10
Jogi ultramafic	40	200	100		20				30			60	300	300	600	100	60	8	60
Sev ultramafic	201	3000	300	50	60				10			60	20	100		8	8	100	
Sev ultramafic	203	6000	2000	100	50				60			100	8	200	100	30	20	100	
Sev hy carbonatite	204		300		30				20			200	30	200	200	200	60		20
Sev hy carbonatite	206	600	600		60				60			100	100	100	600	100	60	20	60
Sev oligoclasite	219	1000	1000	60	100				10			300	60		30	10	10	20	
Jogi carbonatite	318	4000	1000	20	60				60			100	100	100	200	30	30	30	30
Apatite magnetite rock	460	100	100		8							200	300	100	300	30	30		8
Sev pyroxenite	789	300			400				10				100	60		60	60		60
Jogi ultramafic	493		790			170	145	90	950			50			100	175	75		675
Jogi ultrapot syenite	39	475	2300			880	380	790	2630			315	280	670	320	375	560	260	390
Jogi melanite sovite	990	120	4913		1115	250	165	123	733			78	73	110	73	73	160	133	610
Jogi shonkinite	561	322	5960		708	136	142	76	534			76	24	64	76	70	50	262	74

Table 3. Trace elements' distribution in carbonatites and associated alkaline rocks

10. Conclusion

The dolerites are hypabyssal equivalents of gabbro and basalts, which are common products of the partial melting of mantle / crustal rocks at depth. The overlapping petrographic

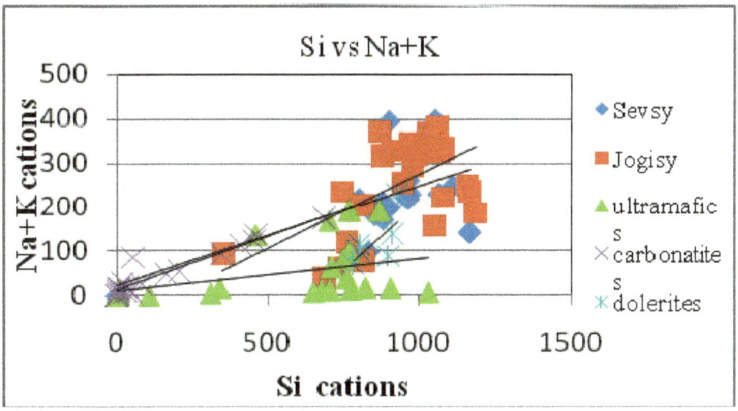

a)

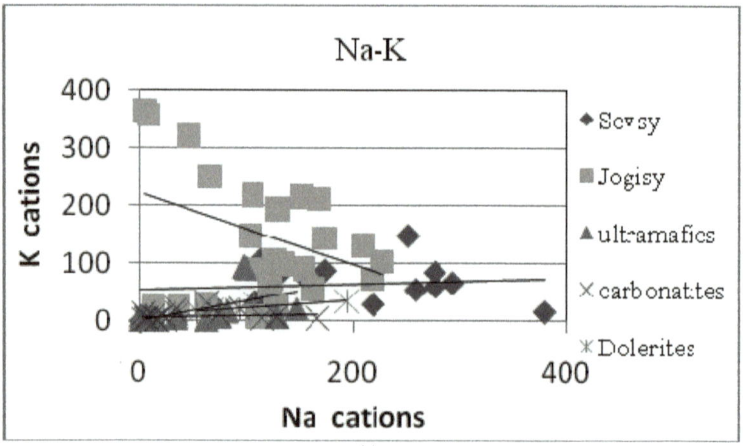
Na-K
K cations
400
300
200
100
0
0
200
400
Na cations
Sevsy
Jogisy
ultramafics
carbonatites
Dolerites

b)

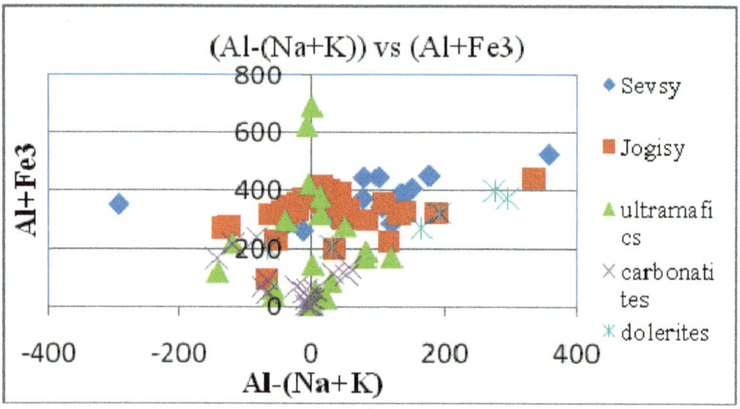
(Al-(Na+K)) vs (Al+Fe3)
Al+Fe3
800
600
400
200
0
-400
-200
0
200
400
Al-(Na+K)
Sevsy
Jogisy
ultramafics
carbonatites
dolerites

c)

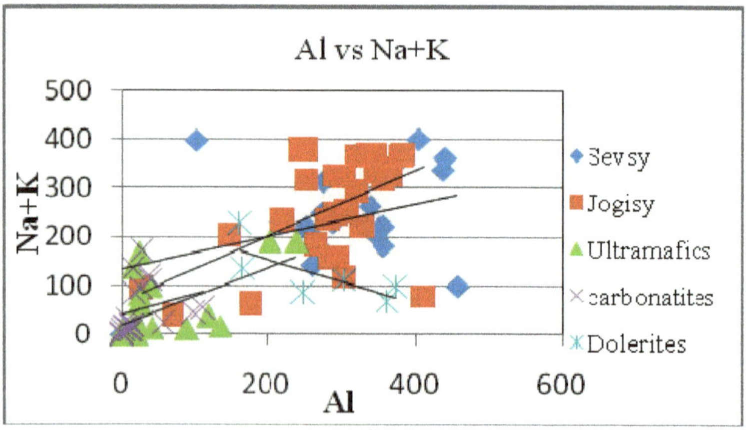
Al vs Na+K
Na+K
500
400
300
200
100
0
0
200
400
600
Al
Sevsy
Jogisy
Ultramafics
carbonatites
Dolerites

d)

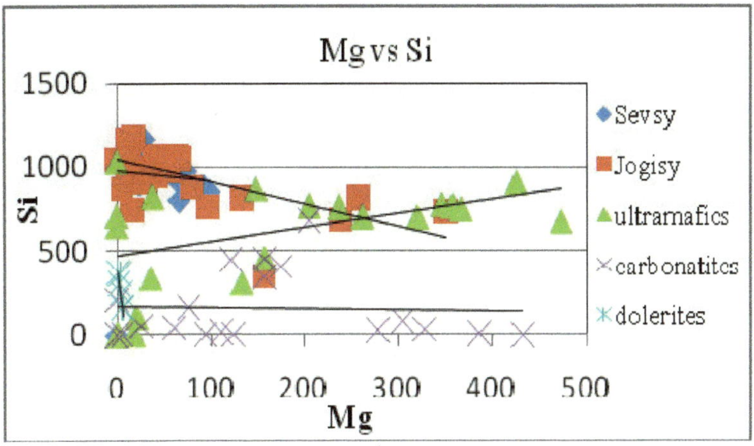
Mg vs Si
1500
1000
500
0
Si
0
100
200
300
400
500
Mg
Sevsy
Jogisy
ultramafics
carbonatites
dolerites

e)

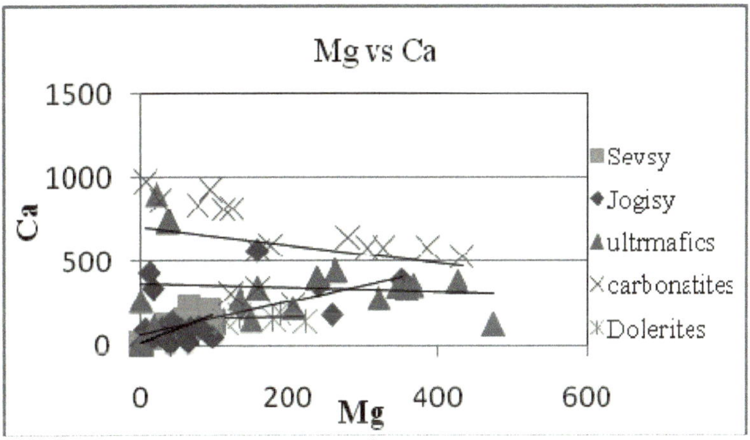
Mg vs Ca
1500
1000
500
0
Ca
0
200
400
600
Mg
Sevsy
Jogisy
ultrmafics
carbonatites
Dolerites

f)

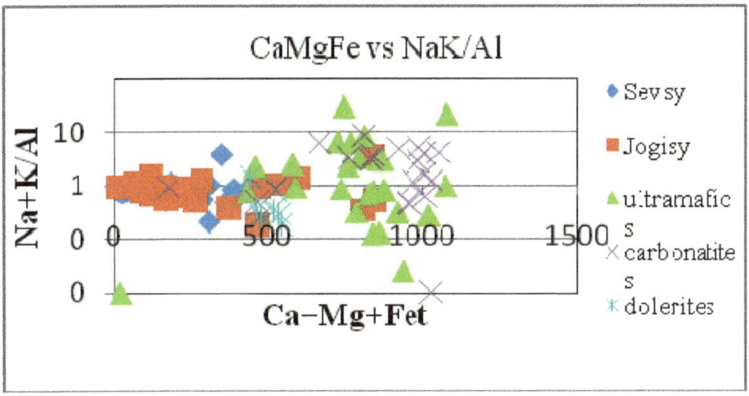
CaMgFe vs NaK/Al
10
1
0
0
Na+K/Al
0
500
1000
1500
Ca-Mg+Fet
Sevsy
Jogisy
ultramafics
carbonatites
dolerites

g)

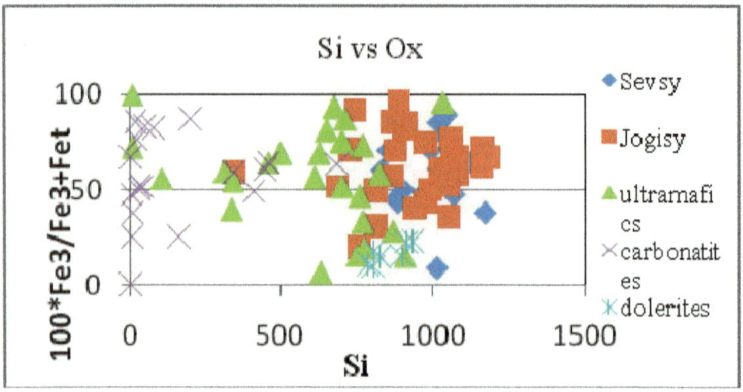

h)

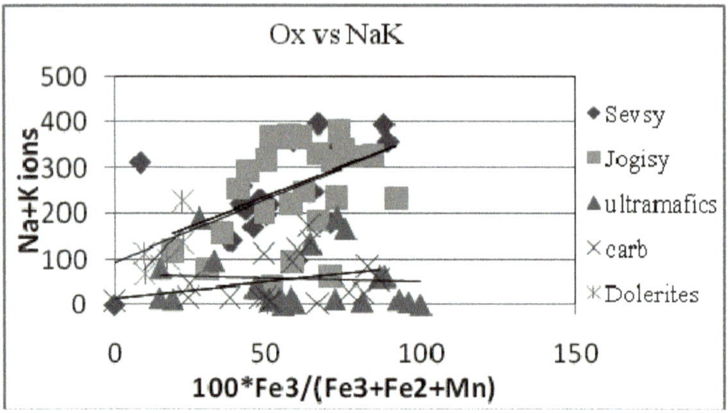

i)

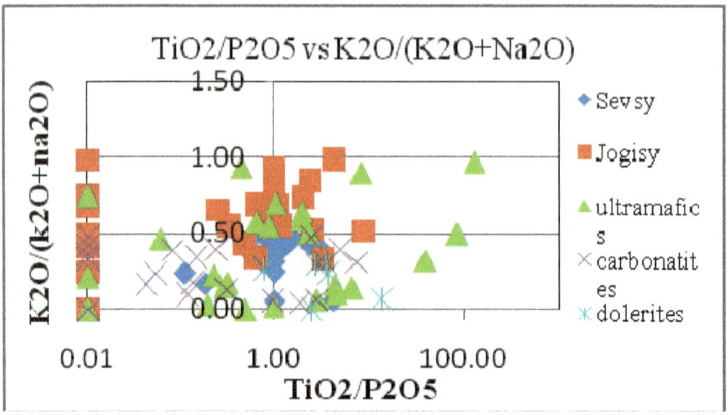

j)

Fig. 5. a-j show linear trends of the magmatic evolution of different types of rocks from common parent magma variations of all these rocks indicate that they are all derived from the same mantle source by different degrees of partial melting. The composition of the parent magma for this complex appears to be very close in composition to that of shonkinite magma, which might have been derived by liquid fractionation and separation from the low degree of the partial melt of the mantle material. The partial melt might have initially been a crystallized olivine and calcic plagioclase; therefore, the residual shonkinitic parent magma so evolved is impoverished in alumina and in silica and acts as a primary magma in a closed magmatic chamber under volatile enriched conditions. Later on, with early crystallizations of calcium-rich clinopyroxenes, the residual magma is impoverished in silica with the enrichment of volatile constituents such as H_2O and CO_2. In an ascending convection current which motivated a low viscous and high temperature shonkinite magma, clinopyroxenes are crystallized with the enrichment of Ca, Mg, Fe, Ti^{iv} and Al^{iv} and a depletion in Si, Al^{vi}, Na and K in low pressure zones towards the top of the magmatic column. On the other hand, from the top of the magmatic column, sinking clinopyroxenes liberate Ca, Mg, Fe, Ti^{iv} and Al^{iv} and accumulate Na and Al^{vi} in their subsolidus crystal lattices during descending convection currents and towards the bottom of the chamber with the crystallization of subsolidus aegirine. Thus, both salic and mafic constituents in the form of ions are concentrated at the top of the magmatic column in different portions. In this closed magmatic chamber under high P_{H2O} and P_{CO2}, an immiscible separation of camafic carbonate liquids and alkali silicate liquids are derived at depth (Saravanan and Ramasamy, 1995). The separated alkali silicate magmas are emplaced in sequences of co-magmatic bodies of syenites, first in the Sevvattur Basin and subsequently emplaced in the Jogipatti basin followed by carbonate magmas in both (Fig. 1 a-d). It is quite a complex matter to study the impact of all these processes by tracing the history of the magmatic evolution inscribed on the rocks over the course of millions of years with the help of detailed petrographic and field investigations.

11. References

Barth, T.F.W. (1952) Theoretical Igneous Petrology, Ms Book and Mineral Company, 2nd Edn, New York, p. 416

Borodin, I.S., Gopal, V., Maralev, V.M. and Subramanian, V. (1971) Precambrian carbonatites of Tamil Nadu, South India, Journal of the Geological Society of India, v. 12, pp. 101-112

Bowen N.L. (1956) The Evolution of Igneous Rocks, Dover, New York, 2nd Edn, p. 332

Deer, W.A., Howie, R.A. and Zussman, J. (1965) Rock Forming Minerals, v. 1 to v. 5, Longmans, London

Dickson,J.A.D. (1965) A modified staining technique for carbonates in thin sections, Nature, v. 205, p. 587

Grady, J.C. (1971) Deep main faults in South India, Journal of the Geological Society of India, v. 12, pp. 56-62

von Eckermann, H. (1966) Wollastonite in carbonatite rocks, Geochimica et Cosmochimica Acta, v. 31 (11), pp. 2253-2254

Schieicher, H., Kramm, U., Pericka, E., Schidlowski, M., Schmidt, F., Subramanian, V., Todt, W. and Viladkar, S.G. (1998) Enriched sub-continental upper mantle beneath Southern India: Evidence from Pb, Nd, Sr and C-O Isotopic studies on Tamil Nadu carbonatites, Journal of Petrology, v. 39, (10)m, pp. 1765-85

Krishnan, M.S. (1962) Geology of India and Burma, Higgin Bothams, Chennai

Macintyre, R. (1975) Age and significance of carbonatites and alkaline intrusive, Mem. Noticicas. Mus. Lab. Min. Geol. Univ. Coimbra, v. 80, pp. 99-103

Mysen, B. (1975) Partitioning of iron and magnesium between crystals and partial melts in Peridotites Upper Mantle, Contrib, Mineral Petrol., v. 52, pp. 69-76

Mysen, B.O., Virgo, D. and Seifert, F.A. (1985) Relationships between properties and structure of aluminosilicate melts, Am. Mineralogist, v. 70, pp. 88-105

Naidu, P.R.J. (1958) 4-Axes Universal Stage, Dept. of Geology, Madras University

Ramasamy, R. (1973) Geology of the area South west of Tiruppattur, Madras State (Tamil Nadu), India, Ph.D Thesis, p. 226, University of Madras

Ramasamy, R. and Shapenko, V. (1980) Fluid inclusion studies in carbonatites of Tiruppattur, Tamil Nadu , India (Abst) Int. Geol. Cong. Paris , v. 1, p. 79

Ramasamy, R. (1981) Tectonomagmatic Episodes of the Peninsular India, Sympet, Jaipur, Geol. Surv. India, Abstract L2-L3.

Ramasamy, R. (1982a) A possible paleo-rift system of the eastern Ghats in the Peninsular India, J. Moscow State Univ., Geology, Ser. 4, No. 2, pp. 32-37 (in Russian)

Ramasamy, R. (1982) Structure and tectonics of carbonatite complex of Tiruppattur, Tamil Nadu In: Bhattacharya, A./K. (Ed.) Current Trends in Geology, IV Indian Geological Congress, Today and Tomorrows Publishers, New Delhi, pp. 119-136

Ramasamy, R. (1986) Titanium-bearing garnets from alkaline rocks of carbonatite complex of Tiruppattur, Tamil Nadu, Current Science, v. 55, pp. 1026-29

Ramasamy, R. (1986a) Ca-rich pyroxenes from the carbonatite complex of Tiruppattur, Tamil Nadu, Current Science, v. 55, pp. 981-984

Ramasamy, R. (1987) Reactivation of the Eastern Ghats Paleorift system during Tertiary and other periods, Proc. National, Seminar Tertiary Orogeny, pp. 109-127.

Ramasamy, R., Gwalani, L.G. and Subramanian, S.P. (2001) A note on the occurrence and formation of magnetite in the carbonatites of Sevvattur, North Arcot district, Tamil Nadu, Southern India Journal of Asian Earth Sciences, v. 19, pp. 297-304

Ramasamy, R., Subramanian, S.P. and Sundaravadivelu, R. (2010) Compositional variations of olivine in shonkinite and its associate ultrabasic rock from the carbonatite complex of Tiruppattur, Tamil Nadu, Current Science, v. 99(10), pp. 1428-1433

Rittmann, A. (1973) Stable Mineral Assemblage of Igneous Rocks, Springer, Berlin, p. 251

Saravanan, S. and Ramasamy, R. (1971) A magnesio-riebeckite from Samalpatti area, Tamil Nadu, India, Mineralogical Magazine, v. 38, pp. 376-377

Saravanan, S. and Ramasamy, R. (1995) Geochemistry and petrogenesis of shonkinite and associated alkaline rocks of Tiruppattur carbonatite complex, Tamil Nadu, Journal of the Geological Society of India, v. 46, pp. 235-243

Schairer, J.F. and Yoder, H.S.J.R., (1960) The nature of residual liquids from crystallization data on the system Nepheline-diopside-silica, American Journal of Science, Bradley Volume, v. 258A, pp. 273-83

Udas, G.R. and Krishnamurthy, P. (1970) Carbonatites of Sevvattur and Jogipatti, Madras State, India, Proc. Of the Indian National Academy of Science, 36A, pp. 331-343

Verwoerd, W.J. (1966) South African Carbonatites and their probable mode of origin, Ph.D. Thesis, Univ. of Stellenboch

Viladkar, S.G. and Subramanian, V. (1995) Mineralogy and geochemistry of the Sevvattur and Samalpatti complexes, Tamil Nadu, Journal of the Geological Society of India, v. 45, pp. 505-518

Winchell, A. N. (1945) Elements of Optical Mineralogy, part II Optical properties of Minerals

Yagi, K. (1953) Petrochemical studies on the alkaline rocks of the Morotu District, Sakhalin, GSA Bull, v. 64, 769-810

Yoder, H.S., Stewart, D.B. and Smith, J.R. (1957) Ternary feldspars, Carnegie Inst. Washington Year Book, pp. 206-214

Permissions

The contributors of this book come from diverse backgrounds, making this book a truly international effort. This book will bring forth new frontiers with its revolutionizing research information and detailed analysis of the nascent developments around the world.

We would like to thank Dr. Marcello Rubens Barsi Andreeta, for lending his expertise to make the book truly unique. He has played a crucial role in the development of this book. Without his invaluable contribution this book wouldn't have been possible. He has made vital efforts to compile up to date information on the varied aspects of this subject to make this book a valuable addition to the collection of many professionals and students.

This book was conceptualized with the vision of imparting up-to-date information and advanced data in this field. To ensure the same, a matchless editorial board was set up. Every individual on the board went through rigorous rounds of assessment to prove their worth. After which they invested a large part of their time researching and compiling the most relevant data for our readers. Conferences and sessions were held from time to time between the editorial board and the contributing authors to present the data in the most comprehensible form. The editorial team has worked tirelessly to provide valuable and valid information to help people across the globe.

Every chapter published in this book has been scrutinized by our experts. Their significance has been extensively debated. The topics covered herein carry significant findings which will fuel the growth of the discipline. They may even be implemented as practical applications or may be referred to as a beginning point for another development.

The editorial board has been involved in producing this book since its inception. They have spent rigorous hours researching and exploring the diverse topics which have resulted in the successful publishing of this book. They have passed on their knowledge of decades through this book. To expedite this challenging task, the publisher supported the team at every step. A small team of assistant editors was also appointed to further simplify the editing procedure and attain best results for the readers.

Our editorial team has been hand-picked from every corner of the world. Their multi-ethnicity adds dynamic inputs to the discussions which result in innovative outcomes. These outcomes are then further discussed with the researchers and contributors who give their valuable feedback and opinion regarding the same. The feedback is then collaborated with the researches and they are edited in a comprehensive manner to aid the understanding of the subject.

Apart from the editorial board, the designing team has also invested a significant amount of their time in understanding the subject and creating the most relevant covers. They scrutinized every image to scout for the most suitable representation of the subject and create an appropriate cover for the book.

The publishing team has been involved in this book since its early stages. They were actively engaged in every process, be it collecting the data, connecting with the contributors or procuring relevant information. The team has been an ardent support to the editorial, designing and production team. Their endless efforts to recruit the best for this project, has resulted in the accomplishment of this book. They are a veteran in the field of academics and their pool of knowledge is as vast as their experience in printing. Their expertise and guidance has proved useful at every step. Their uncompromising quality standards have made this book an exceptional effort. Their encouragement from time to time has been an inspiration for everyone.

The publisher and the editorial board hope that this book will prove to be a valuable piece of knowledge for researchers, students, practitioners and scholars across the globe.

List of Contributors

Evgeniy Selivanov and Roza Gulyaeva
Institute of Metallurgy of the Ural Branch of the Russian Academy of Sciences, Russia

Juan Carlos Rendón-Angeles
Research Institute for Advanced Studies of the NPI Saltillo Campus, Mexico

Zully Matamoros-Veloza
Technological Institute of Saltillo, Mexico

Kazumichi Yanagisawa
Research Laboratory for Hydrothermal Chemistry, Kochi University, Japan

Dragica M. Minić and Vladimir A. Blagojević
University of Belgrade, Faculty for Physical Chemistry, Serbia

Dušan M. Minić
Military Technical Institute, Belgrade, Serbia

João F.Cajaiba da Silva, Andréia P. M. da Silva and Rodrigo C. de Sena
Instituto de Química - Universidade Federal do Rio de Janeiro, Brazil

Françoise Bonneté
Institut des Biomolécules Max Mousseron, Université d'Avignon et des Pays de Vaucluse, France

Humphrey A. Moynihan and Dawn M. Kelly
Dept. of Chemistry / Analytical and Biological Chemistry Research Facility, University College Cork, Cork, Ireland

Edgar Alfonso, Jairo Olaya and Gloria Cubillos
Universidad Nacional de Colombia, Colombia

Marco Giulietti and André Bernardo
Chemical Engineering Department Federal University of São Carlos UFSCar, Brazil

Ahamad Mohiddon Mahamad and Ghanashyam Krishna Mamidipudi
School of Physics, University of Hyderabad, Hyderabad, India

J. J. Gervacio Arciniega, E. Prokhorov, F. J. Espinoza Beltran and G. Trapaga
CINVESTAV, Unidad Queretaro, Juriquilla, Querétaro, Mexico

Kuninori Kitahara
Shimane University, Japan

Akito Hara
Tohoku Gakuin University, Japan

Noriah Bidin and Siti Noraiza Ab Razak
Laser Research Group, Advanced Photonic Science Institute, Faculty of Science, Universiti Teknologi Malaysia Johor Malaysia, Malaysia

R. Ramasamy
Ocean Engineering, IITM, Chennai, India